T0202954

Lecture Notes in Computer Science 14431

Founding Editors

Gerhard Goos
Juris Hartmanis

Editorial Board Members

The series Lecture Notes in Computer Science (LNCS), including its subseries Lecture Notes in Artificial Intelligence (LNAI) and Lecture Notes in Bioinformatics (LNBI), has established itself as a medium for the publication of new developments in computer science and information technology research, teaching, and education.

LNCS enjoys close cooperation with the computer science R & D community, the series counts many renowned academics among its volume editors and paper authors, and collaborates with prestigious societies. Its mission is to serve this international community by providing an invaluable service, mainly focused on the publication of conference and workshop proceedings and postproceedings. LNCS commenced publication in 1973.

Qingshan Liu · Hanzi Wang · Zhanyu Ma ·
Weishi Zheng · Hongbin Zha · Xilin Chen ·
Liang Wang · Rongrong Ji
Editors

Pattern Recognition and Computer Vision

6th Chinese Conference, PRCV 2023
Xiamen, China, October 13–15, 2023
Proceedings, Part VII

Springer

Editors
Qingshan Liu (iD)
Nanjing University of Information Science
and Technology
Nanjing, China

Zhanyu Ma (iD)
Beijing University of Posts
and Telecommunications
Beijing, China

Hongbin Zha (iD)
Peking University
Beijing, China

Liang Wang
Chinese Academy of Sciences
Beijing, China

Hanzi Wang (iD)
Xiamen University
Xiamen, China

Weishi Zheng (iD)
Sun Yat-sen University
Guangzhou, China

Xilin Chen (iD)
Chinese Academy of Sciences
Beijing, China

Rongrong Ji (iD)
Xiamen University
Xiamen, China

ISSN 0302-9743   ISSN 1611-3349 (electronic)
Lecture Notes in Computer Science
ISBN 978-981-99-8539-5   ISBN 978-981-99-8540-1 (eBook)
https://doi.org/10.1007/978-981-99-8540-1

This Springer imprint is published by the registered company Springer Nature Singapore Pte Ltd.
The registered company address is: 152 Beach Road, #21-01/04 Gateway East, Singapore 189721, Singapore

Paper in this product is recyclable.

Preface

Welcome to the proceedings of the Sixth Chinese Conference on Pattern Recognition and Computer Vision (PRCV 2023), held in Xiamen, China.

PRCV is formed from the combination of two distinguished conferences: CCPR (Chinese Conference on Pattern Recognition) and CCCV (Chinese Conference on Computer Vision). Both have consistently been the top-tier conference in the fields of pattern recognition and computer vision within China's academic field. Recognizing the intertwined nature of these disciplines and their overlapping communities, the union into PRCV aims to reinforce the prominence of the Chinese academic sector in these foundational areas of artificial intelligence and enhance academic exchanges. Accordingly, PRCV is jointly sponsored by China's leading academic institutions: the Chinese Association for Artificial Intelligence (CAAI), the China Computer Federation (CCF), the Chinese Association of Automation (CAA), and the China Society of Image and Graphics (CSIG).

PRCV's mission is to serve as a comprehensive platform for dialogues among researchers from both academia and industry. While its primary focus is to encourage academic exchange, it also places emphasis on fostering ties between academia and industry. With the objective of keeping abreast of leading academic innovations and showcasing the most recent research breakthroughs, pioneering thoughts, and advanced techniques in pattern recognition and computer vision, esteemed international and domestic experts have been invited to present keynote speeches, introducing the most recent developments in these fields.

PRCV 2023 was hosted by Xiamen University. From our call for papers, we received 1420 full submissions. Each paper underwent rigorous reviews by at least three experts, either from our dedicated Program Committee or from other qualified researchers in the field. After thorough evaluations, 522 papers were selected for the conference, comprising 32 oral presentations and 490 posters, giving an acceptance rate of 37.46%. The proceedings of PRCV 2023 are proudly published by Springer.

Our heartfelt gratitude goes out to our keynote speakers: Zongben Xu from Xi'an Jiaotong University, Yanning Zhang of Northwestern Polytechnical University, Shutao Li of Hunan University, Shi-Min Hu of Tsinghua University, and Tiejun Huang from Peking University.

We give sincere appreciation to all the authors of submitted papers, the members of the Program Committee, the reviewers, and the Organizing Committee. Their combined efforts have been instrumental in the success of this conference. A special acknowledgment goes to our sponsors and the organizers of various special forums; their support made the conference a success. We also express our thanks to Springer for taking on the publication and to the staff of Springer Asia for their meticulous coordination efforts.

We hope these proceedings will be both enlightening and enjoyable for all readers.

October 2023

Qingshan Liu
Hanzi Wang
Zhanyu Ma
Weishi Zheng
Hongbin Zha
Xilin Chen
Liang Wang
Rongrong Ji

Organization

General Chairs

Hongbin Zha   Peking University, China
Xilin Chen   Institute of Computing Technology, Chinese
     Academy of Sciences, China
Liang Wang   Institute of Automation, Chinese Academy of
     Sciences, China
Rongrong Ji   Xiamen University, China

Program Chairs

Qingshan Liu   Nanjing University of Information Science and
     Technology, China
Hanzi Wang   Xiamen University, China
Zhanyu Ma   Beijing University of Posts and
     Telecommunications, China
Weishi Zheng   Sun Yat-sen University, China

Organizing Committee Chairs

Mingming Cheng   Nankai University, China
Cheng Wang   Xiamen University, China
Yue Gao   Tsinghua University, China
Mingliang Xu   Zhengzhou University, China
Liujuan Cao   Xiamen University, China

Publicity Chairs

Yanyun Qu   Xiamen University, China
Wei Jia   Hefei University of Technology, China

Local Arrangement Chairs

Xiaoshuai Sun	Xiamen University, China
Yan Yan	Xiamen University, China
Longbiao Chen	Xiamen University, China

International Liaison Chairs

Jingyi Yu	ShanghaiTech University, China
Jiwen Lu	Tsinghua University, China

Tutorial Chairs

Xi Li	Zhejiang University, China
Wangmeng Zuo	Harbin Institute of Technology, China
Jie Chen	Peking University, China

Thematic Forum Chairs

Xiaopeng Hong	Harbin Institute of Technology, China
Zhaoxiang Zhang	Institute of Automation, Chinese Academy of Sciences, China
Xinghao Ding	Xiamen University, China

Doctoral Forum Chairs

Shengping Zhang	Harbin Institute of Technology, China
Zhou Zhao	Zhejiang University, China

Publication Chair

Chenglu Wen	Xiamen University, China

Sponsorship Chair

Yiyi Zhou	Xiamen University, China

Exhibition Chairs

Bineng Zhong	Guangxi Normal University, China
Rushi Lan	Guilin University of Electronic Technology, China
Zhiming Luo	Xiamen University, China

Program Committee

Baiying Lei	Shenzhen University, China
Changxin Gao	Huazhong University of Science and Technology, China
Chen Gong	Nanjing University of Science and Technology, China
Chuanxian Ren	Sun Yat-Sen University, China
Dong Liu	University of Science and Technology of China, China
Dong Wang	Dalian University of Technology, China
Haimiao Hu	Beihang University, China
Hang Su	Tsinghua University, China
Hui Yuan	School of Control Science and Engineering, Shandong University, China
Jie Qin	Nanjing University of Aeronautics and Astronautics, China
Jufeng Yang	Nankai University, China
Lifang Wu	Beijing University of Technology, China
Linlin Shen	Shenzhen University, China
Nannan Wang	Xidian University, China
Qianqian Xu	Key Laboratory of Intelligent Information Processing, Institute of Computing Technology, Chinese Academy of Sciences, China
Quan Zhou	Nanjing University of Posts and Telecommunications, China
Si Liu	Beihang University, China
Xi Li	Zhejiang University, China
Xiaojun Wu	Jiangnan University, China
Zhenyu He	Harbin Institute of Technology (Shenzhen), China
Zhonghong Ou	Beijing University of Posts and Telecommunications, China

Contents – Part VII

Feature Extraction and Feature Selection

Multimedia Analysis and Reasoning

Optimization and Learning Methods

Document Analysis and Recognition

Feature Enhancement with Text-Specific Region Contrast for Scene Text Detection

Xurui Sun[1,2], Jiahao Lyu[1,2], Yifei Zhang[1,2], Gangyan Zeng[3], Bo Fang[1,2], Yu Zhou[1(✉)], Enze Xie[1,2], and Can Ma[1]

[1] Institute of Information Engineering, Chinese Academy of Sciences, Beijing, China
{sunxurui,lvjiahao,zhangyifei0115,fangbo,zhouyu,xieenze,macan}@iie.ac.cn
[2] School of Cyber Security, University of Chinese Academy of Sciences, Beijing, China
[3] School of Information and Communication Engineering, Communication University of China, Beijing, China
zgy1997@cuc.edu.cn

Abstract. As a fundamental step in most visual text-related tasks, scene text detection has been widely studied for a long time. However, due to the diversity in the foreground, such as aspect ratios, colors, shapes, *etc.*, as well as the complexity of the background, scene text detection still faces many challenges. It is often difficult to obtain discriminative text-level features when dealing with overlapping text regions or ambiguous regions of adjacency, resulting in suboptimal detection performance. In this paper, we propose Text-specific Region Contrast (TRC) based on contrastive learning to enhance the features of text regions. Specifically, to formulate positive and negative sample pairs for contrast-based training, we divide regions in scene text images into three categories, *i.e.*, text regions, backgrounds, and text-adjacent regions. Furthermore, we design a Text Multi-scale Strip Convolutional Attention module, called TextMSCA, to refine embedding features for precise contrast. We find that the learned features can focus on complete text regions and effectively tackle the ambiguity problem. Additionally, our method is lightweight and can be implemented in a plug-and-play manner while maintaining a high inference speed. Extensive experiments conducted on multiple benchmarks verify that the proposed method consistently improves the baseline with significant margins.

Keywords: Scene Text Detection · Contrastive Learning · Feature Enhancement · Lightweight Method

1 Introduction

Scene text conveys valuable information and thus is of critical importance for the understanding of natural scenes. As an essential step prior to many text-related

Supported by the Natural Science Foundation of China (Grant NO 62376266), and by the Key Research Program of Frontier Sciences, CAS (Grant NO ZDBS-LY-7024).

tasks *e.g.*, text recognition [18], text retrieval [5], *etc.*, scene text detection (STD) has received extensive attention from researchers and practitioners alike.

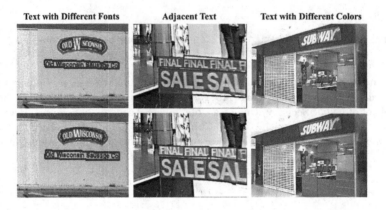

Fig. 1. The left side depicts challenges encountered in scene text detection under various scenarios, while the right side showcases the method's explanation in feature space.

Roughly speaking, most STD methods are inspired by object detection and segmentation paradigms, which can be divided into regression-based methods [12,31] and segmentation-based methods [4,11,17,24,25]. A segmentation-based text detection method has an innate advantage in the detection speed of scene text. Nevertheless, directly applying general segmentation networks for STD suffers from several limitations. First, commonly used semantic segmentation backbones often consist of stacked convolution networks and pooling layers, which can cause a loss of context information and increase false positive samples (e.g., railings misclassified to text category). Second, segmentation-based methods classify images at the pixel level, making adjacent text can not be separated effectively. Third, the larger aspect ratio of scene text makes it difficult for segmentation models designed for general objects to adapt.

Admittedly, there are some efforts to tackle one or more of the problems mentioned above. For example, in targeting the false positive problem, SPC-NET [26] utilizes a text context module to strengthen global information and a re-scoring mechanism to filter out false positives. For adjacent text detection, PixelLink [4] introduces the link prediction scheme, which uses 8 neighbors of each pixel to distinguish different instances. PSENet [24] designs a progressive scale expansion algorithm to adapt to text with large aspect ratios. However, due to the diversity of text foreground and the complexity of the background, these methods still struggle to effectively address the challenges of detecting variable text. Moreover, complex feature-processing blocks or post-processing steps are required, which causes a serious decrease in model efficiency.

In this work, aiming to ensure a fast inference speed, we address the aforementioned problems from a feature enhancement perspective. A method termed Text-specific Region Contrast (TRC) is proposed, which utilizes contrastive learning [19] to obtain more discriminative features for STD. To be specific, different from the common practice [8], we construct a new category of adjacent text samples to form more meaningful contrast pairs accompanied by a novel dynamic region-level sampling strategy. Through these operations, our model aggregates more context information and obtains enhanced features. Furthermore, we design a feature refinement module Text Multi-scale Strip Convolutional Attention (TextMSCA), which utilizes strip convolutions to fit text instances with extreme aspect ratios. As shown in Fig. 1, equipped with the proposed contrastive learning scheme, the detector can obtain more accurate and complete detection results. Extensive experiments conducted on benchmarks show that using TRC and TextMSCA techniques improves the baseline PAN [25] with significant margins. Especially, with the help of SynthText pretraining, our method achieves F-measure performance gains of 2.1%, 2.8%, and 7.5% respectively on the CTW1500 [30], TotalText [3] and MSRA-TD500 [29] datasets while maintaining a high inference speed. The contributions are summarized as follows:

- We propose a novel region-level contrastive learning framework for scene text detection, named Text-specific Region Contrast (TRC), which is able to tackle typical detection challenges via feature enhancement.
- In the proposed framework, text adjacent regions are involved as a new negative category, and a dynamic region-level sampling strategy is implemented based on the weighting of the groundtruth map and score map.
- We design a feature refinement module, $i.e.$, TextMSCA, which could be accustomed to extreme aspect ratios of texts to extract robust representations.
- Extensive experiments demonstrate that the proposed method surpasses the baseline with significant margins and maintains efficient inference overhead.

2   Related Work

Scene Text Detection. Inspired by upstream detection and segmentation frameworks, existing scene text detection methods can be mainly divided into regression-based methods [12,31] and segmentation-based methods [4,11,17, 24,25]. Compared with regression-based methods, segmentation-based methods achieve great success to detect curved text instances. In this paper, we focus on segmentation-based methods with lightweight backbones. PAN [25] and DBNet [11] are two representations of real-time methods, which separately design a pixel aggregation and a differentiable binarization mechanism.

Contrastive Learning. Contrastive learning typically pulls positive samples closer and pushes negative samples away. As a pioneering research, MoCo [7] utilizes data augmentation techniques to obtain positive sample pairs and stores negative samples in a memory bank. SimCLR [2] further simplifies this architecture by using other samples in the batch as negative samples. In addition to being applied in unsupervised scenarios, contrastive learning can effectively

assimilate annotation information. Supervised contrastive learning [10] introduces labels into the contrastive paradigm, and this formulation has been successfully applied in semantic segmentation. Wang et al. [23] raise a pixel-to-pixel
contrastive learning method for semantic segmentation in the fully supervised
setting. Hu et al. [8] form contrastive learning in a region manner.

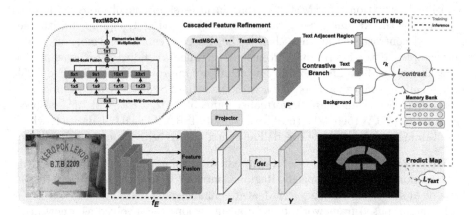

Fig. 2. Illustration of the pipeline. A general segmentation-based scene text detection
framework is placed at the bottom, and a plug-and-play contrastive learning branch is
placed at the top. Details of the strip convolution setting in TextMSCA are shown in
the top left. Text-specific region contrast (TRC) is shown at the top right.

3 Methodology

3.1 Overall Architecture

Previous works [17,24] have demonstrated that segmentation-based text detection methods are effective in handling complex curved text instances. In our
experiments, we mainly utilized lightweight segmentation-based pipelines such
as PAN [25] as baselines. These models share commonalities: a CNN-based backbone for extracting image features, a feature refinement layer for integrating
multi-scale features, and a detection head for classifying dense pixels prior to
post-processing the segmentation map.

To state our method better, we pre-define some abbreviations of the general
model. We formulate f_E as the combination of the CNN-based backbone and
the feature fusion module. Given an input image $x \in \mathbb{R}^{C \times H \times W}$, the feature map
$F \in \mathbb{R}^{D \times H \times W}$ can be extracted by feeding x into f_E. C, H, and W represent the
channel, height, and width of the image, respectively. D represents the feature
dimension of every pixel embedding $F_p \in \mathbb{R}^D, p \in \{p = (i,j)|i = 1, ..., H, j =
1, ..., W\}$ in feature map F. Then a detection head f_{det} maps the feature F into
a normalized score map $Y = \{y_p|p\} = f_{det}(F) \in \mathbb{R}^{H \times W}$, which indicates the
presence of texts on the position of pixel p.

In our methods, we make use of the feature map F as the embedding input in the refinement and contrast branch. We design a cascaded feature refinement module TextMSCA to transfer F into $F^* \in \mathbb{R}^{D \times H \times W}$ to obtain more suitable text-specific features. And the binary groundtruth map $\hat{Y}$ is the supervised signal in contrastive paradigm. With the supervision of score map Y and the groundtruth map $\hat{Y} \in \mathbb{R}^{H \times W}$, we aggregate features of text regions, background, and text adjacent regions and perform contrastive learning in F^*. Details are presented in Sect. 3.2, and the pipeline of our approach is displayed in Fig. 2.

3.2   Sampling in Text-Specific Region Contrast

The cross-image contrastive paradigm includes two sample groups: one from a mini-batch and another from a memory bank. Both groups have samples from the text or background category. We notice that nearby texts are often linked due to proximity and similarity, requiring special consideration. To handle this, we introduce negative samples called text adjacent regions (shown in Fig. 3). This helps address ambiguity in adjacent text connections. Finally, we use a contrastive loss to bring positive samples closer and separate negative samples. In this section, we provide a detailed account of the negative sample construction and sampling strategy. Additionally, we explain how to construct features of text adjacent regions.

1) **Region-Level Sampling.** Region-Level feature construction pays attention to designing an overall feature that comes from the average pooling of pixel embedding belonging to the same class in an image. To make full use of labels, we use predicted score maps and groundtruth masks to distinguish true or false foreground pixel samples and assign them different weights, which means to guide the model focusing on samples easy to misclassify. Given the enhanced feature map $F^* \in \mathbb{R}^{D \times H \times W}$ and score map $Y \in \mathbb{R}^{H \times W}$, firstly we use a predefined threshold t which follows the experiment setting in [25] and groundtruth map $\hat{Y} \in \mathbb{R}^{H \times W}$ to obtain right or wrong anchors. For example, for a specified category $text$: $\{k = 1 | category = text\}$, right pixel map can be denoted as $RMap^1 = \mathbb{1}(Y_p \geq t) \cap \mathbb{1}(\hat{Y}_p \geq t)$, while wrong pixel map can be denoted as $WMap^1 = \mathbb{1}(Y_p < t) \cap \mathbb{1}(\hat{Y}_p \geq t)$. The definition of $text$ region anchor could be set as:

$$r_k = \frac{\sum_p F_p^* \cdot ((1 - y_p) \cdot RMap_p^k + WMap_p^k)}{\sum_p \mathbb{1}|\hat{Y}_p^k|}, \tag{1}$$

where k denotes the class of $text$: $\{k = 1 | category = text\}$ or $background$: $\{k = 0 | category = background\}$ while $\hat{Y}_p \geq t$ denotes Y_p^{text} and $\hat{Y}_p < t$ denotes $Y_p^{background}$. And $\mathbb{1}(\cdot)$ represents the binary classifier for pixel in $\hat{Y}$ and Y and $\mathbb{1}|\cdot|$ denotes the operation employed on corresponding category k, such as when $k = text$, $\mathbb{1}|\hat{Y}_p^k| = \mathbb{1}|\hat{Y}_p^k > t|$.

Regarding the other group of sampling pixels in the memory bank, we update the region anchor vectors within a mini-batch to the corresponding memory

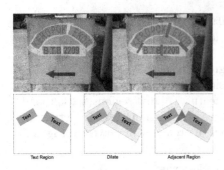

Fig. 3. We visualize and explain the specific design of contrast sampling. As we can see, we add a new category (red areas) of the negative sample to represent the text adjacent region. (Color figure online)

Algorithm 1. Text Adjacent Region Generation

Input: batchsize B, refined feature F^*, dilated coefficient σ, GT map Y, area of text instance set T of each GT map

Output: text adjacent region embedding $r_{adjacent} \in \mathbf{R}^{B \times D}$

1: **for** i-*th* feature and GT map in B **do**
2:    Refined feature F_i^*, GT map Y_i.
3:    **for** k-*th* text instance T_k in Y_i **do**
4:       Dilate region T_k with coef. σ.
5:    **end for**
6:    Get intersection region r_i in Y_i.
7:    Apply Eq. 2 with F_i^* and Y_i.
8: **end for**
9: Get text adjacent region embedding $r_{adjacent}$.

---

queue after computing the training loss. These updated vectors are then used in the subsequent iterations. The complete sampling process is illustrated in the Supplemental Material.

2) Text Adjacent Region Generation. We discovered a false detection issue in some segmentation-based scene text detection models. When two text instances are close to each other, they may merge into one instance. This is difficult to avoid due to dense text and a single category. To address this problem, we creatively constructed negative samples representing adjacent text regions as shown in Fig. 3. By separating these areas from the background, we enable the detection model to recognize the existence of text adjacent areas.

The features of text adjacent regions are calculated separately. We dilate text regions with an offset by the Vatti clipping algorithm and get intersections of dilated regions to represent text adjacent region $TAMap$. The adjacent region feature is calculated by using Eq. 2 and details are shown in Algorithm 1.

$$r_{adjacent} = \frac{\sum_{x,y} F^* \cdot TAMap}{\sum \mathbb{1} |\hat{Y}_{adjacent}|}. \tag{2}$$

3.3   Text-Specific Feature Refinement Module

Features used in contrastive learning mostly arise from general feature fusion layers. Inspired by the MSCA module in SegNeXt [6], we design a text-specific multi-scale strip convolutional attention module (TextMSCA) to refine features F from f_E because text often has strip-like shape and extreme aspect ratios. As shown in Fig. 2, we look into the aspect ratio of text in the dataset and set four special strip convolutions to refine text features.

$$F^* = Conv_{1 \times 1}(\sum_{i=0}^{4} SC_i(DC(F))) \otimes F \tag{3}$$

Here DC denotes depth-wise convolution and SC_i denotes the i-th strip convolution branch in TextMSCA. $\otimes$ is the element-wise matrix multiplication operation. After putting the feature map into this module, TextMSCA can aggregate the strip context and extract better text-specific features. Therefore, F^* represents scene text features better in more directions and aspects.

3.4 Loss Function and Inference

Contrastive Loss. After describing our contrastive sampling and refinement module, we introduce the supervised contrastive formula. This method is employed when data is labeled. Our method is derived from self-supervised contrastive learning, which uses the InfoNCE [16] loss function. For segmentation-based tasks, we need a more fine-grained loss calculation at the region and pixel levels defined below:

$$\mathcal{L}_{contrast} = \frac{1}{|M_r|} \sum_{r^+ \in M_r} -log \frac{exp(r \cdot r^+/\tau)}{exp(r \cdot r^+/\tau) + \sum_{r^-} exp(r \cdot r^-/\tau)} \tag{4}$$

In detail, r denotes the anchor feature, when faced with region-level contrast loss, r^+ means the positive region samples that belong to the same label as the anchor region r, and r^- means the negative region samples. M_r represents a set of positive regions samples in a mini-batch or the memory bank. Note that region embedding r always comes from the average pooling of pixels embedding in corresponding category regions.

Overall Objective. The overall loss function can be formulated as:

$$\mathcal{L} = \mathcal{L}_{text} + \lambda \cdot \mathcal{L}_{contrast} \tag{5}$$

where $\mathcal{L}_{text}$ denotes the original loss and $\mathcal{L}_{contrast}$ represents the contrast loss used in training stage. λ is a hyper-parameter that balances the weights of $\mathcal{L}_{text}$ and $\mathcal{L}_{contrast}$. The contrast branch does not participate in the inference phase and makes the model maintain the original efficiency.

4 Experiment

4.1 Experimental Setting

Datasets. We utilize 5 typical scene text detection datasets for experiments. **ICDAR 2015** [9] includes many dense and small instances. **CTW1500** [30] and **Total-Text** [3] include instances with various shapes such as horizontal, multi-oriented, and curved situations. **MSRA-TD500** [29] is a multi-language dataset including Chinese and English text instances. At last **MLT-2017** [15] is also a large multi-language dataset that includes 9 languages.

Table 1. Ablation of text adjacent regions as negative samples. TRC(w/o) denotes without consideration of text adjacent regions.

Method	CTW1500			TotalText			TD500		
	P	R	F	P	R	F	P	R	F
PAN	84.6	77.7	81	88	79.4	83.5	80.7	77.3	78.9
PAN+TRC(w/o)	85.7	77.2	81.3	88.3	80	83.9	85.8	78.9	82.1
PAN+TRC	**86.7**	**78.0**	**82.1**	**89.3**	**81.3**	**85.1**	**88.2**	**80.7**	**84.3**

Implementation Details. We use ResNet18 as the lightweight backbone by default and PAN [25] is used as the baseline to evaluate the effectiveness of our method due to its stability and universality. The contrastive learning branch is a plug-and-play component, so the detection branch uses the original baseline settings for hyperparameters. PAN is trained from scratch, and we follow default strategies for data augmentation and optimization. The coefficient λ in Eq. 5 is set to 0.025, and the temperature τ in the contrastive loss is set to 0.7 as in [23]. During inference, we use the same method as the original model, and the TextMSCA module and contrastive branch are not involved.

4.2 Ablation

Effectiveness of Text Adjacent Regions. We created a contrastive sampling strategy specifically tailored to the characteristics of the text, with text adjacent regions added to the memory bank as negative samples. We compared the results of using this strategy versus not using it on three datasets with complex text shapes. Table 1 shows that adding text adjacent regions improves the effectiveness of contrastive learning, especially on the MSRA-TD500 dataset with 2.2% improvement. This design helps the model pay attention to ambiguous regions and effectively distinguish text boundaries, resulting in more robust features and alleviating the problem of connecting adjacent texts.

Effectiveness of TextMSCA. We compared our proposed TextMSCA module with the original MSCA [6] module to validate its effectiveness. We conducted experiments on three datasets, setting the contrastive branch's TRC with two types of negative samples, as in the previous ablation experiment. Table 3 shows that TextMSCA outperforms MSCA by up to 3.1% on MSRA-TD500. The heatmap figure in the supplemental materials further demonstrates that our large-scale strip convolution can better extract text features in various aspect ratios, providing more detailed and robust information in images.

The Flexibility of TRC. To justify the flexibility of TRC in a plug-and-play manner, we transfer the contrastive branch into another segmentation-based scene text detection model DB [11] and explore whether it brings improvements

Table 2. Ablation on DB.

Method	IC15	MLT17	CTW1500	TotalText	TD500
DB	82.3	71.0	81.0	82.8	82.8
DB+TPC	82.1	**72.5**	**81.4**	83.2	83.2
DB+TRC	**82.6**	72.1	**81.4**	**84.0**	**84.1**

Table 4. Comparisons on CTW1500. ∗ means adding TextMSCA.

Method	Ext.	P	R	F	FPS
EAST [32]	–	78.7	49.1	60.4	21.2
TextSnake [14]	✓	67.9	85.3	75.6	1.1
TextField [27]	✓	83.0	79.8	81.4	–
TextRay [21]	✓	82.8	80.4	81.6	–
ABCNet [12]	✓	81.4	78.5	81.6	–
DBNet(R18) [11]	✓	84.8	77.5	81.0	55
CTN(R18) [17]	–	85.5	79.2	82.2	40.8
Fuzzy(R18) [22]	✓	84.6	77.7	81.0	35.2
PAN(R18) [25]	–	84.6	77.7	81.0	39.8
PAN(R18)∗+TPC	–	86.9	77.9	82.1	38.2
PAN(R18)∗+TRC	–	86.7	78.0	82.1	34.2
PAN(R18)∗+TRC	✓	87.2	79.4	**83.1**	34.2

Table 3. Ablation of TextMSCA.

Method	CTW1500			TotalText			TD500		
	P	R	F	P	R	F	P	R	F
PAN	84.6	77.7	81	88	79.4	83.5	80.7	77.3	78.9
+MSCA	86.3	77.9	81.9	88.4	80.5	84.3	84.7	78	81.2
+TextMSCA	**86.7**	**78**	**82.1**	**89.3**	**81.3**	**85.1**	**88.2**	**80.7**	**84.3**

or not. We employ the contrastive branch to train a new DB model without TextMSCA module on five classic datasets and results are shown in Table 2. Text Pixel Contrast (TPC) [23] is a pixel-level sampling method for semantic segmentation tasks. We also compare it with TRC in the experiment. Experimental results show that with the attachment of the contrastive branch, DB gains 1.5%, 1.2%, and 1.3% improvements on MLT17, TotalText, and TD500, respectively. Moreover, TRC outperforms TPC on multiple datasets, which demonstrates the effectiveness of our method.

4.3 Performance Comparison

To evaluate the effectiveness of our method, we conduct thorough experiments on three benchmark datasets CTW1500, TotalText, and MSRA-TD500 in both qualitative and quantitative forms.

Table 5. Comparisons on TotalText. ∗ means adding TextMSCA.

Method	Ext.	P	R	F	FPS
TextSnake [14]	✓	82.7	74.5	78.4	–
PSENet-1s [24]	✓	84.0	78.0	80.9	3.9
TextField [27]	✓	81.2	79.9	80.6	–
CRAFT [1]	✓	87.6	79.9	83.6	–
DBNet(R18) [11]	✓	88.3	77.9	82.8	50
ABCNet [12]	✓	87.9	81.3	84.5	–
OPMP [31]	✓	85.2	80.3	82.7	3.7
CTN [17]	–	–	–	85.6	24
Fuzzy [22]	✓	88.7	79.9	84.1	24.3
PAN(R18) [25]	–	88.0	79.4	83.5	39.6
PAN(R18)∗+TPC	–	88.5	80.4	84.3	39.1
PAN(R18)∗+TRC	–	89.3	81.3	85.1	37.9
PAN(R18)∗+TRC	✓	90.7	82.4	**86.3**	37.2

Table 6. Comparisons on MSRA-TD500. ∗ means adding TextMSCA.

Method	Ext.	P	R	F	FPS
MCN [13]	✓	88.0	79.0	83.0	–
PixeLink [4]	✓	83.0	73.2	77.8	3
TextSnake [14]	✓	83.2	73.9	78.3	1.1
MSR [28]	✓	87.4	76.7	81.7	–
CRAFT [1]	✓	88.2	78.2	82.9	8.6
SAE [20]	✓	84.2	81.7	82.9	–
DBNet [11]	✓	91.5	79.2	84.9	32
CTN [17]	–	–	–	83.5	20.5
Fuzzy [22]	✓	89.3	81.6	85.3	–
PAN(R18) [25]	–	80.7	77.3	78.9	30.2
PAN(R18)∗+TPC	–	85.8	78.9	82.1	33.7
PAN(R18)∗+TRC	–	88.2	80.7	84.3	32.8
PAN(R18)∗+TRC	✓	89.8	83.3	**86.4**	32.6

Firstly, we present the results under different contrastive settings qualitatively. To reflect the effectiveness that contrastive learning brings to the scene text detection, we follow the training mode for PAN [25] that only uses the **ResNet18** as the backbone **with or without** pretraining using SynthText. As we can see in Table 4, it brings 1.1% improvements to PAN without pretraining. In Table 5 and Table 6, results show that it brings a maximum of 1.6% improvements on TotalText and 5.4% improvements on MSRA-TD500. We also pretrain the model with SynthText and finetune the model on these datasets, and it brings further improvement. Results from Table 4, Table 5 and Table 6 demonstrate that using cross-image region-level contrastive learning assists PAN to get impressive improvements and it even catches up with most methods using ResNet50 as their backbone. FPS would decrease slightly due to the increasing number of connected areas in pixel aggregation before filtering, but the time consumption is negligible. In addition, Fig. 4 shows our visualization results. Intuitively, the contrastive branch with TRC and TextMSCA solves the problem of text detection errors such as stripe-like patterns. When faced with a line of text with different colors or fonts, it also performs better. The quantitative result and the qualitative visualization prove that our proposed model extracts more robust text features from images instead of relying on colors or shapes.

Fig. 4. Visualization of groundtruth (left), results of the original PAN (middle), and PAN with the proposed contrastive branch (right).

5 Conclusion

In this paper, we propose a text-specific region contrast (TRC) method to enhance the feature of text regions for scene text detection. Then, a text multiscale strip convolutional attention module (TextMSCA) is designed to further refine embedding feature. We conduct extensive experiments and visualizations to demonstrate the effectiveness of contrastive learning in enhancing text detection. Moreover, the proposed contrastive branch is a plug-and-play component without introducing additional computation during the inference phase. In the future, we will extend our method to address other relevant tasks, for example, multi-language text detection.

References

1. Baek, Y., Lee, B., Han, D., Yun, S., Lee, H.: Character region awareness for text detection. In: Proceedings of the IEEE/CVF Conference on Computer Vision and Pattern Recognition, pp. 9365–9374 (2019)
2. Chen, T., Kornblith, S., Norouzi, M., Hinton, G.: A simple framework for contrastive learning of visual representations. In: International Conference on Machine Learning, pp. 1597–1607. PMLR (2020)
3. Ch'ng, C.K., Chan, C.S.: Total-Text: a comprehensive dataset for scene text detection and recognition. In: 2017 14th IAPR International Conference on Document Analysis and Recognition (ICDAR), vol. 1, pp. 935–942. IEEE (2017)
4. Deng, D., Liu, H., Li, X., Cai, D.: PixelLink: detecting scene text via instance segmentation. In: Proceedings of the AAAI Conference on Artificial Intelligence, vol. 32 (2018)
5. Gómez, L., Mafla, A., Rusiñol, M., Karatzas, D.: Single shot scene text retrieval. In: Ferrari, V., Hebert, M., Sminchisescu, C., Weiss, Y. (eds.) Computer Vision – ECCV 2018. LNCS, vol. 11218, pp. 728–744. Springer, Cham (2018). https://doi. org/10.1007/978-3-030-01264-9_43
6. Guo, M.H., Lu, C.Z., Hou, Q., Liu, Z., Cheng, M.M., Hu, S.M.: SegNeXt: rethinking convolutional attention design for semantic segmentation. arXiv preprint arXiv:2209.08575 (2022)
7. He, K., Fan, H., Wu, Y., Xie, S., Girshick, R.: Momentum contrast for unsupervised visual representation learning. In: Proceedings of the IEEE/CVF Conference on Computer Vision and Pattern Recognition, pp. 9729–9738 (2020)
8. Hu, H., Cui, J., Wang, L.: Region-aware contrastive learning for semantic segmentation. In: Proceedings of the IEEE/CVF International Conference on Computer Vision, pp. 16291–16301 (2021)
9. Karatzas, D., et al.: ICDAR 2015 competition on robust reading. In: 2015 13th International Conference on Document Analysis and Recognition (ICDAR), pp. 1156–1160. IEEE (2015)
10. Khosla, P., et al.: Supervised contrastive learning. Adv. Neural. Inf. Process. Syst. **33**, 18661–18673 (2020)
11. Liao, M., Wan, Z., Yao, C., Chen, K., Bai, X.: Real-time scene text detection with differentiable binarization. In: Proceedings of the AAAI Conference on Artificial Intelligence, vol. 34, pp. 11474–11481 (2020)
12. Liu, Y., Chen, H., Shen, C., He, T., Jin, L., Wang, L.: ABCNet: real-time scene text spotting with Adaptive Bezier-Curve network. In: Proceedings of the IEEE/CVF Conference on Computer Vision and Pattern Recognition, pp. 9809–9818 (2020)
13. Liu, Z., Lin, G., Yang, S., Feng, J., Lin, W., Goh, W.L.: Learning Markov clustering networks for scene text detection. arXiv preprint arXiv:1805.08365 (2018)
14. Long, S., Ruan, J., Zhang, W., He, X., Wu, W., Yao, C.: TextSnake: a flexible representation for detecting text of arbitrary shapes. In: Ferrari, V., Hebert, M., Sminchisescu, C., Weiss, Y. (eds.) ECCV 2018. LNCS, vol. 11206, pp. 19–35. Springer, Cham (2018). https://doi.org/10.1007/978-3-030-01216-8_2

15. Nayef, N., et al.: ICDAR 2017 robust reading challenge on multi-lingual scene text detection and script identification-RRC-MLT. In: 2017 14th IAPR International Conference on Document Analysis and Recognition (ICDAR), vol. 1, pp. 1454–1459. IEEE (2017)

16. Oord, A.V.D., Li, Y., Vinyals, O.: Representation learning with contrastive predictive coding. arXiv preprint arXiv:1807.03748 (2018)

17. Sheng, T., Chen, J., Lian, Z.: CentripetalText: an efficient text instance representation for scene text detection. Adv. Neural. Inf. Process. Syst. **34**, 335–346 (2021)

18. Shi, B., Yang, M., Wang, X., Lyu, P., Yao, C., Bai, X.: ASTER: an attentional scene text recognizer with flexible rectification. IEEE Trans. Pattern Anal. Mach. Intell. **41**(9), 2035–2048 (2018)

19. Tian, Y., Krishnan, D., Isola, P.: Contrastive multiview coding. In: Vedaldi, A., Bischof, H., Brox, T., Frahm, J.-M. (eds.) ECCV 2020. LNCS, vol. 12356, pp. 776–794. Springer, Cham (2020). https://doi.org/10.1007/978-3-030-58621-8_45

20. Tian, Z., et al.: Learning shape-aware embedding for scene text detection. In: Proceedings of the IEEE/CVF Conference on Computer Vision and Pattern Recognition, pp. 4234–4243 (2019)

21. Wang, F., Chen, Y., Wu, F., Li, X.: TextRay: contour-based geometric modeling for arbitrary-shaped scene text detection. In: Proceedings of the 28th ACM International Conference on Multimedia, pp. 111–119 (2020)

22. Wang, F., Xu, X., Chen, Y., Li, X.: Fuzzy semantics for arbitrary-shaped scene text detection. IEEE Trans. Image Process. **32**, 1–12 (2022)

23. Wang, W., Zhou, T., Yu, F., Dai, J., Konukoglu, E., Van Gool, L.: Exploring cross-image pixel contrast for semantic segmentation. In: Proceedings of the IEEE/CVF International Conference on Computer Vision, pp. 7303–7313 (2021)

24. Wang, W., et al.: Shape robust text detection with progressive scale expansion network. In: Proceedings of the IEEE/CVF Conference on Computer Vision and Pattern Recognition, pp. 9336–9345 (2019)

25. Wang, W., et al.: Efficient and accurate arbitrary-shaped text detection with pixel aggregation network. In: Proceedings of the IEEE/CVF International Conference on Computer Vision, pp. 8440–8449 (2019)

26. Xie, E., Zang, Y., Shao, S., Yu, G., Yao, C., Li, G.: Scene text detection with supervised pyramid context network. In: Proceedings of the AAAI Conference on Artificial Intelligence, vol. 33, pp. 9038–9045 (2019)

27. Xu, Y., Wang, Y., Zhou, W., Wang, Y., Yang, Z., Bai, X.: TextField: learning a deep direction field for irregular scene text detection. IEEE Trans. Image Process. **28**(11), 5566–5579 (2019)

28. Xue, C., Lu, S., Zhang, W.: MSR: multi-scale shape regression for scene text detection. arXiv preprint arXiv:1901.02596 (2019)

29. Yao, C., Bai, X., Liu, W., Ma, Y., Tu, Z.: Detecting texts of arbitrary orientations in natural images. In: 2012 IEEE Conference on Computer Vision and Pattern Recognition, pp. 1083–1090. IEEE (2012)

30. Yuliang, L., Lianwen, J., Shuaitao, Z., Sheng, Z.: Detecting curve text in the wild: new dataset and new solution. arXiv preprint arXiv:1712.02170 (2017)

31. Zhang, S., Liu, Y., Jin, L., Wei, Z., Shen, C.: OPMP: an omnidirectional pyramid mask proposal network for arbitrary-shape scene text detection. IEEE Trans. Multimedia **23**, 454–467 (2020)

32. Zhou, X., et al.: EAST: an efficient and accurate scene text detector. In: Proceedings of the IEEE Conference on Computer Vision and Pattern Recognition, pp. 5551–5560 (2017)

Learning Efficient Representations for Image-Based Patent Retrieval

Hongsong Wang[1]([✉])[iD] and Yuqi Zhang[2][iD]

[1] Department of Computer Science and Engineering, Southeast University,
Nanjing 210096, China
`hongsongwang@seu.edu.cn`
[2] Baidu Inc., Beijing, China

Abstract. Patent retrieval has been attracting tremendous interest from researchers in intellectual property and information retrieval communities in the past decades. However, most existing approaches rely on textual and metadata information of the patent, and content-based image-based patent retrieval is rarely investigated. Based on traits of patent drawing images, we present a simple and lightweight model for this task. Without bells and whistles, this approach significantly outperforms other counterparts on a large-scale benchmark and noticeably improves the state-of-the-art by 33.5% with the mean average precision (mAP) score. Further experiments reveal that this model can be elaborately scaled up to achieve a surprisingly high mAP of 93.5%. Our method ranks first in the ECCV 2022 Patent Diagram Image Retrieval Challenge.

Keywords: Image-Based Patent Retrieval · Patent Search · Sketch-Based Image Retrieval

1 Introduction

With a large and ever-increasing number of scientific and technical documents on the web every year, diagram image retrieval and analysis [1,33] become crucially important for intelligent document processing and understanding. Different from natural images that contain rich information about color, texture, and intensity, diagram images only involve line and shape. Although existing computer vision methods have achieved significant progress on natural image understanding (e.g., classification, detection and retrieval) [12,16,26,31], diagram image understanding retains a challenging and less developed area.

Image-based patent retrieval [2,11,37], one of the most typical research topics of diagram image understanding, has drawn increasing interest from intellectual property and information retrieval communities. The need for effective and efficient systems of patent retrieval becomes inevitably crucial due to the tremendous amounts of patent data. As visual image plays an essential role in information retrieval, patent image search can help people quickly understand

---

H. Wang and Y. Zhang—Equal Contribution.

Q. Liu et al. (Eds.): PRCV 2023, LNCS 14431, pp. 15–26, 2024.
https://doi.org/10.1007/978-981-99-8540-1_2

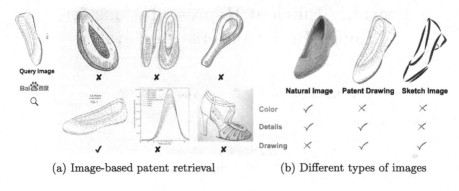

(a) Image-based patent retrieval   (b) Different types of images

Fig. 1. An illustration of image-based patent retrieval.

the patent contents and check differences between patents. However, accurate patent retrieval based on visual content remains an open challenge. Commercial search engines such as Google and Baidu currently fail to retrieve relevant patent images with the query of a patent drawing [15]. An example of this task is illustrated in Fig. 1(a), where many unrelated drawing images with a similar style to the query are retrieved.

One of the well-concerned problems in computer vision most related to image-based patent retrieval is sketch-based image retrieval [9]. Although both are abstract drawings, there are subtle differences between sketch images and patent drawings. As shown in Fig. 1(b), patent drawings contain more detailed contours and patterns about the object when compared with sketch images.

Existing patent search systems mostly use textual and metadata information. Only little published work studies content- and image-based patent retrieval. These approaches represent the patent drawing with relational skeletons [13], edge orientation autocorrelogram [27], contour description matrix [38], adaptive hierarchical density histogram [30], etc. For example, the patent retrieval system PATSEEK [27] applies shape-based retrieval and represents a patent image with an edge orientation autocorrelogram. However, these representations are low-level and handcrafted visual features, which are not discriminative enough for large-scale applications. In addition, previous public patent retrieval datasets such as the `concept` [29] only contains a limited number of ID categories and images. The evaluation protocols of image-based patent retrieval are also not consistent across most existing approaches.

Recently, a large-scale dataset called `DeepPatent` [15] focused on patent drawings is introduced. This dataset makes it possible for large-scale deep learning-based image-based patent retrieval. The PatentNet [15] is currently the only deep learning model designed for large-scale image-based patent retrieval. However, this model is sophisticatedly trained in two stages with different training losses. To the best of our knowledge, there exists no single-stage training model that exploits the unique characteristics of patent drawings and comprehensively studies how to train an effective patent retrieval system from patent images.

To this end, we present a simple yet strong pipeline for image-based patent retrieval which adopts a lightweight backbone and subsequently applies a neck structure to obtain low-dimensional representation for efficient patent retrieval. Training this network is straightforward by only using a classification loss in a geometric angular space and data augmentation without scaling, which is specifically designed for patent drawings according to their characteristics. Experimental results demonstrate that the proposed method significantly outperforms both the traditional and deep learning-based counterparts. In addition, this baseline can be easily scaled up from the current lightweight model to large models which achieve even better performance.

2   Related Work

2.1   Image-Based Patent Retrieval

Patent retrieval [14] is the task of finding relevant patents for a given image query, which can be a sketch, a photo or a patent image. An image and text analysis approach automatically extracts concept information describing the patent image content [29]. A retrieval system uses a hybrid approach that combines feature extraction, indexing and similarity matching to retrieve patent images based on text and shape features [18]. A combined method of classification and indexing techniques retrieves patents for photo queries and evaluates it on a large-scale dataset [10]. PatentNet [15] learns to embed patent drawings into a common feature space using a contrastive loss function, where similar drawings are close and dissimilar drawings are far apart. The diversity and complexity of patent images which contain different objects and the data scarcity and imbalance are main difficulties of image-based patent retrieval.

2.2   Sketch-Based Image Retrieval

Sketch-based image retrieval [32] receives hand-drawn sketch as an input and retrieves images relevant to the sketch. Existing deep learning approaches can be roughly divided into three categories: generative models, cross-modal learning and zero-shot learning. Generative models aim to generate synthetic images or sketches from the input sketch or image using generative neural networks [3,5]. Cross-modal learning approaches learn a common feature space for sketches and images using techniques such as metric learning, domain adaptation and knowledge distillation [4,23]. Zero-shot learning retrieves images that belong to unseen categories using semantic information such as attributes or word embeddings [28,35]. This problem is extremely challenging as the appearance gap between sketches and natural images makes it difficult to learn effective cross-modal representations and style variations among different sketchers introduce noise and ambiguity to the sketch representations. In contrast, image-based patent retrieval does not involve cross-modal learning and variations among patent drawings are smaller than those among sketches.

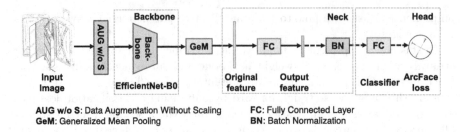

Fig. 2. The network structure of the proposed method. Dotted lines denote modules that only appear in the training pipeline.

3   Method

We present a simple yet effective training pipeline for image-based patent retrieval, which is shown in Fig. 2. Similar to the strong baseline for person re-identification (Re-ID) [17], the network mainly consists of the backbone, the neck and the head. However, training strategies and structures of key components are very different from those of Re-ID. Details are described as follows.

3.1   Backbone Network

EfficientNet-B0 [26] is an effective and efficient convolutional network obtained by neural architecture search [25,39]. Compared with ResNet-50 [12], EfficientNet-B0 significantly reduces network parameters and FLOPS by an order of magnitude. The main building block of this network is the mobile inverted bottleneck called MBConv [24,25], which uses the efficiency inverted structure in the residual block. This architecture can be easily scaled up from dimensions of depth, width, and resolution to obtain a family of models.

Given the output feature maps of the backbone network, we use Generalized Mean (GeM) pooling [22] instead of the commonly used max- or average-pooling to get image features. The pooled feature dimension is the same as the number of channels of feature maps. Let $f \in \mathbb{R}^n$ be the output feature vector of dimension n. For the k-th component of vector f, the formulation of GeM pooling is

$$f_k = \left(\frac{1}{|Z_k|} \sum_{x \in Z_k} x^{p_k}\right)^{1/p_k}, \tag{1}$$

where Z_k is the k-th feature map, and p_k is a learnable pooling parameter, $p_k > 0$. This operation is average-pooling when $p_k = 1$, and approximates max-pooling when $p_k \to \infty$. The GeM pooling focuses on the salient and discriminative features from different spatial positions of the image.

3.2   Neck Structure

The strong baseline for Re-ID presents a neck structure that consists of a batch normalization (BN) layer and empirically analyzes the functions of this BN layer while combining ID classification loss and triplet loss [17]. There exists inconsistency between the classification loss and the triplet loss while combing them in the same feature space, and the BN layer alleviates this inconsistency by separating the two losses into different feature spaces.

Different from previous neck structures [17,34] that involve solely a BN layer for person re-identification, we design a neck structure that comprises a fully connected (FC) layer and a consecutive BN layer. The FC layer maps the pooled feature vector of an image into a compact and low-dimensional space, and, as a result, reduces the parameters of the classifier. In addition, it transforms feature vectors of various dimensions from different backbones to fixed dimensions.

During training, the BN layer normalizes the feature distributions and improves the intra-class compactness and inter-class discrimination, which benefits the ID classification loss. In the inference phase, the output feature of the FC layer is directly used for image retrieval.

3.3 Classification Head

The PatentNet [15] first trains the patent retrieval model with the classification loss and then finetunes the network using either the triplet or contrastive loss. The strong baseline [17] trains the person re-identification network with both the classification and triplet losses. Our approach for image-based patent retrieval is trained in a single stage and only uses classification loss.

Let x_i denote the retrieval feature vector of the i-th image sample, and y_i be the patent ID label. Suppose there are m different ID categories and N samples, the widely used cross-entropy loss or the Softmax loss L_c is

$$L_c = -\frac{1}{N} \sum_{i=1}^{N} \log \frac{e^{W_{y_i}^T x_i + b_{y_i}}}{\sum_{j=1}^{m} e^{W_j^T x_i + b_j}}, \qquad (2)$$

where W and b are the weight and bias parameters of the classifier that consists of an FC layer.

The ArcFace loss [8] first performs L2 normalization upon the feature vector and the weight of the classifier, then computes the cosine distance in a geometric angular space. The loss L_a is formulated as

$$L_a = -\frac{1}{N} \sum_{i=1}^{N} \log \frac{e^{s \cdot \cos(\theta_{y_i} + m)}}{e^{s \cdot \cos(\theta_{y_i} + m)} + \sum_{j=1, j \neq y_i}^{m} e^{s \cdot \cos \theta_j}}, \qquad (3)$$

where s and m are scale and angular margin penalty hyperparameters, respectively. As discussed in [8], the ArcFace loss could enhance similarity for intra-class samples and increase diversity for inter-class samples, which is particularly suitable for large-scale retrieval scenarios due to the huge intra-class appearance variations.

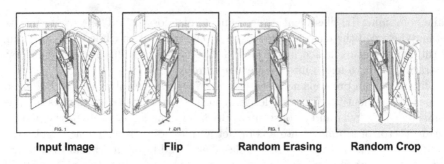

| **Input Image** | **Flip** | **Random Erasing** | **Random Crop** |

Fig. 3. Typical data augmentations without scaling transformation for image-based patent retrieval. For `Random Crop`, the cropped image is not resized to the original image resolution.

3.4 Data Augmentation Without Scaling

Scaling Transformation is a prevalent data augmentation technique for natural image recognition and analysis. As the concerned object in images varies considerably in different sizes, scaling transformation makes the model learn scale-invariant features from images. For image understanding with deep learning, scaling transformation is almost inevitably included in any data augmentations.

Different from natural images, patent drawings span a relatively fixed line width in the document, regardless of the content, and so objects in the drawings vary little in size. For this reason, we present data augmentation without scaling (`AUG w/o S`) for image-based patent retrieval. Typical `AUG w/o S` transformations are shown in Fig. 3. Empirically studies of data augmentations are provided in experiments.

4   Experiments

4.1 Datasets and Implementation Details

The DeepPatent [15] is a recent large-scale dataset for image-based patent retrieval. It consists of a total of 45,000 different design patents and 350,000 drawing images. The object in a patent is presented in multiple drawing images from different views. The training set has 254,787 images from 33,364 patents, and the validation set has 44,815 images from 5,888 patents. This dataset is dedicated to visual patent retrieval based on detailed and abstract drawings. The mean Average Precision (mAP) and ranked accuracy (Rank-N) are used to evaluate the performance of retrieval methods.

The input image resolution is 256×256 if not otherwise specified. The output feature dimension of the FC layer in the neck structure is 512. The scale and margin hyperparameters s and m in the ArcFace loss are 20 and 0.5, respectively. The network is trained using the AdamW optimizer with an initial learning rate of 0.001 and weight decay of 0.0005. The batch size is 128, and the maximum number of iterations is 20,000 (Table 1).

Table 1. Comparison with the state-of-the-art approaches for large-scale patent retrieval on the DeepPatent dataset.

Method	mAP	Rank-1	Rank-5	Rank-20
HOG [7]	8.3	27.2	31.7	35.9
SIFT FV [6]	9.2	20.6	28.9	37.5
LBP [19]	6.9	21.0	25.2	34.3
AHDH [30]	9.5	28.8	34.3	39.9
VisHash [20]	9.3	27.4	34.0	40.2
Sketch-a-Net [36]	13.5	36.1	45.1	53.6
PatentNet [15]	37.6	69.1	78.4	84.1
Ours	71.2	88.9	95.8	98.1
Ours w/o ArcFace	55.4	82.3	91.3	95.2

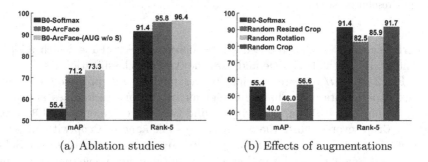

(a) Ablation studies   (b) Effects of augmentations

Fig. 4. Effects of key components of the proposed method.

4.2 Experimental Results

As image-based patent retrieval is a relatively new research topic and traditional general techniques of computer vision are likely to perform well, we compare our approach with both deep learning and traditional approaches.

Representative traditional methods are HOG [7], SIFT FV [6,21], AHDH [30], and VisHash [20]. Deep learning-based comparison methods are Sketch-a-Net [36] and PatentNet [15]. The proposed method significantly outperforms both traditional and deep learning approaches. It beats the recently proposed deep learning-based PatentNet [15] by 33.5% and 19.8% for the scores of mAP and Rank-1, respectively.

4.3 Analysis and Visualization

Ablation studies of key components of the proposed method are shown in Fig. 4(a). ArcFace loss significantly improves the results over Softmax loss for large-scale image-based patent retrieval, as it increases the value of mAP from 55.4% to 71.2%. Data augmentation without scaling (AUG w/o S) further boosts

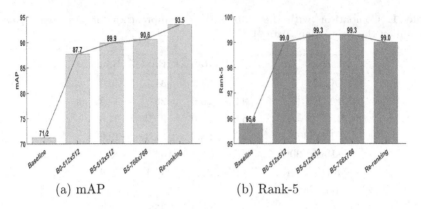

(a) mAP   (b) Rank-5

Fig. 5. Scalability analysis of the proposed method. For the sake of simplicity, B0 and B5 denote EfficientNet-B0 and EfficientNet-B5, respectively, and $m \times m$ means that the input resolution is $m \times m$.

the performance over the proposed strong baseline B0-ArcFace which adopts EfficientNet-B0 as the backbone and uses the ArcFace classification loss.

In Fig. 2, AUG w/o S are specifically designed in the proposed method. Here we experimentally evaluate different augmentations techniques and provide results in Fig. 4(b). Surprisingly, we find that the commonly used Random Resized Crop severely hurts the performance while Random Crop without scaling could improve performance. The results confirm our hypothesis in Subsect. 3.4 that scaling transformation is not suitable for image-based patent retrieval. Another observation is that Random Rotation also dramatically damages the performance. One possible reason is that drawing images are horizontally placed in the document without any geometric distortion.

Although state-of-the-art performance has been reached, the proposed strong baseline can be scaled up to achieve even stronger performance. As a trial experiment, we extend the model from three aspects: increasing input resolution, using a large backbone, and post-processing with re-ranking. The results are summarized in Fig. 5. Notably, the mAP score is first boosted from 71.2% of the baseline to 87.7% with the input image resolution of 512×512, then gets 89.9% using EfficientNet-B5, and finally reaches 93.5% by combining these three expanding techniques. Notably, the mAP score is astonishingly boosted from 71.2% of the baseline to 93.5% by combining these three expanding techniques. The final extended approach wins the first place in the Image Retrieval Challenge of an International Conference.

To further investigate the retrieval performance, we visualize the search results of different approaches in Fig. 6. We find that our approach is robust to viewpoints. For the query patent (b) which is an aerial view of a patent object, our approach successfully retrieves the patent images from other different views while the recent state-of-the-art PatentNet [15] fails to find this patent for the top-4 retrieved results. For the query patent (c) which contains apparent texture

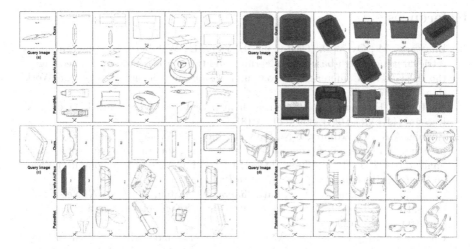

Fig. 6. Comparison of image-based patent retrieval results of different approaches of ours and the state-of-the-art PatentNet [15]. Given a query image, we only show the top-5 retrieved images. Symbols ✓ and ✗ denote correct and wrong hits, respectively.

patterns, the PatentNet [15] find other patents with similar shapes by mistake while ours correctly retrieves the right patents from other views.

5 Conclusion

In this paper, we empirically study image-based patent retrieval and present a simple, lightweight yet strong model which consists of a backbone network, neck structure, and training head. We find that the loss function plays an important role in this task, and deliberately choose the ArcFace classification loss. We also find that the classic crop and resize data augmentation greatly hurts the performance, and accordingly devise data augmentation without scaling which benefits the performance. To study the scalability of the model, we scale this baseline to large models by several expanding techniques such as increasing input resolution and post-processing with re-ranking. Astonishingly, such techniques significantly boost the retrieval performance, which demonstrates that the proposed model has good scalability. In the future, we will investigate more robust losses for image-based patent retrieval.

Acknowledgement. This work was supported by the Start-up Research Fund of Southeast University under Grant RF1028623063. This work is also supported by the Big Data Computing Center of Southeast University.

References

1. Bhattarai, M., Oyen, D., Castorena, J., Yang, L., Wohlberg, B.: Diagram image retrieval using sketch-based deep learning and transfer learning. In: Proceedings of the IEEE/CVF Conference on Computer Vision and Pattern Recognition Workshops, pp. 174–175 (2020)
2. Bhatti, N., Hanbury, A.: Image search in patents: a review. Int. J. Doc. Anal. Recogn. (IJDAR) **16**(4), 309–329 (2013)
3. Bhunia, A.K., Chowdhury, P.N., Sain, A., Yang, Y., Xiang, T., Song, Y.Z.: More photos are all you need: semi-supervised learning for fine-grained sketch based image retrieval. In: Proceedings of the IEEE/CVF Conference on Computer Vision and Pattern Recognition, pp. 4247–4256 (2021)
4. Chaudhuri, U., Banerjee, B., Bhattacharya, A., Datcu, M.: CrossATNet-a novel cross-attention based framework for sketch-based image retrieval. Image Vis. Comput. **104**, 104003 (2020)
5. Chen, W., Hays, J.: SketchyGAN: towards diverse and realistic sketch to image synthesis. In: Proceedings of the IEEE Conference on Computer Vision and Pattern Recognition, pp. 9416–9425 (2018)
6. Csurka, G., Renders, J.M., Jacquet, G.: XRCE's participation at patent image classification and image-based patent retrieval tasks of the Clef-IP 2011. In: CLEF (Notebook Papers/Labs/Workshop), vol. 2. Citeseer (2011)
7. Dalal, N., Triggs, B.: Histograms of oriented gradients for human detection. In: IEEE Conference on Computer Vision and Pattern Recognition, vol. 1, pp. 886–893. IEEE (2005)
8. Deng, J., Guo, J., Xue, N., Zafeiriou, S.: ArcFace: additive angular margin loss for deep face recognition. In: Proceedings of the IEEE/CVF Conference on Computer Vision and Pattern Recognition, pp. 4690–4699 (2019)
9. Eitz, M., Hildebrand, K., Boubekeur, T., Alexa, M.: Sketch-based image retrieval: benchmark and bag-of-features descriptors. IEEE Trans. Visual Comput. Graphics **17**(11), 1624–1636 (2010)
10. Gong, G., Guo, M.: Image based design patent retrieval with classification and indexing. In: International Conference on Information Technology and Computer Application (ITCA), pp. 481–488. IEEE (2020)
11. Hanbury, A., Bhatti, N., Lupu, M., Mörzinger, R.: Patent image retrieval: a survey. In: Proceedings of the Workshop on Patent Information Retrieval, pp. 3–8 (2011)
12. He, K., Zhang, X., Ren, S., Sun, J.: Deep residual learning for image recognition. In: Proceedings of the IEEE Conference on Computer Vision and Pattern Recognition, pp. 770–778 (2016)
13. Huet, B., Kern, N.J., Guarascio, G., Merialdo, B.: Relational skeletons for retrieval in patent drawings. In: International Conference on Image Processing, vol. 2, pp. 737–740. IEEE (2001)
14. Krestel, R., Chikkamath, R., Hewel, C., Risch, J.: A survey on deep learning for patent analysis. World Patent Inf. **65**, 102035 (2021)
15. Kucer, M., Oyen, D., Castorena, J., Wu, J.: DeepPatent: large scale patent drawing recognition and retrieval. In: Proceedings of the IEEE/CVF Winter Conference on Applications of Computer Vision, pp. 2309–2318 (2022)
16. Liu, C., et al.: Efficient token-guided image-text retrieval with consistent multimodal contrastive training. IEEE Trans. Image Process. **32**, 3622–3633 (2023)
17. Luo, H., et al.: A strong baseline and batch normalization neck for deep person re-identification. IEEE Trans. Multimedia **22**(10), 2597–2609 (2019)

18. Mogharrebi, M., Ang, M.C., Prabuwono, A.S., Aghamohammadi, A., Ng, K.W.: Retrieval system for patent images. Procedia Technol. **11**, 912–918 (2013)
19. Ojala, T., Pietikainen, M., Maenpaa, T.: Multiresolution gray-scale and rotation invariant texture classification with local binary patterns. IEEE Trans. Pattern Anal. Mach. Intell. **24**(7), 971–987 (2002)
20. Oyen, D., Kucer, M., Wohlberg, B.: VisHash: visual similarity preserving image hashing for diagram retrieval. In: Applications of Machine Learning, vol. 11843, pp. 50–66. SPIE (2021)
21. Perronnin, F., Sánchez, J., Mensink, T.: Improving the fisher kernel for large-scale image classification. In: Daniilidis, K., Maragos, P., Paragios, N. (eds.) ECCV 2010. LNCS, vol. 6314, pp. 143–156. Springer, Heidelberg (2010). https://doi.org/10.1007/978-3-642-15561-1_11
22. Radenović, F., Tolias, G., Chum, O.: Fine-tuning CNN image retrieval with no human annotation. IEEE Trans. Pattern Anal. Mach. Intell. **41**(7), 1655–1668 (2018)
23. Sain, A., Bhunia, A.K., Yang, Y., Xiang, T., Song, Y.Z.: StyleMeUp: towards style-agnostic sketch-based image retrieval. In: Proceedings of the IEEE/CVF Conference on Computer Vision and Pattern Recognition, pp. 8504–8513 (2021)
24. Sandler, M., Howard, A., Zhu, M., Zhmoginov, A., Chen, L.C.: MobileNetV2: inverted residuals and linear bottlenecks. In: Proceedings of the IEEE Conference on Computer Vision and Pattern Recognition, pp. 4510–4520 (2018)
25. Tan, M., et al.: MnasNet: platform-aware neural architecture search for mobile. In: Proceedings of the IEEE/CVF Conference on Computer Vision and Pattern Recognition, pp. 2820–2828 (2019)
26. Tan, M., Le, Q.: EfficientNet: rethinking model scaling for convolutional neural networks. In: International Conference on Machine Learning, pp. 6105–6114. PMLR (2019)
27. Tiwari, A., Bansal, V.: PATSEEK: content based image retrieval system for patent database. In: International Conference on Electronic Business, pp. 1167–1171. Academic Publishers/World Publishing Corporation (2004)
28. Verma, V.K., Mishra, A., Mishra, A., Rai, P.: Generative model for zero-shot sketch-based image retrieval. In: IEEE/CVF Conference on Computer Vision and Pattern Recognition Workshops (CVPRW), pp. 704–713. IEEE (2019)
29. Vrochidis, S., Moumtzidou, A., Kompatsiaris, I.: Concept-based patent image retrieval. World Patent Inf. **34**(4), 292–303 (2012)
30. Vrochidis, S., Papadopoulos, S., Moumtzidou, A., Sidiropoulos, P., Pianta, E., Kompatsiaris, I.: Towards content-based patent image retrieval: a framework perspective. World Patent Inf. **32**(2), 94–106 (2010)
31. Wang, H., Liao, S., Shao, L.: AFAN: augmented feature alignment network for cross-domain object detection. IEEE Trans. Image Process. **30**, 4046–4056 (2021)
32. Xu, P., Hospedales, T.M., Yin, Q., Song, Y.Z., Xiang, T., Wang, L.: Deep learning for free-hand sketch: a survey. IEEE Trans. Pattern Anal. Mach. Intell. (2022)
33. Yang, L., Gong, M., Asari, V.K.: Diagram image retrieval and analysis: challenges and opportunities. In: Proceedings of the IEEE/CVF Conference on Computer Vision and Pattern Recognition Workshops, pp. 180–181 (2020)
34. Ye, M., Shen, J., Lin, G., Xiang, T., Shao, L., Hoi, S.C.: Deep learning for person re-identification: a survey and outlook. IEEE Trans. Pattern Anal. Mach. Intell. **44**(6), 2872–2893 (2021)
35. Yelamarthi, S.K., Reddy, S.K., Mishra, A., Mittal, A.: A zero-shot framework for sketch based image retrieval. In: Ferrari, V., Hebert, M., Sminchisescu, C., Weiss,

Y. (eds.) ECCV 2018. LNCS, vol. 11208, pp. 316–333. Springer, Cham (2018). https://doi.org/10.1007/978-3-030-01225-0_19

36. Yu, Q., Yang, Y., Liu, F., Song, Y.Z., Xiang, T., Hospedales, T.M.: Sketch-a-Net: a deep neural network that beats humans. Int. J. Comput. Vision **122**(3), 411–425 (2017)

37. Zhang, Y., Qian, Q., Wang, H., Liu, C., Chen, W., Wan, F.: Graph convolution based efficient re-ranking for visual retrieval. IEEE Trans. Multimedia (2023)

38. Zhiyuan, Z., Juan, Z., Bin, X.: An outward-appearance patent-image retrieval approach based on the contour-description matrix. In: Joint Workshop on Frontier of Computer Science and Technology, pp. 86–89. IEEE (2007)

39. Zoph, B., Le, Q.V.: Neural architecture search with reinforcement learning. In: International Conference on Learning Representations. OpenReview.net (2017)

HelixNet: Dual Helix Cooperative Decoders for Scene Text Removal

Kun Liu, Guangtao Lyu, and Anna Zhu[✉]

School of Computer Science and Artificial Intelligence,
Wuhan University of Technology, Wuhan, China
{miracle,297903,annazhu}@whut.edu.cn

Abstract. Scene text removal aims to remove scene text from images and fill the resulting gaps with plausible and realistic content. Within the context of scene text removal, two potential sub-tasks exist, i.e., text perception and text removal. However, most existing methods have ignored this premise or only divided this task into two consecutive stages, without considering the interactive promotion relationship between them. By leveraging some transformations, better segmentation results can better guide the process of text removal, and vice versa. These two sub-tasks can mutually promote and co-evolve, creating an intertwined and spiraling process similar to the double helix structure of Deoxyribonucleic acid (DNA) molecules. In this paper, we propose a novel network, HelixNet, incorporating Dual Helix Cooperative Decoders for Scene Text Removal. It is an end-to-end one-stage model with one shared encoder and two interacted decoders for the text segmentation and text removal sub-tasks. Through the use of dual branch information interaction, we can fuse complementary information from each sub-task, achieving interaction between scene text removal and segmentation. Our proposed method is extensively evaluated on publicly available and commonly used real and synthetic datasets. The experimental results demonstrate the promotion effect of the specially designed decoder and also show that HelixNet can achieve state-of-the-art performance.

Keywords: Scene text removal · Text segmentation · One-stage

1 Introduction

Scene text removal (STR) is a task of removing the text and meanwhile filling the text regions with plausible content precisely. In recent years, STR has attracted increasing research interests because it can help avoid privacy leaks by hiding personal messages such as ID numbers, addresses, and license plate numbers. Additionally, it has valuable applications in image text editing [1], visual translation [2], and document restoration.

Generally, the existing STR research can be classified into two categories: text segmentation-agnostic and text segmentation-based. Text segmentation-agnostic STR methods [3,4] generally utilize single encoder-decoder architectures to process scene text images and generate the text-removed output result directly

Q. Liu et al. (Eds.): PRCV 2023, LNCS 14431, pp. 27–38, 2024.
https://doi.org/10.1007/978-981-99-8540-1_3

without text detection or segmentation explicitly. These models are known for their efficiency and fast execution. However, due to the combined learning of text detection and inpainting within a single network, the accuracy of text localization is compromised, leading to an unpredictable text-erasing process.

Text segmentation-based approaches [5] typically employ two or more encoder-decoder architectures. In the initial step, they detect scene text or text strokes, and consider them as masks. In the subsequent step, they use image inpainting techniques [6] to fill the masked regions. These kinds of methods are required to extract text implicitly and then guide the model for text removal. However, they are implemented with two-stage or multi-stage structures, tending to be inefficient and requiring complex training. Moreover, these methods have limitations due to their heavy reliance on text detection or the initial coarse removal results at the front-end.

Within the context of scene text removal, there exist two potential sub-tasks: text perception and text removal. However, most existing methods have either overlooked this premise or simply divided the task into two consecutive stages, without considering the mutually beneficial relationship between them.

The analysis presented above prompts us to rethink the network architecture and effect of text segmentation for scene text removal. *Is it possible to use a single network for both text segmentation and STR?* EraseNet [7] gives us the answer it is possible. It used a coarse-to-fine strategy for STR. The coarse stage was designed by one encoder and two parallel decoders, with one for text segmentation and the other for STR. However, the results are unsatisfactory using only the coarse network. FETNet [8] injects the text segmentation results into STR decoder, boosting the STR performance. *Is the STR result boost the text segmentation performance?* In this way, the two sub-tasks can work in interactive way to mutually reinforce and evolve, creating an intertwined and spiraling process akin to the double helix structure of DNA molecules. Drawing inspiration from these two concepts, we propose a novel approach called HelixNet, which stands for Dual Helix Cooperative Decoder for STR. HelixNet is an end-to-end one-stage model featuring a shared encoder and two interconnected decoders for the two sub-tasks. By facilitating dual-branch information interaction, we can integrate complementary information from each sub-task, enabling effective collaboration between scene text removal and segmentation. We introduce the interactive information acquisition modules that are simple, efficient, and parameter-free, enabling the acquisition of valuable auxiliary information. Additionally, a interactive information fusion module is proposed for combining the acquired auxiliary interactive information. Our proposed approach, HelixNet, has been extensively evaluated on two widely used benchmark datasets: the real-world dataset SCUT-EnsText [7] and the synthetic dataset SCUT-Syn [4]. Through extensive experiments on both real and synthetic datasets, HelixNet demonstrates remarkable performance, surpassing state-of-the-art (SOTA) methods in terms of quantitative and qualitative results. Ablation studies further validate the effectiveness of the proposed interactive module.

The contributions of this work can be summarized as follows:

- We first introduce the concept that text removal and text segmentation can mutually reinforce each other in STR task, and propose a novel network, HelixNet, containing Dual Helix Cooperative Decoders for STR task.
- We propose the simple and efficient interactive information acquisition modules that require zero parameters, enabling the acquisition of valuable auxiliary information.
- Extensive experimental results demonstrate the effectiveness of the proposed interactive module and the superior performance of HelixNet, surpassing existing SOTA methods on many evaluation metrics.

2 Related Works

Scene Text Segmentation. Scene text segmentation is a traditional task in Optical Character Recognition (OCR) [9] field. It aims to accurately segment the text regions in the image and distinguish them from the background.

The traditional method is based on threshold and edge detection [10], which is used to process document text images. However, their effectiveness is limited when dealing with complex natural background scene texts. The first CNN-based scene text segmentation work [3] was built on a three-stage model that utilized the edges and regions of text to fuse shallow information from convolution networks and deep information from deconvolution networks. This approach aims to detect, refine, and filter candidate text regions. Afterwards, a lot of deep learning-based methods were proposed for this task, e.g., SMANet [11], PGTSNet [12].

Image Inpainting. Image inpainting aims to repair the damaged part of an image. This technology can be applied to image editing [13], image completion [14], etc. Traditional inpainting methods include diffusion-based [15] and patch-based [13]. The diffusion-based method propagates contextual pixels from the boundary to holes along the iso-photometric direction. In the patch-based method, missing areas are filled using neighborhood appearances or similar patches. However, these two methods may have shortcomings when completing a large number of missing areas in complex scenes that require semantic repair.

With the development of deep learning, inpainting models based on deep learning [16] can synthesize more reasonable content for complex scenes. Many STR works adopted the image inpainting methods, considering the text detection regions and text segmentation results as masks. They regraded text as normal objects and did not use the characteristics of text for inpainting.

Scene Text Removal. The current methods for scene text removal can be roughly divided into text segmentation-agnostic and text segmentation-based methods.

Text segmentation-agnostic STR methods regard the task as an image conversion task, where the input natural scene text image is converted to the corresponding non-text natural image. These methods are normally built on a single-stage architecture without text detection and segmentation explicitly. Nakamura et al. [3] proposed the first STR model to automatically erase, and fill natural scene text images without text detection and segmentation. Further, EnsNet [4] and FETNet are proposed with several designed loss functions and feature transformation to improve the performance.

Text segmentation-based approaches generally extract text regions with text removal task in parallel [7] or cascaded [5] ways. These kinds of method implemented STR in two-stage networks, using the segmented text as guidance information and then applying image inpainting pipelines to reconstruction the text regions. Some methods adopted iterative algorithms, employing multiple erasure iterations to refine the text removal process [17].

3   Proposed Method

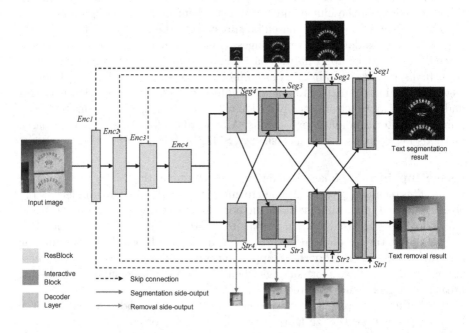

Fig. 1. The overall architecture of HelixNet.

3.1   Overall Pipeline

In this section, we present the details of the proposed HelixNet. As shown in Fig. 1, the pipeline of our proposed model consists of a shared encoder and two task-specific decoders. The basic block is based on ResNet [18] block. In the encoder, each downsampling process reduces the size of the feature maps by half and doubles the number of feature channels. There are two parallel branches in the decoder to generate segmentation and erasure results respectively. To better utilize the information from the two sub-tasks, we propose an interactive mechanism and apply it to each layer of the two decoders. Through this interaction mechanism, we expect the two sub-tasks to promote each other and co-evolve. Additionally, each layer of the two decoders outputs the result of text segmentation and text removal, as multi-scale interaction is able to better promote the model to pay attention to text information at different scales.

3.2   Interactive Mechanism

In this section, we introduce the information interaction mechanism between text removal and text segmentation decoder, considering the impact of each layer of decoder output on subsequent results, as well as the relationship between segmentation results and erasure results. A simple information interaction block is designed to obtain the complementary features of two sub-tasks. In the removal branch, the interaction block in each decoder layer outputs the difference and absolute value of the corresponding encoder features Enc_i and segmentation features Seg_i, denoting as $Inter(Enc_i, Seg_i) = |Enc_i - Seg_i|$. Its purpose is to find the precise non-text features that can help restore the text regions, which are required in the text erasure branch. Similarly, by taking the difference and absolute value of the layer-wise encoder features Enc_i and erasure features Str_i in the interaction block of the segmentation branch (denoting as $Inter(Enc_i, Str_i)$ $= |Enc_i - Str_i|$), we can obtain information about which areas are erased by the erasure features compared to the encoding features. These areas correspond to the regions where the text is located. This process can be beneficial for the text segmentation branch.

Through these two simple interaction blocks, the non-text features and text features are highlighted for text removal and segmentation respectively. The interactive features are fused with the specified decoder features and encoder features by skip-connection as follows:

$$Seg_i = Dec_{seg}(reduce_{seg}([Seg_i, Enc_i]), Inter(Enc_i, Str_i)) \qquad (1)$$

$$Str_i = Dec_{str}(reduce_{str}([Str_i, Enc_i]), Inter(Enc_i, Seg_i)) \qquad (2)$$

where $[\cdot,\cdot]$ denotes concatenation. $Reduce_{seg}$ and $Reduce_{str}$ are convolution operation to combine features with encoder features Enc_i, $Inter$ is an information inter-module used to highlight corresponding features. Dec_{seg} and Dec_{str} are information fusion modules to combine reduce features and inter features. Seg_i and Str_i correspond to the output of two sub-tasks of the decoder layer.

We use the last layer of decoder features and simple upsampling to obtain the final text segmentation features and text erasure features. Additionally, we also use side-output decoders to deeply supervise the decoder networks to learn inpainting text features for STR and extracting text features for text segmentation.

3.3 Loss Functions

In our network training process, we use L_1 Loss for text removal and Dice loss for text segmentation respectively. We monitor the results of text erasure and segmentation output from each layer of the decoder. Given the input scene text image I, text-removed Ground Truth (GT) I_{gt} and the binary text GT mask M_{gt} with 0 for non-text regions and 1 for text regions, the STR output is denoted as I_o and text segmentation results as M_o. For the side-output of i_{th} layer, all the above images and GTs are resized to the corresponding down-scaling sizes.

Removal Loss. First, we would like to force both backgrounds and removed text regions to approach the real GT I_{gt}, using a standard L_1 loss for each decoder layer:

$$L_{str}^i = ||(I_o^i - I_{gt})^i||_1 \tag{3}$$

Segmentation Loss. For the learning of text segmentation module, we use dice loss [19] for i_{th} layer as (4), which considers the contour similarity between the prediction results in M_o and GT M_{gt}.

$$L_{seg}^i = 1 - \frac{2\sum_{x,y}(M_o^i(x,y) \times M_{gt}^i(x,y))}{\sum_{x,y}(M_o^i(x,y)^2 + M_{gti}(x,y)^2)} \tag{4}$$

The overall loss function is the addition of the balanced above two losses, representing as:

$$L_{all} = \sum_{i=1}^{n}(L_{str}^i + \lambda L_{seg}^i) \tag{5}$$

where λ is set to 1 experimentally.

4 Experiments

4.1 Dataset and Evaluation Metrics

We evaluated our proposed HelixNet on the following two benchmark datasets.

SCUT-Syn. SCUT-Syn is a synthetic dataset generated by text synthesis technology and only includes English text instances, with 8,000 images for training and 800 images for testing. The images are collected from ICDAR 2013 and ICDAR, MLT-2017, etc. The details for generating and using these synthetic text images can be found in [20] and [4].

SCUT-EnsText. SCUT-EnsText It contains 2,749 training images and 813 test images which are collected in real scenes with various text properties. All text in each image is carefully removed to maintain consistency in the text removal area and surrounding textures. More descriptions about this dataset refer to the work [7].

A text detector CRAFT [21] is employed to detect text on the output STR images by measuring the precision, recall, and F-score. The lower their values, the better text removal performance. Six image-level evaluation metrics are also adopted for measurement of the generated STR images with GT images, i.e., Peak Signal Noise Ratio (PSNR), Mean Square Error (MSE), Mean Structural Similarity (MSSIM), Average Gray-level Error (AGE), percentage of Error Pixels (pEPs), and percentage of Clustered Error Pixels (pCEPS) [4]. A higher MSSIM and PSNR, and a lower AGE, pEPs, pCEPS, and MSE indicate better results.

4.2 Implementation Details

We train HelixNet on the training sets of SCUT-EnsText, and SCUT-Syn and evaluate them on their corresponding testing sets. The masks are generated by subtracting the input images and the corresponding STR GT, i.e., text removal images. We follow [7] to apply data augmentation during training. The model is optimized using the Adam algorithm. The initial learning rate is set to 0.0001, alternating training for 1000 periods in a single NVIDIA A40 GPU, with a batch size of 8 and an input image size of 256×256.

4.3 Ablation Study

In this section, We set three comparison experiments as follows: removing interaction (baseline), using the addition function to combine interact features (add), and using subtraction and absolute value function to combine interact features (abs_diff). * means only keeping the text region of the erased result in the text segmentation result and using the origin input in other regions. We analyze the impact of different components within the network on the SCUT-EnsText dataset.

Baseline. For the baseline model, we adopt a simple encoder-decoder structure without the text segmentation and text erasure interaction module. The experimental results yield a PSNR of 34.20 and SSIM of 96.95.

Different Interaction Schemes. As we know, introducing extra parameters in the model may improve the performance. However, that is unfair since the improvement might be brought by the larger model. Our interaction blocks are **parameter-free**. This point ensures fairness in comparative experiments. We investigate other different simple interaction schemes without inducing extra parameters (i.e. add operation). Results are presented in Table 1.

Table 1. Ablation study results of different modules effect on SCUT-Ensnet.

Method	PSNR↑	MSSIM↑	MSE↓	AGE↓	pEPs↓	pCEPs↓
- baseline	34.20	96.95	0.0015	2.0938	0.0145	0.0086
- baseline*	35.04	97.14	0.0014	1.7227	**0.0130**	**0.0076**
- add	34.84	96.90	0.0018	2.1363	0.0165	0.0104
- add*	35.12	97.09	0.0015	1.8009	0.0142	0.0086
- abs_diff(Our model)	34.77	97.02	0.0023	2.0400	0.0156	0.0100
- abs_diff*(Our model)	**35.15**	**97.18**	**0.0014**	**1.7215**	0.0133	0.0079

Table 2. Comparison with SOTA methods and proposed method on SCUT-EnsText. R: Recall; P: Precision; F: F-score.

Method	Image-level Evaluation						Detection-Eval (%)		
	PSNR↑	MSSIM↑	MSE↓	AGE↓	pEPs↓	pCEPs↓	P↓	R↓	F↓
Original Images	–	–	–	–	–	–	79.8	69.7	74.4
Pix2pix [22]	26.75	88.93	0.0033	5.842	0.048	0.0172	71.3	36.5	48.3
STE [3]	20.60	84.11	0.0233	14.4795	0.1304	0.0868	52.3	14.1	22.2
EnsNet [4]	29.54	92.74	0.0024	4.1600	0.2121	0.0544	68.7	32.8	44.4
MTRNet++ [23]	27.34	95.20	0.0033	8.9480	0.0953	0.0619	56.1	3.6	6.8
EraseNet [7]	32.30	95.42	0.0015	3.0174	0.0160	0.0090	53.2	4.6	8.5
PSSTRNet [17]	34.65	96.75	0.0014	**1.7161**	0.0135	**0.0074**	**47.7**	5.1	9.3
FETNet [8]	34.53	97.01	**0.0013**	1.7539	0.0137	0.0080	51.3	5.8	10.5
Helixnet (Ours)	**35.15**	**97.18**	0.0014	1.7215	**0.0133**	0.0079	49.4	**1.7**	**3.3**

Table 3. Comparison with SOTA methods and proposed method on SCUT-Syn.

Method	Category	PSNR↑	MSSIM↑	MSE↓	AGE↓	pEPs↓	pCEPs↓	Parameters↓
Pix2pix	one-stage	25.16	87.63	0.0038	6.8725	0.0664	0.0300	54.4M
STE	one-stage	24.02	89.49	0.0123	10.0018	0.0728	0.0464	89.16M
EnsNet	one-stage	37.36	96.44	0.0021	1.73	0.0069	0.0020	12.4M
MTRNet++	two-stage	34.55	98.45	0.0004	–	–	–	18.7M
EraseNet	two-stage	38.32	97.67	0.0002	1.5982	0.0048	0.0004	19.74M
PSSTRNet	multi-stage	39.25	98.15	0.0002	1.2035	0.0043	0.0008	**4.88M**
FETNet	one-stage	39.14	97.97	0.0002	1.2608	0.0046	0.0008	8.53M
HelixNet (Ours)	one-stage	**41.77**	**98.64**	**0.0001**	**1.0151**	**0.0018**	**0.0002**	18.17M

Two STR results are obtained. One is the direct output of the text removal branch. The other is the output combination of two branches. Since we could get the text segmentation results in the two branch decoder structures, we replace the non-text regions of text removed output with the corresponding original input image regions based on the segmentation results.

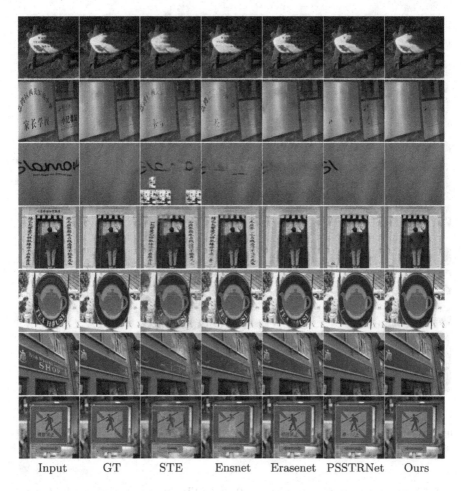

Input   GT   STE   Ensnet   Erasenet   PSSTRNet   Ours

Fig. 2. Qualitative comparison of all methods on SCUT-EnsText. From left to right: input image, GT, STE, Ensnet, Erasenet, PSSTRNet, Ours (HelixNet).

Observing the experimental results, it is evident that the addition of the interaction module has led to overall improvement. Our proposed non-parameter-induced interaction block, i.e., abs_diff, is able to generate better STR results than adding interaction operation. The narrowing performance gap between the direct STR output and combination STR output indicates that our model has effectively enhanced its ability to perceive text by integrating segmentation and erasure of information (Table 2).

4.4 Comparing with SOTA

Qualitative Comparison. We compare our proposed HelixNet with other SOTA approaches on both real-world and synthetic text images.

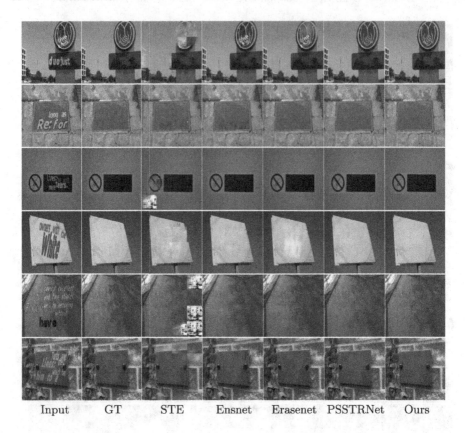

Input   GT   STE   Ensnet   Erasenet PSSTRNet   Ours

Fig. 3. Qualitative comparison of all methods on SCUT-Syn. From left to right: input image, GT, STE, Ensnet, Erasenet, PSSTRNet, Ours (HelixNet).

As shown in the first column of Fig. 2, the STR results of other methods tend to preserve a significant portion of the text within the shadows. By introducing the mutual supervision of text segmentation and erasure, HelixNet exhibits enhanced sensitivity to textual information and achieves complete text removal while generating plausible backgrounds, even under complex lighting conditions.

Our approach appears to be particularly effective in handling Chinese characters in scene text images, as evident in the second, third, and seventh columns of Fig. 2. It demonstrates our model has strong generalization capabilities in erasing scene text in images.

In the sixth column of Fig. 2, HelixNet accurately distinguishes the differences between background patterns and text of the same color, and can predict more realistic text region textures. It significantly reduces the restoration inconsistencies, generating surprising erasure results that surpass the ground truth. This showcases the semantic excellence of HelixNet.

The qualitative results on the SCUT-Syn dataset as shown in Fig. 3 also demonstrate the SOTA performance of our model. It could better distinguish between text and background, even those with easily confused colors. Moreover, our erasure results are more refined, so there is no problem of text remaining.

Quantitative Comparison . According to the experimental results in Table 3, the HelixNet has reached a leading level in the synthesis of the dataset. It also gets the highest scores on PSNR and MSSIM metrics on the SCUT-EnsText dataset. The trainable parameters of HelixNet are 18.17M. Compared with other networks of similar scales, our model exhibits superior performance.

Overall, both the qualitative and quantitative results demonstrate that our approach achieves excellent text removal and background restoration effects on both real-world and synthetic text images.

5 Limitations

Although our model achieves great performance in most scenarios, it exhibits limitations in handling certain text scenarios, such as stereoscopic characters and large characters as shown in Fig. 4, due to complex text effect and CNN's limited global perception ability. As a result, there is still room for improvement in our model's ability to acquire global information and accurately represent character text information.

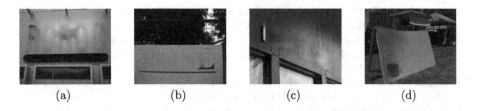

| (a) | (b) | (c) | (d) |

Fig. 4. Some failure cases of our method for STR.

6 Conclusion

This paper proposes a lightweight one-stage network called HelixNet for scene text removal. It is based on the U-net encoder-decoder structure, which processes scene text removal into two sub-tasks: text segmentation and text erasure. These two sub-tasks use two decoders, sharing a common encoder. We utilized the relationship between two sub-tasks that can mutually supervise and enhance each other during the decoding process and proposed a simple information exchange module without additional parameters. Experiments have shown that our HelixNet outperforms the most advanced STR methods.

Acknowledgement. This work is supported by the Open Project Program of the National Laboratory of Pattern Recognition (NLPR) (No. 202200049).

References

1. Yang, Q., Jin, H., Huang, J., Lin, W.: SwapText: image based texts transfer in scenes. In: CVPR (2020)
2. Singh, A., Pang, G., Toh, M., Huang, J., Hassner, T.: TextOCR: towards large-scale end-to-end reasoning for arbitrary-shaped scene text. In: CVPR (2021)
3. Nakamura, T., Zhu, A., Yanai, K., Uchida, S.: Scene text eraser. In: ICDAR (2017)
4. Zhang, S., Liu, Y., Jin, L., Huang, Y., Lai, S.: EnsNet: ensconce text in the wild. In: AAAI (2019)
5. Tursun, O., Rui, Z., Denman, S., Sridharan, S., Fookes, C.: MTRNet: a generic scene text eraser. In: ICDAR (2019)
6. Yu, T., et al.: Inpaint anything: segment anything meets image inpainting. arXiv preprint arXiv:2304.06790 (2023)
7. Liu, C., Liu, Y., Jin, L., Zhang, S., Wang, Y.: EraseNet: end-to-end text removal in the wild. IEEE Trans. Image Process. **29**, 8760–8775 (2020)
8. Lyu, G., Liu, K., Zhu, A., Uchida, S., Iwana, B.K.: FETNet: feature erasing and transferring network for scene text removal. Pattern Recognit. **140**, 109531 (2023)
9. Nguyen, N., et al.: Dictionary-guided scene text recognition. In: CVPR (2021)
10. Nobile, N., Suen, C.Y.: Text segmentation for document recognition. In: Doermann, D., Tombre, K. (eds.) Handbook of Document Image Processing and Recognition, pp. 257–290. Springer, London (2014). https://doi.org/10.1007/978-0-85729-859-1_8
11. Bonechi, S., Bianchini, M., Scarselli, F., Andreini, P.: Weak supervision for generating pixel level annotations in scene text segmentation. Pattern Recogn. Lett. **138**, 1–7 (2020)
12. Xixi, X., Qi, Z., Ma, J., Zhang, H., Shan, Y., Qie, X.: BTS: a bi-lingual benchmark for text segmentation in the wild. In: CVPR (2022)
13. Barnes, C., Shechtman, E., Finkelstein, A., Goldman, D.B.: PatchMatch: a randomized correspondence algorithm for structural image editing. ACM Trans. Graph. **28**(3), 24 (2009)
14. Iizuka, S., Simo-Serra, E., Ishikawa, H.: Globally and locally consistent image completion. ACM Trans. Graph. **36**(4), 1–14 (2017)
15. Ballester, C., Bertalmio, M., Caselles, V., Sapiro, G., Verdera, J.: Filling-in by joint interpolation of vector fields and gray levels. IEEE Trans. Image Process. **10**(8), 1200–1211 (2001)
16. Pathak, D., Krahenbuhl, P., Donahue, J., Darrell, T., Efros, A.A.: Context encoders: feature learning by inpainting. In: CVPR (2016)
17. Lyu, G., Zhu, A.: PSSTRNet: progressive segmentation-guided scene text removal network. In: ICME (2022)
18. He, K., Zhang, X., Ren, S., Sun, J.: Deep residual learning for image recognition. In: CVPR (2016)
19. Milletari, F., Navab, N., Ahmadi, S.A.: V-Net: fully convolutional neural networks for volumetric medical image segmentation. In: 3DV (2016)
20. Gupta, A., Vedaldi, A., Zisserman, A.: Synthetic data for text localisation in natural images. In: CVPR (2016)
21. Baek, Y., Lee, B., Han, D., Yun, S., Lee, H.: Character region awareness for text detection. In: CVPR (2019)
22. Isola, P., Zhu, J.Y., Zhou, T., Efros, A.A.: Image-to-image translation with conditional adversarial networks. In: CVPR (2017)
23. Tursun, O., Denman, S., Zeng, R., Sivapalan, S., Sridharan, S., Fookes, C.: MTRNet++: one-stage mask-based scene text eraser. Comput. Vis. Image Underst. **201**, 103066 (2020)

Semantic-Information Space Sharing Interaction Network for Arbitrary Shape Text Detection

Hua Chen, Runmin Wang[✉][iD], Yanbin Zhu, Zhenlin Zhu, Jielei Hei, Juan Xu, and Yajun Ding

School of Information Science and Engineering, Hunan Normal University, Changsha 410081, China
runminwang@hunnu.edu.cn

Abstract. Arbitrary shape text detection is a challenging task due to significant variations in text shapes, sizes, and aspect ratios. Previous approaches relying on single-level feature map generated through a top-down fusion of different feature levels have limitations in harnessing high-level semantic information and expressing multi-scale features. To address these challenges, this paper introduces a novel arbitrary shape scene text detector called the Semantic-information Space Sharing Interaction Network (SSINet). The proposed network leverages the Semantic-information Space Sharing Module (SSM) to generate a single-level feature map capable of expressing multi-scale features with rich semantic and prominent foreground, enabling effective processing of text-related information. Experimental evaluations on three benchmark datasets, namely CTW-1500, MSRA-TD500, and ICDAR2017-MLT, validate the effectiveness of our method. The proposed SSINet achieves impressive results with an F-score of 86.0% on CTW-1500, 89.1% on MSRA-TD500, and 72.4% on ICDAR2017-MLT. The code will be available at https://github.com/123cjjjj/SSINet.

Keywords: Scene text detection · Arbitrary shape text · Multi-scale feature representation

1 Introduction

Texts contained within images serve as concise depictions of pertinent image details, thereby enhancing human comprehension. Notably, the advent of deep learning methodologies has profoundly advanced the progression of scene text detection, yielding remarkable outcomes [9,21]. Such advancements have engendered numerous pragmatic applications across domains like autonomous driving, online education, and video retrieval. Nevertheless, the formidable nature of

This work was supported in part by the Natural Science Foundation of Hunan Province (No. 2020JJ4057), the Key Research and Development Program of Changsha Science and Technology Bureau (No. kq2004050), and the Scientific Research Foundation of the Education Department of Hunan Province of China (No. 21A0052).

Q. Liu et al. (Eds.): PRCV 2023, LNCS 14431, pp. 39–51, 2024.
https://doi.org/10.1007/978-981-99-8540-1_4

scene text detection tasks persists, attributable to intricate textual environments within images, the prevalence of extreme aspect ratios, and the assortment of text structures and orientations, thus conferring immense challenges.

Texts exhibit distinctive characteristics compared to generic objects, primarily stemming from their significant variances in terms of shapes, sizes, and aspect ratios. In order to tackle the challenge of scale discrepancies in scene texts, conventional approaches have predominantly embraced a top-down methodology, gradually integrating feature maps originating from diverse levels [18, 27]. Nonetheless, the prevailing fusion techniques fail to fully exploit the wealth of high-level semantic information and the inter-layer correlations present within the feature maps. Consequently, there exists a deficiency in expressing multi-scale features with utmost efficacy within a single-level feature map, particularly when confronted with highly extensive variations in scene texts with large scales. Hence, for tasks reliant upon a single-level feature map for text detection, it becomes crucial to endow said feature map with rich semantic and foreground emphasis, thereby empowering it with the capacity for multi-scale feature expression.

Therefore, to address the above issues, we propose a arbitrary shape text detector called Semantic-information Space Sharing Interaction Network (SSINet). This model mainly consists of a Semantic-information Space Sharing Module, which is composed of three sub-modules with distinct functionalities. Specifically, the Semantic-information Space Sharing Module makes full use of multi-scale high-level semantic to guide the interaction of information between high-level and low-level feature maps, and enhances the perception of foreground text targets, making it easier to process text information in subsequent steps.

In summary, this paper has three main contributions:

1. We propose a Semantic-information Space Sharing Module (SSM) to generate a single-level feature map capable of representing multi-scale feature with rich semantic and prominent foreground.
2. We introduce a multi-scale feature interaction fusion module based on cross-level feature correlations to generate multi-scale fused feature maps.
3. Extensive experiments on three publicly available datasets (i.e., CTW-1500, MSRA-TD500, and ICDAR2017-MLT) demonstrate the superiority of our method.

2   Related Works

In the domain of text detection, scholarly investigations have witnessed a notable transition in research priorities, wherein the emphasis has shifted from conventional horizontal texts to encompass more intricate forms such as multi-oriented and arbitrary shape texts [7]. Presently, deep learning-based approaches in text detection can be broadly categorized into several distinctive methodologies, including regression-based methods, segmentation-based methods, and hybrid approaches that amalgamate both regression and segmentation methods.

2.1 Regression-Based Methods

Drawing inspiration from object detection methodologies, regression-based approaches in text detection primarily encompass boundary box-based regression and pixel-based regression techniques. In these methodologies, the text is regarded as a distinct class of objects, and popular object detection frameworks, such as Faster R-CNN [14] and SSD [10], are employed to localize text boxes by predicting anchor or pixel offsets. For instance, by building upon the foundations of Faster R-CNN [14], the work [12] devises the Rotation Region Proposal Network (RRPN), specifically tailored to detect text instances with arbitrary orientations. Another notable approach, EAST [28], leverages Fully Convolutional Networks (FCN) [11] to directly predict pixel-to-text box offsets, thereby streamlining the process of text detection.

2.2 Segmentation-Based Methods

Segmentation-based text detection methods predominantly rely on the foundational principles established by Fully Convolutional Networks (FCN) [11]. These methods generate text masks through pixel-level semantic segmentation, subsequently employing intricate post-processing algorithms to derive accurate boundary boxes encapsulating the text regions. Remarkable progress has been achieved in detecting texts of arbitrary orientations and shapes through such approaches. For instance, PseNet [16] addresses the challenge of adjacent texts by generating diverse size of kernels for text instances and progressively expanding these kernels to effectively segment adjacent text entities. Another noteworthy method, Textmountain [29], leverages a combination of a center boundary probability map and a text center direction map to generate finely delineated text instance regions through text region segmentation.

2.3 Methods Based on the Combination of Regression and Segmentation

These methods usually include text region detection based on semantic segmentation and text box prediction based on regression. For example, in Contour-Net [17], an adaptive-RPN is proposed to generate proposal boxes, and a set of contour points of text regions are learned by orthogonal convolutions to locate texts with arbitrary shapes. TextFuseNet [20] proposes a novel detector that utilizes both detection and mask branches to obtain character-wise and word-wise features, and uses semantic segmentation to obtain global features.

Despite the notable advancements achieved by these existing methods, there remains a dearth of extensive research dedicated to the generation of a single-level feature map that encapsulates abundant semantic information and prominent foreground. Recognizing this research gap, we put forth a novel text detection framework, namely SSINet, that operates on a single-level feature map enriched with substantial semantic details and foreground emphasis. Our proposed SSINet aims to address the aforementioned limitations and empower

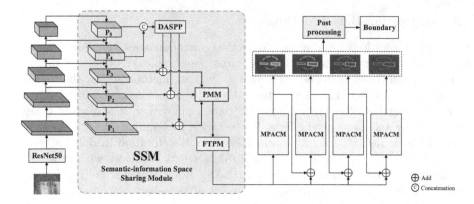

Fig. 1. The overall framework of our SSINet.

text detection systems with enhanced feature expression capabilities, ultimately improving the accuracy and robustness of text localization.

3   Proposed Method

The framework of SSINet is visually depicted in Fig. 1. Motivated by TextPMs [25], the overall architecture of the SSINet is meticulously devised to effectively represent discriminative features, leveraging the Semantic-information Space Sharing Module (SSM) in training a robust scene text detection model. The SSINet exhibits a cohesive coarse-to-fine framework, which encompasses pivotal components such as the feature extraction module, the feature fusion module, the text feature representation module, and the iteratively optimized probability maps generation module, etc. These meticulously designed modules synergistically contribute to the accurate detection of arbitrary shape scene texts.

Our SSINet leverages ResNet50 [6] as the underlying backbone network. Initially, the input text image is propagated through the backbone network employing a Feature Pyramid Network (FPN) architecture, which yields a multi-scale feature pyramid. These feature maps, denoted as P_1, P_2, P_3, P_4, P_5, correspond to stages 1 to 5 of the network. Within the Semantic-information Space Sharing Module (SSM), the high-level semantic features derived from the multi-scale feature pyramid are amalgamated with cross-level feature correlations, facilitating the integration of semantic and spatial information between high-level and low-level features. This module enhances the expressive capacity of low-level features in terms of semantic information, thereby enhancing the model's perception of foreground text targets. Subsequently, the Multi-branch Parallel Asymmetric Convolution Module (MPACM) is employed to process the feature maps, enabling the acquisition of text-specific characteristics while mitigating the introduction of background noise. This tailored processing of feature maps serves to facilitate subsequent stages of text information analysis.

The output branch of our SSINet is designed similar to TextPMs [25] to minimize the impact of noise caused by inaccurate text annotations. The final output is represented as more reasonable text pixel probabilities, and a set of probability maps are iteratively optimized to describe distributions. Finally, the watershed algorithm and the threshold filtering post-processing algorithm are used to reconstruct and filter the text instances to obtain detection results.

3.1   Semantic-Information Space Sharing Module

In this subsection, we devote our focus to introducing the Semantic-information Space Sharing Module (SSM). The SSM embodies a comprehensive framework, comprising three essential sub-modules, namely the multi-scale semantic information extraction sub-module, the semantic-to-space information guidance sub-module, and the foreground text perception sub-module. These sub-modules synergistically operate to facilitate the exchange and integration of critical information between the semantic and spatial domains. The multi-scale semantic information extraction sub-module serves to distill rich semantic cues from the multi-scale context, enabling the model to capture and comprehend intricate textual semantics. Subsequently, the semantic-to-space information guidance sub-module establishes a robust connection between the extracted high-level semantic features and the corresponding spatial locations, facilitating the alignment of semantics with spatial characteristics. Finally, the foreground text perception sub-module augments the model's capability to discern and perceive foreground text elements, thus enabling enhanced comprehension and accurate processing of text-centric information.

Multi-scale Semantic Information Extraction Sub-module. Based on the fact that high-level feature maps contain a large amount of semantic information beneficial to classification tasks, the ASPP [2] optimized by depthwise separable convolutions (DASPP) is used to process the high-level feature maps P_5 and P_4, generating multi-scale semantic information. Moreover, the depth separable convolutions significantly reduce the computational complexity of this module while maintaining similar (or better) performance. Specifically, downsampling is performed on feature map P_4, and then it is concatenated with the high-level feature map P_5 to obtain the high-level semantic fusion feature map P_5'.

Since features at different scales typically focus on different semantic information, in order to facilitate the guidance of semantic information on low-level spatial information, the high-level semantic fusion feature map P_5' is processed using the DASPP module to obtain the multi-scale semantic information P_{ms}.

$$P_5' = Concat(P_4, P_5) \tag{1}$$

$$P_{\mathrm{ms}} = DASPP(P_5') \tag{2}$$

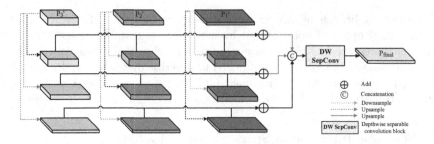

Fig. 2. The structure of PMM.

Semantic-to-Space Information Guidance Sub-module. To compensate for the inherent deficiency of high-level semantic information within the current feature fusion procedure, we employ an additional fusion method to supplement the low-level feature maps, imbuing them with additional semantic information. By incorporating this fusion technique, the resulting feature maps attain a more comprehensive representation that combines both the rich spatial information from the low-level features and the crucial semantic context from the high-level features.

$$P_1' = P_{\mathrm{ms}} + P_1 \tag{3}$$

$$P_2' = P_{\mathrm{ms}} + P_2 \tag{4}$$

$$P_3' = P_{\mathrm{ms}} + P_3 \tag{5}$$

To address the issue of large-scale variations in size, the cross-layer fusion strategy is proposed to realize the guidance of high-level semantic information to low-level spatial information. Specifically, the interaction of cross-layer features to obtain the fusion feature map is achieved via the parallel merge module (PMM), and the specific details of the PMM are shown in Fig. 2.

The work [30] elaborates that the pyramid level features are mainly associated with two adjacent levels and feature maps with large differences in levels may impair the performance. The PMM mainly targets the feature maps of the two adjacent levels of P_2' and feature map P_2'. To be more specific, the resolutions of feature map P_1', P_2' and P_3' are used as the benchmark resolutions of the three single branches respectively. For every single branch, the feature map with the benchmark resolution is kept unchanged, and the remaining two feature maps are upsampled or downsampled to the benchmark resolution. Then, the single branch fusion feature map is obtained by addition. Finally, the fused feature maps from the three benchmark resolutions are concatenated along the channel dimension, and a depthwise separable convolution block for dimension reduction is used for feature fusion to obtain the fused feature map P_{final}.

Foreground Text Perception Sub-module. The foreground attention map is obtained by performing 3×3 and 1×1 convolutions on the fused feature map P_{final}, and an exponential activation operation is applied to it to increase the feedback difference between foreground and background. Finally, the foreground attention map is applied to the fused feature map P_{final} to obtain the semantic-rich and foreground-prominent single-level feature map.

The dice loss function is computed for the generated foreground attention map with foreground truth annotations as an auxiliary loss of the overall model, constraining the foreground attention map to be more accurate.

3.2   Loss Function

The loss function of our SSINet is formulated as Eq. (6).

$$L_{\text{total}} = L_{\text{reg}} + \beta\, L_{\text{att}} \qquad (6)$$

where L_{reg} represents the text pixel probability regression loss, L_{att} represents the loss of foreground attention map. β represents the balance parameter and is empirically set to 1 in our work.

For the calculation of the L_{reg}, the same ground truths and MSE loss function as in TextPMs [25] are used for processing.

$$L_{\text{reg}} = \sum_{i=1}^{N} L_i \qquad (7)$$

L_i is the probability map loss for the i-th prediction, and N represents the total number of predictions. In our experiments, we set N to 4 as in TextPMs [25]. Note that L_{att} and L_{reg} share the same ground truth annotation set.

4   Experiment

4.1   Datasets and Evaluation Method

In our empirical analysis, we select four public benchmark datasets, i.e., Total-Text [4], ICDAR2017-MLT [13], CTW-1500 [22], and MSRA-TD500 [19], that are widely acknowledged for assessing the efficacy of natural scene text detection methodologies. To evaluate the performance of our proposed approach, we employ standard evaluation metrics, namely Precision (P), Recall (R), and F-score (F). The corresponding evaluation code, which facilitates the rigorous quantification of these performance indicators, can be conveniently downloaded from the designated website[1].

4.2   Implementation Details

We utilize a pretrained ResNet50 [6] as the backbone network. Our SSINet is pre-trained on ICDAR2017-MLT dataset for 300 epochs using the Adam optimizer, with a batch size of 6 and an initial learning rate of 1e−2. During the fine-tuning

---

[1] https://rrc.cvc.uab.es/.

process, we first randomly crop text regions and resize them to 640×640. Then, we resize them again to 800×800. For both stages, the model is trained with a batch size of 6 using the SGD optimizer. The initial learning rate is set to $1e-2$ and multiplied by 0.9 every 100 epochs. All experiments are conducted on a single GPU (RTX-3090) using PyTorch 1.7.0.

4.3   Ablation Experiments

Ablation Study for SSM. To verify the effectiveness of the proposed SSM, multiple ablation experiments are conducted on Total-text without any additional pre-training. The overall model is trained by removing different submodules from SSM to validate their effectiveness.

Specifically, SSM w/o DASPP means no multi-scale semantic information learning for high-level semantic information. SSM w/o PMM indicates using channel-wise concatenation for feature fusion. FTPM means the foreground text perception sub-module, and SSM w/o FTPM indicates no foreground text perception attention weights are learned. The quantitative results are shown in Table 1, demonstrating the complementary effects among the components in SSM and confirming that the design of SSM is valid in our SSINet.

Table 1. Results of the ablation experiments for the SSM and its different sub-modules on Total-text. The symbol * means our implementation by using the code downloaded from the website (https://github.com/GXYM/TextPMs).

Method	R(%)	P(%)	F(%)
TextPMs* [25] (Baseline)	84.08	82.84	83.46
Baseline + SSM w/o DASPP	80.50	85.78	83.06
Baseline + SSM w/o PMM	80.83	88.08	84.30
Baseline + SSM w/o FTPM	82.48	87.87	85.09
Baseline + SSM	83.27	87.13	85.15

The impact of different sub-modules in the SSM is shown in Table 1. The results demonstrate the varying degrees of influence and importance of the DASPP sub-module in the multi-scale semantic information extraction submodule, the PMM in the semantic-to-space information guidance sub-module, and the foreground text perception sub-module (FTPM). Among them, DASPP of the multi-scale semantic information extraction sub-module has a greater impact on our SSINet, compared to the PMM and FTPM. SSM achieves 4.29% and 1.69% performance gains in Precision and F-score, respectively, compared to the baseline method TextPMs [25].

FED-Net [3] recognizes that features of different resolutions have semantic differences, and directly adding the high-level semantic fusion feature map P_5' to the low-level feature map P_1, P_2 and P_3 for fusion can result in semantic discrepancies, negatively affecting subsequent cross-layer feature fusion. Therefore, the design of DASPP addresses this issue by extracting multi-scale receptive field semantic information to alleviate the semantic differences among feature maps at different levels, thereby significantly influencing the subsequent feature fusion interaction learning and text information processing.

4.4   Comparisons with State-of-the-Art Methods

To fully validate the performance of our SSINet, we selectively choose different types of datasets and conduct experiments, comparing them with previous methods on several publicly available datasets, including CTW-1500, MSRA-TD500, and ICDAR2017-MLT. The specific quantitative results are shown in Table 2.

Curved Text Detection. To validate the robustness of our SSINet on curved texts, experiments are conducted on CTW-1500. In the testing phase, we scale the short side of the test images to 512, the long side is no larger than 1024. As shown in Table 2, our method achieves 86.0% F-score accuracy, improving F-score, Recall, and Precision by 0.5%, 0.8% and 0.1%, respectively, compared to the baseline model. Moreover, in terms of F-score metric, our SSINet's performance surpasses most state-of-the-art methods. With the assistance of the SSM, our SSINet effectively enhances the ability of single-level feature map to express multi-scale features with rich semantic and prominent foreground. Here are some representative detection results of our SSINet on CTW-1500, as shown in Fig. 3(a).

Table 2. Experimental results on CTW-1500, MSRA-TD500, and ICDAR2017-MLT. The symbol * means our implementation by using the code downloaded from the website (https://github.com/GXYM/TextPMs), and Ext means extra data pretraining. The best results of all metrics are indicated in bold.

Method	Backbone	CTW-1500			MSRA-TD500			ICDAR2017-MLT			Ext
		R(%)	P(%)	F(%)	R(%)	P(%)	F(%)	R(%)	P(%)	F(%)	
R-Net [18]	VGG16	71.0	74.6	72.8	79.7	83.7	81.7	64.5	70.4	67.3	✓
LOMO [23]	Res50	69.6	**89.2**	78.4	–	–	–	60.6	78.8	68.5	✓
DBNet [8]	Res18-DCN	77.5	84.8	81.0	76.3	90.4	82.8	63.8	81.9	71.7	✓
PSENet-1s [16]	Res50	79.7	84.8	82.2	–	–	–	**68.2**	73.8	70.9	✗
DBNet [8]	Res50-DCN	80.2	86.9	83.4	79.2	91.5	84.9	67.9	83.1	**74.7**	✓
CRAFT [1]	VGG16	81.1	86.0	83.5	78.2	88.2	82.9	**68.2**	80.6	73.9	✓
ContourNet [17]	Res50	**84.1**	83.7	83.9	–	–	–	–	–	–	–
DRRG [26]	VGG16	83.0	85.9	84.5	82.3	88.1	85.1	61.0	75.0	67.3	✗
PCR [5]	DLA34	82.3	87.2	84.7	83.5	90.8	87.0	–	–	–	–
TextBPN [27]	Res50	83.6	86.5	85.0	84.5	86.6	85.6	–	–	–	–
FewNet [15]	Res50	82.4	88.1	85.2	84.8	91.6	88.1	–	–	–	–
DBNet++ [9]	Res50-DCN	82.8	87.9	85.3	83.3	91.5	87.2	–	–	–	–
TextBPN++ [24]	Res50	83.8	87.3	85.5	85.4	89.2	87.3	65.7	80.5	72.3	✗
TextPMs* [25]	Res50	82.2	89.1	85.5	**87.3**	89.9	88.6	64.7	80.1	71.6	✗
Ours	Res50	83.0	**89.2**	**86.0**	85.6	**92.9**	**89.1**	63.8	**83.6**	72.4	✗

(a) CTW-1500   (b) MSRA-TD500   (c) ICDAR2017-MLT

Fig. 3. Visualization detection results of our SSINet. Images in column 1–3 are sampled from CTW-1500 [22], MSRA-TD500 [19], and ICDAR2017-MLT [13], respectively.

Multi-oriented Text Detection. To validate the robustness of our SSINet on multi-oriented texts, experiments are conducted on MSRA-TD500. In the testing phase, we scale the long side of the test images to 832, and there is no limit on the short side. As shown in Table 2, our SSINet achieves 89.1% F-score accuracy, improving F-score and Precision by 0.5% and 3.0% respectively, compared to the baseline model. Furthermore, our SSINet outperforms several state-of-the-art methods on this dataset. Some representative detection results are shown in Fig. 3(b).

Multi-language Text Detection. To validate the robustness of our SSINet on multi-language texts, experiments are conducted on ICDAR2017-MLT. In the testing phase, we scale the short side of the test images to 256, the long side is no larger than 1920. As shown in Table 2, on ICDAR2017-MLT, our SSINet achieves 72.4% F-score accuracy without any extra data pretraining, improving

F-score and Precision by 0.8% and 3.5% respectively, compared to the baseline model. Furthermore, our SSINet outperforms several state-of-the-art methods. Some representative detection results are shown in Fig. 3(c).

5 Conclusion

In this paper, we propose a novel end-to-end arbitrary shape scene text detector SSINet. The proposed SSM effectively improves the model's multi-scale feature representation of the semantic-rich and foreground-prominent single-level feature map, which combines high-level semantic with correlations between cross-level features. Meanwhile, the DASPP in the multi-scale semantic information extraction sub-module plays an important role in reducing the semantic differences between different-level features and facilitating the fusion of cross-layer features. Extensive experiments on three benchmark datasets demonstrate the effectiveness of the proposed method. Further research on this work will attempt to extend it to real-time arbitrary shape scene text detection and end-to-end text recognition systems.

References

1. Baek, Y., Lee, B., Han, D., Yun, S., Lee, H.: Character region awareness for text detection. In: Proceedings of the IEEE/CVF Conference on Computer Vision and Pattern Recognition, pp. 9365–9374 (2019)
2. Chen, L.-C., Zhu, Y., Papandreou, G., Schroff, F., Adam, H.: Encoder-decoder with atrous separable convolution for semantic image segmentation. In: Ferrari, V., Hebert, M., Sminchisescu, C., Weiss, Y. (eds.) ECCV 2018. LNCS, vol. 11211, pp. 833–851. Springer, Cham (2018). https://doi.org/10.1007/978-3-030-01234-2_49
3. Chen, X., Zhang, R., Yan, P.: Feature fusion encoder decoder network for automatic liver lesion segmentation. In: 2019 IEEE 16th International Symposium on Biomedical Imaging (ISBI 2019), pp. 430–433. IEEE (2019)
4. Ch'ng, C.K., Chan, C.S.: Total-Text: a comprehensive dataset for scene text detection and recognition. In: 2017 14th IAPR International Conference on Document Analysis and Recognition (ICDAR), vol. 1, pp. 935–942. IEEE (2017)
5. Dai, P., Zhang, S., Zhang, H., Cao, X.: Progressive contour regression for arbitrary-shape scene text detection. In: Proceedings of the IEEE/CVF Conference on Computer Vision and Pattern Recognition, pp. 7393–7402 (2021)
6. He, K., Zhang, X., Ren, S., Sun, J.: Deep residual learning for image recognition. In: Proceedings of the IEEE Conference on Computer Vision and Pattern Recognition, pp. 770–778 (2016)
7. Kang, J., Ibrayim, M., Hamdulla, A.: Overview of scene text detection and recognition. In: 2022 14th International Conference on Measuring Technology and Mechatronics Automation (ICMTMA), pp. 661–666. IEEE (2022)
8. Liao, M., Wan, Z., Yao, C., Chen, K., Bai, X.: Real-time scene text detection with differentiable binarization. In: Proceedings of the AAAI Conference on Artificial Intelligence, vol. 34, pp. 11474–11481 (2020)
9. Liao, M., Zou, Z., Wan, Z., Yao, C., Bai, X.: Real-time scene text detection with differentiable binarization and adaptive scale fusion. IEEE Trans. Pattern Anal. Mach. Intell. **45**(1), 919–931 (2022)

10. Liu, W., et al.: SSD: single shot multibox detector. In: Leibe, B., Matas, J., Sebe, N., Welling, M. (eds.) Computer Vision-ECCV 2016: 14th European Conference, Amsterdam, The Netherlands, 11–14 October 2016, Proceedings, Part I 14, vol. 9905, pp. 21–37. Springer, Cham (2016). https://doi.org/10.1007/978-3-319-46448-0_2

11. Long, J., Shelhamer, E., Darrell, T.: Fully convolutional networks for semantic segmentation. In: Proceedings of the IEEE Conference on Computer Vision and Pattern Recognition, pp. 3431–3440 (2015)

12. Ma, J., et al.: Arbitrary-oriented scene text detection via rotation proposals. IEEE Trans. Multimedia **20**(11), 3111–3122 (2018)

13. Nayef, N., et al.: ICDAR 2017 robust reading challenge on multi-lingual scene text detection and script identification-RRC-MLT. In: 2017 14th IAPR International Conference on Document Analysis and Recognition (ICDAR), vol. 1, pp. 1454–1459. IEEE (2017)

14. Ren, S., He, K., Girshick, R., Sun, J.: Faster R-CNN: towards real-time object detection with region proposal networks. In: Advances in Neural Information Processing Systems, vol. 28 (2015)

15. Tang, J., et al.: Few could be better than all: feature sampling and grouping for scene text detection. In: Proceedings of the IEEE/CVF Conference on Computer Vision and Pattern Recognition, pp. 4563–4572 (2022)

16. Wang, W., et al.: Shape robust text detection with progressive scale expansion network. In: Proceedings of the IEEE/CVF Conference on Computer Vision and Pattern Recognition, pp. 9336–9345 (2019)

17. Wang, Y., Xie, H., Zha, Z.J., Xing, M., Fu, Z., Zhang, Y.: ContourNet: taking a further step toward accurate arbitrary-shaped scene text detection. In: Proceedings of the IEEE/CVF Conference on Computer Vision and Pattern Recognition, pp. 11753–11762 (2020)

18. Wang, Y., Xie, H., Zha, Z., Tian, Y., Fu, Z., Zhang, Y.: R-Net: a relationship network for efficient and accurate scene text detection. IEEE Trans. Multimedia **23**, 1316–1329 (2020)

19. Yao, C., Bai, X., Liu, W., Ma, Y., Tu, Z.: Detecting texts of arbitrary orientations in natural images. In: 2012 IEEE Conference on Computer Vision and Pattern Recognition, pp. 1083–1090. IEEE (2012)

20. Ye, J., Chen, Z., Liu, J., Du, B.: TextFuseNet: scene text detection with richer fused features. In: IJCAI, vol. 20, pp. 516–522 (2020)

21. Yu, W., Liu, Y., Hua, W., Jiang, D., Ren, B., Bai, X.: Turning a clip model into a scene text detector. In: Proceedings of the IEEE/CVF Conference on Computer Vision and Pattern Recognition, pp. 6978–6988 (2023)

22. Yuliang, L., Lianwen, J., Shuaitao, Z., Sheng, Z.: Detecting curve text in the wild: new dataset and new solution. arXiv preprint arXiv:1712.02170 (2017)

23. Zhang, C., et al.: Look more than once: an accurate detector for text of arbitrary shapes. In: Proceedings of the IEEE/CVF Conference on Computer Vision and Pattern Recognition, pp. 10552–10561 (2019)

24. Zhang, S.X., Yang, C., Zhu, X., Yin, X.C.: Arbitrary shape text detection via boundary transformer. IEEE Trans. Multimedia 1–14 (2023)

25. Zhang, S.X., Zhu, X., Chen, L., Hou, J.B., Yin, X.C.: Arbitrary shape text detection via segmentation with probability maps. IEEE Trans. Pattern Anal. Mach. Intell. (2022). https://doi.org/10.1109/TPAMI.2022.3176122

26. Zhang, S.X., et al.: Deep relational reasoning graph network for arbitrary shape text detection. In: Proceedings of the IEEE/CVF Conference on Computer Vision and Pattern Recognition, pp. 9699–9708 (2020)

27. Zhang, S.X., Zhu, X., Yang, C., Wang, H., Yin, X.C.: Adaptive boundary proposal network for arbitrary shape text detection. In: Proceedings of the IEEE/CVF International Conference on Computer Vision, pp. 1305–1314 (2021)
28. Zhou, X., et al.: EAST: an efficient and accurate scene text detector. In: Proceedings of the IEEE Conference on Computer Vision and Pattern Recognition, pp. 5551–5560 (2017)
29. Zhu, Y., Du, J.: TextMountain: accurate scene text detection via instance segmentation. Pattern Recogn. **110**, 107336 (2021)
30. Zhuang, J., Qin, Z., Yu, H., Chen, X.: Task-specific context decoupling for object detection. arXiv preprint arXiv:2303.01047 (2023)

AIE-KB: Information Extraction Technology with Knowledge Base for Chinese Archival Scenario

Shiqing Bai[✉], Yi Qin, and Peisen Wang

Zhengzhou University, Zhengzhou, Henan, China
{iesqbai,ieyqin,wps28501}@gs.zzu.edu.cn

Abstract. The current visual information extraction (VIE) methods usually focus only on textual features or image features of the document modality, ignoring the segment-level structural information in the document, thus limiting the accuracy of information extraction. To address the problem of information extraction in the scenario of scanned personnel archival images, the paper proposes an Archive Information Extraction model with Knowledge Base (AIE-KB), which injects knowledge into the model in the coarse relationship prediction module (CRP). Our model undergoes several modules, including entity recognition, CRP, and entity linking, after encoding the features for multi-modal fusion. As detailed in Fig. 2. Specifically, we (1) add new intermediate entity types and (2) design a new relationship prediction head to explicitly exploit structural information in scanned images, solving the problem of nested entity relationships specific to personnel archives. To address the lack of a certain amount of annotated data on personnel files, we also introduce a benchmark dataset of personnel archives in which the "answer" types in each scanned archival image are handwritten. The experimental results show that our model AIE-KB achieves excellent results in two tasks with different datasets, demonstrating the effectiveness and robustness of the model. All metrics for both tasks exceeded the baseline model, with the F1 metric for entity recognition on the Visually rich scanned Images in the personnel Archive (VIA) (from 84.36 to 94.61) and entity linking on VIA (from 69.79 to 87.94).

Keywords: information extraction · archival knowledge base · multi-modal fusion

1 Introduction

The specialized audit of personnel archives is a process of appraisal and review of the archives, which can effectively ensure the authenticity, accuracy, and completeness of the archives [1]. The traditional specialized audit of personnel archives adopts the manual identification method, which is not only time-consuming and labor-intensive but also prone to subjective errors and omission of information. The key information extraction technique is aimed at automatically extracting predefined information entities and their relationships from visually rich documents. Applying this technology to the process of digital archive

© The Author(s), under exclusive license to Springer Nature Singapore Pte Ltd. 2024
Q. Liu et al. (Eds.): PRCV 2023, LNCS 14431, pp. 52–64, 2024.
https://doi.org/10.1007/978-981-99-8540-1_5

processing, extract the relevant text pairs from the complex structure scanned archival images, and establish the basic information database of personnel in preparation for the subsequent archival special audit. Such automated processing not only improves work efficiency but also reduces the occurrence of human errors.

In the traditional methods, there are two common methods for the key information extraction task, based on text features and based on multi-modal features. The text-based feature approach usually splits the task into a serial process of text recognition and plain text information mining [2]. However, this approach only considers textual content and cannot handle complex structured scanned archival images. On the other hand, the multi-modal feature-based approach has made significant progress in the field of visual information extraction (VIE) [5–7], which takes into account the visual and layout features of the image and implements cross-modal feature fusion.

In the actual personnel archive scenario, there may be nested relationships between entities, which has not been solved by existing methods. Inspired by the field of VIE, we propose a multi-modal personnel archive information extraction network model and split the task into two subtasks, entity recognition and entity linking [3, 9] to implement without predefined entity types. To solve the problem of nesting relationships between entities, we introduce the intermediate entity type in the entity recognition task. As shown in Fig. 1, e.g. ("工资情况" labeling Question token and "4" labeling Answer token in the bottom-left image). The scanned image may have entities with nested relationships in the archival scenario, which causes the relational linking task to be difficult. We add a new entity type to connect entities with nested relationships e.g. ("职务工资", "档次" which the model classifies as the "qa" token in the bottom-left image).

In the training phase of information extraction, entity recognition (ER) and entity linking (EL) are the main tasks of information extraction [5,6]. Consider that there may be multiple levels of nested relationships between entities, and within this chain of relationships, there are intermediate entities that can be

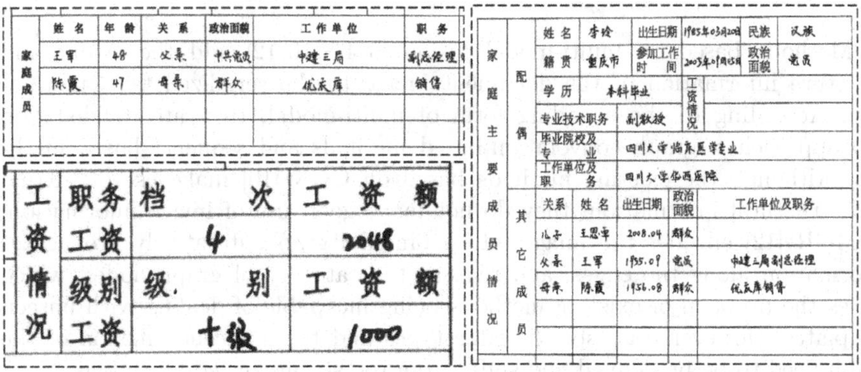

Fig. 1. Snippet of synthetic scanned archival images

treated as both "question" and "answer" types. Therefore, it is necessary to add an entity category, which we mark as "qa" type, to represent entities with nested relations. The introduction of the "qa" entity category has led to more complex relationships between pairs of entities, therefore we add the coarse relation prediction (CRP) module between the ER module and the EL module. The CRP module can roughly classify the relationships between pairs of entities, by training a binary classification model to determine whether a relationship exists between pairs of entities.

Our model proves its effectiveness and robustness in two public datasets (XFUND-zh, SIBR-zh) and the Visually rich scanned Images in the personnel Archive (VIA) constructed by our research group. The contributions of this paper are as follows:

- We propose a model for information extraction in the Chinese personnel archive scenario that takes up the tasks of entity recognition and entity linking. A new entity type: intermediate entity, is added to solve the problem of nested relationships between entities.
- The coarse relationship prediction module is conceived to support the modeling of entity linking types. In addition, the relationship prediction head is designed to explicitly exploit the distance and direction in the layout information, and we set different training strategies for different link types.
- The experimental evaluation shows that AIE-KB outperforms multiple models and that AIE-KB can replace manual extraction of key information to a certain extent, effectively facilitating the automation of the subsequent archival special audit.

2   Related Works

The task of information extraction was first applied in the field of natural language processing, and there has been considerable research. The experience summarized from previous works is that the task is usually divided into two stages, classifying the categories of entity features, and classifying the relations of entity pairs.

Methods based on multi-modal fusion [3,4,8,10,12] add the layout and structure information of the document image to the serialized text information. According to the encoding form of multi-modal, they are divided into two approaches: graph convolution-based methods and sequence-based methods. Although the existing methods based on GCN [10] make use of textual and structural information, they do not make good use of image information. SDMGR [12] encodes the image into a bimodal graph, iteratively propagates information along the edges, and reasons the category of graph nodes, which solves the problem of existing methods being incapable of dealing with unseen templates. However, task-specific edge types and the adjacency matrix of the graph need to be predefined and suffer from too strong an inductive bias.

LayoutXLM [3] is a sequence-based approach, that uses Transformers to fuse textual, layout, and image information in the pre-training stage, and adds a spatial-aware self-attention mechanism to support the model to fuse visual and textual features. However, the layout features learned by the model are still not fine enough due to the limitations of the pre-training task. BROS [4] encodes the relative spatial position of text into BERT to better learn the layout representation. StrucTexT [8] proposes a paired box direction task to model the geometric orientation of text segments before training. However, only the geometric relationship at the pairwise level is explored. GeoLayoutLM [5] expands the relations of multi-pairs and triplets and designs three geometric pre-training tasks, which presents an efficient approach to geometric pre-training for VIE tasks. However, the model's performance might be affected when dealing with intricate visual elements or entities with nest relationships.

Several studies [14–16] have explored how to inject external knowledge into relation extraction (RE) tasks to improve model accuracy. RE-SIDE [14] injects external knowledge into entity types, but only considers limited characteristics of entities. RECON [15] proposes a model for encoding attribute triples and relation triples in a knowledge graph and combines it with sentence embedding. ConceptBert [16] proposes a novel concept-aware representation learning method to enhance the performance of visual question-answering models by combining the concept information in images and questions.

Inspired by these works, we inject archive knowledge into the coarse relation prediction module and set the corresponding training strategy for the relationship prediction head, which enhances the relation feature representation of the EL task and effectively improves the understanding of complex document structure in archival images.

3   Model Architecture

The overall model architecture of the AIE-KB is shown in Fig. 2, given an input image containing the text contents and the corresponding bounding boxes. In the feature embedding stage, various types of information from image, text, and layout, are utilized to generate embeddings for each mode, and these embeddings capture the specific representations of each modality. The embeddings from different modes are then fed into a pre-trained transformer encoder, where they are concatenated together to obtain cross-mode fusion features. Then the entity recognition module implements the prediction of the category of the entity, the coarse relation prediction module filters the relational connections of entities in the inference phase, and the entity linking module is realized through the decoder and the output of the above modules. In this section, the overall framework and highlight components are introduced.

3.1   Multi-modal Feature Embedding

Given a scanned image I, we adopt the DBNet algorithm [11] to obtain the text segment regions B. And the PP-OCRv3 algorithm [9] is used to extract the text content in each B. For the layout embedding, we input the coordinates of

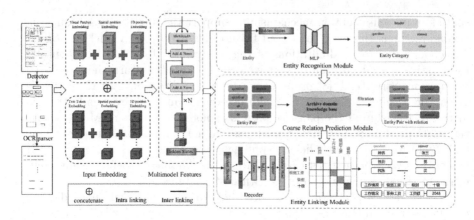

Fig. 2. The overall model architecture of the AIE-KB.

each text segment, encoded as layout embedding L. For the textual embedding, the text sequence S is input to the text encoding layer W_t, and the sorted 1D position text embedding P_t^s and the layout embedding L are added to obtain the text embedding, $T = W_t(S) + P_t^s + L$. For the visual embedding, we adopt ResNet50 as the image feature extractor and extract the visual features of each text segment in combination with the position of the text detection box B. Furthermore, the visual embedding V of the image I is obtained by adding the sorted 1D position visual embedding P_v^s and the layout embedding L, which can be expressed as $V = ResNet(B) + P_v^s + L$. In addition, the S^{id} of the segment embedding is added to represent the different modal features. Finally, the fused cross-modal representation M is obtained by concatenating the embeddings of different modalities as input to the encoder layer.

$$M = Encoder\left(Concat\left(T, V\right) + S^{id}\right). \tag{1}$$

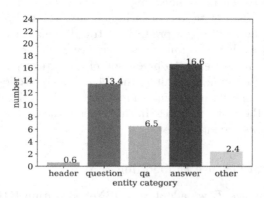

Fig. 3. The average number of entity categories per image on VIA. (Color figure online)

3.2 Feature Enhancement Module

Entity Recognition Module. Entity recognition (ER) task in VIE is usually modeled as a token classification problem, and learning a simple multilayer perceptron (MLP) classifier is effective for ER task [7]. There are only four types of entity types [6,13] in VIE, and the label of entity is categorized as "question", "answer", "header", and "other". To solve the problem of nested relationships between entities, our model adds intermediate entity types for connecting entities with nested relationships and defines the class of all entities: $E = \{e_{header}, e_{question}, e_{answer}, e_{qa}, e_{other}\}$, where e_{qa} represents the intermediate entity type, which can be treated as either a "question" entity or an "answer" entity. For each text segment token, we feed the output vector into the activation function, after that we will get the predicted probabilities $p_{category}$ for each entity class.

$$p_{category} = Softmax\left(MLP\left(M, E\right)\right), \tag{2}$$

where M represents the fused cross-modal representation. And we use the focal loss as the loss function in an entity recognition task, to address the problem of extreme entity type imbalance, which can also be seen in Fig. 3. The formula for the loss $Loss_{ER}$ in the ER task is as follows:

$$Loss_{ER} = FocalLoss(p_{category}) = -\alpha(1 - p_{category})^{\gamma}log(p_{category}), \tag{3}$$

where the parameter α has five different categories of values, which are set according to the proportion of positive and negative samples of each category on VIA, the parameter γ is used to adjust the difficulty level of distinguishing the number of samples, and we set to 5.

Coarse Relationship Prediction Module. In the coarse relationship prediction (CRP) module, the archival knowledge base is used for the preliminary prediction of entity relationships, and the performance of information extraction is enhanced by building semantically informative entity pairs. In this regard, CRP is essentially a binary classification model for determining whether a text pair is related, in which the pre-trained Chinese model ERNIE-Tiny [17] is fine-tuned in the archival dataset. Given two entity segments e, they are classified by the archival corpus, which in turn filters out the entity pairs with semantic information.

$$p_{ij}^{CRP} = Softmax(MLP(e_i, e_j)), \tag{4}$$

which $p_{ij}^{CRP} \in \{0, 1\}$ is the probability value of determining whether i-th entity segment e_i has a semantic association with j-th entity segment e_j, using the simple MLP network: FC and ReLU, and only two nodes in the output layer, this results in a loss function $Loss_{CRP}$ for the CRP module.

$$Loss_{CRP} = BCE(P^{CRP}, L^{CRP}), \tag{5}$$

where L^{CRP} is the label for CRP modeling, and the binary cross-entropy (BCE) function is utilized in CRP loss.

Entity Linking Module. Inspired by SIBR [6], entities with nested relations are considered to be transformed into intra-entity link predictions. For the entity linking module, given all entities in the scanned archival image I, a predefined set of semantic entity categories E and a set of relationship types $R = \{r_{inter}, r_{intra}\}$, the module needs to find the function $F_{re}(I, E, R) \Rightarrow Y$, where Y is the set of predicted semantic relationships.

$$
\begin{aligned}
Y = \{ \ &(link\,[e_{question}, e_{answer}]\,, r_{inter})\,, \\
&(link\,[e_{question}, e_{qa}]\,, r_{intra})\,, \\
&(link\,[e_{qa}, e_{qa}]\,, r_{intra})\,, \\
&(link\,[e_{qa}, e_{answer}]\,, r_{inter})\}\,,
\end{aligned} \tag{6}
$$

where $link[*]$ denotes a collection of entity pairs, r_{inter} denotes an inter-entity linking with a textual question-answering relationship, and r_{intra} denotes an intra-entity linking with a textual entailment relationship.

It is observed that there are such characteristics in scanned archival images: there are entity pairs with intermediate entities, the textual information and layout information are used to design the links between the entities. The direction of the entity is determined by calculating the angle between entity segment i pointing to entity segment j and the vertical line. According to the habit of normal reading order: from left to right, from top to bottom [13], so we only consider two directions: right and down. Finally, the orientation problem between entities and their corresponding entities is transformed into a binary classification problem, and the result p_{ij}^{direct} is the predicted direction probability.

$$
p_{ij}^{direct} = Sigmoid\,(MLP\,(e_i, e_j))\,, \tag{7}
$$

which e_i, e_i represent the i-th and j-th entity tokens respectively, $Sigmoid$ is chosen as the activation function.

When the entity pair does not contain intermediate entities, only one answer will exist. This means that only a one-to-one entity link. To take advantage of this feature, we explicitly include the distance of entity pairs in the layout information. Determine whether the j-th entity segment e_j is closest to the i-th entity segment e_i in its right and down directions, based on the determination of at most one nearest segment by distance. p_{ij}^{dist} is the probability of the nearest entity pair, a bilinear layer is applied in the distance strategy, See the following Eq. (8):

$$
p_{ij}^{dist} = Sigmoid\,(Bilinear\,(e_i, e_j))\,. \tag{8}
$$

For relation prediction head, which is a ReLU-FC-ReLU-FC block followed by a softmax layer.

$$
p_{ij}^{link} = Softmax\,(MLP\,(e_i, e_j))\,, \tag{9}
$$

where p_{ij}^{link} is the probability of the entity pair with the relation between e_i and e_j. Finally, the loss function $Loss_{EL}$ of entity linking task is defined as:

$$
\begin{aligned}
Loss_{EL} = \ &CrossEntropy(P^{link}, L^{link}) + \\
&BCE(P^{direct}, L^{direct}) + BCE(P^{dist}, L^{dist})\,,
\end{aligned} \tag{10}
$$

where L^{link} L^{direct} and L^{dist} are labels for linking direction and distance modeling. The whole modules are jointly trained by minimizing the following loss function $Loss_{total}$ as:

$$Loss_{total} = \lambda Loss_{ER} + (1 - \lambda)Loss_{EL}, \tag{11}$$

where $Loss_{ER}$ and $Loss_{EL}$ are defined in Eq. (3) and Eq. (10) individually, and λ is set to 0.05 for fine-tuning.

4 Experiments

4.1 Datasets

Due to a lack of digital personnel archive data for information extraction, our group established a new dataset called Visually rich scanned Images in the personnel Archive (VIA), which contains 474 images with 18,723 annotated entity instances and 7,206 entity links. Figure 3 shows the average number of entity categories per image. The cyan, orange, green, blue, and gray columns represent entity types for "header", "question", "answer", "qa", and "other", respectively. Figure 4 shows the visualization of labels on VIA. Note that all archival images appearing in this paper are synthetic. The cyan, purple, gray, and green boxes represent entity types for "question", "answer", "qa", and "other", respectively. Moreover, question-answer pairs and qa-answer pairs are linked with green lines, and question-qa pairs, and qa-qa pairs are linked with blue lines.

For training the AIE-KB model, we select image data similar to the layout of personnel archive documents from an online competition on structured text extraction from Visually-Rich Document Images and add 60% of VIA as the original training set. The evaluation dataset contains, in addition to VIA, two publicly available datasets, SIBR-zh [6], which contains 700 Chinese invoices, bills, and receipts, and XFUND-zh [3], which contains form data in Chinese.

For the archive text pairs generation. The annotations for the archival knowledge base dataset training are in the form of triples: ($entity_1$, $entity_2$, $label$), where $entity_1$ and $entity_2$ represent pairs of entities, the label represents the ground truth of the annotation, indicating whether the entity pair is associated or not, and they are separated by "\t". The positive sample triplet is constructed from the data in the personnel archive, and the negative sample triplet is constructed by $entity_1$ randomly selecting the non-corresponding $entity_2$. The final ratio of negative and positive samples is 3:1.

4.2 Implementation Details

Model Setting. AIE-KB is based on the Transformer which consists of a 12-layer encoder with 768 hidden sizes, 12 attention heads, and the sequence of a maximum length of 512 tokens. To take advantage of existing pre-trained models and adapt to scanned document image downstream tasks, we initialize the weight of the AIE-KB model with the pre-trained LayoutXLM [3] model except for the position embedding.

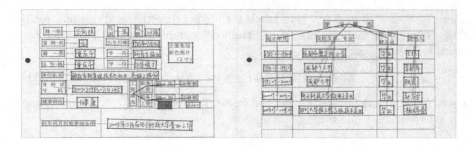

Fig. 4. Example of labels on VIA. (Color figure online)

Experiments Setting. Our model was trained on 2 NVIDIA RTX3060TI GPUs with 16 GB memory and used Adam as the optimizer to minimize the total loss at the training phase. The learning rate is set to 0.0001 and the batch size is 8 over the whole training phase. Our model was trained over 50 epochs, each epoch about 55 min.

4.3 Experiments and Result Analysis

Baseline Comparison. We select the LayoutXLM algorithm as the baseline task, the structure that proved effective in work VIE [3,7]. For the entity recognition (ER) task, we calculate the F1 score as the primary evaluation metric, with precision and recall also reported. For entity linking (EL) tasks, adopt the F1-score as our evaluation metric. As can be seen in Table 1, we compared the F1 for the baseline and AIE-KB on VIA. The experimental results for the ER task showed that AIE-KB outperformed the baseline for all entities and improved the overall F1 score by 10.04%. See from the sixth row, the precision of AIE-KB almost gets a full score. Further analysis shows that the most significant increase in the data on both models was in F1 performance for the "qa" entity type.

Table 1. Performance comparison between baseline method and Ours on VIA.

Entities		header	question	qa	answer	other	Total (micro)
LayoutXLM (baseline)	F1	97.84	83.23	70.07	84.53	93.5	**84.39**
	precision	96.48	80.32	65.28	88.47	91.69	83.32
	recall	99.23	88.54	75.61	80.92	95.38	85.84
AIE-KB (Ours)	F1	99.55	93.82	90.63	95.61	94.95	**94.43**
	precision	99.98	92.45	93.24	94.12	96.59	93.99
	recall	99.12	95.23	88.16	97.15	93.36	94.93

To verify the performance of our proposed model, we add the advanced ESP algorithm [6] and the GeoLayoutLM algorithm [5] in VIE filed on XFUND-zh,

SIBR-zh, VIA datasets, the performance of entity recognition and entity linking tasks is shown in Table 2, which the symbol "-" indicates that the ESP model is not opensource or the GeoLayoutLM model is only applicable to English document images, so the experimental results cannot be obtained. Moreover, AIE-KB outperformed the baseline model by 7.81%, 8.7%, and 10.25% on XFUND-zh, SIBR-zh, and VIA on the ER task, a result that suggests that archival knowledge injection and multimodal feature fusion alignment sorted by position embedding are helpful to ER. For the RL task, The F1 metric of AEI-KB is 13.77% higher than that of the baseline model, and 4.2% higher than that of the latest model GeoLayoutLM on VIA, which indicates the good generality of our training strategy for relational prediction heads. However, it is not surprising that slightly lower than the latest model GeoLayoutLM on XFUND-zh. It proves that our model has a significant effect on relational linking in terms of archival images, and also benefits from the coarse relational prediction module based on the archival knowledge base.

Table 2. Comparisons on Entity Extraction and Entity Linking tasks, and the evaluation metric is F1.

Task	Model	XFUND-zh	SIBR-zh	VIA
ER	LayoutXLM	85.65	71.87	84.36
	ESP	90.3	-	-
	GeoLayoutLM	93.28	-	93.43
	AIE-KB(Ours)	93.46	80.57	**94.61**
EL	LayoutXLM	69.46	62.33	69.79
	ESP	85.96	-	-
	GeoLayoutLM	89.71	-	83.71
	AIE-KB (Ours)	86.48	76.10	**87.94**

The comparison is shown in Fig. 5, the columns from left to right indicate the visualization of ground truth, LayoutXLM, and AIE-KB, among which green rectangles and arrowed lines indicate correct information extraction results, and red rectangles and arrowed lines indicate unidentified or incorrectly identified results. Our model AIE-KB successfully predicts almost all links in an archival image that contains entities with nested relationships. However, LayoutXLM tends to link two entities relying more on their semantics. Most of the false relation links predicted by LayoutXLM are table columns, which is to illustrate that AIE-KB can handle the challenge of entities with nested relationships better than LayoutXLM on VIA, especially in the entity linking task.

To evaluate the contribution of each component of our model, we conduct an ablation study. As shown in Table 3, when we remove the coarse relation prediction module (CRP) from AIE-KB, the most striking observation that emerges from the data comparison is the degraded F1 performance of AIE-KB on VIA.

(1) Ground Truth   (2) AIE-KB   (3) LayoutXLM

Fig. 5. The visualization results of LayoutXLM model and ours on VIA. Note that only entity pairs are shown for better visualization. (Color figure online)

This shows that semantic features play an important role in solving problems with a large number of relations between entities. This result is interpretable because the knowledge base of the specialized domain can provide richer semantic features and richer association text pairs. In addition, the improved relational head algorithm also affects the performance of AIE-KB. Specifically, See from the last row, removing the relation prediction head (RPH) from AIE-KB leads to a decrease in the F1-metric score on both datasets, especially for images containing nested relationships of entities. Thus, the added intermediate entities are good at handling the nested structure of the entities.

Table 3. Ablation study on the entity linking task.

fusion method	Entity Linking	
	XFUND-zh	VIA
AIE-KB (Full model)	**86.5**	**87.9**
w/o AKB	85.9	81.6
w/o RPH	82.1	86.1

Ablation Study

5   Conclusion

This paper constructs a dataset called VIA to address the problem of limited data in the archival domain. Moreover, we propose the model AIE-KB, which identifies entity types more finely than general methods. To explore the relationship between entity pairs, we add the coarse relation prediction module, which introduces the archival knowledge base for relationship extraction.

The AIE-KB model experimented on different domain datasets, including archival, medical, and financial domains. The experimental results show that the AIE-KB model is effective and superior in dealing with undefined entity categories and entity nesting relationships, which provides a reference for future research in the archival field.

References

1. Fan, Y.: Research on the development and utilization of personnel archival resources in the digital era. Adm. Assets Finance **24**, 94–96 (2022)
2. Zhao, X., Wu, Z., Wang, X.: CUTIE: learning to understand documents with convolutional universal text information extractor. arXiv preprint arXiv:1903.12363 (2019)
3. Xu, Y., Lv, T., Cui, L., et al.: LayoutXLM: multimodal pre-training for multilingual visually-rich document understanding. arXiv preprint arXiv:2104.08836 (2021)
4. Hong, T., Kim, D., Ji, M., et al.: BROS: a pre-trained language model focusing on text and layout for better key information extraction from documents. In: Proceedings of the AAAI Conference on Artificial Intelligence, vol. 36(10), pp. 10767–10775 (2022)
5. Luo, C., et al.: GeoLayoutLM: geometric pre-training for visual information extraction. In: Proceedings of the IEEE/CVF Conference on Computer Vision and Pattern Recognition, pp. 7092–7101 (2023)
6. Yang, Z., Long, R., Wang, P., et al.: Modeling entities as semantic points for visual information extraction in the wild. In: Proceedings of the IEEE/CVF Conference on Computer Vision and Pattern Recognition, pp. 15358–15367 (2023)
7. Xu, Y., Xu, Y., Lv, T., et al.: LayoutLMv2: multi-modal pre-training for visually-rich document understanding. arXiv preprint arXiv:2012.14740 (2020)
8. Li, Y., Qian, Y., Yu, Y., et al.: StrucTexT: structured text understanding with multi-modal transformers. In: Proceedings of the 29th ACM International Conference on Multimedia, pp. 1912–1920 (2021)
9. Li, C., et al.: PP-OCRv3: more attempts for the improvement of ultra lightweight OCR system. arXiv preprint arXiv:2206.03001 (2022)
10. Liu, X., Gao, F., Zhang, Q., et al.: Graph convolution for multimodal information extraction from visually rich documents. arXiv preprint arXiv:1903.11279 (2019)
11. Liao, M., et al.: Real-time scene text detection with differentiable binarization. In: Proceedings of the AAAI Conference on Artificial Intelligence, pp. 11474–11481 (2020)
12. Sun, H., Kuang, Z., Yue, X., et al.: Spatial dual-modality graph reasoning for key information extraction. arXiv preprint arXiv:2103.14470 (2021)
13. Gu, Z., Meng, C., Wang, K., et al.: XYLayoutLM: towards layout-aware multimodal networks for visually-rich document understanding. In: Proceedings of the IEEE/CVF Conference on Computer Vision and Pattern Recognition, pp. 4583–4592 (2022)
14. Türker, R., Zhang, L., Alam, M., Sack, H.: Weakly supervised short text categorization using world knowledge. In: Pan, J.Z., et al. (eds.) ISWC 2020. LNCS, vol. 12506, pp. 584–600. Springer, Cham (2020). https://doi.org/10.1007/978-3-030-62419-4_33

15. Bastos, A., et al.: RECON: relation extraction using knowledge graph context in a graph neural network. In: WWW, pp. 1673–1685. ACM, Online (2021)
16. Gardères, F., et al.: ConceptBERT: concept-aware representation for visual question answering. Findings of the Association for Computational Linguistics: EMNLP 2020 (2020)
17. Su, W., Chen, X., Feng, S., et al.: ERNIE-Tiny: a progressive distillation framework for pretrained transformer compression (2021). https://doi.org/10.48550/arXiv. 2106.02241

Deep Hough Transform for Gaussian Semantic Box-Lines Alignment

Bin Yang[1], Jichuan Chen[1,2], Ziruo Liu[1,2], Chao Wang[1], Renjie Huang[1], Guoqiang Xiao[1]([✉]), and Shunlai Xu[2]

[1] College of Computer and Information Science, Southwest University, Chongqing 400715, China
gqxiao@swu.edu.cn
[2] Chongqing Academy of Animal Sciences, Chongqing 402460, China

Abstract. Image correction and trimming for the digitization of paper-based records is a challenging task. Existing models and techniques for object detection are limited and cannot be applied directly. For example, non-directional bounding boxes cannot correct skewed images, paper-based scanned images with complex boundaries (torn pages and damaged edges) cannot locate obvious bounding boxes, resizing high-resolution image leads to accuracy errors, etc. To this end, in this paper, we proposed a Boundary Detection Network (BDNet) based on deep hough transform, which implements semantic boundary detection with geometrical restriction in a coarse-to-fine mean. The model is mainly divided into two stages: coarse location and refined adjustment. The former predicts the boundaries' orientations and positions at the down-sampled image. The latter refines the boundary positions using the image patches sampled from the coarsely-located boundary in the original image. Among them, each stage contains two main modules: The Semantic Box-Lines (SBL) utilizes Gaussian heatmaps to capture extensive boundary semantic information, while the Deep Hough Alignment (DHA) efficiently extracts global line orientations to align semantic boundaries (box or lines). Detailed experiments and metric analysis verify that our proposed model is effective and feasible on the open-source datasets we collected, i.e., for scanned images with a resolution of 2500×3500 pixels, our method can accurately locate content boundaries and achieve 0.95 IoU accuracy.

Keywords: Image correction and trimming · Object detection · Deep hough transform · Semantic boundary · Gaussian heatmap

1 Introduction

The digitization of paper-based records is currently in the ascendant. Image correction and trimming of digital images scanned by high-resolution scanners

This work is funded by NSFC 62172339.
B. Yang and R. Huang—Contribute equally to this work.

Q. Liu et al. (Eds.): PRCV 2023, LNCS 14431, pp. 65–75, 2024.
https://doi.org/10.1007/978-981-99-8540-1_6

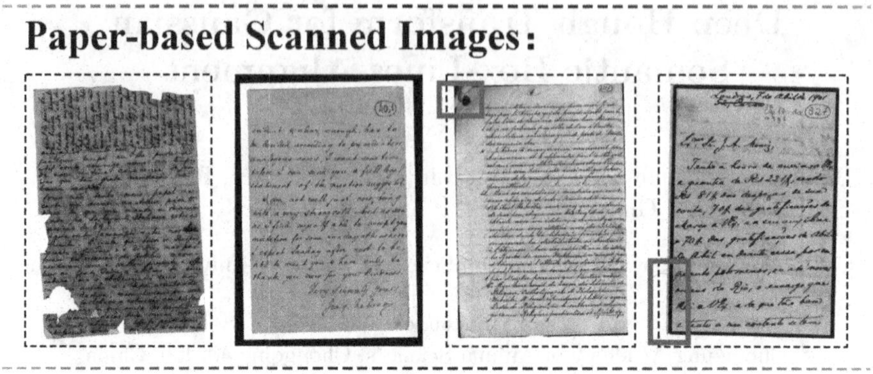

Paper-based Scanned Images:

Fig. 1. High-resolution (about 2500 × 3500 pixels) scanned images.

is an essential task in the digitization process. In earlier work, this is usually modeled as document edge detection [1,2] and skew detection using handcrafted features [3,4] and hough line detection [5], and the processed document object is strict rectangle. Deep neural networks model document detection as keypoint (corner) detection [6,7], contour detection [8], line detection [9–11], bounding box detection [12], and instance segmentation [13].

However, it is difficult to accurately locate the document boundary because of the following characteristics shown in Fig. 1, including the complex non-rectangle edge, skewness, unobvious boundary, long range of scale variations, high resolution and white borders around the scanned content or impurities (such as black dots, black borders, etc.) that affect the image quality caused by the scanning device. All these factors make this task challenging, and it mainly relies on manual cropping in actual projects. Therefore, previous algorithms and techniques can no longer meet the current application requirements. Algorithms and technologies need to be innovated urgently.

Obviously, to implement this task, two critical problems need to be settled. Firstly, it is required to determine the appropriate boundaries constituting the oriented bounding box. Considering that the oriented document boundaries will become more obvious, and the orientation remains unchanged when downsampling the image, it is reasonable to predict the orientation and coarse position of document in a down-sampled image. Secondly, the accurate boundary in the original high-resolution image is required to avoid that the scanned content is trimmed according to the special requirement of processing paper-based records (historical documents).

Based on the above analysis, inspired by semantic line detection [14], we propose the Boundary Detection Network (BDNet) on deep Hough transform [15], a novel task and dataset, which is used to study the image correction and trimming for paper-based records directly and integrally. Our BDNet is a CNN-based two-stage prediction-refinement boundary framework, which employs a coarse-to-fine strategy: (1) In the coarse location stage, predict the semantic box, and perform

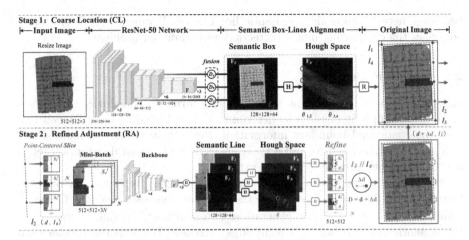

Fig. 2. Schematic illustration of our BDNet pipeline: A digital image is fed into a backbone (ResNet-50) for feature extraction, followed by deconvolution (D) and upsampling fusion operations to generate an intermediate feature map with multi-scale features. The feature map first predicts the Gaussian semantic box F_b under the point supervised L_h interaction in the Hough space F_h, and then locates the result lines (l_1, l_2, l_3, l_4) on the original image by reverse Hough transform (R). Finally, the point-centered slices $(S_1, S_2, S_3, ..., S_n)$ along the four edges further refine the semantic lines F_l through the refinement stage that focuses on the high-resolution original image to regress more precise coordinate information.

point supervision in deep Hough space to align more accurate semantic lines. (2) In the refined adjustment stage, uniform cropping is performed along the up-sampled box (or lines) on original images, and the obtained mini-batch is refined to eliminate regression errors of high-resolution images. The contributions of this paper are as follows:

(1) The proposed BDNet utilizes deep hough transform to exploit the semantic boundaries, which better handles the various edge information of paper-based scanned images.
(2) The coarse-to-fine strategy in BDNet makes our method better deal with the high-resolution scanned images.
(3) A new open-source dataset is collected and a novel metric of evaluating document skewness is proposed to better supervise the optimization process of BDNet.

2   Methodology

In this section, we state the details of BDNet. The proposed model takes a digital image as input and utilizes a novel Semantic Box-Lines (SBL) module and Deep Hough Alignment (DHA) module for classification and regression in a cascaded two-stage network to achieve semantic boundaries alignment, which efficiently addresses the challenging image correction and trimming problem.

2.1   Semantic Box-Lines

Since the input image content is broken and torn, its boundaries are always blurred, making it impossible to determine a clear bounding box. To well balance the contextual boundary information, the transition can be made slowly in a gentle manner, and we use a two-dimensional Gaussian distribution [6] to approximate the semantic boundary extent.

We let the parameters $(cx, cy, w, h, angle)$ uniquely represent an oriented bounding box [16], and all points on the box $\tilde{p}$ are processed using a Gaussian kernel $K(p, \tilde{p}) = e^{-\gamma\|p-\tilde{p}\|^2}$ to generate a Gaussian heatmap that characterizes the semantic box [17], where γ is an object size-adaptive hyperparameter used to represent local effects. Similarly, the mini-batch sliced along the midpoints of each edge in the refinement phase characterizes the semantic lines F_l in the same way. If the two Gaussian coincide at adjacent boundary points, we take the element-wise mean value instead [18].

As shown in Fig. 2, our BDNet feeds a variety of images with varying scale different boundaries into a cascaded two-stage shared backbone. The former is a pre-processed (down-sampled) scanned image $I \in \mathbb{R}^{512 \times 512 \times 3}$, which ignores the original boundaries details at low-resolution levels to roughly predicts the oriented semantic box, and the latter is post-processed slice images $S \in \mathbb{R}^{512 \times 512 \times 3N}$ without losing any high-resolution boundary information, which in turn predicts accurate semantic lines to refine the final boundary. To better capture the broad contextual semantic boundary information in deep features, we improve the Focal loss [19] for pixel-wise logistic regression with class imbalance:

$$L_f = -\frac{1}{N} \sum_{i=1}^{N} \begin{cases} \alpha(1 - \hat{y}_i)^\beta \log(\hat{y}_i) & \text{if } y_i = 1 \\ (1 - \alpha)(\hat{y}_i)^\beta \log(1 - \hat{y}_i) & \text{if } y_i \neq 1 \end{cases} \tag{1}$$

where $\hat{y}_i$ is the confidence level with respect to the ground-truth y_i, and β is used as a modulation factor to focus on hard-to-classify samples. Based on this, we add another modulation factor α to adjust the ratio between positive and negative sample loss. The values of α and β have a range of values that influence each other, and we use $\alpha = 2$ and $\beta = 4$ in all experiments.

2.2   Deep Hough Alignment

Since the SBL module only encodes semantic features with blurred boundaries, but cannot decode the location and orientation information of specific semantic boundaries, it is necessary to use the Hough transform (HT) [5] to model its prediction results in the subsequent process. HT is a common line detection method, which mainly detects lines by counting votes in Hough space.

As shown in Fig. 2, our DHA takes deep semantic box-lines features $F_{b,l} \in \mathbb{R}^{128 \times 128 \times 64}$ as input to generate the transformed Hough space $F_h \in \mathbb{R}^{N_\rho \times N_\theta \times 64}$, where the quantized sizes $[N_\rho, N_\theta]$ are determined by the quantization interval $\Delta\rho$ and $\Delta\theta$. Therefore, the Hough space is discretized into $N_\rho \times N_\theta$ bins. And for an arbitrary bin (ρ, θ), our DHT accumulates the global contextual geometric information for the feature map activations along the semantic boundary direction [20]:

$$H(\rho, \theta) = \sum_i F_{b,l}(x_i, y_i) \qquad (2)$$

where i is the orientation position index, and the pixel (x_i, y_i) and bin (ρ, θ) are mapped by $\rho = x_i \cos\theta + y_i \sin\theta$. Moreover, it is calculated per channel, and we neglect the channel dimension here.

Since a bin in Hough space corresponds to a straight line in image space, line detection can be reduced to point detection. We choose to compute the point-wise supervised loss directly in the Hough space. Thus, the ground-truth boundary lines are first transformed into the same Hough space with a standard Hough transform, and then the ground-truth points are smoothed and augmented with an adaptive Gaussian kernel to rapidly converge the features. Ultimately, we optimize the binary cross-entropy loss to focus on dilated ground-truth points:

$$L_h = -\frac{1}{N} \sum_{i=1}^{N} \hat{Y}_i \log\left(P(Y_i)\right) + (1 - \hat{Y}_i) \log\left(1 - P(Y_i)\right) \qquad (3)$$

where $P(Y_i)$ is the predictive confidence, and $\hat{Y}_i = K * Y_i$ is the smoothed ground truth. The K is a 5×5 Gaussian kernel with the symbol $(*)$ denoting the convolution operation. Local operations in Hough space correspond to global operations in image space, we use two 3×3 convolutional layers in Hough space to aggregate contextual line features, and then use the Reverse Hough transform (R) to remap the $[N_\rho, N_\theta]$ Hough space back to the original space. Our RHT maps bins to the pixel location (x, y) by the average of all voted bins:

$$R(x, y) = \frac{1}{N_\theta} \sum_\theta F_h(x \cos\theta + y \sin\theta, \theta) \qquad (4)$$

where θ is the orientation parameter and we normalize the angular information, $N_\theta = \pi/\Delta\theta$.

2.3 Coarse-to-Fine Strategy

When feeding high-resolution images to the network, the graph memory explodes as the training task is performed. For object detection on large-scale images, such as tele-sensing object detection [21], the conventional approach is to first perform sliding window overlap cropping on large images and then stitch the prediction results for each small image.

Our proposed BDNet pays more attention to the boundary information of the object, and therefore dose not need to perform sliding window cropping on the entire image. But only the training of partial images around the boundary is required. Therefore, we employ a coarse-to-fine strategy to predict semantic boundaries during a refinement process. In the coarse prediction stage, we predict the location and orientation of the oriented bounding box (d, l_θ), slide-window cropping along the l_θ. Then the obtained local images are refine to a high accurate offset location Δd and semantic orientation $l_{\bar\theta}$ through the refinement semantic box-lines alignment process (see Sec.2.1 and Sec.2.2 for details):

$$l_{\bar\theta} = \frac{1}{|l|} \sum_i l_\theta(i) \qquad (5)$$

where $|l|$ is the number of sliced images on l, and the final oriented bounding box is uniquely represented by the parameters $(d + \Delta d, l_{\bar\theta})$, which is directly applied to subsequent image correction and trimming processes.

3   Experiments

3.1   Datasets and Implementation Details

Datasets: We trained and evaluated our models on private datasets provided by partner archives and on our own collected public datasets. The public datasets will be published on our Github. The private datasets D_{pri} contains $1,326$ high-definition scanned images, which are 2480×3508 pixels (the A4 paper size), and the public datasets D_{pub} contains $1,610$ scanned images and $2,000$ close-up images taken with a mobile phone. We partition the D_{pri} and $9/10$ D_{pub} to train our model and test on the remaining $1/10$ D_{pub}.

Implementation Details: Our model is implemented using Pytorch and trained on four RTX 2080 Super. Input images are resized to 512×512 (or 320×320) pixels and the image augmentation consists of horizontal and vertical flipping, rotation, scaling, etc. Our model is trained using the Adam optimizer [22] with a weight decay of 1×10^{-4}. The initial learning rate is set to 4×10^{-4} and is reduced by a factor of 10 after 24 epochs. We train the model for a total of 64 epochs with a batch size of 4.

We evaluate our results based on two heatmap-based metrics, AP^H and F^H, which are used in previous detection tasks [10,23], and two structural-based metrics, sAP and sF, the former is proposed in L-CNN [9] and the latter is designed by ourselves.

3.2   Feasibility Analysis

There are two important assumptions in our proposed method: one is that a 2-D Gaussian distribution can be used to approximate the semantic extent of

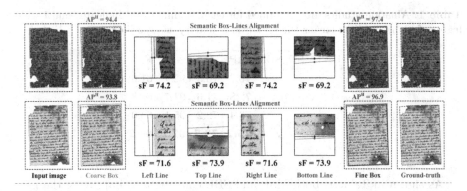

Fig. 3. Visualize the results and the corresponding scores. Two images were randomly selected from 100 high-definition scanned images with complex boundaries to demonstrate the visualizations detected by our BDNet and to calculate evaluation scores.

the blurred boundary, and the other is that a coarse-to-fine strategy can eliminate the scale difference error at the high-resolution level. To validate the above hypothesis, we select 100 high-definition scanned images with complex boundaries and design a novel metric, sF, to analyze the feasibility of our proposed method by measuring the similarity between pairs of lines. Let the prediction $\hat{l}_i$ and the ground-truth l_i be a pair of lines to be measured, and we normalize the image to the unit square to be computed as follows:

$$L_d = 1 - D(\hat{l}_i, l_i), \quad L_\theta = 1 - \frac{\theta(\hat{l}_i, l_i)}{\pi/2} \tag{6}$$

where $D(\hat{l}_i, l_i)$ is the Euclidean distance between midpoints of $\hat{l}_i$ and l_i, and $\theta(\hat{l}_i, l_i)$ is the angular distance. Finally, our proposed metric combines the both distances as follows:

$$sF = (L_d \cdot L_\theta)^2 \tag{7}$$

As shown in Fig. 3, we show some visualization results and the corresponding scores, where the higher the score, the more similar the lines are and the more accurate our method is. When we include all the prediction results, our method is feasible and meets the application requirements.

3.3 Ablation Study

We perform a series of ablation and comparison experiments to analyze the effectiveness of our proposed module. We first build a baseline model based on the naive ResNet-50 and DHA modules, and then individually validate the effectiveness of the multi-scale fusion (MSF) and Semantic Box-Lines (SBL) modules. We attach MSF and SBL to the baseline models and evaluate their performance.

Table 1. Ablation studies are performed for each module. MSF indicates multi-scale feature fusion, and SBL means the semantic boundary as described in Sec.2.1.

DHA	MSF	SBL	sAP	sF	AP^H	F^H
✓			70.7	70.3	92.4	61.6
✓	✓		72.8	71.6	94.7	62.8
✓		✓	73.4	72.2	95.1	63.0
✓	✓	✓	**75.2**	**73.7**	**97.2**	**64.2**

As shown in Table.1, both MSF and SBL contribute to improving the performance of the baseline model. We combine all the modules to form the final complete method that achieves the best performance. Of these, the line structure-based accuracies sAP and sF reach 75.2 and 73.7, which are 4.5 and 3.4 higher than the baseline model. Also, the heatmap-based accuracy of AP^H and F^H relative to the PR curve outperforms the baseline model by 4.8 and 2.6, achieving 97.2 and 64.2. Excellent metrics demonstrate the effectiveness of our approach.

3.4   Comparison Analysis

To verify the effectiveness of our BDNet, as shown in Table.2, we combine our proposed modules with several existing state-of-the-art line segment detection methods for experiments.

Table 2. Quantitative comparison of BDNet with different line segment detection models on public datasets.

BoundaryNet	Input	AIoU	AS	Params(M)	FPS
L-CNN [6]	512	82.6	69.7	9.8	36.6
HAWP [7]	512	**88.7**	**72.3**	18.4	24.9
TP-LSD [8]	320	86.2	71.4	23.9	22.7
BDNet320	320	94.7	73.8	9.9	31.5
BDNet512	512	96.6	72.6	16.5	26.2

Quantitative Comparison. As shown in Table 2, the parameters with the highest scores are marked in bold blue and red in the single model and our BDNet, respectively. We use the detection results to evaluate our model: (1) Calculate the average IOU (AIoU) between the frame-line alignment result box and the GT box. (2) Calculate the average skewness (AS) between the result box and the horizontal line. The experimental results are as follows:

First, comparing three state-of-the-art line segment detection models, the HAWP model has the highest AIoU(88.7) and AS(72.3) scores. Second, our

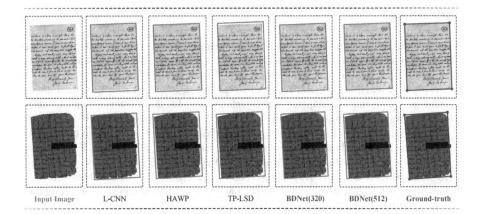

| Input Image | L-CNN | HAWP | TP-LSD | BDNet(320) | BDNet(512) | Ground-truth |

Fig. 4. Qualitative comparison of BDNet with different line segment detection models on public datasets. The top is a relatively regular scanned image, and the bottom is an irregular image.

BDNet512 outperforms HAWP by 7.9, TP-LSD by 10.4 and L-CNN by 14.0 on AIoU. In addition, in terms of AS score, BDNet320 outperforms BDNet512, which is 1.5, 2.4 and 4.1 higher than HAWP, TP-LSD and L-CNN, respectively. Finally, BDNet falls between these three models in terms of both Params and FPS. Therefore, considering both accuracy and performance, BDNet512 is undoubtedly the best model.

Qualitative Comparison. As shown in Fig. 4, the top is a relatively regular scanned image, and the bottom is an irregular image. We employ various line detection models and BDNet to predict regular and irregular input images. In contrast, the detection results based on the BDNet512 model are more refined, the edge error is smaller, and the boundaries of the box-lines alignment results are closer.

4 Conclusion

In this paper, we propose an efficient and feasible two-stage semantic boundary alignment network for paper-based scanned images from coarse to fine. To better evaluate our approach, we design a novel metric that simultaneously measures the distance and orientation between pairs of lines. In addition, an open-source dataset was collected to visualize our detection results, compensating for the fact that the archival data is private. Ultimately, extensive experimental results demonstrate that our method is capable of image correction and trimming tasks for current paper-based digitized images.

References

1. Lyu, P., Yao, C., Wu, W., et al.: Multi-oriented scene text detection via corner localization and region segmentation. In: Proceedings of the IEEE Conference on Computer Vision and Pattern Recognition, pp. 7553–7563 (2018)
2. Mechi, O., Mehri, M., Ingold, R., et al.: Text line segmentation in historical document images using an adaptive u-net architecture. In: 2019 International Conference on Document Analysis and Recognition (ICDAR), pp. 369–374. IEEE (2019)
3. Wu, J.-W., Yin, F., Zhang, Y.-M., Zhang, X.-Y., Liu, C.-L.: Image-to-markup generation via paired adversarial learning. In: Berlingerio, M., Bonchi, F., Gärtner, T., Hurley, N., Ifrim, G. (eds.) ECML PKDD 2018. LNCS (LNAI), vol. 11051, pp. 18–34. Springer, Cham (2019). https://doi.org/10.1007/978-3-030-10925-7_2
4. Von Gioi, R.G., Jakubowicz, J., Morel, J.M., et al.: LSD: a line segment detector. Image Process. Line **2**, 35–55 (2012)
5. Ballard, D.H.: Generalizing the Hough transform to detect arbitrary shapes. Pattern Recogn. **13**(2), 111–122 (1981)
6. Law, H., Deng, J.: Cornernet: detecting objects as paired keypoints. In: Proceedings of the European Conference on Computer Vision (ECCV), pp. 734–750 (2018)
7. Law, H., Teng, Y., Russakovsky, O., et al.: Cornernet-lite: efficient keypoint based object detection. arXiv preprint arXiv:1904.08900 (2019)
8. Arbelaez, P., Maire, M., Fowlkes, C., et al.: Contour detection and hierarchical image segmentation. IEEE Trans. Pattern Anal. Mach. Intell. **33**(5), 898–916 (2010)
9. Zhou, Y., Qi, H., Ma, Y.: End-to-end wireframe parsing. In: Proceedings of the IEEE/CVF International Conference on Computer Vision, pp. 962–971 (2019)
10. Xue, N., Wu, T., Bai, S., et al.: Holistically-attracted wireframe parsing. In: Proceedings of the IEEE/CVF Conference on Computer Vision and Pattern Recognition, pp. 2788–2797 (2020)
11. Huang, S., Qin, F., Xiong, P., Ding, N., He, Y., Liu, X.: TP-LSD: tri-points based line segment detector. In: Vedaldi, A., Bischof, H., Brox, T., Frahm, J.-M. (eds.) ECCV 2020. LNCS, vol. 12372, pp. 770–785. Springer, Cham (2020). https://doi.org/10.1007/978-3-030-58583-9_46
12. Liu, W., Anguelov, D., Erhan, D., Szegedy, C., Reed, S., Fu, C.-Y., Berg, A.C.: SSD: single shot multibox detector. In: Leibe, B., Matas, J., Sebe, N., Welling, M. (eds.) ECCV 2016. LNCS, vol. 9905, pp. 21–37. Springer, Cham (2016). https://doi.org/10.1007/978-3-319-46448-0_2
13. Bolya, D., Zhou, C., Xiao, F., et al.: Yolact: real-time instance segmentation. In: Proceedings of the IEEE/CVF International Conference on Computer Vision, pp. 9157–9166 (2019)
14. Lee, J.T., Kim, H.U., Lee, C., et al.: Semantic line detection and its applications. In: Proceedings of the IEEE International Conference on Computer Vision, pp. 3229–3237 (2017)
15. Zhao, K., Han, Q., Zhang, C.B., et al.: Deep hough transform for semantic line detection. IEEE Trans. Pattern Anal. Mach. Intell. **44**(9), 4793–4806 (2021)
16. Ding, J., Xue, N., Long, Y., et al.: Learning ROI transformer for oriented object detection in aerial images. In: Proceedings of the IEEE/CVF Conference on Computer Vision and Pattern Recognition, pp. 2849–2858 (2019)
17. Zhou, X., Wang, D., Krähenbühl, P.: Objects as points. arXiv preprint arXiv:1904.07850 (2019)

18. Dong, Z., Li, G., Liao, Y., et al.: Centripetalnet: pursuing high-quality keypoint pairs for object detection. In: Proceedings of the IEEE/CVF Conference on Computer Vision and Pattern Recognition, pp. 10519–10528 (2020)
19. Lin, T.Y., Goyal, P., Girshick, R., et al.: Focal loss for dense object detection. In: Proceedings of the IEEE International Conference on Computer Vision, pp. 2980–2988 (2017)
20. Lin, Y., Pintea, S.L., van Gemert, J.C.: Deep hough-transform line priors. In: Vedaldi, A., Bischof, H., Brox, T., Frahm, J.-M. (eds.) ECCV 2020. LNCS, vol. 12367, pp. 323–340. Springer, Cham (2020). https://doi.org/10.1007/978-3-030-58542-6_20
21. Han, J., Ding, J., Li, J., et al.: Align deep features for oriented object detection. IEEE Trans. Geosci. Remote Sens. **60**, 1–11 (2021)
22. Kingma, D.P., Ba, J.: Adam: a method for stochastic optimization. arXiv preprint arXiv:1412.6980 (2014)
23. Huang, K., Wang, Y., Zhou, Z., et al.: Learning to parse wireframes in images of man-made environments. In: Proceedings of the IEEE Conference on Computer Vision and Pattern Recognition, pp. 626–635 (2018)

Chinese-Vietnamese Cross-Lingual Event Causality Identification Based on Syntactic Graph Convolution

Enchang Zhu[1,2], Zhengtao Yu[1,2(✉)], Yuxin Huang[1,2], Yantuan Xian[1,2], Yan Xiang[1,2], and Shuaishuai Zhou[1,2]

[1] Faculty of Information Engineering and Automation, Kunming University of Science and Technology, Kunming 650500, China
[2] Yunnan Key Laboratory of Artificial Intelligence, Kunming University of Science and Technology, Kunming 650500, China
ztyu@hotmail.com

Abstract. The Chinese-Vietnamese cross-lingual event causality identification aims to identify the cause and effect events from the news text describing the event information and present them in a structured form. The existing event causality extraction model faces the following two challenges: 1) The research work related to event causality extraction is mainly focused on resource-rich monolingual scenarios, and the performance of resource-scarce languages needs to be further improved; 2) Existing event causality identification methods are not good at capturing implicit causal semantic relations. Therefore, we propose a novel Chinese-Vietnamese Cross-lingual Event Causality Identification Based on Syntactic Graph Convolution. Firstly, the Chinese-Vietnamese word vectors are mapped into the same semantic space through pre-trained cross-lingual word embeddings. Then the syntactic graph convolutional neural network is used to capture the deep semantic information of the event sentence. Finally, the in-depth semantic features of event sentences in different languages are obtained by combining the cross-attention mechanism of event types. Experiment results on a self-built dataset that the proposed method outperforms the state-of-the-art models.

Keywords: Event causality identification · Cross-language · Syntactic graph convolution · Cross-attention mechanism

1 Introduction

In the context of globalization, exchanges between countries have become more frequent. The number of events of common interest and mutual influence is also increasing. Timely access to the correlations between events on news websites of different countries is of great value in promoting communication and cooperation between countries. The main types of correlation between news events include causal, chronological and co-referential relationships. Event causality identification (ECI) aims to identify causal relations between events in texts, which can provide crucial clues for NLP tasks, such as Reading Comprehension [1] and logical reasoning [2]. This task is usually modeled as a classification problem, i.e. determining whether there is a causal relation between two events in a sentence. This paper focuses on the causal relationships between bilingual Chinese and Vietnamese news events. The purpose of causality recognition is to determine whether there is a causal relationship between events in Chinese and Vietnamese news texts. The core problem is how to map Chinese and Vietnamese bilingual texts into the same semantic space for effective computation.

Currently, in event causality identification, researchers have proposed rule-based methods and machine learning-based methods [3]. However, rule-based and machine-learning-based methods require a large amount of expert knowledge, poor domain migration and low coverage of event causality recognition. End-to-end neural network-based models have achieved good results [4]. Deep learning-based approaches have better results on languages with large-scale annotated data (Chinese and English) [5]. However, there is much room for improvement in performance in small languages with little or no annotated data. The above methods have been studied in monolingual and resource-rich language scenarios, and less in multilingual scenarios, especially in bilingual Chinese and Vietnamese scenarios. Research in multilingual event analysis has focused on how to address the semantic divide between languages, including the use of machine translation, alignment-based resources, and cross-lingual word embedding. Through these approaches, different languages are represented in a unified space for computation to solve the cross-linguistic computational problem. The performance of all these approaches relies on large-scale alignment resources, and good results have been achieved in resource-rich languages such as Chinese and English. However, compared to English and Chinese, Vietnamese is a typical non-common language due to its relatively small usage and population, and non-common languages are mainly characterized by the lack of linguistic knowledge and linguistic resources, especially the scarcity of alignment resources for the non-common language, Chinese, under such a premise, it is difficult to compute different language text representations in the same semantic space, and it is difficult to carry out multilingual It is still difficult to carry out multilingual event causal analysis for these small languages. In summary, Chinese-Vietnamese cross-lingual event causality identification currently faces the following two problems: 1) end-to-end neural network-based news event causality extraction methods rely heavily on large-scale and high-quality annotated corpus, while Chinese-Vietnamese event alignment corpus is scarce, making it difficult to map the two languages into the same semantic space; 2) news event causality extraction requires The existing models based on end-to-end deep learning have achieved certain results in event causality extraction, but there are still problems such as insufficient capture of deep semantic information within event sentences, which leads to ambiguous recognition of implicit causality boundaries.

Inspired by Subburathinam [6] and Lu [7], this paper will make use of language-independent lexical and syntactic features for cross-linguistic structural migration. Therefore, we propose a syntactic graph convolution-based method for causality extraction of bilingual news events in Chinese and Vietnamese. The method first pre-trains a bilingual word embedding. Then a graph convolutional neural network is used to learn the contextual representation of nodes in the syntactic dependency tree to capture the deep semantic information within the event sentence. Finally, a cross-attention mechanism is used to simultaneously model the deep causality semantics of event sentences in both languages, which is finally transformed into an event causality classification problem. Experiments on a self-built Chinese-Vietnamese cross-lingual event causality identification dataset demonstrate that our proposed method outperforms the traditional baseline approach, verifying that the incorporation of syntactic dependency information enhances the model's ability to capture deep semantic information in news event sentences.

2   Related Work

Event causality extraction is one of the tasks of event association relationship extraction. Existing research on event causality extraction falls into three main categories: pattern matching-based methods, methods based on a combination of pattern matching and machine learning, and methods based on deep learning. Among them, pattern-matching-based methods use semantic and lexical-symbolic [8], and other features for causal relationship extraction through pattern matching. Causal relations and then used semantic constraints to classify candidate event pairs as causal or non-causal pairs. Methods that rely entirely on pattern-matching rules are usually poorly adapted across domains and may require extensive domain knowledge in solving domain-specific problems. The combined pattern-based and machine learning approach mainly addresses the causal pair extraction task in a pipelined manner, dividing causal pair extraction into two subtasks, candidate causal pair extraction and relation classification. Luo et al. [9] extracted causal terms from a large-scale web text corpus and then used causal cues based on statistical measures of point-state mutual information to measure the causal strength between web text corpora.

With the increase in computer power and the availability of high-quality datasets, and the powerful representational learning capability of deep neural networks to effectively capture the implicit causal semantic relationships in texts describing events, neural network-based approaches have become the mainstream approach. Li et al. [10] proposed a causal relationship extraction model based on the BiLSTM-CRF model. This model uses a new causal relationship labeling scheme to directly extract the causal relationship. To solve the problem of insufficient data, context embedding is applied to the task of causality extraction. Liu et al. [11] proposed a method to augment event-causal inference using external knowledge. In addition, a masking mechanism was added to the model to allow the model to learn generic inferential information. A gating mechanism was also introduced to balance the contribution of background and semantic knowledge to prevent the external background knowledge from being overweighted and perturbing the original semantics of the sentence. Cao et al. [12] proposed a method to use external knowledge to help event-causal inference, where external knowledge is divided into descriptive and relational knowledge, which is encoded mainly through graph neural networks. [5] proposed a new semantic structure integration model for ECI, which leveraged two kinds of semantic structures, i.e., eventcentric structure and event-associated structure. The above end-to-end deep learning-based event causality extraction methods are mainly related to studies conducted in monolingual scenarios and have not been reported in multilingual, especially in Chinese-Vietnamese bilingual scenarios.

3   Cross-Lingual Mapping

Different languages do not share a joint vector space, which means, the learnt representations for the same thing expressed in different languages vary greatly. This makes event causality identification in multilingual situations challenging. Also, it means sharing knowledge learnt in one language with another is problematic. Our solution for overcoming these issues is to use Cross-language word embedding.

The process begins by training a monolingual embedding model for each language considered. In this paper, we try nonlinear Cross-language word embedding methods. we choose LNMAP [13]. LNMAP is a model that operates independently of isomorphic assumption. It comprises two auto-encoders with non-linear hidden layers for each language. The auto-encoders are first trained independently in a self-supervised way to induce the latent code space of the respective languages. A small seed dictionary is then used to learn the non-linear mappings, which are implemented as feed-forward neural networks with non-linear activation layers between the two learnt latent spaces. As the first non-linear cross-lingual word embedding method, LNMAP has shown outstanding performance with many language pairs including far-distance language pairs.

4   Event Causality Identification

In this section, we describe the architecture of our model in detail. Our model aims at identifying the causal relationship between Chinese-Vietnamese bilingual news events from Chinese-Vietnamese news texts describing event information. Figure 1 shows the overview of our proposed model.

4.1   Word Vector Representation Layer

The first problem to be solved for bilingual Chinese-Vietnamese news event causality recognition is the cross-linguistic problem between Chinese-Vietnamese and Vietnamese. In the case of sparse Chinese-Vietnamese aligned corpus, Chinese-Vietnamese machine translation is ineffective, and direct translation from Vietnamese to Chinese or from Chinese-Vietnamese to Vietnamese will cause error transmission and reduce the accuracy of the model. Therefore, this

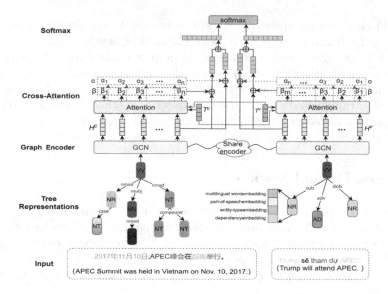

Fig. 1. Structure diagram of the event causality identification model for Chinese-Vietnamese bilingual news events based on syntactic graph convolution.

paper will use pre-trained cross-lingual word embeddings to vectorize Chinese-Vietnamese news texts.

It has been demonstrated that incorporating linguistic features such as lexical, syntactic, and entity features in the model can more effectively characterize event information [14,15]. Given a Chinese sentence vector $C = \{c_1, c_2, \cdots c_m\}$, where m is the length of the words of the sentence. For each word in C, c_i is labeled with POS p_i and entity type e_i. Then each word, its corresponding POS, entity, and entity type are then vectorized separately to obtain M_{c_i}, M_{p_i}, and M_{e_i}. Finally, the word vector, POS vector, and entity vector are stitched together as the final vector representation M_i of c_i:

$$M_i = [M_{c_i}; M_{p_i}; M_{e_i}] \tag{1}$$

Representing each word c_i in C as a vector M_i in the above manner and performing the splicing operation in the vector direction using the $\oplus$ operator, the semantic representation matrix X^C of the sentence C is:

$$X^C = M_1 \oplus M_2 \oplus \cdots M_m \tag{2}$$

Similarly, we can get the Vietnamese word representation N_i and sentence representation X^V:

$$N_i = [N_{v_i}; N_{p_i}; N_{e_i}] \tag{3}$$

$$X^V = N_1 \oplus N_2 \oplus \cdots N_n \tag{4}$$

4.2 Syntactic Graph Convolutional Layer

As mentioned in the introduction, the syntactic graph convolution module captures language-independent lexical and syntactic information through a shared encoder between Chinese and Vietnamese to achieve cross-linguistic structural migration [6,7], thereby alleviating the corpus sparsity problem and capturing deep semantic information within event sentences. In this paper, LTP [16] and the Vietnamese open-source dependency parsing tool [17] are used to construct the dependency syntactic relationship between Chinese and Vietnamese. Separately, sentences in different languages are represented as undirected graphs $\zeta = (\gamma, \varepsilon)$, where γ and ε are the set of nodes and the set of edges, respectively. Using Vietnamese as an example, in γ each v_i denotes a node of w_i, each edge $(v_i, v_j) \in \varepsilon$ is a directed syntactic arc from word w_i to word w_j, and the type label of the edge is $K(w_i, w_j)$. To allow information to flow in the opposite direction, the reverse edge $K'(w_i, w_j)$ with type label (v_j, v_i) is added. In addition, the self-loop edges of all nodes are added.

At layer k of the syntactic graph convolution network module, compute the graph convolution vector $\bar{h}_{v_i}^{k+1}$ for node v:

$$\bar{h}_{v_i}^{k+1} = f(\sum_{u \in N(v)} (W_{K(u,v)}^{(k)} \bar{h}_{v_i}^{(k)} + b_{K(u,v)}^{(k)})) \tag{5}$$

where $K(u,v)$ denotes the type label of edge (u,v), $W_{K(u,v)}^{(k)}$ and $b_{K(u,v)}^{(k)}$ are the weight matrix and deviation, respectively, for type label $K(u,v)$, $N(v)$ is the set of neighbours of node v, and f is the non-linear activation function. It should be noted that considering the aim to reduce model training costs, we opted to use the traditional GCN in our experiments.

4.3 Cross-Attention Layer

Event trigger words and event types can effectively improve the performance of event causality extraction. In addition to capturing deep semantic information about events, a central issue in event causality extraction is the use of the captured event information for event causality extraction. Based on this, we introduce a cross-attention layer in the model, which is a representation of Chinese (Vietnamese) news event types to Vietnamese (Chinese) news event mutual attention to capture the correlations between events, so as to improve the performance of Chinese-Vietnamese cross-lingual event causality identification. Specifically, Chinese news event types to Vietnamese news event text attention to get a better representation of Vietnamese news events; Vietnamese news event types to Chinese news event text attention to allow the model to get a better representation of Chinese news events.

When encoding Chinese event sentences, the event types corresponding to Vietnamese event sentences are used as an attention mechanism with Chinese event sentences to obtain deeper semantic information in Chinese. The specific formula is as follows:

$$K_1 = \tanh(W_{XH} \begin{bmatrix} T^V \\ H^C \end{bmatrix}^T) \tag{6}$$

$$\alpha = \text{softmax}(W_K K_1) \tag{7}$$

$$r_1 = \alpha H^C \tag{8}$$

$$g^C = \tanh(r_1) \tag{9}$$

where T^V is the vector of event types of Vietnamese event sentences, ($W_{XH} \in R^{2d*2d}$) is the weight vector of joint T^V and H^C; $\alpha = \{\alpha_1, \alpha_2, \alpha_3, ..., \alpha_n\}$ is the weight vector $\alpha \in R^n$ of the attention mechanism and $W_K W_K \in R^{2d}$ is the weight matrix of K. $r_1(r_1 \in R^d)$ is the weight vector of Chinese event sentences corresponding to the event types of Vietnamese news sentences, and $g^C(g^C \in R^d)$ is the vector of deep Chinese event semantic features mined by the cross-attention mechanism.

Similarly, we can obtain the Vietnamese event semantic representation $g^V(g^V \in R^d)$. Finally, the semantic information obtained by the cross-attention mechanism is used as the deep semantic information of event sentences for the final classification.

4.4 Classification Layer

The classification layer utilizes a softmax classifier to identify event causality.

$$y = \text{softmax}(W[g^C; g^V] + b) \tag{10}$$

where g^C and g^V denotes the causal semantic features of Chinese event sentences and Vietnamese event sentences, respectively, and y is the output of the model. Its loss function is a binary cross-entropy loss.

For optimization method, we adopt mini-batch stochastic gradient descent (SGD) [18] to minimize the objective function. For regularization, we also employ dropout on the penultimate layer, which can prevent the coadaptation of hidden units by randomly dropping out a proportionof p the hidden units during forward and backpropagation.

5   Experiments and Analysis of Results

5.1   Datasets and Evaluation Metrics

Datasets: To our knowledge, there is no publicly available dataset for causal extraction of multilingual news events to date. Firstly, we crawled 813 Vietnamese news articles and 4065 Chinese news articles respectively. Then manually labeled 7 event types and 1 non-event type, which included 30177 event sentences in set $D_i = \{e_1, e_2, ..., e_n\}$, where n is the number of events. Further, the event sentence set D_i was annotated to form the Chinese-Vietnamese bilingual news event causality dataset.

Evaluation Metrics: For evaluation, we adopt widely used Precision (P), Recall (R), and F1-score (F1) as evaluation metrics. Considering the small size of the Vietnamese dataset, a significance test was conducted and the level of significance was set at $\rho = 0.05$.

5.2   Analysis of Experimental Results

Baselines: In order to prove the effectiveness of this work, we compare the proposed method with the state-of-the-art models as follows: **RB:** proposed by [19], which raises a rule-based method for ECI; **DD:** proposed by [19], a data driven approach based on machine learning; **VR-C:** proposed by [20], a method with data filtering and enhanced causal signals based on verb rules; **CNN-softmax:** the model uses CNNs to encode event sentences to aggregate contextual information about the text, and the model consists of two parts: a CNN encoder and a softmax classifier; **BiLSTM-softmax:** proposed by [21], end-to-end model based on Bi-LSTM, which consists of two parts: Bi-LSTM encoder and softmax classifier; **CCNN-BiLSTM-CRF:** proposed by [22], neural sequence model based on CNN and Bi-LSTM hierarchical coding; **SERC:** proposed by [23], an event causality extraction model based on LSTM and incorporating word properties (POS), dependencies and word order information; **KMMG:** proposed by [11], a BERT-based model with mention masking generalization and external knowledge; **LearnDA:** proposed by [4], which creates a learnable knowledge-guided data augmentaion method for this task; **GenECI:** GenECI [24] are the methods that formulate ECI as a generation problem; **mBERT-GCN-softmax and XMLR-GCN-softmax:** the models uses multilingual-BERT (mBERT) [25] and XLM-RoBERTa(XML-R) [26] respectively, instead of the pre-trained Chinese-Vietnamese bilingual word vector proposed.

Table 1. Performances of all comparison methods.

Methods	P	R	F1
RB	41.50	19.82	26.83
DD	75.40	30.90	43.92
VR-C	**79.21**	35.84	49.30
CNN-softmax	69.43	61.74	65.36
BiLSTM-softmax	71.65	64.38	67.82
CCNN-BiLSTM-CRF	73.81	66.92	70.20
SERC	74.09	67.18	70.46
KMMG	74.73	66.81	70.56
LearnDA	75.46	68.20	71.64
GenECI	76.83	67.93	72.11
mBERT-GCN-softmax	76.21	68.49	72.14
XMLR-GCN-softmax	76.87	69.15	72.81
Ours	78.67	**70.49**	**74.36***

Main Results: From the experimental results in Table 1, it can be seen that the syntactic graph convolution-based causality extraction method for bilingual Chinese and Vietnamese news events proposed in this paper outperforms the other baseline methods. Where $*$ represents the significance level of $\rho = 0.05$. Comparing with the models CCNN-BiLSTM-CRF and SERC, it is found that the method proposed in this paper has significantly improved in accuracy (P), recall (R), and F1 values, the performance of ours drops 4.16% and 3.5% on F1-score, respectively. This indicates the effectiveness of the syntactic graph convolution and cross-attention mechanism proposed in this paper, which we believe is since the syntactic dependency tree of the event sentence enhances the semantic relatedness between words in the sentence through syntactic edges, and aggregates the semantic information of the neighboring points through GCN, which is more conducive to the model capturing the causal semantic information within the event sentence. Through comparative experiments, we can also observe that the performance of the method in this paper outperforms that of mBERT-GCN-softmax and XLM-R, indicating the effectiveness of our propose pre-training method and that the trained Chinese-Vietnamese bilingual word vectors are more applicable to the Chinese-Vietnamese bilingual news event causality extraction task. In addition, the performance of the Bi-LSTM-based model is significantly better than that of the CNN-based model. The possible reason for this is that the Bi-LSTM can capture contextual information more effectively and learn the semantic representation of causal relationships.

Ablation Study: We perform ablation experiments to understand how each part of our model contributes to the task improvements. In the following sections, we will explore each part in detail.

(1) Effect of GCN

We conduct an experiment to demonstrate that the performance of our proposed models, namely on the encoder side, we compare the impact of GCN, Tree-LSTM [27], and LSTM on the model performance respectively. We perform this experiment on the Self-built dataset and present the result in Fig. 2.

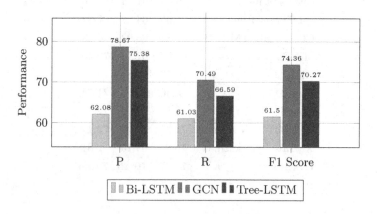

Fig. 2. Performance for Bi-LSTM, GCN and Tree-LSTM.

From Fig. 2, We can see that both GCN and Tree-LSTM encoding structures significantly outperform the Bi-LSTM baseline, and the GCN encoder generally outperforms Tree-LSTM. It is because that structural information is important for event causality recognition tasks and GCN-based model benefits more from the explicit information derived from universal dependency parsing.

(2)Effect of GCN layers

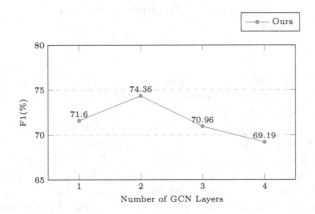

Fig. 3. The impacts of the number of GCN layers.

In order to verify the effect of the number of syntactic graph convolution layers on the performance of the model in extracting causal relationships between Chinese and Vietnamese bilingual news events, we investigated the use of different number of graph convolution layers in the model for ablation experiments. The F1 values using different number of layers of GCN in the model are illustrated by Fig. 3. We can observe that increasing the number of falling asleep GCN layers does not always improve the model performance, and for the Chinese-Vietnamese bilingual news event causality extraction task, the best results are achieved when the number of GCN layers is 2.

(3) Effect of cross-attention mechanisms and feature combinations
We exploit the effects of cross-attention mechanisms and feature(POS, entity and dependency relation) combinations. In order to verify the effectiveness of our method, experiments are carried out in a multilingual environment.
The experiment results are displayed in Table 2. We can draw the following conclusions:

1) From the experimental results, without cross-attention mechanism, the performance of model drops 3.68% in terms of F1-score. We believe that the reason for this phenomenon is that using one language event type as a cross-attention mechanism with news event sentences in another language allows for a more effective characterisation of news event sentences. In addition,

Table 2. Performance on cross-attention mechanism and features combination.

Methods	F1
W/o cross-attention	70.68
Multilingual Word Embedding	69.60
+POS Embedding	71.52
+Entity Embedding	72.28
+Dependency Relation Embedding	73.46
All	**74.36***

according to the multilingual consistency [28] and multilingual complementarity [29] hypotheses, performing cross-attention can effectively alleviate the sparse data problem in Vietnamese.

2) According to the results, POS embedding, entity embedding, and dependency relation embedding provide complement improvement in the model. It seems that, dependency relation embedding is more effective than POS embedding and entity embedding (by 1.56% on average). An intuition explanation is that: It is because that structural information is important for capturing deep semantic information of event sentences and GCN-based model benefits more from the explicit information derived from universal dependency parsing.

6   Conclusion

In this paper, we propose a novel Chinese-Vietnamese Cross-lingual Event Causality Identification Based on Syntactic Graph Convolution. This method first uses cross-lingual word embedding to solve the semantic gap problem; then, the deep semantic information in the event sentence and the causal semantic relationship between event sentences are captured through syntactic graph convolution and cross-attention mechanism, to improve the model's causal relationship recognition performance. Experiments on a self-built Chinese-Vietnamese cross-lingual event causality identification dataset showed that the F1 value of the proposed method improved by 4.16%. This further validates the effectiveness and superiority of the method for event causality extraction tasks, especially in low-resource and multi-lingual scenarios. In the next research work, we will design more advanced models and investigate multilingual event causality extraction that incorporates external knowledge such as multilingual knowledge graphs and multilingual matter graphs.

Acknowledgement. This work was supported by the National Natural Science Foundation of China(U21B2027, 61972186, 62266027, 62266028); Yunnan provincial major science and technology special plan projects(202302AD080003, 202202AD080003); Yunnan Fundamental Research Projects(202301AT070393, 202301AT070471); Kunming University of Science and Technology's"Double First-rate" construction joint project(202201BE070001-021).

References

1. Oh, J. H., Torisawa, K., Hashimoto, C., Sano, M., De Saeger, S., Ohtake, K. Why-question answering using intra-and inter-sentential causal relations. In: Proceedings of the 51st Annual Meeting of the Association for Computational Linguistics, vol. 1: Long Papers, pp. 1733–1743 (2013)
2. Girju, R.: Automatic detection of causal relations for question answering. In: Proceedings of the ACL 2003 Workshop on Multilingual Summarization and Question Answering, pp. 76–83 (2003)
3. Kim, H., Joung, J., Kim, K.: Semi-automatic extraction of technological causality from patents. Comput. Ind. Eng. **115**, 532–542 (2018)
4. Zuo, X., et al.: Learnda: learnable knowledge-guided data augmentation for event causality identification. arXiv preprint arXiv:2106.01649 (2021)
5. Hu, Z., et al.: Semantic structure enhanced event causality identification. arXiv preprint arXiv:2305.12792 (2023)
6. Subburathinam, A., Lu, D., Ji, H., May, J., Chang, S.F., Sil, A., Voss, C.: Cross-lingual structure transfer for relation and event extraction. In: Proceedings of the 2019 Conference on Empirical Methods in Natural Language Processing and the 9th International Joint Conference on Natural Language Processing (EMNLP-IJCNLP), pp. 313–325 (2019)
7. Lu, D., Subburathinam, A., Ji, H., May, J., Chang, S.F., Sil, A., Voss, C.: Cross-lingual structure transfer for zero-resource event extraction. In: Proceedings of The 12th Language Resources and Evaluation Conference, pp. 1976–1981 (2020)
8. Ittoo, A., Bouma, G.: Extracting explicit and implicit causal relations from sparse, domain-specific texts. In: Muñoz, R., Montoyo, A., Métais, E. (eds.) NLDB 2011. LNCS, vol. 6716, pp. 52–63. Springer, Heidelberg (2011). https://doi.org/10.1007/978-3-642-22327-3_6
9. Luo, Z., Sha, Y., Zhu, K.Q., Hwang, S.W., Wang, Z.: Commonsense causal reasoning between short texts. In: Fifteenth International Conference on the Principles of Knowledge Representation and Reasoning (2016)
10. Li, Z., Li, Q., Zou, X., Ren, J.: Causality extraction based on self-attentive bilstm-crf with transferred embeddings. Neurocomputing **423**, 207–219 (2021)
11. Liu, J., Chen, Y., Zhao, J.: Knowledge enhanced event causality identification with mention masking generalizations. In: Proceedings of the Twenty-Ninth International Conference on International Joint Conferences on Artificial Intelligence, pp. 3608–3614 (2021)
12. Cao, P., et al.: Knowledge-enriched event causality identification via latent structure induction networks. In: Proceedings of the 59th Annual Meeting of the Association for Computational Linguistics and the 11th International Joint Conference on Natural Language Processing, vol. 1: Long Papers, pp. 4862–4872 (2021)
13. Mohiuddin, T., Bari, M.S., Joty, S.: Lnmap: departures from isomorphic assumption in bilingual lexicon induction through non-linear mapping in latent space (2020)
14. Hsi, A., Carbonell, J.G., Yang, Y.: Modeling event extraction via multilingual data sources. In: TAC (2015)
15. Hsi, A., Yang, Y., Carbonell, J.G., Xu, R.: Leveraging multilingual training for limited resource event extraction. In: Proceedings of COLING 2016, the 26th International Conference on Computational Linguistics: Technical Papers, pp. 1201–1210 (2016)

16. Che, W., Li, Z., Liu, T.: LTP: a Chinese language technology platform. In: Coling 2010: Demonstrations, pp. 13–16 (2010)
17. Kočiský, T., Hermann, K.M., Blunsom, P.: Learning bilingual word representations by marginalizing alignments. arXiv preprint arXiv:1405.0947 (2014)
18. Ruder, S.: An overview of gradient descent optimization algorithms. arXiv preprint arXiv:1609.04747 (2016)
19. Mirza, P., Tonelli, S.: An analysis of causality between events and its relation to temporal information. In: Proceedings of COLING 2014, the 25th International Conference on Computational Linguistics: Technical Papers, pp. 2097–2106 (2014)
20. Mirza, P.: Extracting temporal and causal relations between events. In: Proceedings of the ACL 2014 Student Research Workshop, pp. 10–17 (2014)
21. Wang, P., Qian, Y., Soong, F.K., He, L., Zhao, H.: Part-of-speech tagging with bidirectional long short-term memory recurrent neural network. arXiv preprint arXiv:1510.06168 (2015)
22. Ma, X., Hovy, E.: End-to-end sequence labeling via bi-directional lstm-cnns-crf. arXiv preprint arXiv:1603.01354 (2016)
23. Venkatachalam, K., Mutharaju, R., Bhatia, S.: Serc: syntactic and semantic sequence based event relation classification. In: 2021 IEEE 33rd International Conference on Tools with Artificial Intelligence (ICTAI), pp. 1316–1321. IEEE (2021)
24. Man, H., Nguyen, M., Nguyen, T.: Event causality identification via generation of important context words. In: Proceedings of the 11th Joint Conference on Lexical and Computational Semantics, pp. 323–330 (2022)
25. Lee, J.D.M.C.K., Toutanova, K.: Pre-training of deep bidirectional transformers for language understanding. arXiv preprint arXiv:1810.04805 (2018)
26. Conneau, A., et al.: Unsupervised cross-lingual representation learning at scale. In: Proceedings of the 58th Annual Meeting of the Association for Computational Linguistics(ACL2020), pp. 8440–8451 (2020)
27. Tai, K.S., Socher, R., Manning, C.D.: Improved semantic representations from tree-structured long short-term memory networks. arXiv preprint arXiv:1503.00075 (2015)
28. Liu, S., Liu, K., He, S., Zhao, J.: A probabilistic soft logic based approach to exploiting latent and global information in event classification. In: Proceedings of the AAAI Conference on Artificial Intelligence, vol. 30 (2016)
29. Liu, J., Chen, Y., Liu, K., Zhao, J.: Event detection via gated multilingual attention mechanism. In: Proceedings of the AAAI Conference on Artificial Intelligence, vol. 32 (2018)

MCKIE: Multi-class Key Information Extraction from Complex Documents Based on Graph Convolutional Network

Zhicai Huang[1,3], Shunxin Xiao[2,3], Da-Han Wang[2,3(✉)], and Shunzhi Zhu[2,3]

[1] College of Information and Smart Electromechanical Engineering,
Xiamen Huaxia University, Xiamen 361024, China
huangzc@hxxy.edu.cn
[2] School of Computer and Information Engineering,
Xiamen University of Technology, Xiamen 361024, China
{wangdh,szzhu}@xmut.edu.cn
[3] Fujian Key Laboratory of Pattern Recognition and Image Understanding,
Xiamen 361024, China

Abstract. The majority of key information extraction in document analysis work relies on simple layout scenes with few classes, such as the date and amount on invoices or receipts. However, many document applications entail sophisticated layouts with various types of elements. The mortgage contract agreement, for example, often has over ninety entities, and it is expensive to design template for each document. To this end, we propose an efficient multi-class key information extraction (MCKIE) method based on graph convolutional network. In detail, we design a graph construction strategy to generate a text box graph. Then, MCKIE utilizes the message-passing mechanism to learn node representation by aggregating information from neighborhoods. Besides, we compare the performance of various graph construction methods and verify the effectiveness of MCKIE on the realistic contract document dataset. Extensive experimental results show that the proposed model significantly outperforms other baseline models.

Keywords: Key Information Extraction · Complex Document Image · Graph Convolutional Network

1 Introduction

Optical character recognition (OCR) methods have lately shown significant success as computer vision development has improved. Specifically, extracting key information from recognized text demonstrates larger application scenarios citesroie, such as layout analysis and robotic process automation. Most traditional key information extraction (KIE) works [3,14,17] usually designed a template to extract document key information. They used recognized text and position information to match entities with the help of regex. Nevertheless, these methods have low expandability because designing hand-crafted feature for each document template takes time and requires lots of prior knowledge. Hence many

Q. Liu et al. (Eds.): PRCV 2023, LNCS 14431, pp. 89–100, 2024.
https://doi.org/10.1007/978-981-99-8540-1_8

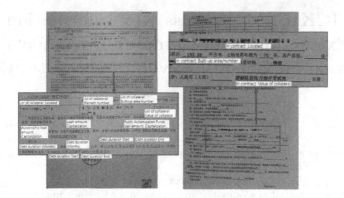

Fig. 1. Examples of key information extraction from scanned contract documents.

researchers utilized the named entity recognition (NER) technology to solve the KIE task by considering it as a sequence tagging problem. However, NER suffers from ambiguous feature representation since it only uses text information. Therefore, many researchers [6,9] attempt to fuse the additional information to enrich the document feature representation. In particular, the document KIE tasks with multiple classes, such as contract documents, require more information and clarified features to distinguish them. Generally, these solutions include the transformer-based methods and the graph-based methods. The transformer-based [19] model usually perform better than traditional recurrent neural network (RNN) networks. Many studies [8,21] turn to embed more information into the transformer-based model to learn document features. For instance, LayoutLM [21,22] pre-trained the BERT [4] model on text and layout of documents and applied the pre-trained model on downstream document understanding tasks. LayoutLM performs well on KIE tasks with rich visual information. However, training the BERT model is expensive and requires extensive document data, which is hard to collect in practical applications.

The graph-based models [5,12,15,23] generate a graph for each document page. Qian et al. [15] proposed the GraphIE model, which constructed a document graph with plain text or text boxes and learns embedding by aggregating node information. It needs some effort to mark the type of predetermined edge in GraphIE. The work [12] also employed text boxes to form a fully linked graph and used a graph convolutional network (GCN) model to learn graph embedding. It directly designed edge properties without the requirement to designate edge kinds. However, the model may add redundant information to nodes due to the fully links. Then Yu et al. [23] proposed the PICK model to automatically learn textual and visual elements of documents by utilizing a soft adjacent matrix for comprehending node relationships rather than artificially specified graph edge features. Further study indicate these methods could not be better for complex document KIE tasks due to three main factors. To begin, each node in the graph has a connection to others. The work [23] analyzes that fully linked graph nodes will bring redundant information to graph representation, and then PICK develops automatically acquiring graph connection information. Inspired by it, we hold that local information on complex documents is more significant than other remote information. The second premise is that the layout structure of the evaluation document stays fixed or varies little, such as purchase receipts or tickets. However, the graph layout is changing in the contract document. Third, the previously mentioned methods must have at least two items to construct a document graph. However, the contract documents vary in that their entity numbers span from one to many. If there only exists one text box on a document page, the earlier techniques are unable to generate a document graph and fail to operate. These factors cause contract document KIE tasks more difficult.

In our research, we primarily handle mortgage contract documents with multiple categories. Figure 1 shows two instances of key information extraction from scanned contract documents. To this end, we propose a multi-class key information extraction (MCKIE) model based on graph convolutional network. The structure of the MCKIE model is depicted in Fig. 2. We design a graph construction strategy, named constant node context (CNC), to generate text box graph, which is inspired by the human eye's scope of having a continuous context in a glance when reading a document. the proposed CNC strategy preserves local invariance by focusing on local rather than remote information. Because most contract key elements are numbers, names, or dates, they interact less with remote nodes but are more relevant to their neighbors. For example, the neighbor node of a date-type entity with the text "signed date" has more helpful information than other remote nodes. However, the proposed CNC graph-building technique would render the graph nodes local and independent, therefore, additional holistic features are required to enrich graph embedding. To capture global features, we add global and relative position information to graph edges. Such a solution can learn the local information and filter the redundant remote features. Furthermore, certain entities are similar that may impact the outcome, such as debt maturity, debt maturity initiation, and debt maturity end. Inspired by SEED model [16], our method leverages an augmented semantic network to supervise graph node embedding in order to learn semantic features. Furthermore, the independence of CNC may result in only one text box in the document graph. To solve it, we add a self-loop operation to each node to improve the representation of graph node properties. The following are the contributions of this paper:

1. We present an efficient graph-building method that concentrates on the local information of the graph. Then we explore global position characteristics and link them to edge properties to improve graph node embedding. A self-loop is also introduced to each node to investigate self-features.
2. To exploit the semantical features of graph nodes, we utilized an augmented semantic network to supervise graph node embedding, which can learn richer representations.
3. We are the first work to study the multi-class mortgage contract document KIE task. Extensive experimental results show that our proposed approach

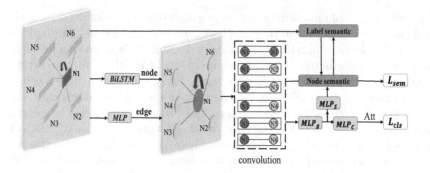

Fig. 2. Structure of MCKIE model.

outperforms baseline methods regarding efficacy and robustness while processing real mortgage contract documents.

2   Related Work

Earlier techniques treated KIE as a sequence labeling problem and solved it using Markov random fields [10], conditional random field (CRF) [13], NER [11], or other RNN-based algorithms. These approaches only focus on plain text and disregard important non-sequence information such as visual, location, and layout information. Recently, several works try to introduce visual features into KIE models [1,2,7]. For example, Guo et al. [7] proposed an end-to-end model to extract visual image features, whereas it only uses visual information and applies to fixed layouts with a small number of entities. Later, combining multimodal features became the cutting-edge research on the KIE task, with two primary directions: graph-based and transformer-based approaches. The LayoutLM [21] was designed to tackle document comprehension tasks by integrating textual, layout, and visual aspects. Recently, Tang et al. [18] proposed the UDOP model, which fuses text, image and layout modalities to understand and generate documents using a unified representation; UDOP has achieved SOTA results on several document AI tasks. However, these methods have low interpretability and lack explicit semantic relationship modeling. Wei et al. [20] used GCN to pre-trained language models by encoding text information with the BERT model and utilizing GCN to encode other layout information. However, training BERT models are costly, and the models' weights are large, increasing the difficulty of application deployment. Hence several researchers prefer to utilize the graph approaches for addressing the KIE challenge, in which GCN can explicitly connect document entities and fuse visual and textual features. For example, the methods of Qian et al. [15] and Wei et al. [12] employed graph models to capture non-local characteristics and extract entities. In detail, Liu et al. [12] constructed a network that treated text box graphs as fully linked regardless of the intricacy of the layout document, leading to aggregating redundant information between nodes. To filter the redundant information and address the multiple categories KIE task, we present the MCKIE model with a CNC graph construction approach to learn a better representation of the node embedding.

3 Methodology

In this work, we mainly focus on processing the mortgage contract documents, which contain a set of text boxes generated by OCR system, and each text box consists of coordinates and texts. When building the document graph, our model regards each text box as a graph node and connects the nodes following the local connecting strategy. We define the graph with N nodes as G, where $G = (V, E)$, the $V = \{v_1, v_2, \dots, v_N\}$ is the set of text box in single page. $E = \{v_i \times r_{ij} \times v_j | i, j \in N, r_{ij} \in R\}$ is the set of directed edges of graph and R is the connection matrix where r_{ij} is the edge connection from node v_i to node v_j. The document graph model will first construct triple pairs by linking text boxes in the document picture. Second, we embed layout information into graph edges by using the relative and absolute locations of text boxes as layout characteristics. Then, our model utilizes graph convolution to learn each node's embedding by aggregating the information from each node's triple pairs. Third, the graph embedding will interact with node labeled textual information to enhance the model capacity of semantic features learning. Finally, the learned embedding is used to predict the labels of entities.

Continuous Node Context. The proposed continuous node context (CNC) graph construction strategy allows MCKIE focus on the local information of text boxes. Since most of the contract document entities are relevant to their neighbors, reflecting the remote nodes less influence document entities. Accordingly, we develop two methodologies based on CNC to create document graphs: N-neighbors connection and Max-distance connection. Our approach utilizes each text box's central coordinates as the graph node's centroid point. For N-neighbors connection, it selects the N nodes in closest proximity when constructing the document graph. Note that, N-neighbors connection may extend to distant nodes. Conversely, the Max-distance connection chooses neighbor nodes within a predefined limit range, ensuring their distances are less than the specified threshold.

Global Edge Position Embedding. Due to the localization of CNC strategy, our approach learn edge embedding with global position information. In this study, the origin point is placed in the top-left corner of the document picture. The edge embedding of the relationship between nodes v_i and v_j is as follows:

$$r_{ij} = [\frac{x_i}{W}, \frac{y_i}{H}, \frac{x_j}{W}, \frac{y_j}{H}, \frac{w_i}{h_j}, \frac{h_j}{h_i}, \frac{w_j}{h_i}], \tag{1}$$

where W, H are width and height of contract document, respectively. Besides, (x_i, y_i) is the coordinate point of node v_i with horizontal and vertical distance respectively. w_i and h_i are the width and height of the text box used for node v_i. Concretely, in the Eq. 1, the elements $[\frac{x_i}{W}, \frac{y_i}{H}, \frac{x_j}{W}, \frac{y_j}{H}]$ are the absolute position ratio in the image, and they indicate absolute position features of nodes. The others elements $[\frac{w_i}{h_j}, \frac{h_j}{h_i}, \frac{w_j}{h_i}]$ are the aspect ratio of node v_i, relative height and breadth of node v_j, which illustrate inner relationship of connected nodes.

Self-Loop Node. Our model is built on graph triple pairs, requiring the presence of at least two nodes and its link edge. In practice, the document image may consist of only one text box. Additionally, the localization of CNC may produce separate nodes while creating the document graph model. In some instances, such a situation cannot create a graph to train GCN. To address this, we apply a self-loop operation to each node in the graph to resolve node isolation problems and enrich graph features.

Graph Learning. The proposed MCKIE model adopts the GCN to learn the graph embedding by combining textual and layout information. We define graph triples as (v_i, r_{ij}, v_j). For node v_i, graph convolution aggregates features of each neighbor v_j to generate graph hidden features $h_{ij} = [v_i \| r_{ij} \| v_j]$. If only one node exists, node v_j could be v_i due to the self-loop operation. Furthermore, we employ multi-layer perceptron MLP_g to extract graph embedding U_i:

$$U_i = MLP_g(h_{ij}). \tag{2}$$

Semantic Enhanced with Graph Embedding. Document KIE task has not only complex layouts but also include multiple kinds of entities. Some categories, in particular, have a remarkable resemblance, such as debt maturity, the beginning of debt maturity, and the end of debt maturity. In these cases, it is insufficient to use text embedding to learn node representation, and we must investigate additional textual information to classify these comparable categories. We introduce an upgraded semantic network inspired by [16] to examine more of the node's textual content. The augmented semantic task is an auxiliary branch in our model to help graph embedding interact with node labeled text semantics and obtain better graph representation. The procedure is as follows:

$$L_{sem} = \sum_i COSIM(MLP_s(U_i), E_i). \tag{3}$$

The total of node text embedding is $E_i = \sum_m e_{im}$, where e_{im} is the text sequence of the node v_i with sequence length m. The cosine similarity loss $COSIM$ computes the similarity between node text embedding and graph embedding features produced by MLP_s. The final semantic similarity loss for each document page is the total cosine similarity loss of all nodes.

Self-attention Network. In contrast to full connection with all nodes in the document graph, as illustrated in the lower right of Fig. 2, MCKIE applies self-attention network to learn node feature of its neighbors, then obtains the node graph embedding to predict the node category. The self-attention module is capable of expressing the following:

$$v_i' = \sigma(\sum_{j \in \{0,...,m\}} \alpha_{ij} h_{ij}), \tag{4}$$

$$\alpha_{ij} = \frac{exp(ReLU(w_a^T h_{ij}))}{\sum_{j \in \{0,...,m\}} exp(ReLU(w_a^T h_{ij}))}. \tag{5}$$

Because of the self-loop node, j counts from 0, and the index 0 reflects the node itself. The attention coefficients are represented by α_{ij}, while the activation function is represented by σ and $ReLU$ is rectified linear unit. The parameters for shared attention weighting are denoted by w_a. The graph embedding feeds the MLP_c network, after which the attention module predicts the entity's categories. The classification loss is as follows:

$$Loss_{cls} = \sum_{i \in \{1,...,N\}} CE(MLP_c(v_i'), t_i), \tag{6}$$

$$Loss_{model} = \lambda Loss_{cls} + (1 - \lambda) Loss_{sem}. \tag{7}$$

The CE represents cross-entropy loss, and t_i represents the ground truth of node i. Equation 7 is the final loss, and λ is a hyperparameter employed to change the classification loss ratio and improve semantics.

4 Experiments

4.1 Datasets

We conduct experiments on real-world mortgage contract data (MCD) with 2,534 contracts from eighteen banks. Each bank has multiple templates with different layouts. As shown in Fig. 1, each set of contracts contains a cover, general terms, specific terms, and other documents. MCD was scanned and manually labeled with ground truth information, which defines 98 entity categories, including entities related to borrowers, mortgagees, and loans, among others.

4.2 Model Evaluation and Ablation Study

Model Evaluation. We compare the KIE task performance of the proposed model under different conditions: the baseline model [12], the baseline with an attention module, the baseline with a semantic enhancement module, and the full MCKIE model that integrates both the attention module and the semantic enhancement branch. We also compare it with other models like PICK and layoutXLM. The results are shown in Table 1, where the case of the baseline model without CNC represents fully connected graph nodes. We implemented the N-neighbors and Max-distance connections and compared them with a fully linked graph node. We empirically fixed the number of neighbors to six for the N-neighbors connection. We set the Max-distance connection's limit distance D to 250. All other conditions for these models are kept the same to guarantee consistency and comparability of results. The F1 score is utilized to evaluate model performance.

Table 1. F1-score results of all compared models.

Model	without CNC	CNC	
		N-neighbors $(N = 6)$	Max-distance $(D < 250)$
PICK [23]	72.37	–	–
LayoutXLM [22]	75.12	–	–
Baseline [12]	77.98	78.88	72.88
Baseline+ Attention	80.98	81.47	81.58
Baseline+ Semantic	79.05	79.23	73.48
Full Model	82.12	**82.84**	81.70

As shown in Table 1, the MCKIE model outperforms other models under different graph construction strategies. Moreover, the MCKIE model also outperform existing multi-modal approaches PICK and LayoutXLM. The results reveal that the model with the attention module outperformed the baseline model significantly. Besides, the model with the expanded semantic network shows slight improvement. In detail, the attention module improves model performance by 3.8%, 3.3%, and 11.9% for different graph construction approaches, demonstrating that the attention mechanism has a more significant impact on key information extraction. Furthermore, N-neighbors and max-distance connections outperform the fully connected method without CNC, demonstrating that the CNC technique can effective addresses the document KIE task.

Model Robustness Study. The contract document KIE task is built based on text detection. However, the detecting branch might miss vital text boxes. In this section, we conduct an experiment in which eliminated document text boxes at random with a certain probability to evaluate the robustness of MCKIE. Our experiment is carried out on full model with a removal probability of 0.5, while all other model configurations remained unchanged.

In this experiment, we randomly delete document text boxes to imitate real-world scenarios text detection. As shown in Table 2, the experimental results

Table 2. Robust Performance of MCKIE model.

Model	Full Connection	N-neighbors ($N = 6$)	Max-distance ($D < 250$)
MCKIE	71.21	**77.91**	74.16

indicate that the robustness and resilience of our proposed MCKIE model. Concretely, the N-neighbors connection surpasses the other two methods, and both graph construction strategies outperform the fully connected approach due to the locality of the proposed CNC method. The outcome indicates the robust of our model for processing KIE tasks in documents.

Edge Attribute Study. The proposed MCKIE model introduces absolute and relative positional features in edge attributes to efficiently extract the layout features of documents. To validate the effectiveness of the edge property design in MCKIE model, we set three different graph edge attributes. Firstly, Edge = 7 refers to the edge property in Eq. 1, including the global and inner relation features. Second, we established a new edge property, Edge = 6, which only comprises absolutely position embedding, inspired by the predefined edge in work [12], which is written in Eq. 8:

$$r_{ij} = [x_{ij}, y_{ij}, \frac{x_i}{W}, \frac{y_i}{H}, \frac{x_j}{W}, \frac{y_j}{H}], \tag{8}$$

where x_{ij} and y_{ij} are the horizontal and vertical distances between two boxes, respectively. Thirdly, Edge = 5 is the edge characteristic mentioned in work [12].

Table 3. Experimental results of MCKIE with different edge attributes.

Model	Edge = 7	Edge = 6	Edge = 5
Full Model	**82.84**	81.89	81.52

As showed in Table 3, the edge attribute Edge = 7 with absolute and relative position information performs better than the other edge attributes. It demonstrate that relative feature is vital in layout information. Furthermore, relative features represent the inner relationship of nodes. In the Edge = 7, MCKIE model can capture spatial characteristics by integrating global and inner relation features. It alleviates the limitations of local features and learn more global features.

Node Self-loop Study. Due to the independence and locality of CNC, we designed a self-loop method in which combine nodes to form triple pairs. To verify its effectiveness, we conduct an experiment with and without self-loop on the

Table 4. Impact of self-loop on the N-neighbors model.

Model	w. self-loop	w/o. self-loop
Full Model	**82.84**	81.98

N-neighbors model with N = 6. The Table 4 show that the model with self-loop outperforms the model without self-loop. The outcome confirms the effectiveness and feasibility of this structure in MCKIE model. It also indicates that the graph nodes requires further mining self-feature to richer their embeddings.

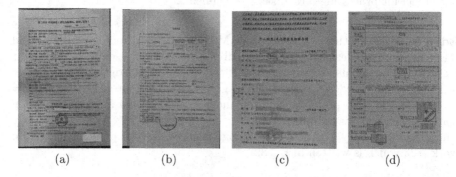

    (a)   (b)   (c)   (d)

Fig. 3. Visual example of KIE task in MCD dataset, the image has been anonymized. The correct classification text box is green, and the misclassified text box is red. (Color figure online)

4.3 Visualization and Analysis

Figure 3 display KIE result visualization of MCKIE model on the MCD dataset. Figure 3(a) and 3(b) show MCKIE can accurately extract key information from various layout document. However, the statistics on error instances demonstrate that the majority of error occur in borrower and mortgagee related categories, where usually encounter multi-value or a multi-label classification problem, as shown in Fig. 3(c). Furthermore, as seen in Fig. 3(d), long text boxes have a significant effect on graph construction due to the huge variance in aspect ratio.

5 Conclusions and Future Work

Through constructing document graphs followed by the text box graph strategy, the MCKIE model can aggregate neighbors' features and utilize global information to boost graph node embedding by self-loop operation and the semantic-enhanced network. Comprehensive experiments show the benefit of the MCKIE model in processing multi-class complex documents. In the future, we plan to better integrate multi-modal features for contract document KIE tasks, including addressing multi-label problems and improving model generalization.

Acknowledgements. This work is supported by National Natural Science Foundation of China (No. 61773325), Natural Science Foundation of Xiamen, China (No. 3502Z20227319), Industry-University Cooperation Project of Fujian Science and Technology Department (No. 2021H6035), Fujian Key Technological Innovation and Industrialization Projects (No. 2023XQ023), and Fu-Xia-Quan National Independent Innovation Demonstration Project (No. 2022FX4).

References

1. Boroş, E., et al.: A comparison of sequential and combined approaches for named entity recognition in a corpus of handwritten medieval charters. In: 2020 17th International Conference on Frontiers in Handwriting Recognition (ICFHR), pp. 79–84. IEEE (2020)
2. Carbonell, M., Fornés, A., Villegas, M., Lladós, J.: A neural model for text localization, transcription and named entity recognition in full pages. Pattern Recogn. Lett. **136**, 219–227 (2020)
3. D'Andecy, V.P., Hartmann, E., Rusinol, M.: Field extraction by hybrid incremental and a-priori structural templates. In: Proceedings of the 13th IAPR International Workshop on Document Analysis Systems (DAS 2018), pp. 251–256. Institute of Electrical and Electronics Engineers Inc. (2018)
4. Devlin, J., Chang, M.W., Lee, K., Toutanova, K.: Bert: pre-training of deep bidirectional transformers for language understanding. In: Proceedings of the 2019 Conference of the North American Chapter of the Association for Computational Linguistics: Human Language Technologies (NAACL HLT 2019), vol. 1, pp. 4171–4186. Association for Computational Linguistics (ACL) (2019)
5. Gemelli, A., Biswas, S., Civitelli, E., Lladós, J., Marinai, S.: Doc2Graph: a task agnostic document understanding framework based on graph neural networks. In: Karlinsky, L., Michaeli, T., Nishino, K. (eds.) ECCV 2022, vol. 13804, pp. 329–344. Springer, Cham (2023). https://doi.org/10.1007/978-3-031-25069-9_22
6. Gui, T., et al.: A lexicon-based graph neural network for chinese ner. In: Proceedings of the 2019 Conference on Empirical Methods in Natural Language Processing and the 9th International Joint Conference on Natural Language Processing (EMNLP-IJCNLP), pp. 1040–1050 (2019)
7. Guo, H., Qin, X., Liu, J., Han, J., Liu, J., Ding, E.: Eaten: entity-aware attention for single shot visual text extraction. In: Proceedings of the International Conference on Document Analysis and Recognition, ICDAR. pp. 254–259. IEEE Computer Society (2019)
8. Hong, T., Kim, D., Ji, M., Hwang, W., Nam, D., Park, S.: Bros: a pre-trained language model focusing on text and layout for better key information extraction from documents. Proc. AAAI Conf. Artif. Intell. **36**, 10767–10775 (2022)
9. Hwang, W., et al.: Post-OCR parsing: building simple and robust parser via bio tagging. In: Workshop on Document Intelligence at NeurIPS 2019 (2019)
10. Kumar, S., Gupta, R., Khanna, N., Chaudhury, S., Joshi, S.D.: Text extraction and document image segmentation using matched wavelets and MRF model. In: IEEE Transactions on Image Processing, vol. 16, pp. 2117–2128. Institute of Electrical and Electronics Engineers Inc. (2007)

11. Lample, G., Ballesteros, M., Subramanian, S., Kawakami, K., Dyer, C.: Neural architectures for named entity recognition. In: Proceedings of the 2016 Conference of the North American Chapter of the Association for Computational Linguistics: Human Language Technologies (NAACL HLT 2016), pp. 260–270. Association for Computational Linguistics (ACL) (2016)

12. Liu, X., Gao, F., Zhang, Q., Zhao, H.: Graph convolution for multimodal information extraction from visually rich documents. In: Proceedings of the 2019 Conference of the North American Chapter of the Association for Computational Linguistics: Human Language Technologies (NAACL HLT 2019), vol. 2, pp. 32–39. Association for Computational Linguistics (ACL) (2019)

13. Ma, X., Hovy, E.: End-to-end sequence labeling via bi-directional LSTM-CNNS-CRF. In: 54th Annual Meeting of the Association for Computational Linguistics (ACL 2016) - Long Papers. vol. 2, pp. 1064–1074. Association for Computational Linguistics (ACL) (2016)

14. Medvet, E., Bartoli, A., Davanzo, G.: A probabilistic approach to printed document understanding. Int. J. Document Anal. Recogn. **14**(4), 335–347 (2011). https://doi.org/10.1007/s10032-010-0137-1

15. Qian, Y., Santus, E., Jin, Z., Guo, J., Barzilay, R.: Graphie: a graph-based framework for information extraction. In: Proceedings of the 2019 Conference of the North American Chapter of the Association for Computational Linguistics: Human Language Technologies (NAACL HLT 2019), vol. 1, pp. 751–761. Association for Computational Linguistics (ACL) (2019)

16. Qiao, Z., Zhou, Y., Yang, D., Zhou, Y., Wang, W.: Seed: semantics enhanced encoder-decoder framework for scene text recognition. In: Proceedings of the IEEE Computer Society Conference on Computer Vision and Pattern Recognition, pp. 13525–13534. IEEE Computer Society (2020)

17. Rusinol, M., Benkhelfallah, T., Dandecy, V.P.: Field extraction from administrative documents by incremental structural templates. In: Proceedings of the International Conference on Document Analysis and Recognition (ICDAR), pp. 1100–1104. IEEE Computer Society (2013)

18. Tang, Z., et al.: Unifying vision, text, and layout for universal document processing. In: 2023 IEEE/CVF Conference on Computer Vision and Pattern Recognition (CVPR), pp. 19254–19264 (2023)

19. Vaswani, A., et al.: Attention is all you need **30** (2017)

20. Wei, M., He, Y., Zhang, Q.: Robust layout-aware IE for visually rich documents with pre-trained language models. In: Proceedings of the 43rd International ACM SIGIR Conference on Research and Development in Information Retrieval (SIGIR 2020), pp. 2367–2376. Association for Computing Machinery, Inc. (2020)

21. Xu, Y., Li, M., Cui, L., Huang, S., Wei, F., Zhou, M.: Layoutlm: pre-training of text and layout for document image understanding. In: Proceedings of the ACM SIGKDD International Conference on Knowledge Discovery and Data Mining, pp. 1192–1200. Association for Computing Machinery (2020)

22. Xu, Y., et al.: Layoutxlm: multimodal pre-training for multilingual visually-rich document understanding (2021)

23. Yu, W., Lu, N., Qi, X., Gong, P., Xiao, R.: Pick: processing key information extraction from documents using improved graph learning-convolutional networks (2020)

A Pre-trained Model for Chinese Medical Record Punctuation Restoration

Zhipeng Yu[1,2], Tongtao Ling[1,2], Fangqing Gu[1(✉)], Huangxu Sheng[1], and Yi Liu[2]

[1] Guangdong University of Technology, Guangzhou, China
fqgu@gdut.edu.cn
[2] IMSL Shenzhen Key Lab, Shenzhen, China

Abstract. In the medical field, text after automatic speech recognition (ASR) is poorly readable due to a lack of punctuation. Even worse, it can lead to the patient misunderstanding the doctor's orders. Therefore, punctuation after ASR to enhance readability is an indispensable step. Most recent work has been fine-tuning downstream tasks on pre-trained models, but the models lack knowledge of relevant domains. Furthermore, most of the research is based on the English language, and there is less work in Chinese and even less in the medical field. Based on this, we thought of adding Chinese medical data to the model pre-training stage and adding the task of punctuation restoration in this work. From this, we proposed the Punctuation Restoration Pre-training Mask Language Model (PRMLM) task in the pre-training stage and used contrastive learning at this stage to enhance the model effect. Then, we propose a Punctuation Prior Knowledge Fine-tuning (PKF) method to play the role of contrast learning better when fine-tuning the downstream punctuation restoration task. In our medical field dataset, we performed a series of comparisons with existing algorithms to verify the proposed algorithm's effectiveness.

Keywords: Punctuation restoration · Automatic speech recognition · Pre-training mask language model · Supervised contrast learning

1 Introduction

The significant advantage of electronic medical record systems (EMRs) [9] is that they can store patient records by using ASR [17, 19] software instead of typing them out manually. Although ASR systems trained on a large corpus can accurately transcribe human speech, the resulting spoken speech records usually exclude punctuation. It makes it difficult for users to understand the doctor's orders. Thus, punctuation restoration is a necessary post-processing step in ASR, which poses a challenge in enhancing the human readability of ASR-generated transcripts or many text-processing applications.

This research was in part supported by the Natural Science Foundation of Guangdong Province (2021A1515011839), and in part supported by Shenzhen Science and Technology Program (No: JCYJ20210324135809025).

Q. Liu et al. (Eds.): PRCV 2023, LNCS 14431, pp. 101–112, 2024.
https://doi.org/10.1007/978-981-99-8540-1_9

In recent years, pre-trained language models (e.g., BERT [6], RoBERTa [15], GPT [21], and BART [12]) have gradually become the mainstream building block for a wide range of Natural Language Processing (NLP) tasks. Their performance continues to improve as the size of the model increases. These pre-trained models can be divided into two branches. One is the autoencoder language model, such as the BERT model; One is an autoregressive language model, such as the GPT model. The autoencoder language model randomly masks the words in the input sentences, while the pre-training process predicts the masked words according to the context. The advantage of this kind of model is that the encoding process is bidirectional. Therefore, the context can be well understood. The disadvantage is that the masked words are replaced with [MASK] in the pre-training process. However, [MASK] does not appear when the downstream task is fine-tuned, which leads to a mismatch between the pre-training task and the downstream task, and it is not suitable to do the generative downstream task. The autoregressive language model is to predict the next possible word according to the above content. The advantage of this model is that it is suitable for the downstream generation task, and the disadvantage is that it can only use the above or the following information but cannot use the above and the following details simultaneously. These large transformer-based models [8, 28] are usually pre-trained on massive amounts of unlabelled text data to learn general knowledge and representation of sentences, which allows the model to perform well on a variety of downstream NLP tasks such as text classification tasks [13], knowledge inference tasks [14], sequence labeling tasks, question answer.

Recently, contrast learning (CL) [18] is a new kind of machine learning widely used in computer vision, natural language processing, and other fields. These algorithms can be roughly divided into self-supervised contrast learning [4] and supervised contrast learning (SCL) [10]. Self-supervised contrast learning does not have real labeled data and usually sets some metrics or heuristic functions. The training goal of the model is to make these indicators reach an extreme convergence. Supervised contrast learning has labeled data. The training aims to make the label or score predicted by the training data close to the real label. It compares the samples of the current category with all the samples of the other categories. A suitable loss function is designed to narrow the distance between the features of the same category and pull the distance between the features of different categories so that the categories can easily find the classification boundary. However, SCL is rarely involved in research on applying Chinese medical Punctuation Restoration.

While most previous research has focused on speech transcription in the general domain, our work focuses on real-world patient records. However, applying this conventional approach to the medical field could emerge some problems: (1) Lack of domain knowledge: while these language models have pre-trained on the general domain, there is a lack of Chinese medical knowledge; (2) Lack of attention to the colons: most of the previous work focuses on commas, periods, and question marks, but in Chinese medicine, colons occur far more often than question marks; (3) Word segmentation problem: conventional Chinese pre-trained language models are usually segmented into characters by using WordPiece-based approach, but there is no punctuation between words according to Chinese grammar.

In this work, we focus on Chinese medical punctuation restoration and propose some methods to solve these problems and improve accuracy:

- First, we capture a large amount of Chinese medical data and conducted incremental pre-training on RoBERTa to make the model a certain knowledge understanding in the Chinses medical field.
- Second, we propose the Punctuation Restoration Pre-training Mask Language Model (PRMLM) task in the pre-training stage based on Whole Word Masking (WWM) [5].
- Third, after RoBERTa outputs the representation of the mask position, we add SCL to narrow the representation of the same category in the features space and pull the different categories away from each other.
- Fourthly, the experimental results verify that the proposed Punctuation Prior Knowledge Fine-tuning (PKF) is more suitable for downstream ASR tasks.

The rest of this paper is organized as follows. Section 2 reviews the related works and describes some of the research backgrounds. Section 3 carefully describes the implementation details of the proposed algorithm. Section 4 provides experimental design, experimental process, and experimental results. Finally, Sect. 5 draws a conclusion.

2 Related Work and Background

2.1 The Recent Research and Development of Punctuation Restoration

The punctuation restoration task is an indispensable step after ASR, and the existing research has obtained many effective results, but most focus on English, and few in the Chinese field. The task of punctuation restoration can be classified into three categories: First, some researchers treated it as a token-level classification problem. The research will label each token to restore punctuation. For example, X.Che et al. proposed a training model of Deep Neural Network(DNN) or Convolutional Neural Network (CNN) [26] to classify whether a punctuation mark should be inserted after the third word of a 5-word sequence and which kind of punctuation mark the inserted one should be [2]. Second, some scholars regard this task as a machine translation task [20]. The machine translation method traditionally builds an encoder-decoder architecture that inputs text without punctuation marks and outputs the result of punctuation restoration. For example, O Klejch et al. made a sequence-to-sequence model which replaced the traditional encoder with a hierarchical encoder to do machine learning for punctuation restoration [11]. To further utilize pre-trained language models, Makhija et al. used BERT as a base encoder and LSTM [7,23] network with a random condition field (CRF) for punctuation restoration [16]. The others take it as a sequence labeling task, assigning a punctuation mark to each word by probability. J Yi et al. proposed a self-attention-based model to predict punctuation marks for word sequences [24]. Salloum et al. used bidirectional recurrent neural network (B-RNN) as the backbone and viewed punctuation restoration in medical reports as a sequence tagging problem [22].

On the other hand, current punctuation restoration models use text feature alone, text and speech features together, or flexibly choose according to the scene. O Klejch et al. combined lexical and acoustic features for punctuation restoration [11]. And the model proposed by J Yi et al. was trained using word and speech embedding features obtained from the pre-trained Word2Vec and Speech2Vec, respectively [24]. And then, J Yi et al. also try to use the part-of-speech (POS) tags to help train the punctuation restoration task [25]. Zhu Yaoming proposed a single modal and multimodal punctuation restoration framework: unified multimodal punctuation restoration framework (UniPunc) [29]. UniPunc has three main components: lexicon encoder, acoustic assistant, and coordinate bootstrapper. The three components learn lexical features, possible acoustic properties, and cross-modal hybrid features. However, due to data reasons, there is less data corresponding to speech and text for the Chinese medical field, so the all work in this paper is based on text only.

2.2   Problem Definition of Punctuation Restoration

In this paper, punctuation restoration is formalized as a sequence labeling problem. Generally, given a non-punctuation input sequence $X = (x_1, x_2, ..., x_N)$ and punctuation marks $Y = (y_1, y_2, ..., y_N)$, where N is the length of the input sequence. The punctuation restoration model needs to output a sequence $\hat{Y} = (\hat{y_1}, \hat{y_2}, ..., \hat{y_N})$ in $[O,$ *COMMA, PERIOD, COLON*[1]$]$, where O denotes the label is not a punctuation mark. An ideal punctuation restoration model aims to predict whether each token x_i in the input sequence X is followed by a punctuation mark. During the inference stage, the punctuation restoration model automatically adds the predicted mark to the position after the current token.

2.3   Whole Word Mask (WWM)

In the original BERT series, BPE-based coding techniques were used in the WordPiece tokenizer, which often divided a word into smaller fragments. However, in Chinese, many words are often fixed collocations, so after the pre-training MLM stage, the effect of the wordpiece model in the Chinese field is not as good as that in the English field. Therefore, this WWM technology is proposed for the Chinese field. It divides the text into several words together with Chinese Word Segmentation (CWS) [1, 27], and carries out mask MLM pre-training on the words as a whole. In this paper, the base model is the RoBERTa (Robustly Optimized BERT Pretraining Approach)-wwm. But the RoBERTa model removes the next sentence prediction (NSP) step in pre-training.

3   Method

In this paper, we have the following steps for punctuation restoration: 1. Masks of whole word and Punctuation Restoration Pre-training Mask Language Model (PRMLM) task in the pre-training stage; 2. Learn by contrast during PRMLM pre-straining stage; 3. Punctuation Prior Knowledge Fine-tuning (PKF) for downstream tasks.

---

[1] Unlike general domain, QUESTION marks are hardly ever seen in medical domain, but COLON marks appear frequently. We focused on three punctuation marks: COMMA, PERIOD, and COLON.

The purpose of step 1 is to enhance the model's understanding ability of the current Chinese data set and add the punctuation prediction task to the pre-training stage, which is closer to the downstream task. The meaning of step 2 is to narrow the features representation belonging to the same class of samples in the vector space and pull the hidden vectors of different class samples away. The role of step 3 is because the pre-training phase is the initial expectation with punctuation, while in the downstream task, the text generated by ASR is not punctuated, and we propose a Punctuation Prior Knowledge Fine-tuning (PKF) to fine-tune it in the downstream task.

3.1 Mask Classifier During Pre-straining Stage

Formally, given a pair of sequences $X = (x_1, x_2, ..., x_N)$ and punctuation marks label $Y = (y_1, y_2, ..., y_N)$, each label falls into one of four categories which are [O, $COMMA$, $PERIOD$, $COLON$]. Then, RoBERTa-wwm converts X into a contextualized representation $H^L \in \mathbb{R}^{N \times d}$ through an embedding layer (which consists of word embedding, positional embedding, and token type embedding), and a consecutive L-layer transformer according to Eqs. (2) and (3), where N is the maximum sequence length, and d is the dimension of hidden layers as shown in Fig. 1.

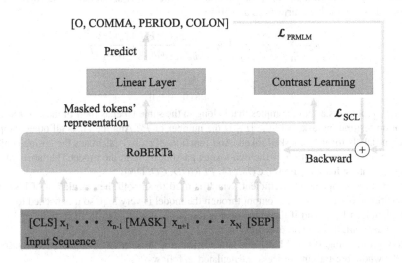

Fig. 1. The PRMLM for pre-training stage

$$\mathbf{X} = [CLS]x_1..x_{n-1}[MASK]x_{n+1}..x_N[SEP] \tag{1}$$

$$\mathbf{H}^{(0)} = \mathbf{Embedding}(\mathbf{X}) \tag{2}$$

$$\mathbf{H}^{(l)} = \mathbf{Transformer}(\mathbf{H}^{l-1}) \tag{3}$$

The rule of masking is according to the WWM and the I is the set of all masked tokens. Therefore, sentences with [MASK] will input the embedding layer and transformer layer, output $\mathbf{H}^{(L)}$. And we only use the contextualized representation of [MASK] to predict the punctuation.

We collect a subset with these contextualized representation of the masked token, forming the replaced representation $\mathbf{H}_I \in \mathbb{R}^{k \times d}$, where k is the number of whole masked token in X. The $\mathbf{H}_i$ is used to predict the probability distribution p of the punctuation set [O, $COMMA$, $PERIOD$, $COLON$]. So we initialize a linear layer with weight $\mathbf{W}_{pr} \in \mathbb{R}^{4 \times d}$, $\mathbf{b}_{pr} \in \mathbb{R}^{1 \times d}$ to get the probability distribution of the i th token by Eq. (4).

$$\mathbf{p}_i = \mathbf{H}_i \mathbf{W}_{\mathbf{pr}}^{\mathbf{T}} + \mathbf{b}_{pr} \tag{4}$$

And we take the cross-entropy as loss function to optimize the punctuation restoration in PRMLM task by Eq. (5).

$$\mathcal{L}_{PRMLM} = -\frac{1}{k} \sum_{i=1}^{k} \mathbf{y}_i \log \mathbf{p}_i \tag{5}$$

3.2   Learn by Contrast During Pre-straining Stage

We added supervised contrast learning to the pre-training process, extracted $\mathbf{H}$ from Eq. (3), and used this $\mathbf{H}$ for contrast learning, so as to enhance the model's representation of punctuation restoration problems in the input sequence. The formula for supervised comparative learning is as follows:

$$\mathcal{L}_{SCL} = \sum_{i \in I} \frac{-1}{|P_i|} \sum_{p \in P_i} \log(\frac{exp(\varphi(\mathbf{H}_i, \mathbf{H}_p)/\tau)}{\sum_{k \in A_i} exp(\varphi(\mathbf{H}_i, \mathbf{H}_k)/\tau)}) \tag{6}$$

$$\varphi(\mathbf{s}_1, \mathbf{s}_2) = \frac{\mathbf{s}_1 \cdot \mathbf{s}_2}{||\mathbf{s}_1|| \cdot ||\mathbf{s}_2||} \tag{7}$$

The P_i represents a set of samples that belong to the same label as the i th masked token in punctuation restoration task. And the $|P_i|$ is the number of P_i. And the A_i is all other samples of different labels for the i masked token. And function $\varphi(\mathbf{s}_1, \mathbf{s}_2)$ calculates the cosine similarity of the vector representation of the two tokens passing through the model. The labels of this supervised contrast learning are [O, $COMMA$, $PERIOD$, $COLON$].

The τ is the temperature coefficient, which is used to smooth the calculation of loss. The representational dimension of the output through the model is very high so it is needed to adjust the calculation of loss. And if the value of τ is too large, it will be difficult to train, and if the values is too small, it is easy to over-fit.

In our pre-training, PRMLM task and contrast learning are carried out simultaneously, so the loss in the whole pre-training process is calculated as follows:

$$\mathcal{L}_{pr} = \lambda \mathcal{L}_{PRMLM} + (1 - \lambda)\mathcal{L}_{SCL} \tag{8}$$

The λ is the weight factor that weighs the loss of MLM task and contrast learning in pre-training.

3.3   Punctuation Prior Knowledge Fine-Tuning (PKF) for Downstream Task

Because in the downstream practical task, the text output by ASR is completely without any punctuation mark, and in the pre-training, the random mask mechanism is still used, and the input sentence may have punctuation marks, so we also need to fine-tune in the actual downstream task.

More importantly, because in the pre-training process, we use contrast learning to make representations belonging to the same category closer in the representation space, and representations belonging to different categories pull away in the representation space, then we consider that the three punctuation marks in the category are also in the vocabulary, and according to the grammar rules, these three punctuation marks do not appear consecutive to each other. Therefore, when the three punctuation marks in the input sequence are masked, their classification is always [O], so we use this part of the prior knowledge.

Based on the above analysis, the PKF was proposed as shown in Fig. 2. The represtantion after the three punctuation marks in the embedding layer of the pre-training model is added to the downstream fine-tuning training so that we can make better use of this prior knowledge in the training process.

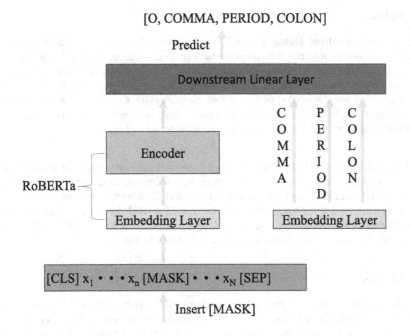

Fig. 2. The method of PKF for the downstream task

3.4   Overview

In addition, WWM mechanism is adopted in the pre-training stage, while in the downstream task, after CWS word segmentation of the plain text output by ASR, [MASK] was inserted only according to the word, not according to the token.

Then we can predict the classification of the masks' position and complete the downstream punctuation restoration task.

We set the location of all inserted masks to set J, forming the representation $\mathbf{H}_J \in \mathcal{R}^{v \times d}$, where v is the number of the whole inserted mask. Then we initialize a linear layer with weight $W_{ft} \in \mathbb{R}^{4*3d}, \mathbf{b}_{pr} \in \mathbb{R}^{1 \times d}$ in fine-tuning stage. For the $j \in J$ th position inserting [MASK], the probability distribution can be calculated by Eq. (9). The $\mathbf{H}_{COMMA}^{emb}, \mathbf{H}_{PERIOD}^{emb}, \mathbf{H}_{COLON}^{emb} \in \mathbb{R}^{1 \times d}$ are the three punctuation marks' representation by embedding layer in RoBERTa. And the loss function of downstream task training, we take the cross-entropy as loss function by Eq. (10). The truth label of the j th position is y_j

$$\mathbf{p}_j = (\mathbf{H}_j, \mathbf{H}_{COMMA}^{emb}, \mathbf{H}_{PERIOD}^{emb}, \mathbf{H}_{COLON}^{emb})\mathbf{W}_{ft}^{\mathbf{T}} + \mathbf{b}_{ft} \tag{9}$$

$$\mathcal{L}_{ft} = -\frac{1}{v}\sum_{j=1}^{v}\mathbf{y}_j \log \mathbf{p}_j \tag{10}$$

4   Experiments

4.1   Datasets

Ours Chinese Medical Dataset. The Chinese medical data set used in the experiment is the data we captured from major medical public websites, including medical knowledge, doctor's advice, patient's chief complaint, and other data. More precisely, this dataset consists of 48,844,038 words in the training set, 498,915 words in the validation set, and 502,375 words in the test set. Each word is labeled by one of four categories: [*O, COMMA, PERIOD, COLON*], and punctuation marks that do not belong to period, colon, and comma will be classified as [*O*]. The detail of the label in this dataset distribution can be found in Table 1.

IWSLT2012. IWSLT TED Talk dataset [3] is a commonly used dataset for punctuation restoration tasks. This dataset consists of 2.1M words in the training set, 295k words in the validation set from the IWSLT20212 machine translation track, and 12.6k words (reference transcripts) in the test set from the IWSLT2011 ASR track. Each word is labeled by one of four classes: [*O, COMMA, PERIOD, QUESTION*]. The ratios of these four labels are 85.7%, 7.53%, 6.3%, 0.47% respectively. The detail of the data distribution can be found in Table 2.

4.2   Metrics

We evaluated the predictive results of three punctuation recoveries on the test set, and we used precision (P), recall (R), and F_1 score (F_1) metrics to measure the specific effects of our model. Since this is a punctuation restoration task, we only evaluated the [*COMMA, PERIOD, COLON*] three punctuation restoration effects.

4.3   Experiment Setup

We performed the above pre-training and downstream task fine-tuning based on the ROBERTA model. We set the maximum length of the input sequence to 512, use an A100 for training, and set batch size to 64. We use AdamW optimizer for gradient backward, and the learning rate is set to 1×10^{-5}. The dropout rate in our experiment is 0.1.

In the pre-training, we annotate the dataset according to the label category, use WWM rule for mask operations, and then train the PRMLM task. We set the weight τ in contrast learning loss to 0.6. After training, we save the weights and use them to fine-tune downstream tasks.

In the PKF step, we processed the previous dataset. In order to be closer to the ASR task, we retained the annotated data, then removed all punctuation marks in the dataset, and then fine-tuned the downstream task, and performed loss calculation based on the saved annotated data. Since it is a downstream task, we only take the first 50000 text samples of the training set in the pre-training task in the training set. And then we still only trained 5 epochs.

Table 1. Label distributions of the Chinese Medical dataset

	Train	Validation	Test
EMPTY	43,348,327	442,505	44,5973
COMMA	4,337,312	44,477	44,538
PERIOD	828,325	8,415	8,510
COLON	330,074	3,518	3,354

Table 2. Label distributions of the IWSLT2012

	Train	Validation	Test
EMPTY	1,801,727	252,922	10,943
COMMA	158,392	22,451	830
PERIOD	132,393	18,910	807
QUESTION	9,905	1,517	46

Table 3. Model comparisons on our Chinese medical dataset. P, R, F_1 denote the Precision, Recall, and F_1 Score on test dataset respectively

Model (RoBERTa-wwm)	COMMA			PERIOD			COLON			OVERALL		
	P	R	F_1	P	R	F_1	P	R	F_1	P	R	F_1
+Linear	0.952	0.959	0.955	**0.962**	0.928	0.945	0.907	0.895	0.901	0.954	0.944	0.949
+CRF	**0.968**	0.965	0.961	0.945	0.951	0.948	0.930	0.881	0.904	0.956	0.948	0.952
+LSTM	0.955	0.966	0.960	0.949	0.945	0.947	0.928	0.882	0.904	0.956	0.946	0.951
+PRMLM (w/o SCL)	0.955	0.968	0.961	0.949	0.956	**0.952**	0.939	0.871	0.904	0.959	0.947	0.953
+PRMLM($\lambda = 0.5$)	0.959	0.965	0.962	0.941	**0.958**	0.949	0.952	0.859	0.903	0.962	0.944	0.952
+PRMLM ($\lambda = 0.9$)	0.952	**0.970**	0.961	0.958	0.946	**0.952**	0.962	0.853	0.904	0.966	0.940	0.953
+PRMLM+LSTM	0.949	0.963	0.956	0.936	0.951	0.944	0.951	0.854	0.900	0.958	0.940	0.948
+PRMLM+CRF	0.959	0.967	**0.963**	0.950	0.952	0.951	0.916	**0.897**	0.907	0.955	**0.952**	0.954
+PKF (w/o PMLM & SCL)	0.960	0.965	0.962	0.949	0.952	0.950	0.926	0.892	**0.909**	0.957	0.951	**0.954**
+PMLM+PKF	0.954	0.965	0.960	0.959	0.938	0.948	**0.968**	0.845	0.902	**0.969**	0.936	0.951

Table 4. Model comparisons on IWSLT2011 Ref dataset. P, R, F_1 denote the Precision, Recall, and F_1 Score on test dataset respectively

Models	COMMA			PERIOD			QUEATION			OVERALL		
	P	R	F_1	P	R	F_1	P	R	F_1	P	R	F_1
BERT$_{Based}$+Linear	0.713	0.705	0.709	0.843	0.821	0.832	0.826	0.775	0.800	0.794	0.767	0.780
BERT$_{Based}$+PKF (Ours)	0.712	0.708	0.710	0.852	0.813	0.832	0.869	0.769	0.816	0.811	0.763	0.786
ALBERT+Linear	0.621	0.686	0.652	0.836	0.771	0.802	0.652	0.638	0.645	0.703	0.698	0.700
ALBERT+PKF (Ours)	0.650	0.666	0.658	0.817	0.790	0.803	0.695	0.653	0.673	0.721	0.703	0.712
BERT$_{Large}$+Linear	0.712	0.705	0.708	0.857	0.828	0.842	0.804	0.804	0.804	0.791	0.779	0.785
BERT$_{Large}$+PKF (Ours)	0.734	0.709	0.721	0.8401	0.837	0.8385	0.869	0.740	0.799	0.814	0.762	0.786
RoBERTa$_{Based}$+Linear	0.696	0.719	0.707	0.841	0.842	0.841	0.869	0.727	0.792	0.802	0.763	0.780
RoBERTa$_{Based}$+PKF (Ours)	0.683	0.715	0.698	0.851	0.818	0.834	0.891	0.773	0.828	0.808	0.769	0.787
RoBERTa$_{Large}$+Linear	0.754	0.722	0.737	0.838	0.879	0.858	0.847	0.866	0.857	0.813	0.822	0.817
RoBERTa$_{Large}$+PKF (Ours)	0.749	0.735	0.742	0.856	0.855	0.855	0.891	0.836	0.863	0.832	0.809	0.820

4.4 Results

During the experiment, we conducted ablation a series of experiments, hyperparameter comparison experiments, and comparison experiments with other algorithms to verify the effectiveness of the proposed algorithm.

As shown in Table 3, we performed a series of comparisons in our Chinese medical dataset based on the RoBERTa-wwm pre-trained model. First, to verify the effectiveness of PRMLM, we conducted an ablation experiment. We only trained RoBERTa-wwm's model with PRMLM, without adding comparative learning. The results show that adding only PRMLM in the pre-training stage can outperform the RoBERTa-wwm+Linear model on all overall indexes. Moreover, the overall P index and the average F_1 index both exceed the RoBERTa-wwm+CRF and RoBERTa-wwm+LSTM models.

Secondly, to verify the effectiveness of adding SCL, we also set up an ablation experiment, and we compared adding SCL and not adding SCL in the pre-training stage. The hyperparameter a was also analyzed. When $\lambda = 0.5$, the RoBERTa-wwm+PRMLM model added to SCL increased its overall P index by nearly 0.3%, while when $\lambda = 0.9$, the RoBERTa-wwm+PRMLM model added to SCL increased its overall P index by nearly 0.8%. And after we conducted several groups of experiments, we found that the model had the best effect when λ was 0.9. In order to further verify the effectiveness of the PRMLM technology added to SCL, we added it to the RoBERTa-wwm+LSTM and RoBERTa-wwm+CRF models, and we found that the PRMLM technology added to SCL improved the two models to varying degrees. RoBERTa-wwm+PRMLM+LSTM increased by 0.9% in P index compared with RoBERTa-wwm+LSTM. RoBERTa-wwm+PRMLM+CRF increased by 0.4% in R index compared with RoBERTa-wwm+CRF and achieved the state-of-the-art result in all experiments. Not only that, it also achieves a state-of-the-art result in the F_1 index in the comma punctuation restoration and in the R index in the colon punctuation restoration.

Then, we verified the effectiveness of the proposed PKF and conducted ablation experiments. We directly applied PKF to the original RoBERTa pre-trained model and found that the RoBERTa+PKF model achieved a state-of-the-art result in F_1 indexes of all experiments. Moreover, the RoBERTa+PRMLM+PKF model added to SCL achieved a state-of-the-art result in the P indexes of overall and colon punctuation restoration.

Moreover, in order to verify the validity of PKF, we also conducted a series of comparative experiments on the IWSLT2012 dataset. We conducted comparative experiments on different pre-trained models, such as BERT, ALBERT, and RoBERTa models, against our proposed PKF. As shown in Table 4, PKF shows a different level of improvement in the effectiveness of the punctuation recovery task on these different models, which once again demonstrates the effectiveness of PKF. The results highlighted in yellow are the index of the improvement brought by PKF.

5   Conclusion

In this paper, we first added the work of punctuation restoration to MLM pre-training of RoBERTa model and add contrast learning in this step to enhance the model's ability to the represent punctuation domain. According to the principle and function of contrastive learning, the embedding representation of the punctuation marks will be added to the linear layer after the downstream task when fine-tuning the original text entered by ASR. The experimental results and ablation experiments validate the effectiveness of the proposed PRMLM and PKF in punctuation restoration tasks. Moreover, this work also provides a dataset on punctuation recovery in the medical field. In future work, we can try to explore and study the use of the ChatGPT large model in punctuation restoration tasks.

References

1. Che, W., Feng, Y., Qin, L., Liu, T.: N-ltp: an open-source neural language technology platform for chinese. arXiv preprint arXiv:2009.11616 (2020)
2. Che, X., Wang, C., Yang, H., Meinel, C.: Punctuation prediction for unsegmented transcript based on word vector. In: Language Resources and Evaluation (2016)
3. Che, X., Wang, C., Yang, H., Meinel, C.: Punctuation prediction for unsegmented transcript based on word vector. In: Proceedings of the Tenth International Conference on Language Resources and Evaluation (LREC 2016), pp. 654–658 (2016)
4. Chen, T., Kornblith, S., Norouzi, M., Hinton, G.: A simple framework for contrastive learning of visual representations. In: International Conference on Machine Learning, pp. 1597–1607. PMLR (2020)
5. Cui, Y., et al.: Pre-training with whole word masking for chinese bert. In: Institute of Electrical and Electronics Engineers (IEEE) (2021)
6. Devlin, J., Chang, M.W., Lee, K., Toutanova, K.: BERT: pre-training of deep bidirectional transformers for language understanding. In: Proceedings of the 2019 Conference of the North American Chapter of the Association for Computational Linguistics: Human Language Technologies (NAACL-HLT), pp. 4171–4186 (2019). https://doi.org/10.18653/v1/N19-1423
7. Graves, A., Graves, A.: Long short-term memory. In: Supervised Sequence Labelling with Recurrent Neural Networks, pp. 37–45 (2012)
8. Hentschel, M., Tsunoo, E., Okuda, T.: Making punctuation restoration robust and fast with multi-task learning and knowledge distillation. In: 2021 IEEE International Conference on Acoustics, Speech and Signal Processing (ICASSP 2021), pp. 7773–7777 (2021). https://doi.org/10.1109/ICASSP39728.2021.9414518
9. Kalogriopoulos, N.A., Baran, J., Nimunkar, A.J., Webster, J.G.: Electronic medical record systems for developing countries. In: 2009 Annual International Conference of the IEEE Engineering in Medicine and Biology Society, pp. 1730–1733. IEEE (2009)
10. Khosla, P., et al.: Supervised contrastive learning. Adv. Neural Inf. Process. Syst. 33, 18661–18673 (2020)
11. Klejch, O., Bell, P., Renals, S.: Sequence-to-sequence models for punctuated transcription combining lexical and acoustic features. In: 2017 IEEE International Conference on Acoustics, Speech and Signal Processing (ICASSP 2017) (2017)
12. Lewis, M., et al.: Bart: denoising sequence-to-sequence pre-training for natural language generation, translation, and comprehension. arXiv preprint arXiv:1910.13461 (2019)
13. Ling, T., Chen, L., Lai, Y., Liu, H.L.: Evolutionary verbalizer search for prompt-based few shot text classification. arXiv preprint arXiv:2306.10514 (2023)
14. Ling, T., Chen, L., Sheng, H., Cai, Z., Liu, H.L.: Sentence-level event detection without triggers via prompt learning and machine reading comprehension. arXiv preprint arXiv:2306.14176 (2023)
15. Liu, Y., et al.: Roberta: a robustly optimized bert pretraining approach. arXiv preprint arXiv:1907.11692 (2019)
16. Makhija, K., Ho, T.N., Chng, E.S.: Transfer learning for punctuation prediction. In: 2019 Asia-Pacific Signal and Information Processing Association Annual Summit and Conference (APSIPA ASC), pp. 268–273 (2019). https://doi.org/10.1109/APSIPAASC47483.2019.9023200
17. Malik, M., Malik, M.K., Mehmood, K., Makhdoom, I.: Automatic speech recognition: a survey. Multim. Tools Appl. 80, 9411–9457 (2021)
18. Oord, A.v.d., Li, Y., Vinyals, O.: Representation learning with contrastive predictive coding. arXiv preprint arXiv:1807.03748 (2018)

19. O'Shaughnessy, D.: Automatic speech recognition: history, methods and challenges. Pattern Recogn. **41**(10), 2965–2979 (2008)
20. Peitz, S., Freitag, M., Mauser, A., Ney, H.: Modeling punctuation prediction as machine translation. In: Proceedings of the 8th International Workshop on Spoken Language Translation: Papers, pp. 238–245 (2011)
21. Radford, A., Narasimhan, K., Salimans, T., Sutskever, I., et al.: Improving language understanding by generative pre-training (2018)
22. Salloum, W., Finley, G., Edwards, E., Miller, M., Suendermann-Oeft, D.: Deep learning for punctuation restoration in medical reports. In: BioNLP 2017, pp. 159–164 (2017). https://doi.org/10.18653/v1/W17-2319
23. Xu, K., Xie, L., Yao, K.: Investigating LSTM for punctuation prediction. In: 2016 10th International Symposium on Chinese Spoken Language Processing (ISCSLP), pp. 1–5. IEEE (2016)
24. Yi, J., Tao, J.: Self-attention based model for punctuation prediction using word and speech embeddings. In: 2019 IEEE International Conference on Acoustics, Speech and Signal Processing (ICASSP 2019) (2019)
25. Yi, J., Tao, J., Bai, Y., Tian, Z., Fan, C.: Adversarial transfer learning for punctuation restoration. arXiv preprint arXiv:2004.00248 (2020)
26. Żelasko, P., Szymański, P., Mizgajski, J., Szymczak, A., Carmiel, Y., Dehak, N.: Punctuation prediction model for conversational speech. arXiv preprint arXiv:1807.00543 (2018)
27. Zhang, Q., Liu, X., Fu, J.: Neural networks incorporating dictionaries for chinese word segmentation. In: Proceedings of the AAAI Conference on Artificial Intelligence, pp. 5682–5689 (2018)
28. Zhang, Z., Liu, J., Chi, L., Chen, X.: Word-level bert-CNN-RNN model for chinese punctuation restoration. In: 2020 IEEE 6th International Conference on Computer and Communications (ICCC) (2020)
29. Zhu, Y., Wu, L., Cheng, S., Wang, M.: Unified multimodal punctuation restoration framework for mixed-modality corpus. In: 2022 IEEE International Conference on Acoustics, Speech and Signal Processing (ICASSP 2022), pp. 7272–7276. IEEE (2022)

Feature Extraction and Feature Selection

English and Spanish Bilinguals' Language Processing: An ALE-Based Meta-analysis of Neuroimaging Studies

Linqiao Liu[1,2], Yan He[2(✉)], and Aoke Zheng[1,2]

[1] College of English Studies, Xi'an International Studies University, Xi'an 710128, China
[2] Key Laboratory for Artificial Intelligence and Cognitive Neuroscience of Language, Xi'an International Studies University, Xi'an 710128, China
1465176791@qq.com

Abstract. The neural mechanisms underlying language processing in English and Spanish bilinguals remain a topic of intense investigation. No neuroimaging meta-analyses about English and Spanish bilinguals' language processing, nevertheless, appeared in the past. In this meta-analysis, Activation Likelihood Estimation (ALE) technique was used to synthesize findings from 13 neuroimaging studies of total 231 Spanish-English or English-Spanish bilinguals. Results showed that there existed 6 peaks in 4 activated brain regions during Spanish processing and 13 peaks in 11 activated brain regions during English processing. In addition, results revealed two distinct peaks of two brain regions commonly activated during language processing across both languages: right insula and left precentral gyrus which may become potential markers for individuals who are bilingual in both English and Spanish. In addition, we found no significant difference in activation between English and Spanish, indicating that these two languages may engage similar neural pathways during language processing.

Keywords: Spanish · English · Bilinguals · Meta-analysis · ALE

1 Introduction

The role of English as a global language cannot be overstated, as it serves as a vital medium for communication and the maintenance of international relationships across various domains, including science, technology, business, education, and tourism [1]. According to Rohmah [2] "English is now a world lingua franca. Nowadays, English is used by people in almost every part of the world" (pp. 106-107). Meanwhile, Spanish stands as the second most widely spoken native language on the planet. According to Green's publication in 2003, Spanish claims the foremost spot as the most extensively utilized modern Romance language, holding a significant lead. Moreover, both languages are part of the Indo-European language family [3], a factor that aids learners in mastering them. Consequently, there exists a multitude of bilingual individuals proficient in English and Spanish, making research and studies pertaining to these languages exceptionally valuable and enlightening.

Q. Liu et al. (Eds.): PRCV 2023, LNCS 14431, pp. 115–125, 2024.
https://doi.org/10.1007/978-981-99-8540-1_10

Nowadays, bilingualism has been widely studied in terms of combination of any two languages or more than two languages. Bilingualism is prevalent, and most of the global population is believed to be bilingual [4]. Bilingual speakers have a greater range of options to express themselves compared to monolingual speakers, which means that bilinguals can select from a variety of speech styles within each language they speak and also switch between the two languages or code systems [5]. In the past, there exist plenty of studies about the brain areas for bilinguals. Nonetheless, disparate studies have separate results in terms of divergent research designs and the different types of stimuli. At the same time, single studies have comparatively rare participants, causing limitations about studies for bilingual neural mechanisms in processing these corresponding languages. Therefore, a growing range of meta-analyses about bilingual have appeared. The term "meta-analysis" delineates the process of scrutinizing multiple analyses [6]. It is employed to denote the statistical examination of an extensive assemblage of analytical outcomes procured from discrete studies, with the aim of effecting a synthesis of findings [6]. This approach embodies a methodological rigor that presents itself as a meticulous substitute for casual and narrative discussions in research studies [6].

A body of meta-analytical research has investigated various dimensions of bilingualism, specifically focusing on microlinguistic and external factors. Sulpizio et al. [7] employed activation likelihood estimation to scrutinize fMRI studies on bilingual language processing and revealed that lexico-semantics processing in the native language (L1) engages a complex network of regions, especially when the second language (L2) is acquired later while L2 processing involves regions beyond the semantic network, linked to executive control. Cargnelutti et al. [8] explored linguistic divergence's impact on neural substrates, highlighting that regardless of linguistic distance or L2 proficiency, L2 usage recruits regions linked to higher-order cognitive functions, with group-specific differences. Sebastian et al. [9] found proficiency influenced brain activation patterns, while Comstock and Oliver [10] discerned task-related differences across bilingual language networks. Liu and Cao [11] synthesized microlinguistics and external factors, revealing early bilinguals exhibit distinct brain activation patterns. Altogether, these studies illuminate complexities within bilingual language processing, revealing the interplay between proficiency, cognitive functions, task dynamics, and language acquisition age.

Dozens of studies about meta-analyses focus on the bilingualism. No neuroimaging meta-analyses about specific English and Spanish bilinguals' language processing, however, appeared in the past. According to the above-mentioned limitation, this meta-analysis mainly aims to solve the following issues: 1) Which brain areas are activated for English and Spanish bilinguals during English processing? 2) Which brain areas are activated for English and Spanish bilinguals during Spanish processing? 3) Which brain areas are commonly activated for dealing with Spanish and English in the English and Spanish bilinguals included in this study? In spite of the results about the above-mentioned issues showing in some single studies, the number of participants is not sufficient and the evidence is not powerful.

To systematize the existing neuroimaging evidence about the brain areas of these two languages, a meta-analytic study has been performed to identify the brain areas of the English and Spanish bilinguals during processing the two languages. In this study, a widely used meta-analytic technique for imaging data—activation likelihood estimation—has been adopted for providing the specific neuroimaging meta-analysis about English and Spanish bilinguals' language processing. ALE methods estimate the probability of spatial convergence of the reported peaks of activations through combining the results of multiple neuroimaging studies with published coordinates [12].

2 Method

2.1 Paper Selection

In order to answer the research questions in this paper, the following inclusion criteria were applied for the selection of pertinent studies and participants: 1) In line with the research topic, the keywords (i.e., English and Spanish and bilingual and fMRI or neuroimaging) were employed for searching relevant published papers in the database of Google Scholar before February 13th, 2023. In total, 6380 records were shown in this database; 2) 6344 papers were excluded since they were not related to this meta-analysis; 3) To align with the study's focus, which revolves around healthy individuals, three records related to disease were removed; 4) Given the meta-analysis nature of this paper, review studies were ineligible for inclusion, leading to the exclusion of an additional four records; 5) With the purpose of exploring the activated brain areas for English and Spanish bilinguals when using English and Spanish, we need to take the whole-brain analysis studies into consideration. One paper has been deleted due to not considering the activation in the whole brain of participants and only exploring particular brain areas; 6) To enable the mapping of brain regions in the Ginger ALE software, 10 articles without mentioning coordinates have also been excluded; 7) 5 papers have been deleted owing to the irrelevant coordinates which are not about English or Spanish. In other words, we need coordinates representing brain areas when processing tasks connected to English and Spanish for those English and Spanish bilinguals. What needs to be noted is that the English and Spanish bilinguals in the current study refer to bilinguals whose first language is Spanish or English.

In total, 13 studies [13–25] related to English coordinates or Spanish coordinates activated by English or Spanish tasks in the Spanish-English bilinguals or English-Spanish bilinguals. The Spanish-related number of foci is 124 and the English-related number of foci is 202. Due to the necessary normalization of all coordinates, we put all coordinates in the Talairach Space due to most of coordinates in the Talaraich space. The rest coordinates in other space were needed to be converted to the corresponding coordinates in the Talaraich space.

2.2 Activation Likelihood Estimation

This study uses the Ginger ALE 3.0.2 software (http://brainmap.org/ale/) to perform the ALE analyses in the standard space of Talaraich in accordance with reported coordinates in these included articles. The present analyses rest upon a coordinate-based Activation Likelihood Estimation (ALE) algorithm, which endeavors to ascertain consistency in functional coordinates across the contrasts delineated in the literature [26–29]. This algorithm employs a random-effects model, employing a mechanism that accounts for spatial uncertainty by treating the reported foci as focal points for the distribution of three-dimensional Gaussian probabilities. The accompanying probability distribution maps, duly weighted by subject count, delineate the likelihood of a given focal point falling within a particular voxel [8]. The visualization of results is executed using Mango 4.1 software (http://mangoviewer.com/mango.html).

In this study, the coordinates about English and Spanish have firstly been analyzed separately in the form of single analyses which mean that Spanish-related coordinated and English-related coordinates are analyzed separately. The uncorrected P value method is employed. According to the manual's recommendation for Ginger ALE 2.3, the very conservative threshold of $P<0.001$ and a minimum cluster size of $250\,mm^3$ were chosen. Then, the conjunction analysis and the contrast analyses have been completed about English and Spanish. Contrast analysis compares and contrasts two ALE datasets and a conjunction image demonstrates the similarity between the datasets. Probability values at $p < 0.001$ were thresholded, with 10,000 p-value permutations and minimum cluster size of $0\,mm^3$.

3   Results

3.1   The ALE Analysis About Spanish

According to the ALE analysis, four clusters have been identified during performing tasks related to Spanish for English and Spanish bilinguals. The two peaks of the first cluster are in right extra-nuclear and right insula (BA 13). The two peaks of the second cluster are in left precentral gyrus. The peak of the third cluster is in right medial frontal gyrus (BA 32) and the peak of the fourth cluster in right middle temporal gyrus (BA 39). The peaks have been demonstrated in Fig. 1.

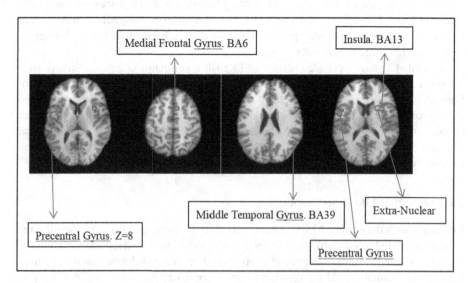

Fig. 1. Significant clusters for Spanish.

3.2 The ALE Analysis About English

The ALE analysis related to English tasks has identified eleven clusters in the English and Spanish bilinguals. The first cluster peaks in left precentral gyrus (BA 44) and left inferior frontal gyrus (BA 47). The two peaks of the second cluster are in right superior frontal gyrus and right middle frontal gyrus. The third cluster peaks in left precuneus (BA 7). The fourth peaks in left declive and the fifth peaks in right insula (BA 13). Peaks of the sixth and seventh clusters are in left precentral gyrus and right insula separately. The peak of the eighth is in left superior frontal gyrus (BA 6). The ninth peaks in left supramarginal gyrus and the tenth peaks in left lateral ventricle. The peak of the eleventh cluster is in left inferior parietal lobule. Figure 2 is about the peaks of significant clusters for the analysis of English.

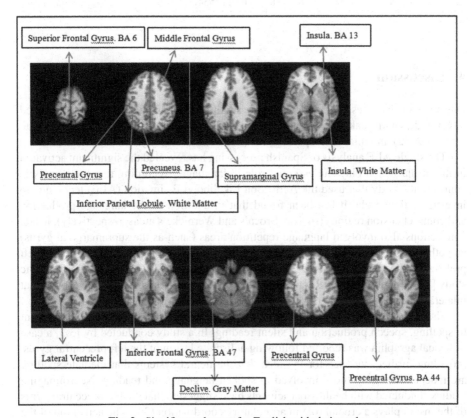

Fig. 2. Significant clusters for English with their peaks.

3.3 Common Brain Regions Activated by Spanish and English in the Bilinguals

Two clusters have been confirmed based on the conjunction analysis. The first cluster is the area from left precentral gyrus to left insula with left precentral gyrus as its peak.

The second cluster is the area from right insula to right extra-nuclear with right insula as its peak. Table 1 is about peaks of significant clusters for the conjunction of Spanish and English. The comparison of English greater than Spanish and Spanish greater than English revealed no significant difference.

Table 1. Peaks of significant clusters for the conjunction of Spanish and English

Cluster	Volume	x	y	z	ALE	Hemisphere	Peaks of clusters
1	144	−46	6	8	0.011791449	L	Precentral Gyrus
2	104	36	8	8	0.011553464	R	Insula

4 Discussion

During Spanish processing, four clusters are significantly activated based on the included target studies with peaks of extra-nuclear, insula (BA 13), precentral gyrus, medial frontal gyrus (BA 32) and middle temporal gyrus (BA 39) individually.

The single ALE analysis of Spanish processing has revealed the significant activaton in the insula which plays an essential role in language. According to a meta-analytic connectivity study that used the Activation Likelihood Estimation (ALE) technique to investigate the insula, it has been found that in addition to the language production and comprehension regions (such as Broca's and Wernicke's areas, respectively), insula connections also involved language repetition areas (such as the supramarginal gyrus) and other linguistic functions, including complex language processes in BA9 of the left prefrontal lobe and lexico-semantic associations in BA37 [30]. Another meta-analytic study [31] containing 42 fMRI studies with healthy participants in adulthood found that bilateral anterior insula was involved in speech perception.

The activation of the precentral gyrus in processing tasks related to Spanish is relevant to spelling, speech production and silent reading. In a study conducted by [32], a case of lexical agraphia was reported following a discrete lesion of the left precentral gyrus, providing further evidence for the idea that the neural systems that mediate spelling are independent from those involved in spoken language and reading. Neuroimaging studies conducted with healthy participants have suggested that the left precentral gyrus of the insula plays a crucial role in articulatory coordination [33]. The activation of the left precentral gyrus was found to be significant in the current meta-analysis since the included studies involved language-related tasks.

The activation of left precentral gyrus (BA 44), left inferior frontal gyrus (BA 47), right superior frontal gyrus, right middle frontal gyrus, left precuneus (BA 7), left declive, right insula (BA 13), left precentral gyrus, right insula, left superior frontal gyrus (BA 6), left supramarginal gyrus, left lateral ventricle and left inferior parietal lobule has been observed in the English and Spanish bilinguals during English processing. Most of these peaks can contain several aspects of language processing. Thus, the areas of activation during English processing for the target participants can be proved by certain relevant literatures.

Several aspects of processing English have been displayed in the activated left inferior frontal gyrus. The inferior frontal gyrus has been demonstrated to be covered in semantic processing, phonological decoding, orthographic processing, and articulatory recoding of print [34–39]. There are clinical cases to demonstrate the function of the left inferior frontal gyrus in language. Marangolo et al. [40] found transcranial direct current stimulation (tDCS) over the left inferior frontal gyrus (IFG) can significantly improve the aphasic symptom of three participants with the long-term effect. Written naming and word writing under dictation tasks can be recovered significantly in subjects 1 and 2 and oral production tasks including word repetition and reading can be improved significantly in subjects 2 and 3 [40]. Damage to the left posterior inferior frontal gyrus can cause an acquired motor speech disorder also called apraxia of speech (AOS) [41–43].

Two clusters are identified by the conjunction analysis with the volume of 144 and 104 separately under the condition of minimum cluster size of o mm3 in the contrast analysis. Their peaks are in left precentral Gyrus and right insula dividedly in order. The right insula and left precentral gyrus may become potential markers for bilinguals who can speak English and Spanish no matter which language is their first language, since these two peaks are overlapped in the analysis of Spanish, English and the conjunction of Spanish and English. The left precentral gyrus and right insula are connected to complicated functions. Thereupon, it is reasonable that they have been found existing in the common brain regions during English and Spanish processing in the studied participants to some extent. There exist no significant difference about the contrast of English greater than Spanish or Spanish greater than English, partly owing to these languages belonging to the same language family and a range of similarities about the learning grammar and vocabulary. In addition, the limited studies about English and Spanish bilinguals included in this meta-analysis is another reason. The neuroimaging research conducted by Fedorenko et al. [44] and others has shown that brain regions involved in language tend to be language-specific, as opposed to being activated during non-linguistic tasks. In this meta-analysis, participants engaged in language-oriented tasks that activated specific regions in the brain, which could explain the existence of common brain regions activated during tasks related to Spanish and English in bilingual individuals analyzed through Ginger ALE. Malik-Moraleda et al. [45] conducted an investigation involving 45 languages and 12 language families, which revealed the existence of a universal language network underlying the organization of all human languages which demonstrates the lateralization to the left hemisphere. The activation likelihood of the first cluster with the peak of the left precentral gyrus in the conjunction analysis can further verify the above-mentioned investigation.

5 Conclusion

On account of the previous limitation, this study is the neuroimaging meta-analysis about English and Spanish bilinguals' language processing for the purpose of answering three questions regarding brain areas during English processing and Spanish processing separately and the common brain areas coping with the two languages. A significant finding of this research is that the right insula and left precentral gyrus may serve as potential indicators for individuals who are bilingual in both English and Spanish, regardless of which language they learned first. In this current study, right extra-nuclear, right insula (BA13), left precentral gyrus and right middle temporal gyrus (BA 39) were engaged for processing tasks pertinent to Spanish in Spanish and English bilinguals; Left precentral gyrus (BA 44), left inferior frontal gyrus (BA 47), right superior frontal gyrus, right middle frontal gyrus, right insula (BA 13), left precentral gyrus, right insula, left superior frontal gyrus (BA 6), left supramarginal gyrus, left lateral ventricle, left inferior parietal lobule were involved in tasks relevant to English in Spanish and English bilinguals; The analysis of common brain regions identified left precentral gyrus and right insula separately as peaks of two clusters with small volume.

In accordance with the tenets of American structuralism, as manifested during the Boas and Sapir epoch, Sapir [46] expounded upon the definition of language, characterizing it as a solely human and non-instinctual mechanism for conveying ideas, sentiments, and wishes through the utilization of a system comprised of intentionally generated symbols. Sapir [46] astutely observed the pervasive attributes inherent in language structures. While variations in formal attributes exist among distinct languages, their fundamental frameworks, encompassing discrete phonetic configurations and concrete amalgamations of sounds and meanings, evince remarkable sophistication [46]. This phenomenon elucidates, to a certain extent, the presence of shared neural substrates activated in both Spanish and English bilinguals when employing each language independently. Notwithstanding the idiosyncrasies characterizing different languages, there subsists a foundational likeness in the realms of phonological articulation and semantic expression.

The empirical discoveries furnished by this inquiry furnish substantiation that bolsters the aforementioned linguistic theory, drawing from an assemblage of thirteen neuroimaging investigations, thereby holding implications for the comprehension of pertinent linguistic paradigms.

Several potential sources of bias warrant consideration: Firstly, in the course of literature retrieval employing specific keywords, the search was confined to published studies. Conceivably, unpublished inquiries exploring the linguistic performance of English-Spanish bilinguals within designated language tasks might have eluded inclusion due to various reasons. Secondly, the equipoise between female and male participants was not rigorously factored into the study selection process, engendering the prospect of influencing the outcomes of this meta-analysis. Thirdly, the age distribution of the participants was regrettably omitted from the deliberations. Lastly, although the investigations encompassed English and Spanish languages, the tasks administered to the participants exhibited divergence, potentially introducing an additional confounding element.

English and Spanish, both members of the Indo-European language family, share a multitude of commonalities. Nevertheless, it is vital to note that the scope of this paper is confined to a limited compilation of only 13 relevant studies, which in turn introduces certain limitations to the study. In the future, with the increasing appearance of neuroimaging studies about English and Spanish bilinguals, the types of tasks, age period of subjects, gender difference can be regarded as sub-groups to perform more specific results. In that case, the relevant results may be more concrete, objective and meaningful.

Acknowledgments. This work was partly supported by the National Natural Science Foundation of China (Grant No. 62207021), in part by the Basic Research Program for the Natural Science of Shaanxi Province of China under Grant 2022JM134, and in part by the start-up foundation from Xi'an International Studies University (No. KYQDF202138).

References

1. Rao, P.S.: The role of English as a global language. Res. J. Engl. **4**(1), 65–79 (2019)
2. Rohmah, Z.: English as a global language: its historical past and its future. Jurnal Bahasa Seni **33**(1), 106–117 (2005)
3. Clackson, J.: Indo-European Linguistics: an Introduction. Cambridge University Press, Cambridge (2007)
4. Ansaldo, A.I., Marcotte, K., Scherer, L., Raboyeau, G.: Language therapy and bilingual aphasia: clinical implications of psycholinguistic and neuroimaging research. J. Neurolinguist. **21**(6), 539–557 (2008)
5. Becker, K.R.: Spanish/English bilingual codeswitching: a syncretic model. Bilingual Rev./La Revista Bilingüe **22**(1), 3–30 (1997)
6. Glass, G.V.: Primary, secondary, and meta-analysis of research. Educ. Res. **5**(10), 3–8 (1976)
7. Sulpizio, S., Del Maschio, N., Fedeli, D., Abutalebi, J.: Bilingual language processing: a meta-analysis of functional neuroimaging studies. Neurosci. Biobehav. Rev. **108**, 834–853 (2020)
8. Cargnelutti, E., Tomasino, B., Fabbro, F.: Effects of linguistic distance on second language brain activations in bilinguals: an exploratory coordinate-based meta-analysis. Front. Human Neurosci. **783**, 744489 (2022)
9. Sebastian, R., Laird, A.R., Kiran, S.: Meta-analysis of the neural representation of first language and second language. Appl. Psycholinguist. **32**(4), 799–819 (2011)
10. Comstock, L., Oliver, B.: A meta-analysis of task-based differences in bilingual L1 and L2 language networks. bioRxiv, 2021-12 (2021).
11. Liu, H., Cao, F.: L1 and L2 processing in the bilingual brain: a meta-analysis of neuroimaging studies. Brain Lang. **159**, 60–73 (2016)
12. Turkeltaub, P.E., Eden, G.F., Jones, K.M., Zeffiro, T.A.: Meta-analysis of the functional neuroanatomy of single-word reading: method and validation. Neuroimage **16**(3), 765–780 (2002)
13. Brignoni-Perez, E., Jamal, N.I., Eden, G.F.: An fMRI study of English and Spanish word reading in bilingual adults. Brain Lang. **202**, 104725 (2020)
14. Rammell, C.S., Cheng, H., Pisoni, D.B., Newman, S.D.: L2 speech perception in noise: an fMRI study of advanced Spanish learners. Brain Res. **1720**, 146316 (2019)
15. Andrews, E., Harshbarger, T., Rammell, C.S.: Multilingual listening and reading: an fMRI study of Russian/English and Spanish/English bilinguals. Glossos (14) (2019)
16. Jamal, N.I., Piche, A.W., Napoliello, E.M., Perfetti, C.A., Eden, G.F.: Neural basis of single-word reading in Spanish-English bilinguals. Hum. Brain Mapp. **33**(1), 235–245 (2012)
17. Meschyan, G., Hernandez, A.E.: Impact of language proficiency and orthographic transparency on bilingual word reading: an fMRI investigation. Neuroimage **29**(4), 1135–1140 (2006)

18. Kovelman, I., Baker, S.A., Petitto, L.A.: Bilingual and monolingual brains compared: a functional magnetic resonance imaging investigation of syntactic processing and a possible "neural signature" of bilingualism. J. Cogn. Neurosci. **20**(1), 153–169 (2008)

19. Stasenko, A., Hays, C., Wierenga, C.E., Gollan, T.H.: Cognitive control regions are recruited in bilinguals' silent reading of mixed-language paragraphs. Brain Lang. **204**, 104754 (2020)

20. Kovelman, I.: Bilingual and monolingual brains compared: a fMRI study of semantic processing. Dartmouth College (2006).

21. Ferreira, R.A., Vinson, D., Dijkstra, T., Vigliocco, G.: Word learning in two languages: neural overlap and representational differences. Neuropsychologia **150**, 107703 (2021)

22. Waldron, E.J., Hernandez, A.E.: The role of age of acquisition on past tense generation in Spanish-English bilinguals: an fMRI study. Brain Lang. **125**(1), 28–37 (2013)

23. Hernandez, A.E., Meschyan, G.: Executive function is necessary to enhance lexical processing in a less proficient L2: Evidence from fMRI during picture naming. Bilingualism Lang. Cogn. **9**(2), 177–188 (2006)

24. Hernandez, A.E.: Language switching in the bilingual brain: what's next? Brain Lang. **109**(2–3), 133–140 (2009)

25. Hernandez, A.E., Woods, E.A., Bradley, K.A.: Neural correlates of single word reading in bilingual children and adults. Brain Lang. **143**, 11–19 (2015)

26. Turkeltaub, P.E., Eickhoff, S.B., Laird, A.R., Fox, M., Wiener, M., Fox, P.: Minimizing within-experiment and within-group effects in activation likelihood estimation meta-analyses. Hum. Brain Mapp. **33**(1), 1–13 (2012)

27. Eickhoff, S.B., Laird, A.R., Grefkes, C., Wang, L.E., Zilles, K., Fox, P.T.: Coordinate-based activation likelihood estimation meta-analysis of neuroimaging data: a random-effects approach based on empirical estimates of spatial uncertainty. Hum. Brain Mapp. **30**(9), 2907–2926 (2009)

28. Laird, A.R., et al.: ALE meta-analysis workflows via the brainmap database: progress towards a probabilistic functional brain atlas. Front. Neuroinf. **3**, 598 (2009)

29. Laird, A.R., Eickhoff, S.B., Li, K., Robin, D.A., Glahn, D.C., Fox, P.T.: Investigating the functional heterogeneity of the default mode network using coordinate-based meta-analytic modeling. J. Neurosci. **29**(46), 14496–14505 (2009)

30. Ardila, A., Bernal, B., Rosselli, M.: Participation of the insula in language revisited: a meta-analytic connectivity study. J. Neurolinguist. **29**, 31–41 (2014)

31. Oh, A., Duerden, E.G., Pang, E.W.: The role of the insula in speech and language processing. Brain Lang. **135**, 96–103 (2014)

32. Rapcsak, S.Z., Arthur, S.A., Rubens, A.B.: Lexical agraphia from focal lesion of the left precentral gyrus. Neurology **38**(7), 1119–1119 (1988)

33. Murphy, K., et al.: Cerebral areas associated with motor control of speech in humans. J. Appl. Physiol. **83**(5), 1438–1447 (1997)

34. Bookheimer, S.: Functional MRI of language: new approaches to understanding the cortical organization of semantic processing. Ann. Rev. Neurosci. **25**(1), 151–188 (2002)

35. Cutting, L.E., et al.: Differential components of sentence comprehension: Beyond single word reading and memory. NeuroImage **29**(2), 429–438 (2006)

36. Fiez, J.A.: Phonology, semantics, and the role of the left inferior prefrontal cortex. Hum. Brain Mapp. **5**(2), 79–83 (1997)

37. Poldrack, R.A., Wagner, A.D., Prull, M.W., Desmond, J.E., Glover, G.H., Gabrieli, J.D.: Functional specialization for semantic and phonological processing in the left inferior prefrontal cortex. Neuroimage **10**(1), 15–35 (1999)

38. Price, C.J.: A review and synthesis of the first 20 years of PET and fMRI studies of heard speech, spoken language and reading. Neuroimage **62**(2), 816–847 (2012)

39. Pugh, K.R., et al.: Neurobiological studies of reading and reading disability. J. Commun. Disord. **34**(6), 479–492 (2001)

40. Marangolo, P., et al.: Electrical stimulation over the left inferior frontal gyrus (IFG) determines long-term effects in the recovery of speech apraxia in three chronic aphasics. Behav. Brain Res. **225**(2), 498–504 (2011)
41. Hillis, A.E., Work, M., Barker, P.B., Jacobs, M.A., Breese, E.L., Maurer, K.: Re-examining the brain regions crucial for orchestrating speech articulation. Brain **127**(7), 1479–1487 (2004)
42. Richardson, J.D., Fillmore, P., Rorden, C., LaPointe, L.L., Fridriksson, J.: Re-establishing Broca's initial findings. Brain Lang. **123**(2), 125–130 (2012)
43. Ochfeld, E., et al.: Ischemia in broca area is associated with broca aphasia more reliably in acute than in chronic stroke. Stroke **41**(2), 325–330 (2010)
44. Fedorenko, E., Behr, M.K., Kanwisher, N.: Functional specificity for high-level linguistic processing in the human brain. Proc. Natl. Acad. Sci. **108**(39), 16428–16433 (2011)
45. Malik-Moraleda, S., et al.: An investigation across 45 languages and 12 language families reveals a universal language network. Nat. Neurosci. **25**(8), 1014–1019 (2022)
46. Sapir, E.: Language: An Introduction to the Study of Speech. Harcourt Brace Jovanovich, Inc., New York (1921)

Robust Subspace Learning with Double Graph Embedding

Zhuojie Huang, Shuping Zhao, Zien Liang, and Jigang Wu[✉]

School of Computer Science and Technology,
Guangdong University of Technology, Guangzhou 510006, China
asjgwucn@outlook.com

Abstract. Low-rank-based methods are frequently employed for dimensionality reduction and feature extraction in machine learning. To capture local structures, these methods often incorporate graph embedding, which requires constructing a zero-one weighted neighborhood graph to extract local information from the original data. However, these methods are incapable of learning an adaptive graph that reveals intricate relationships among distinct samples within noisy data. To address this issue, we propose a novel unsupervised feature extraction method called Robust Subspace Learning with Double Graph Embedding (RSL_DGE). RSL_DGE incorporates a low-rank graph into the graph embedding process to preserve more discriminative information and remove noise simultaneously. Additionally, the $l_{2,1}$-norm constraint is also imposed on the projection matrix, making RSL_DGE more flexible in selecting feature dimensions. Several experiments demonstrate that RSL_DGE achieves competitive performance compared to other state-of-the-art methods.

Keywords: Low-rank representation · Graph embedding · Feature extraction · Subspace learning

1 Introduction

Feature extraction has attracted lots of attention in recent years [6,12,20,29]. In practical applications, original data usually have large dimensions and redundant information, which influence the performance of subsequent tasks. In the past several decades, many feature extraction have been proposed to reduce the dimensionality of high-dimensional data. New representation with low dimension not only improves efficiency but also reduces computational time [7,21].

Since the convenience of obtaining the unlabeled data, unsupervised feature extraction methods have been proposed. The most famous work is principal component analysis (PCA) [1], which identifies the directions of maximum variance to achieve projection. In this way, the main information of the original data is preserved by the low-dimensional representation. Locality preserving projections (LPP) [10] and neighborhood preserving embedding (NPE) [9] are also feature

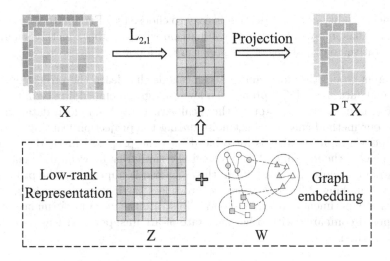

Fig. 1. The general framework of our RSL_DGE.

extraction methods, which preserve the local relationships of data. The above methods can be viewed as graph-based methods and be unified into unify graph embedding framework [24].

In recent years, low-rank based methods [4, 14, 30] have attracted much attention due to their effectiveness and robustness. However, these methods cannot deal with new samples that not include in the training stage. To solve this problem, Latent LRR [15] learns a low-rank projection to extract salient features from original data. Yin et al. [26] proposed a method called double LRR (LRR) that integrates the principle component recovery and salient feature extraction term into a unify model. But the above methods fail to reduce the dimensionality of the original data. To address this issue, low-rank preserving projections (LRPP) [17] and low-rank preserving embedding (LRPE) [28] learn a projection matrix from data. However, the local structure cannot be preserved by the above methods. Low-rank preserving projection via graph regularized reconstruction (LRPP_GRR) [23] preserve introduce graph regularization term to capture the local structure. Low-rank adaptive graph embedding (LRAGE) [16] learn an adaptive graph to preciously preserve the local structure. However, the above method cannot accurately learn the similarity graph from noise data, so the local structure cannot be well preserved.

To solve the problem mentioned above, we propose a method named robust subspace learning with double graph embedding (RSL_DGE). Different from the previous methods, RSL_DGE exploits a low-rank graph projection term to preserve the local structures, eliminating the noise from the original data and learning more accurate local structures during the projection learning. Meanwhile, we further impose the $l_{2,1}$-norm constraint on the projection matrix avoiding to chose the optimal dimensionality of the low-rank projection. By adding these two regularization terms, the local structure of data can be better preserved and the dimensionality of the projection is easy to choose (see Fig. 1). To summarize, RSL_DGE has many differences in comparison with the above methods and the representative contributions of the paper are as follows:

1) Different from other low-rank based methods that learn the projection from original data, RSL_DGE proposed a low-rank graph embedding term to eliminate the noise and to capture the local structures from clean data. In this way, our method can be robust noise during the projection learning.
2) Adaptive graph learning is introduced into the proposed method, in which the neighborhood graph W can reveal the similarity between different samples more accurately. In addition, the $l_{2,1}$-norm is imposed on the projection matrix, making DSL_DGE more easy to select the feature dimensions.
3) Several experiments are conducted to illustrate the good performance of the proposed compared with other the-state-of-art unsupervised feature extraction methods.

For the rest of this paper, Sect. 2 introduces some preliminaries. In Sect. 3, the proposed RSL_DGE, optimization process. In Sect. 4, some experiments are conducted on three real-world databases. The conclusion is finally summarized in Sect. 5

2   Preliminary

In this section, we briefly introduce the low-rank representation and the graph embedding.

2.1   Low-Rank Representation

LRR [5,14] aims to learn a lowest-rank representation from original data. This representation reveals multiple independent subspaces of the sample. The objective function of LRR is the following:

$$\min_{Z} \ \|Z\|_* + \lambda \|E\|_{2,1} \quad \text{s.t.} \quad X = XZ + E \tag{1}$$

where $Z \in R^{n \times n}$ is the low-rank matrix and $E \in R^{m \times n}$ denotes the noisy matrix. $\| \cdot \|_*$ and $\| \cdot \|_{2,1}$ denote the nuclear norm and the $l_{2,1}$-norm, respectively.

2.2   Graph Embedding

Graph embedding [24] first constructs a neighborhood graph W from original data X. And then the projection matrix $P \in R^{m \times d}$ to map the original data X into a low-dimensional latent embedding space. The optimal projection is obtained by minimizing the following objection function:

$$\min_{P} \sum_{ij}^{n} \|P^T x_i - P^T x_j\|_2^2 w_{ij} \quad \text{s.t.} \quad P^T XDX^T P = I \tag{2}$$

where D is a diagonal matrix satisfying $D_{ii} = \sum_{i \neq j} W_{ij}$. We defined $N_k(x_j)$ is a set of k nearest neighbor samples of sample x_j. If $x_i \in N_k(x_j)$ or $x_j \in N_k(x_i)$. The graph W are obtained as follows:

1) **Binary representation:**

$$w_{ij} = 1 \tag{3}$$

2) **Heat kernel weighting:**

$$w_{ij} = e^{-\frac{\|x_i - x_j\|^2}{\rho}} \tag{4}$$

3) **Dot-product weighting:**

$$w_{ij} = \frac{X_i^T X_j}{\max X^T X} \tag{5}$$

3 Proposed Method

In this section, we first introduce the motivation for the proposed method. In addition, we formulate the model of RSL_DGE and design an iterative algorithm for solving our model.

3.1 Motivation

LRR-based methods [13, 22, 27] can explore the hidden structures of data to further improve the performance of the learning model. By preserving the manifold structure information, graph embedding is integrated into the LRR-based method. However, the noisy data degrades the performance of graph embedding. Therefore, it is a problem that how to remove the noise from the original data to improve the robustness of the learning model.

On the other hand, most graph embedding methods first construct a zero-one weighted neighborhood graph W. However, the graph W cannot exactly reveal the similarity relationship between samples. Therefore, adaptive graph learning is also introduced into our model to learn a more precise similarity relationship for feature extraction.

In addition, how to select an optimal dimension for the projection matrix is also a problem. Therefore, we need to impose $l_{2,1}$-norm constraint on the projection matrix since the $l_{2,1}$ can remove the redundant information and enhance the robustness.

In summary, we aim to develop a novel unsupervised feature extraction method, which has the following properties. First, the noise is removed from the original data and the discriminative information can be preserved by the projection matrix. Second, the neighborhood graph is obtained adaptively during projection learning. Third, it is easy to choose a dimension for the projection matrix. We will present how to design this model in the next subsection.

3.2   Problem Formulation and Learning Model

In LRR, the global structures of samples are preserved while ignoring the local structures. Moreover, LRR cannot extract the feature from the new sample which does not contain in the training set. To solve this problem, we integrate LRR and graph embedding into a unify model as follows:

$$\min_{P,Z} \sum_{ij}^{n} \|P^T x_i - P^T x_j\|_2^2 w_{ij} + \lambda_1 \|Z\|_* \quad \text{s.t.} \quad P^T P = I, \quad X = XZ \quad (6)$$

where $P \in R^{d \times m}$ is projection matrix and d is the selected number of the projected dimension $(d < n)$. $Z \in R^{n \times n}$ is a low-rank representation matrix and z_{ij} denotes the similarity between x_i and x_j.

In model (6), the graph W is difficult to set the optimal number of neighbors, and cannot accurately discover the relationship during projection. In addition, the power of capturing local structures is damaged when the training data is polluted by noise. To solve the above problem, we first introduce the low-rank representation Z into the graph regularization term to eliminate the noise from the original data. Moreover, the F-norm is imposed on the graph W, in which a more precise graph W is learned adaptively. The objective function of our model is as follows:

$$\min_{P,Z,W} \sum_{ij}^{n} \|P^T X z_i - P^T X z_j\|_2^2 w_{ij} + \lambda_1 \|Z\|_* + \lambda_2 \|W\|_F^2$$

$$\text{s.t.} \quad P^T P = I, \quad X = XZ, \quad w_{ij} \geq 0, \quad \sum_j w_{ij} = 1 \quad (7)$$

To make the projection matrix P selecting the most discriminative information for feature extraction, we impose $l_{2,1}$-norm on the projection matrix P as follows:

$$\min_{P,Z,W} \sum_{ij}^{n} \|P^T X z_i - P^T X z_j\|_2^2 w_{ij} + \lambda_1 \|Z\|_* + \lambda_2 \|W\|_F^2 + \lambda_3 \|P\|_{2,1}$$

$$\text{s.t.} \quad P^T P = I, \quad X = XZ, \quad w_{ij} \geq 0, \quad \sum_j w_{ij} = 1 \quad (8)$$

where λ_1, λ_2 and λ_3 are the penalty parameters. The $l_{2,1}$-norm is defined as $\|P\|_{2,1} = \sum_i \|p_i\|_2$, p_i is the i-th row vector of matrix P. We will analyze the model (8) in the next subsection.

3.3   Optimization

Since there are three variables in the proposed model, we cannot optimize these three variables simultaneously. To solve this problem, we use the alternating direction method of multipliers (ADMM) [2] to solve the optimization problem (8). We first convert Eq. (8) into the following formula by imposing variable U:

$$\min_{P,Z,W,U} \sum_{ij}^{n} \|P^T X z_i - P^T X z_j\|_2^2 w_{ij} + \lambda_1 \|U\|_* + \lambda_2 \|W\|_F^2 + \lambda_3 \|P\|_{2,1} \tag{9}$$

$$\text{s.t.} \quad P^T P = I, \quad X = XZ, \quad Z = U, \quad w_{ij} \geq 0, \quad \sum_j w_{ij} = 1$$

To simplify the above model, the first term of (9) is rewritten as

$$\sum_{ij}^{n} \|P^T X z_i - P^T X z_j\|_2^2 w_{ij} = Tr(P^T X Z L Z^T X^T P) \tag{10}$$

where $Tr(\cdot)$ denotes the trace operator and L is the Laplacian matrix. Then Eq. (9) can be transformed into the following augmented Lagrangian function:

$$L(P,Z,W,U,C_1,C_2) = \min_{P,Z,W,U} Tr(P^T X Z L Z^T X^T P) + \lambda_1 \|U\|_*$$
$$+ \lambda_2 \|W\|_F^2 + \lambda_3 \|P\|_{2,1} + \frac{\mu}{2} (\|X - XZ + \frac{\mu}{C_1}\|_F^2 + \|Z - U + \frac{\mu}{C_2}\|_F^2) \tag{11}$$

By alternately solving Eq. (11), all variables can be solved.

Step 1 (update Z and fixing other variables). The problem of Eq. (11) can be rewritten:

$$L(Z) = Tr(P^T X Z L Z^T X^T P) + \frac{\mu}{2} (\|X - XZ + \frac{C_1}{\mu}\|_F^2 + \|Z - U + \frac{C_2}{\mu}\|_F^2) \tag{12}$$

Define $M_1 = X + \frac{C_1}{\mu}$ and $M_2 = U - \frac{C_2}{\mu}$, then let $\partial L(Z)/\partial Z = 0$. We can obtain the function as follows:

$$\frac{2}{\mu} X^T P P^T X Z L + (X^T X + I) Z = (X^T M_1 + M_2) \tag{13}$$

Z can be obtained by solving the Sylvester equation (13).

Step 2 (update U and fixing other variables). The minimization function of variable U is as following:

$$L(U) = \lambda_1 \|U\|_* + \frac{\mu}{2} \|Z - U + \frac{\mu}{C_2}\|_F^2 \tag{14}$$

Then U is obtained by using the singular value thresholding (SVT) shrinkage operator [15] as follows:

$$U = \Theta_{\frac{\lambda_1}{\mu}} \left(Z + \frac{C_1}{\mu} \right) \tag{15}$$

where Θ is the SVT shrinkage operator.

Step 3 (update W and fixing other variables). Let $g_{ij} = \|P^T X z_i - P^T X z_j\|_2^2$ and each g_{ij} is the element of the matrix G, then we can obtain W by minimizing the following subproblem:

$$
\begin{aligned}
&\min_{w_j \geq 0, \sum w_j = 1} \sum_i^n g_{ij} w_{ij} + \lambda_2 \|w_j\|_2^2 \\
&= \min_{w_j \geq 0, \sum w_j = 1} 2 \frac{\sum_i^n g_{ij} w_{ij}}{2\lambda_2} + \|w_j\|_2^2 \\
&= \min_{w_j \geq 0, \sum w_j = 1} \|w_j + \frac{g_j}{2\lambda_2}\|_2^2
\end{aligned}
\tag{16}
$$

where w_j and g_j is j-th column of W and G, respectively. w_j can be obtained by the application of **Lemma 1** [16] and its optimal solution is

$$
w_j = \left(\frac{1 + \sum_i^N \widetilde{g_{ij}}}{N} 1 - g_j\right)_+
\tag{17}
$$

where N denotes the number of nonzero elements in g_j. The operator $(\cdot)_+$ imposes the negative elements to zero. The elements of $\widetilde{g_{ij}}$ are those of g_j with the ascending order.

Step 4 (update P and fixing other variables). We can obtain P by minimizing the following problem:

$$
\min_P Tr(P^T(XZLZ^T X^T + \lambda_3 Q)P)
\tag{18}
$$

where $Q = diag(1/\|p_{1,\cdot}\|_2, 1/\|p_{2,\cdot}\|_2, \ldots, 1/\|p_{n,\cdot}\|_2)$. The minimize ratio trace problem (18) can be solve by eigendecomposition:

$$
(XZLZ^T X^T + \lambda_3 Q)P = P\Lambda
\tag{19}
$$

where Λ is eigenvalue matrix. The optimal solution of P is the eigenvectors corresponding to the first d smallest eigenvalues.

Step 5 (update C_1, C_2 and μ). Lagrangian multipliers C_1, C_2, and μ are updated by the following functions:

$$
\begin{aligned}
C_1 &= C_1 + \mu(X - XZ) \\
C_2 &= C_2 + \mu(Z - U) \\
\mu &= \min(\rho\mu, \mu_{max})
\end{aligned}
\tag{20}
$$

where ρ and μ_{max} are constants. The details of the optimization are summarized in Algorithm 1.

3.4   Convergency Analysis

In this subsection, we experimentally illustrate the convergency of the algorithm through a series of experiments including COIL20 [18], Yale B [8], LFW [11]. Figure 2 shows that the proposed method obtains the stable point within about 20 iterations. The experiment result shows that the proposed method has good convergence property.

Algorithm 1. RSL_DGE

Input: data matrix X, parameters λ_1, λ_2 and projected dimension d, nearest neighbor graph W.

Output: P, Z

Initialization: $P = \arg\max\limits_{P^T P=I} Tr(P^T(\Lambda)P)$, where Λ is the data covariance in advance. $Z = W, C_1 = 0, C_2 = 0, \mu = 0.1, \rho = 1.01, \mu_{max} = 10^8$, where ρ and μ_{max} are constants.

while not converged **do**

1.   Update Z solving (13);
2.   Update U solving (14);
3.   Update W using (17);
4.   Update P solving (19);
5.   Update C_1, C_2 and μ by (20);

end while

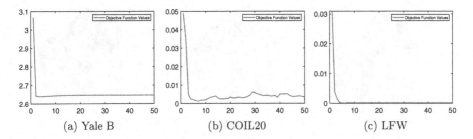

(a) Yale B   (b) COIL20   (c) LFW

Fig. 2. Objective function values versus the number of iterations using the proposed method on Yale B, COIL20 and LFW databases.

4   Experiments

In this section, we conduct several experiments to illustrate the effectiveness of the proposed method. In the experiments, several related unsupervised methods are compared with RSL_DGE. Such as PCA [1], LPP [10], OLPP [3], NPE [9], SPP [19], CRP [25], LatLRR [15], LRPP_GRR [23] and LRAGE [16]. The above methods are performed on three real-world databases, including COIL20 image database [18], Yale B face database [8] and LFW face database [11]. In each experiment, we choose some samples (No.) from each class as the training set and treat the remaining samples as the test samples. In order to improve the efficiency of computation, we perform PCA on all samples and reduce their dimensions by preserving 98% of energy for comparison. The nearest neighbor classifier is selected to obtain the final classification accuracies (%). For fair comparison, we perform the above methods 20 times and report the mean classification accuracies with standard deviation.

4.1 Parameters Selection of RSL_DGE

In Eq. (8), we can find that there are three parameters that need to be turned in advance. To obtain the best performance of the proposed method, we first draw some figures to show the relationship between the classification accuracy with these three parameters, in which the three parameters are selected from a candidate set $\{10^{-5}, 10^{-4}, 10^{-3}, 10^{-2}, 10^{-1}, 1, 10^1, 10^1, 10^2, 10^3, 10^4, 10^5\}$. From Fig. 3, it is obvious that the classification accuracy (ACC) is insensitive to λ_1 and λ_2 on these three databases. However, RSL_DGE obtains a good performance on Yale B and LFW when λ_3 is set to a relatively small number. We can observe that the bast combinations of parameters λ_1, λ_2 and λ_3 are $\{10^3, 10^2, 10^{-3}\}$ for Yale B, $\{10^{-4}, 10^2, 10^{-5}\}$ for COIL20 and $\{10^{-4}, 10^2, 10^{-5}\}$ for LFW, respectively.

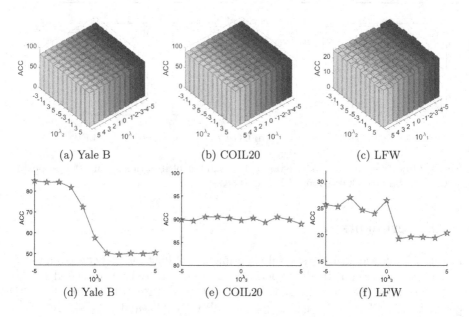

(a) Yale B   (b) COIL20   (c) LFW

(d) Yale B   (e) COIL20   (f) LFW

Fig. 3. The relationship between parameters and ACC (%) of Yale B, COIL20, LFW

4.2 Results and Analysis

From Table 1, we can find that the proposed method obtain the best result than other compared methods. In addition, the results of RSL_DGE are better than LRPP_GRR, since the adaptive graph regularization term can better preserve the local structure to improve the performance of feature extraction. The result on the COIL20 is reported in Table 2. From Table 2, it is obvious that all classification accuracy is very high, while our method is obtain the highest accuracy. The performance of LRPP_GRR, LRAGE, and RSL_DGE is better than other compared methods since the local structure captured by graph embedding is helpful for feature extraction. For LFW, Table 3 shows that the proposed performs much better than other methods while the result of other compared methods is very low. It is because the LFW database is collected under the unconstrained scenario and difficult to learn an optimal projection matrix for feature extraction. From the above experimental result, we can find that our method obtains the best results in three databases.

Table 1. ARR (%) of different methods on the Yale B database

No.	PCA	LPP	OLPP	NPE	SPP	CRP	LatLRR	LRPP	LRPP_GRR	LRAGE	RSL_DGE
10	50.88	56.53	45.26	61.46	85.08	84.03	48.52	50.88	85.23	49.37	**85.69**
	(1.09)	(2.19)	(1.35)	(1.28)	(0.87)	(0.84)	(1.14)	(1.09)	(0.91)	(1.92)	**(0.80)**
15	59.70	64.74	54.05	69.89	88.88	89.16	56.44	59.70	89.28	57.66	**89.29**
	(1.20)	(1.81)	(0.98)	(1.29)	(0.95)	(0.84)	(1.22)	(1.20)	(0.75)	(2.76)	**(0.77)**
20	65.23	70.77	59.66	75.62	91.10	91.34	61.58	65.23	90.88	62.86	**91.44**
	(1.39)	(1.37)	(1.32)	(1.59)	(0.83)	(0.70)	(1.36)	(1.39)	(0.78)	(2.27)	**(0.49)**
25	69.41	74.72	63.55	78.46	92.37	92.58	65.46	69.41	91.58	67.15	**92.72**
	(0.79)	(1.32)	(0.95)	(1.10)	(0.56)	(0.65)	(1.21)	(0.79)	(0.89)	(2.03)	**(0.71)**

Table 2. ARR (%) of different methods on the COIL20 database

No.	PCA	LPP	OLPP	NPE	SPP	CRP	LatLRR	LRPP	LRPP_GRR	LRAGE	RSL_DGE
10	89.87	79.15	90.11	81.45	82.36	87.41	89.62	89.87	90.27	87.38	**90.73**
	(1.09)	(1.92)	(1.03)	(1.42)	(1.32)	(1.29)	(1.17)	(1.09)	(1.17)	(1.79)	**(1.56)**
15	93.24	86.79	93.54	88.78	89.33	90.09	93.09	93.24	93.71	92.76	**93.76**
	(0.82)	(0.98)	(0.88)	(0.99)	(0.96)	(0.88)	(0.82)	(0.82)	(0.87)	(1.01)	**(1.01)**
20	94.95	90.22	95.26	91.47	91.48	92.04	94.80	94.95	95.74	95.62	**95.90**
	(0.73)	(1.05)	(0.72)	(1.00)	(0.84)	(0.99)	(0.75)	(0.73)	(0.92)	(0.75)	**(0.88)**
25	96.42	92.28	96.56	93.61	93.65	94.41	96.39	96.42	96.97	96.81	**97.09**
	(0.43)	(0.73)	(0.44)	(0.56)	(0.65)	(0.51)	(0.45)	(0.43)	(0.74)	(0.89)	**(0.55)**

Table 3. ARR (%) of different methods on the LFW database

No.	PCA	LPP	OLPP	NPE	SPP	CRP	LatLRR	LRPP	LRPP_GRR	LRAGE	RSL_DGE
5	19.96	10.96	19.31	12.95	22.72	13.29	19.62	19.96	21.23	15.35	**25.42**
	(1.18)	(1.25)	(1.17)	(1.01)	(0.87)	(1.08)	(1.32)	(1.18)	(1.14)	(1.52)	**(1.38)**
6	21.27	12.47	20.51	14.45	24.80	13.86	20.80	21.27	22.85	16.60	**27.29**
	(1.00)	(1.24)	(1.01)	(0.94)	(0.88)	(1.71)	(0.85)	(1.00)	(1.58)	(1.56)	**(1.28)**
7	22.51	14.17	22.04	15.90	25.63	15.45	22.09	22.51	24.17	17.40	**28.90**
	(1.41)	(1.47)	(1.03)	(1.09)	(1.40)	(1.31)	(1.50)	(1.41)	(0.96)	(1.47)	**(1.16)**
8	23.38	15.55	22.76	17.78	28.28	16.28	23.20	23.38	22.84	18.15	**31.39**
	(1.13)	(1.01)	(1.20)	(0.96)	(1.56)	(1.08)	(1.20)	(1.13)	(1.85)	(1.61)	**(1.77)**

5   Conclusion

In this paper, we presented a novel unsupervised feature extraction method named RSL_DGE, which introduces a low-rank graph into graph embedding to learn a more accurate neighborhood graph adaptively. Moreover, the $l_{2,1}$-norm are also integrated into the RSL_DGE, which makes the selection of feature dimensions more flexible and further improve the performance and robustness of the proposed RSL_DGE. Compared with other methods, the proposed method can learn a more discriminative projection for feature extraction. Experimental results demonstrate that the proposed method achieves a competitive performance compared to other compared projection learning methods. The proposed method cannot quickly process datasets with large sample sizes and cannot obtain the global optimal solution. The large-scale version of RSL_DGE is also worth in-depth research, which is beyond the scope of this paper and will be explored in our future work.

Acknowledgements. This work was supported in part by Huangpu International Sci&Tech Cooperation Fundation of Guangzhou, China (2021GH12) and Guangdong Basic and Applied Basic Research Foundation under Grant No. 2021B1515120010. It was also supported by the Natural Science Foundation of China under Grant 62106052.

References

1. Abdi, H., Williams, L.J.: Principal component analysis. Wiley Interdisc. Rev. Comput. Stat. **2**(4), 433–459 (2010)
2. Boyd, S., Parikh, N., Chu, E., Peleato, B., Eckstein, J., et al.: Distributed optimization and statistical learning via the alternating direction method of multipliers. Found. Trends® Mach. Learn. **3**(1), 1–122 (2011)
3. Cai, D., He, X.: Orthogonal locality preserving indexing. In: Proceedings of the 28th Annual International ACM SIGIR Conference on Research and Development in Information Retrieval, pp. 3–10 (2005)
4. Candès, E.J., Li, X., Ma, Y., Wright, J.: Robust principal component analysis? J. ACM (JACM) **58**(3), 1–37 (2011)
5. Chen, J., Yang, J.: Robust subspace segmentation via low-rank representation. IEEE Trans. Cybern. **44**(8), 1432–1445 (2013)
6. Chen, M.S., Wang, C.D., Lai, J.H.: Low-rank tensor based proximity learning for multi-view clustering. IEEE Trans. Knowl. Data Eng. **35**, 5076–5090 (2022)
7. Du, S., Liu, B., Shan, G., Shi, Y., Wang, W.: Enhanced tensor low-rank representation for clustering and denoising. Knowl.-Based Syst. **243**, 108468 (2022)
8. Georghiades, A.S., Belhumeur, P.N., Kriegman, D.J.: From few to many: illumination cone models for face recognition under variable lighting and pose. IEEE Trans. Pattern Anal. Mach. Intell. **23**(6), 643–660 (2001)
9. He, X., Cai, D., Yan, S., Zhang, H.J.: Neighborhood preserving embedding. In: Tenth IEEE International Conference on Computer Vision (ICCV 2005), vol. 2, pp. 1208–1213. IEEE (2005)
10. He, X., Niyogi, P.: Locality preserving projections. Adv. Neural Inf. Process. Syst. **16**, 1–8 (2003)
11. Huang, G.B., Mattar, M., Berg, T., Learned-Miller, E.: Labeled faces in the wild: a database for studying face recognition in unconstrained environments. In: Workshop on faces in 'Real-Life' Images: Detection, Alignment, and Recognition (2008)

12. Huang, Z., Zhao, S., Fei, L., Wu, J.: Weighted graph embedded low-rank projection learning for feature extraction. In: ICASSP 2022–2022 IEEE International Conference on Acoustics, Speech and Signal Processing (ICASSP), pp. 1501–1505. IEEE (2022)
13. Khan, G.A., Hu, J., Li, T., Diallo, B., Zhao, Y.: Multi-view low rank sparse representation method for three-way clustering. Int. J. Mach. Learn. Cybern. **13**, 233–253 (2022)
14. Liu, G., Lin, Z., Yan, S., Sun, J., Yu, Y., Ma, Y.: Robust recovery of subspace structures by low-rank representation. IEEE Trans. Pattern Anal. Mach. Intell. **35**(1), 171–184 (2012)
15. Liu, G., Yan, S.: Latent low-rank representation for subspace segmentation and feature extraction. In: 2011 International Conference on Computer Vision, pp. 1615–1622. IEEE (2011)
16. Lu, J., Wang, H., Zhou, J., Chen, Y., Lai, Z., Hu, Q.: Low-rank adaptive graph embedding for unsupervised feature extraction. Pattern Recogn. **113**, 107758 (2021)
17. Lu, Y., Lai, Z., Xu, Y., Li, X., Zhang, D., Yuan, C.: Low-rank preserving projections. IEEE Trans. Cybern. **46**(8), 1900–1913 (2015)
18. Nene, S.A., Nayar, S.K., Murase, H., et al.: Columbia object image library (coil-20) (1996)
19. Qiao, L., Chen, S., Tan, X.: Sparsity preserving projections with applications to face recognition. Pattern Recogn. **43**(1), 331–341 (2010)
20. Tang, C.: Feature selective projection with low-rank embedding and dual laplacian regularization. IEEE Trans. Knowl. Data Eng. **32**(9), 1747–1760 (2019)
21. Wang, M., Wang, Q., Hong, D., Roy, S.K., Chanussot, J.: Learning tensor low-rank representation for hyperspectral anomaly detection. IEEE Trans. Cybern. **53**(1), 679–691 (2022)
22. Wang, S., Xiao, S., Zhu, W., Guo, Y.: Multi-view fuzzy clustering of deep random walk and sparse low-rank embedding. Inf. Sci. **586**, 224–238 (2022)
23. Wen, J., Han, N., Fang, X., Fei, L., Yan, K., Zhan, S.: Low-rank preserving projection via graph regularized reconstruction. IEEE Trans. Cybern. **49**(4), 1279–1291 (2018)
24. Yan, S., Xu, D., Zhang, B., Zhang, H.J., Yang, Q., Lin, S.: Graph embedding and extensions: a general framework for dimensionality reduction. IEEE Trans. Pattern Anal. Mach. Intell. **29**(1), 40–51 (2006)
25. Yang, W., Wang, Z., Sun, C.: A collaborative representation based projections method for feature extraction. Pattern Recogn. **48**(1), 20–27 (2015)
26. Yin, M., Cai, S., Gao, J.: Robust face recognition via double low-rank matrix recovery for feature extraction. In: 2013 IEEE International Conference on Image Processing, pp. 3770–3774. IEEE (2013)
27. Zhang, G.Y., Huang, D., Wang, C.D.: Facilitated low-rank multi-view subspace clustering. Knowl.-Based Syst. **260**, 110141 (2023)
28. Zhang, Y., Xiang, M., Yang, B.: Low-rank preserving embedding. Pattern Recogn. **70**, 112–125 (2017)
29. Zhao, S., Wu, J., Zhang, B., Fei, L.: Low-rank inter-class sparsity based semi-flexible target least squares regression for feature representation. Pattern Recogn. **123**, 108346 (2022)
30. Zhuang, L., Wang, J., Lin, Z., Yang, A.Y., Ma, Y., Yu, N.: Locality-preserving low-rank representation for graph construction from nonlinear manifolds. Neurocomputing **175**, 715–722 (2016)

Unsupervised Feature Selection via Nonlinear Representation and Adaptive Structure Preservation

Aihong Yuan[✉], Lin Lin, Peiqi Tian, and Qinrong Zhang

The College of Information Engineering, Northwest A&F University, Xianyang, China
ahyuan@nwafu.edu.cn

Abstract. Unsupervised feature selection has attracted increasing attention for its promising performance on high dimensional data with higher dimensionality and more expensive labeling costs. Existing unsupervised feature selection methods mostly assume that linear relationships can interpret all feature associations. However, data with exclusively linear relationships are rare and impractical. Moreover, the quality of the similarity matrix significantly affects the effectiveness of conventional spectral-based methods. Real-world data contains lots of noise and redundancy, making the similarity matrix built using the raw data unreliable. To address these problems, we propose a novel and robust method for feature selection over a novel nonlinear mapping function, aiming to mine the nonlinear relationships among features. Furthermore, we incorporated manifold learning into our training process, embedded with adaptive graph constraints based on the principle of maximum entropy, to maintain the intrinsic structure of the data and simultaneously capture more accurate information. An efficient and effective algorithm was designed to perform our method. Experiments with eight benchmark datasets from face images, biology, and time series outperformed nine state-of-the-art algorithms, validating the superiority and effectiveness of our method. The source code is available at https://github.com/aasdlaca/NRASP.

Keywords: Unsupervised feature selection · Nonlinear representation · Adaptive graph constraint · Structure preserving

1 Introduction

With the accelerated advancement of technology, high-dimensional data are becoming more prevalent in machine learning. Undoubtedly, a large amount of information can be obtained from higher-dimension data. However, it also contains more redundancy and even noise, which require more resources such as memory and time to store, utilize, and transmit. To address these issues, feature

This work was supported in part by the National Natural Science Foundation of China under Grant 62306244, and in part by the Key Project of Shaanxi Provision-City Linkage under Grant 2022GD-TSLD-53.

selection [22] and feature extraction [23] are the two main techniques used to reduce the dimensionality of data. Feature selection reduces dimensionality by selecting the most discriminative subset from the original feature space, whereas feature extraction emphasizes combining the original features linearly or nonlinearly to obtain a new feature collection [16]. Hence, feature selection has gained more attention as it preserves the original structure of the data while reducing the dimensions.

Unsupervised feature selection methods have attracted increasing attention in real applications because accurately labeled data is costly and sometimes difficult to obtain [22]. Among the unsupervised feature selection methods, the embedded-based method has excellent performance and interpretability [14,20,21,24]. It addresses feature selection as an optimization problem. In embedded-based methods frequently utilize spectral analysis and sparse constraints to improve performance. Besides, to preserve the intrinsic structure of the raw data, the manifold learning method is also integrated as part of the optimization process [18].

However, the nonlinear relationship between data cannot be ignored in real life. This is also evidenced by the increasing popularity of deep learning methods, which attempt to mine nonlinear relationships among data. These methods [14,20,21,24] project data in a latent low-dimensional space linearly via a transform matrix, which only focuses on the linear relationships among features. They perform poorly for nonlinear features than processes with linear data, ignoring the relationship among the nonlinear features [8]. To learn the nonlinear relationships among data, several nonlinear learning methods have been applied to solve feature selection problems [2,19]. They utilize a sophisticated structure to discover the intrinsic association between data and have achieved promising and outstanding results compared with conventional methods. However, the model complexity and low interpretability are two major problems that cannot be neglected with these methods based on nonlinear learning.

Furthermore, several embedding-based feature selection methods use spectral analysis to leverage the data's geometric structure. Traditional spectral-based methods [3,5,11] typically involve two steps. First, construct a similarity matrix with the k-nearest neighbors method. Subsequently, obtain the feature selection matrix by optimizing the objective function with sparse constraints. These methods treat manifold learning and feature selection separately. Alternatively, the similarity matrix is fixed after it is constructed from the original data, which makes the feature selection matrix suboptimal owing to the noise and redundancy contained in the raw data [10], thus affecting the effect of feature selection. Intuitively, based on the similarity matrix definition, the principle of maximum entropy can be used to construct the similarity matrix dynamically using the probability distribution; thus, the most informative distribution can be obtained adaptively in each iteration during feature selection [9].

To address the aforementioned problems, we propose a novel and robust feature selection method using a novel nonlinear mapping function with sparse constraints and manifold learning, which embeds an adaptive graph constraint called Unsupervised Feature Selection via Nonlinear Representation and Adaptive Structure Preservation (NRASP).

The principal contributions are summarized as follows:

1. We propose a novel nonlinear mapping function with group sparsity regularization to perform feature selection, which explores the nonlinear relationship among features and reduces the noise and redundancy of the original data.
2. An effective adaptive graph constraint founded on the principle of maximum entropy is introduced into manifold learning to preserve the local geometrical structure of the data and capture more accurate information from the data simultaneously.
3. An efficient algorithm is designed for our method, and extensive experiments on eight benchmark datasets demonstrate its effectiveness and superiority over nine other state-of-the-art methods.

The remainder of this paper will be organized in the following way. We will provide an overall review of the related work in Sect. 2. The detailed model of the proposed method and the corresponding optimization method are introduced in Sect. 3. The experimental results and analysis are demonstrated in Sect. 4. The conclusions are presented in Sect. 5. Detailed derivations, experimental setups, more comprehensive experiments, and results are all presented in the appendix, which can be accessed at https://drive.google.com/file/d/1CDItHadqVzJ4lSj38bkdHYy87JcQlix7.

2   Related Work

Embedded methods have recently attracted considerable attention owing to their promising performance in various areas. These methods perform unsupervised feature selection by treating it as an optimization problem. During the optimization process, strategies of spectral analysis and sparse constraints are usually utilized to achieve better performance metrics. Multi-cluster feature selection (MCFS) [3] selects features using l_1-regularized regression model with spectral analysis, which can preserve the clustering structure effectively of data. Unsupervised discriminative feature selection (UDFS) [17] exploits discriminative information using combining discriminative analysis and $l_{2,1}$-norm regularization within a unified framework. To deal with noise in the raw data, robust unsupervised feature selection (RUFS) [15] was developed, which leverages nonnegative matrix factorization (NMF) and $l_{2,1}$-norm regularization to learn pseudo-cluster labels and performs feature selection simultaneously. Nonnegative discriminative feature selection (NDFS) [11] also utilizes discriminative information of data, but what sets it apart is its incorporation of clustering label learning into feature selection and its use of a nonnegative constraint for more accurate cluster labels. Structure optimal graph feature selection (SOGFS) [14] was offered to adaptively learn the local geometrical structure and perform feature selection simultaneously. Generalized uncorrelated regression with the adaptive graph for unsupervised feature selection (URAFS) [10] uses an uncorrelated regression model while incorporating the data's geometric structure into the manifold learning process to selects features and performs manifold learning simultaneously. Considering the correlation of feature groups, the multi-group adaptive graph representation (MGAGR) [20] constructs the global similarity matrix by linearly combining each group's similarity matrix.

Generally, the aforementioned methods perform well in selecting features and have excellent interpretability. However, these methods have a restriction of handling the data with a linear low-dimensional manifold. To address this problem, several nonlinear self-representation based methods have been proposed. Autoencoder feature selection (AEFS) [6] combines an autoencoder network and group LASSO to explore the linear and nonlinear correlations between features. Abid et al. proposed a concrete autoencoder for differentiable feature selection and reconstruction (CAE) [2]. They selected features by utilizing the concrete selector layer and converged to discrete K features during model training. An adaptive autoencoder with redundancy control (AARC) [4] was proposed to control the redundancy of the subset of selected features. They combined structural optimization and redundancy control to improve the performance of sparse learning. Subsequently, Zahra et al. exploited trained sparse denoising autoencoders and used a sparse evolution strategy for training. Afterwards, they use the neuron's strength to quantify each feature's importance and named this method QuickSelection (QS) [1].

The aforementioned methods contain almost one of two main problems: ignoring the nonlinear relationships between features or the lack of effective manifold learning. In this study, we select features using a novel nonlinear mapping function with creative and effective manifold learning, which can explore the features' linear and nonlinear correlation and preserve the data's intrinsic structure.

3 Our Method

We applied nonlinear self-representation to explore the nonlinear relationship between the raw data and latent low-dimensional spaces. Specifically, we projected the features into a latent space using a linear combination of the original features in the first step. Next, we project the linearly combined features from the previous step into a new space using nonlinear transformation to learn the nonlinear relationship. In the training process, we add an $l_{2,1}$-norm to the feature selection matrix as a penalty to ensure it is sparse. The above implies that original data with less effect on the transformed feature space would be removed as redundancy by the group sparsity regularization. Finally, to retain the data's local geometric structure, we employ the manifold learning method and dynamic similarity matrix construction using the principle of maximum entropy.

3.1 Problem Formulation

Given an unlabeled data matrix $X \in R^{m \times n}$ with samples number m, and features dimensionality n, that is, the number of features. First, we preprocess the high-dimensional data matrix X using a projection matrix $W \in R^{n \times d}$ to project the X into a low-dimensional space. d is a hyperparameter that must be adjusted. Subsequently, we utilize the nonlinear self-representation to uncover the underlying nonlinear relationships between the preprocessed features. Then, we reconstruct the data from latent space with the objective of minimization the error of the reconstructed data and the original data.

We measured our method's performance using the sum of the squared errors. Subsequently, the base objective function is defined as:

$$\mathcal{L} = \ \|X - f\left(g\left(XW\right)\right)\|_F^2 \tag{1}$$

where g represents the encoder mapping, f the decoder mapping, $\|\cdot\|_F$ the Frobenius norm.

The aim of feature selection (FS) is to identify the most significant features of the original data, which are usually sparse relative to others. Therefore, the sparsity constraint is always used in the FS methods. Group LASSO regularization ($l_{2,1}$ norm) can make the entire matrix sparse, which is suitable for our proposed method. Its constraint on the projection matrix W can guarantee the row sparsity of W. The original problem becomes:

$$\min_{W,\Theta} \frac{1}{2} \|X - f\left(g\left(XW\right)\right)\|_F^2 + \alpha \|W\|_{2,1} \tag{2}$$

where α is a hyperparameter that must be adjusted to balance the two terms in the current cost function; furthermore, Θ represents the parameters of the nonlinear self-representation component.

Moreover, the data's local manifold structures tend to include information about the discrimination between adjacent data points. It presumes that nearby data points have similar representations. Intuitively, nearby data points should reasonably maintain their neighboring data points after projection, which means that if they are close in the original data space, the projected points $W^T x_i$ and $W^T x_j$ should also have a small distance. Subsequently, we formulated the manifold structure preservation item as (The detailed derivation are shown in Appendix A.1):

$$\min_W \ Tr(W^T X^T L_S X W) \tag{3}$$

Most previous studies first constructed the similarity matrix S at the beginning of the training using the original data matrix X, which makes the similarity matrix S fixed throughout the whole process. In our method, S is obtained dynamically by maximizing the information entropy. For the similarity matrix, the following constraints are required to be satisfied: 1. $s_{ij} \geq 0$ and 2. $\sum_j s_{ij} = 1$. Each row of S appears to fit the probability distribution well with these constraints. Therefore, to obtain the most information, we maximize the information entropy of S distribution [9]. Information entropy is expressed as follows

$$-\sum_{i=1}^m \sum_{j=1}^m s_{ij} \log s_{ij} \quad \text{s.t.} \quad s_{ij} \geq 0, \ \sum_{j=1}^m s_{ij} = 1. \tag{4}$$

In conjunction with the above analysis, the final objective function for FS was expressed as:

$$\min_{W,\Theta,S} \mathcal{L}(W;\Theta;S) = \|X - f(g(XW))\|_F^2 + \alpha\|W\|_{2,1} + \beta Tr(W^T X^T L_S XW)$$

$$+ \gamma \sum_{i=1}^{m}\sum_{j=1}^{m} s_{ij} \log s_{ij} \quad \text{s.t.} \quad s_{ij} \geq 0, \quad \sum_{j=1}^{m} s_{ij} = 1. \tag{5}$$

where Θ represents the parameters of the nonlinear self-representation part, α, β and γ are the hyperparameters that must be adjusted.

By optimizing Eq. (5), we can obtain the projection matrix W, then utilize it to implement the FS. We use $\|w_i\|_2$ to measure the importance of i-th features. For the i-th feature of the raw data input to our method, there is a corresponding weight vector $w_i = [w_{i1}, w_{i2}, \ldots, w_{id}]^T$ directly affecting it. If this feature is more informative than other features, then every component of w_i is likely to be greater than the other weight vectors. Therefore, $\|w_i\|_2$ is reasonably utilized as the index for FS, and it generally reflects the importance of the i-th feature. Afterward, based on the $\|w_i\|_2$, we select the top k features as the feature subset, where k is the number of features that must be selected. To verify the effectiveness of the proposed method, the clustering performance with the selected top features is used as the evaluation criteria to compare with previous methods and other combinations of the proposed method in the experiments.

3.2 Optimization

It is difficult to solve Eq. (5) directly, which involves three different variables: parameters in the nonlinear self-representation part Θ, similarity matrix S, and projection matrix W. Because the optimization of S is independent of the nonlinear self-representation part, we solve the S first and then optimize W and Θ.

For the optimization of W and Θ, we first present the gradient in Eq. (5) w.r.t. W and Θ. We then choose the limited-memory Broyden-Fletcher-Goldfarb-Shanno (L-BFGS) algorithm [12] for optimization as there is no closed-form solution for our method.

Solve S. The part of Eq. (5) related S optimization is presented in the following formula:

$$\min_{S} \quad \mathcal{L}_1 = \beta Tr(W^T X^T L_S XW) + \gamma \sum_{i=1}^{m}\sum_{j=1}^{m} s_{ij} \log s_{ij}$$

$$= \frac{1}{2}\beta \sum_{i=1}^{m}\sum_{j=1}^{m} \|W^T x_i - W^T x_j\|_2^2 s_{ij} + \gamma \sum_{i=1}^{m}\sum_{j=1}^{m} s_{ij} \log s_{ij} \tag{6}$$

$$\text{s.t.} \quad s_{ij} \geq 0, \quad \sum_{j=1}^{m} s_{ij} = 1.$$

By applying the Lagrangian multiplier method with the KKT condition (The detailed derivation are shown in Appendix A.2), we can obtain the optimal solution of s_{ij}, which is

$$s_{ij} = \frac{\exp\{c_{ij}\}}{\sum_{j=1}^{m} \exp\{c_{ij}\}} \tag{7}$$

where $c_{ij} = -\frac{\beta}{2\gamma} \left\| W^T x_i - W^T x_j \right\|_2^2$. According to Eq. (7), the solution of s_{ij} satisfies the concept of a similar matrix well. Given two data points x_i and x_j, it is evident that when the x_i and x_j are near, the s_{ij} is close to 1. In contrast, when the s_{ij} is close to 0, x_i and x_j are far apart, which will not be considered similar.

Solve Θ. The part of Eq. (5), related Θ optimization, is the error term of the nonlinear self-representation part:

$$\min_{\Theta} \quad \mathcal{L}_2 = \|X - f(g(XW))\|_F^2 \tag{8}$$

To ensure easy optimization, we consider the mapping process as the forward propagation in neural network. Subsequently, the calculation of the partial derivatives with respect to Θ can be obtained using the idea of the backpropagation method. Therefore, we ignore the details and provide the following results:

$$\frac{\partial \mathcal{L}_2}{\partial W^{(i)}} = 2a^{(i)} \left(\delta^{(i+1)}\right)^T \tag{9}$$

$$\frac{\partial \mathcal{L}_2}{\partial b^{(i)}} = 2\delta^{(i)} \tag{10}$$

where $W^{(i)}(i = 0, 1, 2)$ (specifically, $W^{(0)} = W$) denotes the weight matrix between the i-th layer, $b^{(i)}(i = 1, 2)$ denotes the i-th layer's bias vector. $\delta^{(i)}(i = 1, 2, 3)$ denotes the error signal backpropagated from the output layer to the certain i-th layer. $a^{(i)}(i = 0, 1, 2)$ denotes the active signal of the i-th layer. The $0, 1, 2, 3$ layers represent the input layer, the first and the second hidden layer, and the output layer individually.

Solve W. With fixed S, Eq. (5) is transformed into

$$\min_{W} \quad \mathcal{L}_3 = \|X - f(g(XW))\|_F^2 + \alpha \|W\|_{2,1} + \beta Tr(W^T X^T L_S XW) \tag{11}$$

Recalling the aforementioned, the linear preproessing can also be naturally considered as a stage of forward propagation. Thus the partial derivatives of the nonlinear self-representation part error $\mathcal{L}_2 = \|X - f(g(XW))\|_F^2$ w.r.t. W can be obtained using Eq. (9), with $i = 0$. The derivation of $\mathcal{L}_3$ w.r.t. W is as follows:

$$\frac{\partial \mathcal{L}_3}{\partial W} = 2a^{(0)} \left(\delta^{(1)}\right)^T + 2\alpha WQ + 2\beta X^T L_S XW \tag{12}$$

Algorithm 1. Algorithm to NRASP

Input:

Data matrix X, first hidden layer size d_1, second hidden layer size d_2, regularization parameter α, β and γ, number to select feature k.

Initialize the parameters of the nonlinear self-representation process.

Repeat

1: Calculate the value of S by Eq. (7).

2. Calculate Laplacian matrix $L_S = D - \left(\frac{S+S^T}{2}\right)$, where the D is a diagonal matrix and the i-th element can be calculated by $\sum_{j=1}^{m} \frac{(s_{ij}+s_{ji})}{2}$.

3: Update the W and Θ by optimizing Eq. (5) with L-BFGS algorithm.

Until converge.

Output:

Calculate scores $\|w_i\|_2$ for each i-th feature and sort them in descending order, then select the top k features.

---

where $Q \in R^{d \times d}$ is a diagonal matrix whose component is calculated by:

$$Q_{ii} = \frac{1}{2\sqrt{w_i^T w_i}}(i = 1, 2, \cdots, d) \tag{13}$$

where w_i denotes the i-th row's transpose of W.

Combining the above results, we can obtain the gradient of the cost function (Eq. (5)) $\nabla \mathcal{L}$ with Eqs. (12), (9) and (10). Subsequently, we used the L-BFGS algorithm to update W and Θ by updating S using Eq. (7) until the objective function converged.

The entire algorithm of our method is shown in Algorithm 1.

4 Experiments

To verify the superiority of the proposed method, eight benchmark datasets from different fields were used, including the image datasets of MNIST, warp-PIE10P, and orlraws10P, the Spoken letter recognition dataset of Isolet, the time-series dataset of HAR, and the biological datasets of lymphoma, TOX_171, GLIOMA.The number of instances, features, classes, and category information for the dataset are presented in Table 2 (included in Appendix B). To evaluate the performance of the FS, we used the selected features of the different methods for K-means clustering. Clustering accuracy (ACC) [20] and normalized mutual information (NMI) [14] were used to evaluate the clustering results.

Due to space constraints, the detailed experimental setup, description of comparative methods, visualization, and ablation study of our method are presented in Appendix B, Appendix C, and Appendix D individually.

Table 1. Clustering performance (ACC% ± std) of 9 FS algorithms on 8 datasets. The bold indicates the best results. (The NMI results are shown in Appendix E).

Dataset	MNIST	lymphoma	Isolet	warpPIE10P	TOX_171	GLIOMA	orlraws10P	HAR
All-Fea	57.96 ± 1.73	75.31 ± 3.47	63.22 ± 2.46	27.76 ± 2.41	43.39 ± 2.49	58.80 ± 7.25	69.60 ± 5.82	59.70 ± 3.22
MCFS	50.66 ± 1.92	76.98 ± 4.27	63.46 ± 1.93	31.81 ± 2.09	44.56 ± 3.34	57.40 ± 3.27	71.90 ± 6.35	60.53 ± 0.82
UDFS	57.84 ± 1.44	74.27 ± 5.38	63.73 ± 2.01	48.38 ± 2.35	47.49 ± 3.61	61.00 ± 5.01	73.60 ± 6.29	58.55 ± 4.36
CAE	58.08 ± 2.92	74.90 ± 5.74	64.17 ± 2.33	27.76 ± 0.64	42.63 ± 2.96	62.80 ± 7.67	76.80 ± 5.31	65.51 ± 3.20
NDFS	60.55 ± 2.10	76.25 ± 5.66	62.70 ± 3.28	30.57 ± 2.14	40.47 ± 5.83	65.00 ± 6.20	73.70 ± 7.02	63.35 ± 4.17
URAFS	62.73 ± 1.55	78.23 ± 4.12	70.88 ± 2.50	40.43 ± 2.24	50.82 ± 1.12	61.00 ± 6.20	76.10 ± 6.19	71.91 ± 2.87
AEFS	60.16 ± 2.20	81.67 ± 2.87	66.91 ± 2.47	28.24 ± 1.29	53.22 ± 0.91	67.20 ± 2.86	77.40 ± 6.92	71.91 ± 5.28
LS	56.97 ± 1.90	72.29 ± 3.93	65.51 ± 2.54	49.57 ± 2.17	44.74 ± 1.33	53.80 ± 3.82	66.50 ± 4.65	62.76 ± 1.14
QS	60.34 ± 2.61	79.06 ± 3.24	67.00 ± 2.90	34.38 ± 3.13	50.82 ± 1.73	68.40 ± 5.48	76.80 ± 4.39	68.77 ± 3.74
Ours	**63.75 ± 4.06**	**83.33 ± 3.29**	**72.04 ± 3.13**	**56.00 ± 4.24**	54.85 ± 4.22	70.40 ± 1.84	80.10 ± 5.69	**73.11 ± 3.90**

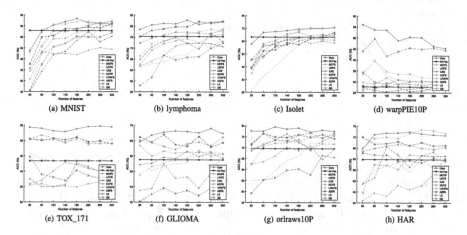

(a) MNIST   (b) lymphoma   (c) Isolet   (d) warpPIE10P

(e) TOX_171   (f) GLIOMA   (g) orlraws10P   (h) HAR

Fig. 1. Clustering accuracy of 9 FS algorithms on 8 datasets with different number of selected feature.

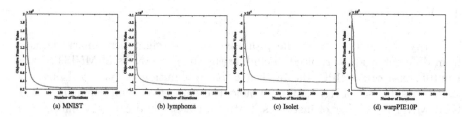

(a) MNIST   (b) lymphoma   (c) Isolet   (d) warpPIE10P

Fig. 2. Convergence curves of our method on 4 different datasets. (Results for an additional 4 datasets are presented in the Appendix E, specifically in Fig. 6)

4.1 Quantitative Analysis and Results

We demonstrate the experimental results in Tables 1 (in Sect. 4) and 5 (in Appendix E) and Figs. 1 (in Sect. 4) and 5 (in Appendix E) of various unsupervised FS methods on the entire datasets. The following conclusions can be drawn in light of the experimental results:

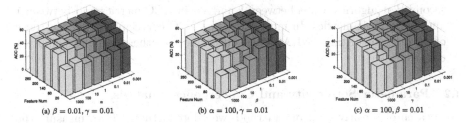

Fig. 3. Clustering ACC with different α, β, γ on MNIST, lymphoma and orlraws10P. The same dataset is indicated in the same row. (a)-(c) are MNIST, (The results for additional datasets are presented in the Appendix E, specifically in Fig. 7). Best view in color.

1. The performance of unsupervised FS methods is superior to that of the baseline method (All-Fea) in general, which shows that FS improves the efficiency of the algorithm and improves its performance. It is necessary for handling noise or redundant information in the original data.
2. Nonlinear self-representation based methods, such as AEFS, QS, and our method, perform better than linear-based methods (MCFS, UDFS, NDFS) in general, especially on biological datasets such as lymphoma, TOX_171, and GLIOMA. Undoubtedly, real-world biological information is extremely complex. This means that linear relations may not be sufficiently effective to describe the relationship between the features [13]. This illustrates that the nonlinear self-representation has not only great effects on dimension reduction and extraction of representative features but also has an excellent ability for nonlinear learning. Therefore, an nonlinear self-representation part is generally effective for feature selection.
3. URAFS can attain suboptimal performance on most datasets, whereas the nonlinear self-representation based method does not perform well in some datasets, such as warpPIE10P. URAFS incorporates adaptive similarity matrix construction into manifold learning, which can capture a more accurate data structure, resulting in an excellent performance. This allows URAFS to achieve better results than traditional spectral analysis methods, including MCFS, UDFS, and NDFS. On this basis, our method also achieves high performance. The above proves that manifold learning with dynamic similarity matrix construction is vital in feature selection.
4. Our proposed method can achieve excellent performance on all datasets, particularly on MNIST, lymphoma, warpPIE10P, TOX_171, GLIOMA, and orlraws10P. The NRASP outperforms the other methods with regard to ACC predominantly. Among the face datasets (warpPIE10P, orlraws10P), NRASP improved ACC by 6.4% on ACC and NMI by 5.7% on NMI over the second-best algorithm LS on warpPIE10P. Moreover, it improved the ACC by 2.7% on ACC over the AEFS and 3.2% on the NMI over the QS on orlraws10P. The ACC curves exceeded those of all other algorithms. For the biological datasets (lymphoma, TOX_171, and GLIOMA), the improvement over the

second-best algorithm was between 2 and 4% in ACC metrics and between 1 and 3% in NMI metrics. In most cases, the ACC curves also exceeded those of the other algorithms. NRASP also outperformed other algorithms on other datasets.

4.2 Parameter Sensitivity and Convergence Analysis

In research of sensitivity of our method, we adopted the ACC to measure the clustering performance under different parameters α, β and γ. The values of each parameter are obtained using a grid search from $\{10^{-3}, 10^{-2}, ..., 10^3\}$. The number of selected features varied between $\{20, 80, 140, 200, 260\}$. The results are shown in Fig. 3. It can be concluded from the figure that the selection of the number of features was significantly more impactful than the selection of parameters. α was not sensitive to the three datasets, whereas β and γ were sensitive to MNIST and orlraws10P, respectively. In addition, we recommend performing a grid search to obtain better results in real applications. The convergence curves of our method on 8 datasets are shown in Fig. 2. The objective function converges at around 100 iterations on MNIST and lymphoma, 30 iterations on warpPIE10P, 50 iterations on Isolet and GLIOMA, and 150 iterations on TOX_171, orlraws10P, and HAR.

5   Conclusion

In this study, we propose a novel nonlinear self-representation based and structure-preserving unsupervised FS method. The above method projects data into a latent lower-dimensional space using an nonlinear mapping, whereas most existing methods use linear transformation. It can identify the correlations among features and tackle with the potential noise in the original data. Meanwhile, the structure preservation constraint was integrated into our method to maintain the local manifold structure of the raw data. Unlike previous spectral analysis methods, we adaptively updated the similarity matrix using the principle of maximum entropy to capture a more accurate structure of the data. Moreover, we designed an efficient algorithm for our method optimization. Experimental results on different datasets demonstrate that our method outperforms nine representative methods.

A   Derivation

A.1   Derivation of the Manifold Structure Preservation Item

Recalling the aforementioned definition of the manifold structure preservation, if the data points are close in the original data space, the projected points $W^T x_i$ and $W^T x_j$ should also have a small distance. Therefore, We can get the manifold structure preservation item as:

$$\min_{W} \frac{1}{2} \sum_{i,j} \left\| W^T x_i - W^T x_j \right\|_2^2 s_{ij} \tag{14}$$

where s_{ij} denotes the similarity between the data x_i and x_j. The value of $\|W^T x_i - W^T x_j\|_2^2$ can be large when the value of s_{ij} is small. Therefore, the neighbor relationship of the original data points can still be maintained in the mapped data points.

We can verify that

$$
\begin{aligned}
\frac{1}{2} \sum_{i,j} \|W^T x_i &- W^T x_j\|_2^2 s_{ij} \\
&= \sum_{i=1}^{n} (W^T x_i)^T W^T x_i D_{ii} - \sum_{i,j=1}^{n} (W^T x_i)^T W^T x_j s_{ij} \\
&= Tr(W^T X^T D X W) - Tr(W^T X^T S X W) \\
&= Tr(W^T X^T L_S X W)
\end{aligned}
\tag{15}
$$

where L_S is a Laplacian matrix. $Tr(\cdot)$ denotes the trace of matrix. L_S is calculated by $L_S = D - (\frac{S+S^T}{2})$, where D is a diagonal matrix, and its elements are defined as:

$$
D_{ii} = \sum_{j=1}^{m} \frac{(s_{ij} + s_{ji})}{2}, i = 1, 2, \cdots, m.
\tag{16}
$$

A.2   Derivation of the KKT Condition

With the Lagrangian multiplier method, Eq. (6) (in main body) is rewritten as:

$$
\begin{aligned}
\mathcal{L}_1' = \frac{1}{2} \beta \sum_{i=1}^{m} \sum_{j=1}^{m} \|W^T x_i &- W^T x_j\|_2^2 s_{ij} + \gamma \sum_{i=1}^{m} \sum_{j=1}^{m} s_{ij} \log s_{ij} \\
&+ \sum_{i=1}^{m} \sum_{j=1}^{m} \lambda_{ij} s_{ij} + \sum_{i=1}^{m} \mu_i \left(\sum_{j=1}^{m} s_{ij} - 1 \right)
\end{aligned}
\tag{17}
$$

where $M = [\mu_1, \mu_2, \ldots, \mu_m]$ and $\Lambda = [\lambda_{ij}]_{m \times m}$ are Lagrangian multipliers. The KKT conditions of Eq. (17) are summarized as

$$
\begin{cases}
\frac{\partial \mathcal{L}_1'}{\partial s_{ij}} = \frac{\beta}{2} \|W^T x_i - W^T x_j\|_2^2 \gamma (\log s_{ij} + 1) + \lambda_{ij} + \mu_i = 0 \\
s_{ij} \geq 0, \quad \lambda_{ij} \geq 0, \quad \lambda_{ij} s_{ij} = 0, \quad \sum_{j=1}^{m} s_{ij} = 1.
\end{cases}
\tag{18}
$$

Based on Eq. (18), we can get the optimal solution of s_{ij} shown in Eq. (7) (in main body).

B   Experiment Setup

For the validation of the effectiveness of our method, we made a comparison with the baseline of all the features and nine representative unsupervised FS methods.

These methods are briefly described as follows.

Table 2. Statistics of used datasets.

dataset	Instances	Features	Classes	Categories
MNIST	3000	784	10	Hand Written Image
lymphoma	96	4026	9	Biological
Isolet	1560	617	26	Spoken letter recognition
warpPIE10P	210	2420	10	Face Image
TOX_171	171	5748	4	Biological
GLIOMA	50	4434	4	Biological
orlraws10P	100	10304	10	Face Image
HAR	10299	561	6	Time Series

1. All-Fea: Use all features for clustering. This method was used as the baseline to verify whether the selected features can outperform all the features in clustering.
2. Laplacian score (LS) [7]: This method measures feature using variance and local structure preservation ability.
3. Multi-cluster feature selection (MCFS) [3]: This method uses l_1-regularized regression model with spectral analysis to select the most important features. It is capable of preserving data's clustering structure.
4. Nonnegative discriminative feature selection (NDFS) [11]: This method utilizes discriminative information of data. It incorporates clustering label learning into FS. In addition, the method used a nonnegative constraint for more accurate cluster labels.
5. Unsupervised discriminative feature selection (UDFS) [17]: This method integrates discriminative analysis with $l_{2,1}$-norm regularization within a unified framework to exploit discriminative information and perform an unsupervised FS problem.
6. Generalized uncorrelated regression with the adaptive graph for unsupervised feature selection (URAFS) [10]: This method selects features and performs manifold learning simultaneously using an uncorrelated regression model while incorporating the data's geometric structure into the manifold learning process.
7. Autoencoder feature selection (AEFS) [6]: This method combines an autoencoder network and group LASSO by excavating the linear and nonlinear information among features to perform FS.
8. Concrete autoencoder (CAE) [2]: This method proposed a concrete autoencoder to implement separable feature selection and reconstruction. CAE uses a concrete selector layer with an effective learning algorithm that can converge to a discrete feature subset.
9. Quick selection (QS) [1]: This method selects features by the strength of the neurons from trained sparse denoising autoencoders with a sparse evolution strategy for training.

It should be noted that the Laplacian score (LS) [7] method does not belong to the embedding-based methods. LS measures features using variance and local structure preservation ability. Due to its efficiency and decent performance, LS remains a popular method for FS. Hence, we include it in our comparative study. Besides, the All-Fea method is a baseline method that includes all features.

To evaluate the performance of various unsupervised methods, we utilized the K-nearest neighbor (KNN) algorithm in LS, MCFS, UDFS, and NDFS, where we set the number of nearest neighbors to five. In addition, the activation functions of the encoder and decoder are set as tanh functions in our method. For the initialization of parameters, we utilized the grid-search strategy from $\{10^{-3}, 10^{-2}, \ldots, 10^3\}$ on parameters to find the optimal parameters in UDFS, URAFS, AEFS, and our method. In the AEFS, we set the size of the neurons in the hidden layer to 256. We selected one hidden layer in CAE, and the activation function was selected as LeakyRelu(0.2). In QS, we utilize the grid-search strategy from $\{0.1, 0.2, 0.3, 0.4, 0.5\}$ for parameter ζ and from $\{2, 5, 10, 13, 20, 25\}$ for parameter ε. In particular, in our proposed method, we also use the grid-search strategy from $\{10^{-3}, 10^{-2}, \ldots, 10^3\}$ for parameters α, β and γ. We selected $k \in \{20, 40, \ldots, 300\}$ features respectively to conduct the experiments.

Considering Eq. (13) (in main body), the W is regularized to be sparse in rows, $\|w_i\|_2$ is most likely zero during training. To avoid this, we add a small positive constant ϵ close enough to infinitesimal, aiming to ensure Q_{ii} to be differentiable. Subsequently, Q is transformed into Q' such that the i-th elements are defined as

$$Q'_{ii} = \frac{1}{2\sqrt{w_i^T w_i + \epsilon}}(i = 1, 2, \cdots, d) \tag{19}$$

Replacing Q with Q', Eq. (12) (in main body) can be written as follows:

$$\frac{\partial \mathcal{L}_3}{\partial W} = 2a^{(0)}\left(\delta^{(1)}\right)^T + 2\alpha W Q' + 2\beta X^T L_S X W \tag{20}$$

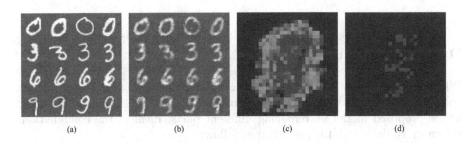

(a)   (b)   (c)   (d)

Fig. 4. Comparison of original and reconstructed images with MNIST and the visualization of the feature importance $\|w_i\|$.

C   Visualization

The training results of the novel nonlinear self-representation are visualized in Fig. 4, where the input sample image is shown in Fig. 4. The reconstructed image is shown in Fig. 4(b). This shows that the nonlinear self-representation model effectively reconstructs the sample by preserving the intrinsic structure and the linear and nonlinear relationships among the original features. Figure 4(c) presents the importance $\|w_i\|$ of each feature i, which is reshaped by the shape of one sample in MNIST, such as that in the top left of Fig. 4(a). The 40 most important features of the MNIST dataset are presented in Fig. 4(d). From Fig. 4(c) and Fig. 4(d), the meaningful features are primarily concentrated in the middle of the image rather than the edges, which is highly consistent with the practical situation. Meanwhile, the region composed of features with high importance is similar to the shape of the number 3. These may imply that the top, bottom, middle, and right parts are the key to distinguishing between different numbers.

Table 3. Clustering performance (ACC% ± std) of different parts combination of our method on 8 datasets. The bold indicates the best results.

Dataset	MNIST	lymphoma	Isolet	warpPIE10P	TOX_171	GLIOMA	orlraws10P	HAR
Base Linear Model	59.40 ± 1.27	72.71 ± 3.39	71.55 ± 2.56	49.00 ± 4.28	42.69 ± 5.29	62.80 ± 3.29	76.40 ± 7.07	68.34 ± 5.02
Nonlinear Self-Representation Part (b)	61.85 ± 1.66	77.92 ± 4.70	69.46 ± 3.33	38.05 ± 2.74	49.94 ± 1.81	69.00 ± 3.30	76.10 ± 5.88	70.37 ± 6.51
b + β	63.54 ± 1.15	82.08 ± 2.94	70.17 ± 2.76	54.24 ± 4.07	56.43 ± 4.25	69.20 ± 4.54	80.00 ± 3.30	**75.43 ± 0.52**
NRASP (b + β + γ)	**63.75 ± 4.06**	**83.33 ± 3.29**	**72.04 ± 3.13**	**56.00 ± 4.24**	54.85 ± 4.22	**70.40 ± 1.84**	**80.10 ± 5.69**	73.11 ± 3.90

Table 4. Clustering performance (NMI% ± std) of different parts combination of our method on 8 datasets. The bold indicates the best results.

Dataset	MNIST	lymphoma	Isolet	warpPIE10P	TOX_171	GLIOMA	orlraws10P	HAR
Base Linear Model	50.51 ± 1.65	60.65 ± 5.18	80.45 ± 1.28	52.11 ± 3.37	18.08 ± 5.77	56.07 ± 2.59	83.68 ± 2.63	65.74 ± 2.56
Nonlinear Self-Representation Part (b)	51.33 ± 1.17	63.28 ± 3.80	78.95 ± 1.39	39.72 ± 3.58	28.11 ± 1.27	55.06 ± 3.90	83.13 ± 3.26	67.28 ± 3.00
b + β	53.08 ± 1.28	67.48 ± 2.36	79.04 ± 1.69	55.34 ± 3.01	33.65 ± 5.86	54.79 ± 6.37	86.47 ± 3.84	**69.93 ± 0.04**
NRASP (b + β + γ)	**53.15 ± 2.41**	**68.68 ± 2.45**	**81.10 ± 1.80**	**57.71 ± 3.28**	**34.87 ± 2.26**	**58.01 ± 7.62**	**87.73 ± 5.32**	69.18 ± 3.04

D   Ablation Study

When it is necessary to understand the function of each part of the model, ablation experiments are commonly performed. We studied the role of each part of the proposed model by removing different parts. Each of the combinations shown in Tables 3 and 4 is presented as follows:

1. Base linear model: The linear self-representation model with sparse constraints is as follows:

$$\mathcal{L} = \|X - XW\|_F^2 + \lambda\|W\|_{2,1} \tag{21}$$

2. Nonlinear self-representation part (b): Basic model of nonlinear mapping based on the method of self-representation and added by sparse constraints. The loss function is

$$\min_{W,\Theta}\quad \mathcal{L}(W;\Theta) = \|X - f(g(XW))\|_F^2 + \alpha\|W\|_{2,1} \tag{22}$$

3. $b + \beta$: Incorporated base nonlinear self-representation model with manifold learning that has a fixed similarity matrix.

$$\min_{W,\Theta}\quad \mathcal{L}(W;\Theta) = \|X - f(g(XW))\|_F^2 + \alpha\|W\|_{2,1} + \beta Tr(W^T X^T L_S XW) \tag{23}$$

4. NRASP $(b + \beta + \gamma)$: The whole proposed model by Eq. (5) (in main body).

As shown in Tables 3 and 4, we first consider the linear self-representation model as the basis on which we introduce nonlinear mapping method adopting the self-representation idea for comparison. It shows comparatively superior performance on most biological datasets, such as lymphoma, TOX_171, and GLIOMA, corroborating the better nonlinear learning ability of the nonlinear self-representation model than the linear model. However, the basic nonlinear self-representation model underperformed on the warpPIE10P image dataset but demonstrated even better performance when combined with manifold learning, which was conducted on all datasets. This indicates that for the information in the data, the nonlinear self-representation model can capture the nonlinear information well, while the structural information of the data may be omitted. Furthermore, when the similarity matrix is dynamically updated for manifold learning, it achieves excellent performance on all but the TOX_171 and HAR datasets, which shows that the structural information in the data is better captured. By comparing the combination of different parts, the NRASP can balance the different parts well and achieve superior performance.

E   Detailed Results

Table 5. Clustering performance (NMI% ± std) of 9 FS algorithms on 8 datasets. The bold indicates the best results.

Dataset	MNIST	lymphoma	Isolet	warpPIE10P	TOX_171	GLIOMA	orlraws10P	HAR
All-Fea	49.81 ± 1.45	61.55 ± 3.71	76.01 ± 1.51	24.40 ± 2.83	15.01 ± 4.04	47.31 ± 11.31	76.57 ± 4.85	59.46 ± 3.46
MCFS	42.07 ± 2.17	59.09 ± 5.16	75.45 ± 0.94	26.95 ± 5.21	16.52 ± 5.70	38.63 ± 4.14	78.80 ± 4.03	62.21 ± 1.47
UDFS	48.86 ± 1.95	56.30 ± 7.61	76.22 ± 1.50	49.78 ± 1.81	18.88 ± 5.30	43.67 ± 6.99	80.46 ± 2.50	51.33 ± 2.82
CAE	46.67 ± 2.34	59.75 ± 5.81	76.66 ± 1.26	25.76 ± 2.08	14.86 ± 4.53	52.71 ± 4.30	81.86 ± 4.58	62.21 ± 1.82
NDFS	49.42 ± 1.59	60.08 ± 6.67	74.66 ± 1.59	28.31 ± 4.37	12.13 ± 5.67	53.17 ± 6.90	80.48 ± 4.03	60.50 ± 3.05
URAFS	52.95 ± 1.33	65.49 ± 4.32	81.00 ± 1.01	41.36 ± 2.76	26.43 ± 4.29	41.63 ± 9.18	81.34 ± 3.17	68.45 ± 2.24
AEFS	50.34 ± 1.91	68.12 ± 2.99	77.63 ± 0.86	26.72 ± 3.15	31.69 ± 2.67	56.64 ± 4.70	84.27 ± 4.28	67.66 ± 2.96
LS	47.28 ± 1.05	54.58 ± 4.79	76.88 ± 1.16	51.99 ± 2.02	16.95 ± 1.05	27.98 ± 6.46	71.33 ± 2.90	65.58 ± 1.15
QS	50.20 ± 2.06	64.16 ± 4.50	77.53 ± 0.71	34.29 ± 3.86	27.22 ± 2.07	54.43 ± 4.07	84.56 ± 4.46	64.41 ± 2.13
Ours	**53.15 ± 2.41**	**68.68 ± 2.45**	**81.10 ± 1.80**	**57.71 ± 3.28**	**34.87 ± 2.26**	**58.01 ± 7.62**	**87.73 ± 5.32**	**69.18 ± 3.04**

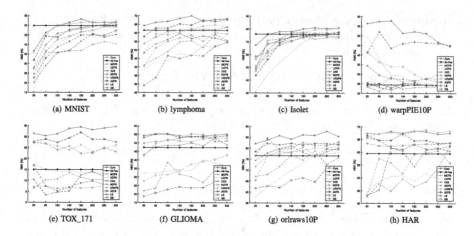

Fig. 5. Clustering NMI of 9 FS algorithms on 8 datasets with different number of selected feature.

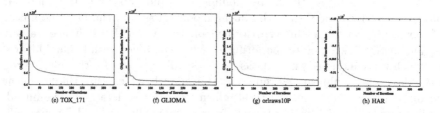

Fig. 6. Convergence curves of our method on 4 additional datasets.

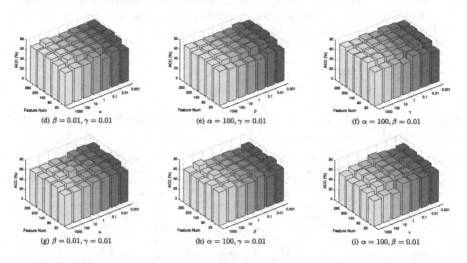

Fig. 7. Additional results of clustering ACC with different α, β, γ on lymphoma and orlraws10P. The same dataset is indicated in the same row. (d)–(f) are lymphoma, (g)–(h) are orlraws10P. Best view in color.

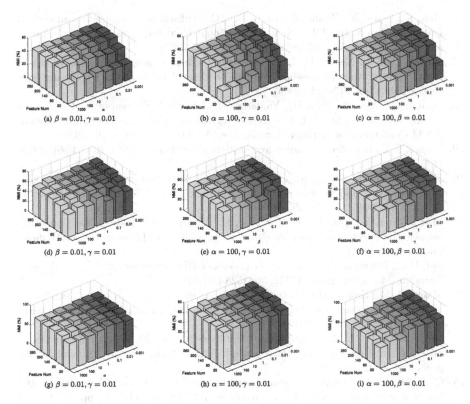

Fig. 8. Clustering NMI with different α, β, γ on MNIST, lymphoma and orlraws10P. The same dataset is indicated in the same row. (a)–(c) are MNIST, (d)–(f) are lymphoma, (g)–(h) are orlraws10P. Best view in color.

References

1. Atashgahi, Z., et al.: Quick and robust feature selection: the strength of energy-efficient sparse training for autoencoders. Mach. Learn. **111**(1), 377–414 (2022)
2. Balın, M.F., Abid, A., Zou, J.: Concrete autoencoders: differentiable feature selection and reconstruction. In: Proceedings of the International Conference on Machine Learning, pp. 444–453 (2019)
3. Cai, D., Zhang, C., He, X.: Unsupervised feature selection for multi-cluster data. In: Proceedings of the 16th ACM SIGKDD International Conference on Knowledge Discovery and Data Mining, pp. 333–342 (2010)
4. Gong, X., Yu, L., Wang, J., Zhang, K., Bai, X., Pal, N.R.: Unsupervised feature selection via adaptive autoencoder with redundancy control. Neural Netw. **150**, 87–101 (2022)
5. Gu, Q., Li, Z., Han, J.: Joint feature selection and subspace learning. In: Proceedings of the International Joint Conference on Artificial Intelligence (IJCAI), pp. 1294–1299 (2011)

6. Han, K., Wang, Y., Zhang, C., Li, C., Xu, C.: Autoencoder inspired unsupervised feature selection. In: 2018 IEEE International Conference on Acoustics, Speech and Signal Processing (ICASSP), pp. 2941–2945 (2018)

7. He, X., Cai, D., Niyogi, P.: Laplacian score for feature selection. In: Advances in Neural Information Processing Systems 18 [Neural Information Processing Systems, NIPS], pp. 507–514 (2005)

8. Huang, Q., Xia, T., Sun, H., Yamada, M., Chang, Y.: Unsupervised nonlinear feature selection from high-dimensional signed networks. In: The Thirty-Fourth AAAI Conference on Artificial Intelligence (AAAI), pp. 4182–4189 (2020)

9. Jaynes, E.T.: Information theory and statistical mechanics. Phys. Rev. **106**(4), 620 (1957)

10. Li, X., Zhang, H., Zhang, R., Liu, Y., Nie, F.: Generalized uncorrelated regression with adaptive graph for unsupervised feature selection. IEEE Trans. Neural Netw. Learn. Syst. **30**(5), 1587–1595 (2019)

11. Li, Z., Yang, Y., Liu, J., Zhou, X., Lu, H.: Unsupervised feature selection using nonnegative spectral analysis. In: Proceedings of the AAAI Conference on Artificial Intelligence (2012)

12. Liu, D.C., Nocedal, J.: On the limited memory BFGS method for large scale optimization. Math. Program. **45**(1), 503–528 (1989)

13. Mahmud, M., Kaiser, M.S., Hussain, A., Vassanelli, S.: Applications of deep learning and reinforcement learning to biological data. IEEE Trans. Neural Netw. Learn. Syst. **29**(6), 2063–2079 (2018)

14. Nie, F., Zhu, W., Li, X.: Unsupervised feature selection with structured graph optimization. In: Proceedings of the Thirtieth AAAI Conference on Artificial Intelligence, pp. 1302–1308 (2016)

15. Qian, M., Zhai, C.: Robust unsupervised feature selection. In: Proceedings of the International Joint Conference on Artificial Intelligence (IJCAI), pp. 1621–1627 (2013)

16. Saberian, M.J., Vasconcelos, N.: Boosting algorithms for simultaneous feature extraction and selection. In: 2012 IEEE Conference on Computer Vision and Pattern Recognition, pp. 2448–2455 (2012)

17. Yang, Y., Shen, H.T., Ma, Z., Huang, Z., Zhou, X.: $l_{2,1}$-norm regularized discriminative feature selection for unsupervised learning. In: Proceedings of the International Joint Conference on Artificial Intelligence (IJCAI), pp. 1589–1594 (2011)

18. You, M., Ban, L., Wang, Y., Kang, J., Wang, G., Yuan, A.: Unsupervised feature selection with joint self-expression and spectral analysis via adaptive graph constraints. Multim. Tools Appl. **82**(4), 5879–5898 (2023)

19. You, M., Yuan, A., He, D., Li, X.: Unsupervised feature selection via neural networks and self-expression with adaptive graph constraint. Pattern Recognit. **135**, 109173 (2023)

20. You, M., Yuan, A., Zou, M., He, D.J., Li, X.: Robust unsupervised feature selection via multi-group adaptive graph representation. In: TKDE, p. 1 (2021)

21. Yuan, A., Huang, J., Wei, C., Zhang, W., Zhang, N., You, M.: Unsupervised feature selection via feature-grouping and orthogonal constraint. In: International Conference on Pattern Recognition (ICPR), pp. 720–726 (2022)

22. Yuan, A., You, M., He, D., Li, X.: Convex non-negative matrix factorization with adaptive graph for unsupervised feature selection. IEEE Trans. Cybern. **52**(6), 5522–5534 (2022)

23. Zhang, Y., et al.: Unsupervised nonnegative adaptive feature extraction for data representation. IEEE Trans. Knowl. Data Eng. **31**(12), 2423–2440 (2019)
24. Zhu, P., Zhu, W., Hu, Q., Zhang, C., Zuo, W.: Subspace clustering guided unsupervised feature selection. Pattern Recogn. **66**(C), 364–374 (2017)

Text Causal Discovery Based on Sequence Structure Information

Yue Li, Donglin Cao, and Dazhen Lin[✉]

Institute of Artificial Intelligence, Xiamen University, Xiamen, Fujian, China
dzlin@xmu.edu.cn

Abstract. Causality forms the basis for reasoning and decision-making in artificial intelligence systems. To take advantage of the vast amount of textual data available today, causal discovery from text has become a significant challenge in recent years. Text data contains rich contextual semantic information. However, traditional causal discovery methods only handle structured data and do not consider serial relationships and semantic relevance between words on textual variables. To address this problem, in this paper, we propose a causal discovery method Text Causal Discovery Based on Sequence Structure Information (TCDSS) discovers strongly correlated text word pairs with semantic relevance and statistical causality and finally constructs lexical causal graphs by introducing sequence-structure information in the causal discovery algorithm. We tested our method TCDSS on the DXY-COVID-19-Data and the Chinese Emergency Corpus (CEC) and compared it with other existing causal discovery methods. The experimental results show that PC, IGCI, RECI, and other forms have improved in precision, recall, and structural Hamming distance (SHD) after the introduction of TCDSS.

Keywords: Causal discovery · Semantic similarity · Syntactic structure

1 Introduction

Causal discovery is an important research direction in natural language processing and artificial intelligence, which can reveal causal connections between events and phenomena and provide critical information for reasoning, prediction, and decision-making. It can be applied to several fields, such as social sciences, medicine, finance, etc. However, accurately capturing and modeling causal relationships in the text is challenging, especially when the semantic complexity of the text is considered.

Traditional causal discovery algorithms are based on statistical methods. The classical algorithms include the PC algorithm based on conditional independent

Supplementary Information The online version contains supplementary material available at https://doi.org/10.1007/978-981-99-8540-1_13.

Q. Liu et al. (Eds.): PRCV 2023, LNCS 14431, pp. 158–169, 2024.
https://doi.org/10.1007/978-981-99-8540-1_13

row constraints [23], the FCI algorithm [20], the GES algorithm based on scoring functions [3], the LiNGAM method based on causal function models [21], the ANM method [10], etc. All these methods are based on structured data, while for unstructured text data, these methods often face some limitations when dealing with text data. For example, although the statistical-based ways can learn a certain degree of semantic information from the data, they do not consider the sequence relationships and semantic relevance between words on textual variables.

Therefore, this paper proposes a causal discovery method based on sequence structure information and semantic correlation. This approach can consider the semantic information between text data in a more examining way. First, our main ideas for this approach are to understand better words' meanings and semantic roles using semantic analysis methods to capture the critical elements of causal relationships accurately. Second, words that may serve as causal variables are added to the initial graph, and by discovering strongly correlated textual word pairs with semantic relevance and statistical causality, constructing lexical causal graphs.

Our main contributions are summarized as follows:

- In response to the problem of not considering sequence relationships when obtaining causal variables, we introduce sequence-structure information in the causal discovery algorithm and add new causal variables based on syntactic structure information.
- To address the problem of not considering semantic relationships in causality, we discover strongly related text word pairs with semantic relevance and statistical causality by using semantic similarity and finally construct lexical causal graphs.

The subsequent sections of this paper are organized as follows: Sect. 2 provides an overview of related work, while Sect. 3 details the proposed causal discovery method that incorporates text semantics. Section 4 presents the experimental data and results, followed by the conclusions in Sect. 5.

2   Related Work

Discovering graphs G from samples of joint distributions P is known as causal structure learning or causal discovery, a fundamental problem in causality [19]. Many approaches have been explored for the task of causal discovery, mainly including three types of causal discovery based on conditional independence constraints, scoring-based causal discovery, and causal function model-based causal discovery.

The causal discovery method based on conditional independence constraints is also called the constraint-based method, which is mainly based on the causal sufficiency assumption and the causal fidelity assumption [7], and first learns the skeleton of the initial causal graph from the observed data, and then uses the conditional independence information to determine the direction of the undirected edges in the skeleton of the causal graph. Some of the classical algorithms are PC algorithm [23], SGS algorithm [22], FCI algorithm [20], RECI algorithm [1], etc. In addition, the invariant causal prediction (ICP) algorithm [9,18] utilizes the causal mechanism to remain invariant under intervention and exclude intervention.

A causal discovery method based on a scoring function searches the space of all possible causal structures intending to optimize a particular metric, based on the assumptions of causal sufficiency and causal fidelity [7], by first setting up a scoring function and searching to find a causal graph structure that maximizes the scoring function. Since the search space of the DAG is hyper-exponential in the number of nodes, many methods rely on greedy search, but the returned graphs are actual equivalence classes [24,25]. For example, GIES [8] repeatedly adds, removes, and flips the orientation of edges in the proposed graph until no higher-scoring diagram is found. The interventional greedy SP (IGSP) algorithm [24] is a hybrid method that uses conditional independence tests in its score function, and the GRaSP algorithm [13] is a method that achieves better results in the case of permutation-based operations in dense graphs and with many variables.

Continuous optimization methods are fraction-based approaches that avoid combinatorial greedy searches for DAGs by using a gradient-based approach. The resulting adjacency matrix is parameterized by representing weights with linear factors or probabilities of having edges between a pair of nodes. This approach's main challenge is restricting the search space to acyclic graphs. A common approach is to consider the search as a constrained optimization problem and deploy an enhanced Lagrangian procedure to solve it, including NOTEARS [26] and DCDI [2]. Alternatively, Ke et al. [12] propose to penalize cyclic graphs using regularization terms while allowing unconstrained optimization. However, the regularizer must be designed and weighted such that the correct, acyclic causal graph is globally optimal for the score function. ENCO [14] is an efficient structural learning method for directed acyclic causal graphs using observation and intervention data.

Causal discovery methods based on causal function models usually consist of sets of random variables and sets of conditions for analyzing multiple variables and describing the causal relationships in a causal graph with a set of structural equations, using the structural equations to qualify the data generation mechanisms between variables and to determine the correct causal direction based on the assumption of mutual independence between the causal and noise variables (equivalently, the causal adequacy assumption). This includes the classical method LiNGAM [21], ANM [10] and IGCI [4].

However, these causal discovery methods are done for structured data and do not consider the serial relationships and semantic correlations between words on textual variables.

3 Methodology

3.1 Problem Statement

Causal discovery aims to infer Structural Causal Model (SCM) from observed data, which models the data generation process. Formally, the fundamental DAG structure learning problem is formulated as follows: we are given a set $D = \{d_1, d_2, ..., d_n\}$, containing a total of n samples, the matrix $X = \{x_1....x_d\} \epsilon R^{n \times d}$ is a matrix consisting of n samples, d variables. A syntactic dependency tree T is obtained from the text data, containing the dependency of each variable on the other variables. Given X and T, we aim to learn a $SCM(P_X, G)$ which encodes causal directed acyclic graphs (DAGs) with a structural equation model (SEM) to reveal the data generated from the distribution of the variable X. Specifically, we denote the DAG by G = (V(G), E(G)), where V(G) is the set of variables and E(G) collects the causal directed edges between the variables. As a result, SEM can be formulated as a structural equation:

$$G = F(f(X) + R(T)) \tag{1}$$

where f is the method we use for causal discovery, R(T) is for syntactic structure analysis, and F denotes the causal discovery performed after adding the syntactic structure.

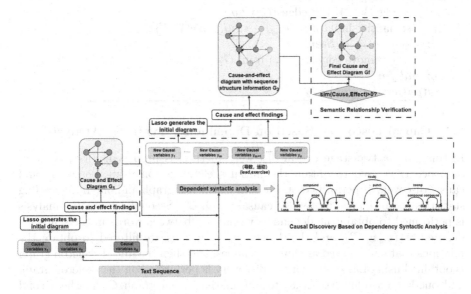

Fig. 1. Structure diagram of text causal discovery model based on sequence structure information

3.2 Overview

To consider text semantic information in the process of causal discovery of text, this part designs a text causal discovery structure based on sequence structure information, as shown in Fig. 1. It mainly contains two parts, one is causal discovery based on sequence structure information, and the other part is to discover strongly related text word pairs with semantic relevance and statistical causality, and finally, to construct lexical causal graphs. Our algorithm is summarized in Algorithm 1.

Algorithm 1. TCDSS Algorithm for DAG Structure Learning

input: A set of text samples $D = \{d_1, d_2, ..., d_n\}$
1: extract the feature matrix $X = \{x_1....x_d\}$
2: extract dependency tree T ← D
3: obtain the initial skeleton graph G_0 ← Glasso(X)
4: *for* each x_i from X *do*
5:     causal discovery:$X_i := f_{(X_i)}(X_{pa}(X_i))$
6:     get the causal graph G_1
7: *end for*
8: filtering to get dependencies R← T
9: extract the new matrix of dependent variables X_{new} ← R
10: feature matrix $\mathbb{X} = concat(X, X_{new})$
11: repeat steps 4-7 to obtain G_2
12: *for* edge(V_i , V_j) in edges(G_2) *do*
13:    semantic similarity:sim_{ij} ← $similarity(V_i, V_j)$
14:    *if* $sim_{ij} \leq \varepsilon$ *then*
15:       remove edge(V_i , V_j) from G_2
16: *end for*
*output:*causal graph G_2

3.3 Causal Discovery Based on Dependency Syntactic Analysis

In this part, a two-stage causal discovery method is proposed. First, the text sequence D is used to obtain the initial skeleton graph using Lasso [6], and causal discovery is performed on the initial skeleton graph and one-hot encoding statistics to obtain the first-stage causal graph G_1. Secondly, syntactic analysis is performed to obtain the dependency relations between words in the sentence, denoted as (v_i, v_j, R), and the dependency relations are filtered to obtain the relations that can be used as causal variables, and then the initial skeleton graph is obtained using Lasso, and causal discovery is performed on the skeleton graph and one-hot encoding statistics to obtain the causal graph G_2 in the second stage. Dependent syntactic analysis is as follows:

In this paper, dependency analysis of the text is applied to determine the degree of dependency and interconnectedness between words. In general, words with causal relationships appear as central components in sentences (e.g., subject, object, etc.), irrelevant modifiers are less likely to be causal variables, and the dependency between the main components of a sentence using syntactic trees improves the quality of causal variable selection. In this paper, sentence constituents are analyzed using Spacy. According to Rui et al. [15], the word-word relationships in the dependency structure of the sentence structure are represented as triads in this paper. The words are encoded using a bidirectional recurrent neural network to obtain the syntactic structure embedding of each word in the dependency tree, and the specific model is shown in Fig. 2. As shown in Fig. 2, take "Long-time and high-intensity exercise can lead to bodily functions dysregulation." as an example. To get the syntactic structure embedding of "intensity", the word embedding vector of its dependent nodes can be stitched together by hiding the state information after the two-way recurrent neural network, which can be expressed as the syntactic structure embedding vector of "intensity". x_i is the word embedding vector, v_i is the hidden state information obtained from the forward recurrent neural network, and u_i is the confidential state information obtained from the reverse recurrent neural network. The final syntactic structure embedding vector is obtained by stitching the two together $(u_0 \| v_5)$.

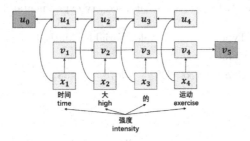

Fig. 2. An example of a syntactic structure embedding model.

The components of the words in the sentences and the semantic relationships with the context are obtained through dependent syntactic analysis. The subject-predicate, indirect object, direct object, and adverbial subordinate clause relations are more likely to be used as causal variables selected for extraction. The extracted components are added to the initial causal graph variables as new ones.

3.4   Semantic Relationship Verification

In this section, we discovered strongly related text word pairs with semantic relevance and statistical causality and finally constructed the lexical causal graph G_f. We perform semantic characterization by using the word2vec [16] and Glove [17] models, respectively, and later verify the causal relationships by calculating the semantic similarity between the candidate causal word pairs. The semantic similarity is calculated by first obtaining the word vectors of cause and effect in the candidate cause-effect word pairs using the textual representation method and calculating the cosine similarity between the word vectors as the semantic similarity. Equation (2) is the calculation of semantic similarity, where V_i and V_j are the word vectors of the two causal variables obtained, respectively.

$$similarity = \frac{V_i \cdot V_j}{||V_i|| \cdot ||V_j||} \qquad (2)$$

4 Experiments

4.1 Experimental Setup

Datasets. In this section, the experiment uses two data, DXY-Data of the New Coronary Pneumonia Epidemic and CEC of the Chinese Emergence Corpus dataset, both of which are Chinese text data used for unstructured text causal discovery studies, and the data details are shown in Table 1. The DXY-Data contains 284 samples, and 1051 variables were processed. The Chinese outbreak corpus dataset CEC includes 332 samples, and 2678 variables were processed.

Table 1. Details of causal discovery experimental data

Data	Type	Sample	Variable	Marked Causal Variable
DXY-Data	Chinese text	284	1051	43
CEC	Chinese text	332	2678	433

Baseline. In this experiment, the baseline used is the following models:

GES [3]: GES is a two-stage greedy search algorithm applied to a specific sparse connected search space, using equivalence classes as states and scoring all operators used in greedy search using local functions of nodes in the domain.

ANM [10]: ANM is an additive noise model that infers causal relationships between nonlinear variables with the help of the mutual independence between the cause and noise variables.

CDS [5]: CDS uses the standard deviation of the standardized conditional distribution as a feature for inferring causality for a given pair of variables, combining developed metrics with standard information theory and statistical metrics to solve causal problems within the framework of Charlene causal pair challenges.

RECI [1]: RECI solves the problem of inferring a causal relationship between two variables by comparing the least squares error of predictions in two possible causal directions under the assumption of independence between the causal function, the conditional noise distribution and the causal distribution.

IGCI [4,11]: IGCI defines independence through orthogonality in the information space, explicitly describing the dependence that occurs between P(Y) and P(X|Y), making the causal hypothesis "Y causes X" implausible.

4.2   Main Results

The experimental results in this chapter are shown in Tables 2 and 3. The best experimental results are shown in bold in the table. The indicators in the experiments are the evaluation of the prediction graphs, and due to the diversity of the graph structure, the indicators Precision, Recall, and SHD in the experiments are only the evaluation of the prediction graphs from different perspectives, and the values of individual indicators do not judge the good or bad results, but express the relative advantages and disadvantages of different methods, and the values of the indicators may be slightly different in different experimental environments, different graph structures or different data.

The experimental results for the DXY-Data dataset and the CEC dataset are shown in Table 2 and Table 3, respectively. The methods in the table are the methods introduced in this part of the baseline. Each method is validated by adding syntactic structure and semantic similarity to each method in this paper. It can be seen from the table that in each method and its improvement methods, the optimal value of each index almost always appears after adding syntactic structure and semantic similarity. However, different methods may achieve the optimal value by adding various semantic similarities. The generally low values of precision and recall in the results are since there are some errors in using these metrics to evaluate the graph structure, so they are only used for relative comparisons between different methods.

The reason why the results of the GES method are worse than other methods on the CEC dataset and the results of this paper are not better after the improvement of the method on GES precisely shows the limitation of the GES method, which is only applicable to the sparse tree graph search of low-dimensional variables, and only fewer edges can be obtained under the CEC dataset containing more than two thousand variables, so there is no way to show the advantage.

Another issue that needs to be clarified is that the DXY-Data data results are at a different level than the CEC data results. CEC's significantly larger SHD value relative to DXY-Data data is because of the target graph nodes and edges. The target graph of DXY-Data data contains only 43 node variables. In comparison, the target graph of CEC data has 433 node variables. SHD is used to assess the gap between the prediction and target graphs, and in this case, the prediction graphs on the CEC data may yield more significant variability. This also shows precisely that the assessment of graph structure in causal discovery tasks is relative and that the indicator values on different experimental settings, graph structures, or data types may be at different levels.

4.3   Ablation Study

To fully demonstrate the effectiveness of the two-stage causal discovery model incorporating syntactic structure and semantic similarity-based validation proposed by the text, ablation experiments are conducted on both datasets in this paper. The results of the ablation experiments on each method in the DXY-Data and CEC datasets are shown in Tables 2 and 3, respectively.

Table 2. Experimental results of different causal discovery methods on DXY-Data dataset

Method	Syntax	Semantics	Precision	Recall	SHD
ANM [10]	-	-	0.1373	0.0409	199
ANM+Syntactic structure	✓	–	0.1702	**0.0468**	188
ANM+Word2Vec	-	✓	0.1333	0.0351	196
ANM+Glove	-	✓	0.1765	0.0351	187
ANM+Syntactic structure+Word2Vec	✓	✓	0.1628	0.0409	192
ANM+Syntactic structure+Glove	✓	✓	**0.2286**	**0.0468**	**184**
CDS [5]	-	-	0.1176	0.0351	200
CDS+Syntactic structure	✓	-	0.16	**0.0468**	199
CDS+Word2Vec	-	✓	0.0889	0.0234	198
CDS+Glove	-	✓	0.1471	0.0292	188
CDS+Syntactic structure+Word2Vec	✓	✓	0.1429	0.0351	195
CDS+Syntactic structure+Glove	✓	✓	**0.2**	0.0409	**188**
RECI [1]	-	-	0.1569	0.0468	198
RECI+Syntactic structure	✓	-	0.1915	**0.0526**	193
RECI+Word2Vec	-	✓	0.1556	0.0409	195
RECI+Glove	-	✓	0.1765	0.0351	187
RECI+Syntactic structure+Word2Vec	✓	✓	0.186	0.0468	191
RECI+Syntactic structure+Glove	✓	✓	**0.2286**	0.0468	**184**
IGCI [4]	-	-	0.1176	0.0351	200
IGCI+Syntactic structure	✓	-	0.1373	**0.0409**	198
IGCI+Word2Vec	-	✓	0.1333	0.0351	196
IGCI+Glove	-	✓	0.0588	0.0117	191
IGCI+Syntactic structure+Word2Vec	✓	✓	**0.1556**	**0.0409**	193
IGCI+Syntactic structure+Glove	✓	✓	0.1111	0.0234	**189**

The experimental results in the table show that the method proposed in this paper performs the best. The best accuracy, recall, and SHD values appear after TCDSS are introduced. They are improved relative to the original method when only syntactic structure and semantic similarity are considered. Still, they are all inferior to the results of fusing syntactic structure and text similarity.

This paper uses two semantic similarity methods for experiments and comparisons. The experiments show that both semantic similarity methods can improve the initial results regardless of which one is used. After adding semantic similarity validation, some redundant edges in the generated causal graph can be removed to improve accuracy, reduce the SHD, and decrease the gap between the prediction graph and the target graph. Glove can give better results than Word2vec in most ways in comparing the two semantic similarity methods.

Table 3. Experimental results of different causal discovery methods on CEC dataset

Method	Syntax	Semantics	Precision	Recall	SHD
GES [3]	-	-	0.1	**0.0017**	599
GES+Syntactic structure	✓	-	0	0	606
GES+Word2Vec	-	✓	0.1111	**0.0017**	599
GES+Glove	-	✓	**0.125**	**0.0017**	**598**
GES+Syntactic structure+word2vec	✓	✓	0	0	604
GES+Syntactic structure+Glove	✓	✓	0	0	602
ANM [10]	-	-	0.0833	0.0285	748
ANM+Syntactic structure	✓	-	0.0942	**0.0352**	753
ANM+Word2Vec	-	✓	0.0936	0.0268	720
ANM+Glove	-	✓	0.0962	0.0251	706
ANM+Syntactic structure+Word2Vec	✓	✓	0.1117	0.0352	720
ANM+Syntactic structure+Glove	✓	✓	**0.1131**	0.0318	**703**
CDS [5]	-	-	0.0718	0.0251	754
CDS+Syntactic structure	✓	-	0.0751	**0.0268**	755
CDS+Word2Vec	-	✓	0.0686	0.0201	727
CDS+Glove	-	✓	0.0875	0.0235	710
CDS+Syntactic structure+Word2Vec	✓	✓	0.082	0.0251	728
CDS+Syntactic structure+Glove	✓	✓	**0.0988**	**0.0268**	**706**
RECI [1]	-	-	0.0766	0.0268	753
RECI+Syntactic structure	✓	-	0.0773	**0.0285**	761
RECI+Word2Vec	-	✓	0.0914	0.0268	723
RECI+Glove	-	✓	0.0812	0.0218	**711**
RECI+Syntactic structure+Word2Vec	✓	✓	0.0895	**0.0285**	734
RECI+Syntactic structure+Glove	✓	✓	**0.093**	0.0268	716
IGCI [4]	-	-	0.067	0.0235	755
IGCI+Syntactic structure	✓	-	0.0773	**0.0285**	761
IGCI+Word2Vec	-	✓	0.0629	0.0184	728
IGCI+Glove	-	✓	0.0875	0.0235	**710**
IGCI+Syntactic structure+Word2Vec	✓	✓	0.0737	0.0235	737
IGCI+Syntactic structure+Glove	✓	✓	**0.0988**	**0.0285**	715

5 Conclusion

In this work, we propose a text causal discovery method based on sequence structure that considers the dependencies between causal variables and the syntactic structure and semantic relationships between them. After introducing our method TCDSS, the original method achieved better text causal discovery results.

Acknowledgements. This work is supported by the National Natural Science Foundation of China (No. 62076210, No. 81973752), the Natural Science Foundation of Xiamen city (No. 3502Z20227188) and the Open Project Program of The Key Laboratory of Cognitive Computing and Intelligent Information Processing of Fujian Education Institutions, Wuyi University(No. KLCCIIP2020203).

References

1. Blöbaum, P., Janzing, D., Washio, T., Shimizu, S., Schölkopf, B.: Cause-effect inference by comparing regression errors. In: International Conference on Artificial Intelligence and Statistics, pp. 900–909. PMLR (2018)
2. Brouillard, P., Lachapelle, S., Lacoste, A., Lacoste-Julien, S., Drouin, A.: Differentiable causal discovery from interventional data. Adv. Neural. Inf. Process. Syst. **33**, 21865–21877 (2020)
3. Chickering, D.M.: Optimal structure identification with greedy search. J. Mach. Learn. Res. **3**, 507–554 (2002)
4. Daniusis, P., et al.: Inferring deterministic causal relations. arXiv preprint arXiv:1203.3475 (2012)
5. Fonollosa, J.A.: Conditional distribution variability measures for causality detection. Cause Effect Pairs in Machine Learning, pp. 339–347 (2019)
6. Friedman, J., Hastie, T., Tibshirani, R.: Sparse inverse covariance estimation with the graphical lasso. Biostatistics **9**(3), 432–441 (2008)
7. Guo, R., Cheng, L., Li, J., Hahn, P.R., Liu, H.: A survey of learning causality with data: problems and methods. ACM Comput. Surv. (CSUR) **53**(4), 1–37 (2020)
8. Hauser, A., Bühlmann, P.: Characterization and greedy learning of interventional markov equivalence classes of directed acyclic graphs. J. Mach. Learn. Res. **13**(1), 2409–2464 (2012)
9. Heinze-Deml, C., Peters, J., Meinshausen, N.: Invariant causal prediction for nonlinear models. J. Causal Inference **6**(2) (2018)
10. Hoyer, P., Janzing, D., Mooij, J.M., Peters, J., Schölkopf, B.: Nonlinear causal discovery with additive noise models. In: Advances in Neural Information Processing Systems 21 (2008)
11. Janzing, D., et al.: Information-geometric approach to inferring causal directions. Artif. Intell. **182**, 1–31 (2012)
12. Ke, N.R., et al.: Learning neural causal models from unknown interventions. arXiv preprint arXiv:1910.01075 (2019)
13. Lam, W.Y., Andrews, B., Ramsey, J.: Greedy relaxations of the sparsest permutation algorithm. In: Uncertainty in Artificial Intelligence, pp. 1052–1062. PMLR (2022)
14. Lippe, P., Cohen, T., Gavves, E.: Efficient neural causal discovery without acyclicity constraints. arXiv preprint arXiv:2107.10483 (2021)
15. Liu, R., Hu, J., Wei, W., Yang, Z., Nyberg, E.: Structural embedding of syntactic trees for machine comprehension. arXiv preprint arXiv:1703.00572 (2017)
16. Mikolov, T., Chen, K., Corrado, G., Dean, J.: Efficient estimation of word representations in vector space. arXiv preprint arXiv:1301.3781 (2013)
17. Pennington, J., Socher, R., Manning, C.D.: Glove: global vectors for word representation. In: Proceedings of the 2014 Conference on Empirical Methods in Natural Language Processing (EMNLP), pp. 1532–1543 (2014)
18. Peters, J., Bühlmann, P., Meinshausen, N.: Causal inference by using invariant prediction: identification and confidence intervals. J. Royal Stat. Soc. Ser. B (Stat. Methodol.), 947–1012 (2016)
19. Peters, J., Janzing, D., Schölkopf, B.: Elements of causal inference: foundations and learning algorithms. The MIT Press (2017)
20. Richardson, T.S.: A discovery algorithm for directed cyclic graphs. arXiv preprint arXiv:1302.3599 (2013)

21. Shimizu, S., Hoyer, P.O., Hyvärinen, A., Kerminen, A., Jordan, M.: A linear non-gaussian acyclic model for causal discovery. J. Mach. Learn. Res. **7**(10) (2006)
22. Spirtes, P., Glymour, C., Scheines, R.: Causality from probability (1989)
23. Spirtes, P., Glymour, C.N., Scheines, R.: Causation, prediction, and search. MIT press (2000)
24. Wang, Y., Solus, L., Yang, K., Uhler, C.: Permutation-based causal inference algorithms with interventions. In: Advances in Neural Information Processing Systems 30 (2017)
25. Yang, K., Katcoff, A., Uhler, C.: Characterizing and learning equivalence classes of causal dags under interventions. In: International Conference on Machine Learning, pp. 5541–5550. PMLR (2018)
26. Zheng, X., Aragam, B., Ravikumar, P.K., Xing, E.P.: Dags with no tears: continuous optimization for structure learning. In: Advances in Neural Information Processing Systems 31 (2018)

MetaSelection: A Learnable Masked AutoEncoder for Multimodal Sentiment Feature Selection

Xuefeng Liang[1,2](✉), Han Chen[2], Huijun Xuan[1], and Ying Zhou[1]

[1] School of Artificial Intelligence, Xidian University, Xi'an, China
xliang@xidian.edu.cn, {xhj,yingzhou}@stu.xidian.edu.cn
[2] Guangzhou Institute of Technology, Xidian University, Guangzhou, China
nopech@stu.xidian.edu.cn

Abstract. Multimodal learning has demonstrated a great advantage in sentimental analysis tasks due to the richer information from different modalities, especially the complementary information. However, our study shows that multimodal data not only provides useful complementary information, but also contains some information that is irrelevant to or conflicts with the task of sentiment prediction. It can degrade the training effectiveness of multimodal models. To tackle this problem, we propose a Learnable Masked AutoEncoder (LMAE) to eliminate the irrelevant or conflicting features of each modality by a learned mask. Afterward, the selected features from modalities are fused by a cross-modal attention. Experiments on samples with conflicting information across modalities and two benchmark datasets, CMU-MOSI and CMU-MOSEI, demonstrate the superiority of our proposal over seven state-of-the-art methods.

Keywords: Multimodal Sentiment Analysis · Modal Feature Selection · Learnable Masked AutoEncoder

1 Introduction

People perceive the world by collaboratively utilizing multiple senses because such multimodal sensing provides more comprehensive information from different aspects [3,13,24]. Recently, multimodal learning significantly improves the performance of Multimodal Sentiment Analysis (MSA) tasks due to the richer information [17,29].

The emerging study [19] stated that different modalities should contain both consistent and complementary sentimental information. And complementary information can considerably boost the performance of multimodal models.

H. Chen and H. Xuan—Contributed equally to this work.
This work was supported in part by the Guangdong Provincial Key Research and Development Programme under Grant 2021B0101410002.

Supplementary Information The online version contains supplementary material available at https://doi.org/10.1007/978-981-99-8540-1_14.

However, our analysis shows some modalities of a sample may also contain task-irrelevant information and even the information conflicting with the given label, which could essentially harm the multimodal learning. Therefore, it will lead to suboptimal results by learning and fusing all features of each modality without selection. To avoid this issue, it is necessary to dynamically evaluate and select only effective features from each modality for fusion and training, meanwhile, excluding irrelevant and conflicting features. Such tasks are defined as Modal Feature Selection tasks (MFS).

Many studies have proposed modal feature selection methods to learn the specific features for each modality. Existing methods can be mainly divided into two categories: 1) *Modality-level selection methods* consider all the features of each modality as a whole to do the evaluation and selection. For example, Mittal et al. [18] proposed M3ER to introduce a check step that used Canonical Correlational Analysis (CCA) [10] to separate ineffective and effective modalities. Sun et al. [21] proposed ICCN, a model which used the outer product of feature pairs along with Deep Canonical Correlation Analysis (DCCA) [1] to study useful multimodal embedding features. Han et al. [6] designed MMIM to maximize the Mutual Information (MI) [22] within the input level and fusion level in order to maintain task-relevant information through multimodal fusion. Yu et al. [26] proposed to jointly train the multimodal and unimodal tasks to learn the consistency and difference between modalities, respectively. However, such methods may ignore the complementary features specific to each modality. 2) *Feature-level selection methods* analyze and select frame-wise features in each modality for a sample, which can be considered as a fine-grained feature selection. Such methods tend to decouple the features into common features and private features between modalities. Inspired by the Domain Separation Network (DSN) [4], Hazarika et al. [7] projected features of each modality to a modality-invariant subspace and a modality-specific subspace, then fused them by Transformer [23]. TAILOR [28] also decoupled modality features into common features and private features, and devised a BERT-like transformer [14] encoder to gradually fuse these features in a granularity descent way. Yang et al. [25] guided the feature decoupling encoders in an adversarial way, and then used cross-modal attention to obtain effective multimodal representations. However, few of above methods consider the irrelevant and conflicting features involved in those private features.

Recently, Masked AutoEncoders (MAE) [8] demonstrated that masking random patches of the input image and reconstructing the missing pixels can preserve effective features, meanwhile, largely eliminate redundancy. Inspired by this work, we propose a feature-level selection method, MetaSelection, to select the effective multimodal features, which are MSA task relevant. Specifically, we design a Learnable Masked AutoEncoder (LMAE) to eliminate the irrelevant or conflicting features of each modality by a mask, in which the mask is dynamically adjusted using the activation values of the modality. To learn the mask, LMAE is jointly optimized by a prediction subtask and a reconstruction task in the training stage. Then, Cross-Modal Attention Fusion is applied to interact the selected features of each modality, thereby achieving a better multimodal feature fusion.

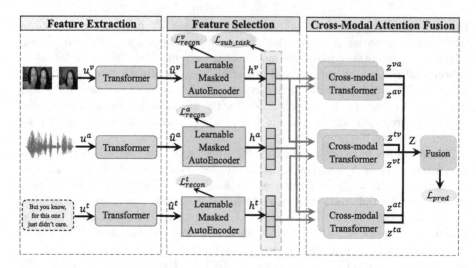

Fig. 1. The framework of MetaSelection, which contains three stages: feature extraction, feature selection and cross-modal attention fusion. First, the unimodal transformer is employed to extract utterance-level unimodal features. The features then are selected by the Learnable Masked AutoEncoder (LAME) to eliminate irrelevant or conflicting information. In the training stage, the output embeddings of LMAE is jointly optimized by a reconstruction loss and a subtask loss, while they are excluded in the test stage. Finally, the selected features are fused by the cross-modal transformer for the sentiment prediction.

The major contributions of this paper are summarized as follows:

(1) We introduce the MetaSelection, including feature extraction stage, feature selection stage, and cross-modal attention fusion stage, for the MSA task.
(2) We propose a novel Learnable Masked AutoEcoder, a precursor to multimodal fusion, which can dynamically select effective sentimental features and eliminate conflicting or irrelevant features.
(3) Experiments on two benchmark datasets CMU-MOSI and CMU-MOSEI demonstrate the effectiveness of our proposed method.

2   Methodology

2.1   Problem Definition

For an utterance u in MFS tasks, the input comprises of three sequences of low-level features from text (t), visual (v) and audio (a) modalities. For a training set with N samples, the input sample is represented as $u = \{u^t, u^a, u^v\}$. For modality $m \in \{t, a, v\}$, $u^m \in \mathbb{R}^{\tau^m \times d^m}$, τ^m represents the sequence length of the sample (i.e., the number of frames), and d^m denotes the feature dimension. Our primary goal is to detect the sentiment orientation of an utterance u from a continuous intensity variable $y \in R$.

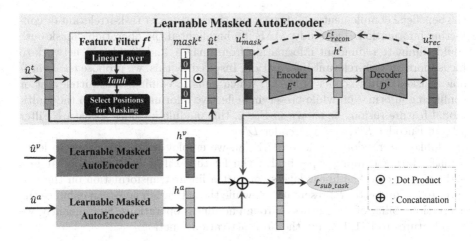

Fig. 2. The feature selection module, which includes a Feature Filter f^m, an Encoder E^m and a Decoder D^m, where $m \in \{t, a, v\}$. The hidden representations h^t, h^a and h^v of three modalities are concatenated for a prediction subtask. The D^m and prediction subtask are only applied during the training stage.

Our proposed MetaSelection model is illustrated in Fig. 1. The key idea is to conduce the multimodal feature selection prior to the feature fusion. Specifically, the Learnable Masked AutoEncoder preserves consistent and complementary information across modalities while eliminating irrelevant or conflicting information. Subsequently, the Cross-Modal Transformer is employed to facilitate information interaction and fusion among these features. Eventually, the fused multimodal features are used for sentiment prediction. The details are given in below.

2.2 Feature Extraction

To capture long-term contextual features from the unimodal data, we utilize a l^m-layer Transformer encoder [23] to enrich the text features, audio features, and visual features, respectively. The Transformer encoder consists of two sub-layers: multi-head self-attention layer and a position-wise feed-forward layer. And the residual connections [9] are adopted, following by a layer normalization. Then the refined embeddings of each modality are obtained and denoted as

$$\hat{u}^m = Transformer(u^m, \theta_t^m) \in \mathbb{R}^{\tau^m \times d}, \tag{1}$$

where θ_t^m are learnable parameters of Transformer, d represents the feature dimension.

2.3 Learnable Masked AutoEncoder for Feature Selection

In the MSA task, the modalities of a sample may contain both consistent and disparate sentiment information. Within the disparate information, apart from the beneficial complementary information, there may exist task-irrelevant or conflicting information. The study, MAE [8], highlighted the operation "mask out" could eliminate redundant information from images, and enable the network to focus more on task-relevant information. Inspired by this, we propose the Learnable Masked AutoEncoder (LMAE), which aims to eliminate task-irrelevant or conflicting information while preserving effective information prior to the multimodal feature fusion. As shown in Fig. 2, this module includes a Feature Filter f^m, an Encoder E^m, and a Decoder D^m.

Unlike the random mask in MAE [8], we employ a Feature Filter to learn a mask during training. Specifically, the Feature Filter consists of linear and activation layers. The linear layers conduct a linear transformation on the features to obtain better representations, while the activation layer can enhance the expression ability of the features through nonlinear operations. Subsequently, we map features to $[-1, 1]$ using the $tanh$ activation function:

$$f^m(\hat{u}^m, \theta_f^m) = tanh(fc(\hat{u}^m)) = [[r_{11}^m, \ldots, r_{1d}^m], \ldots, [r_{\tau^m 1}^m, \ldots, r_{\tau^m d}^m]], \quad (2)$$

where $fc(\cdot)$ denotes the linear layer and θ_f^m is a set of parameters of $fc(\cdot)$.

To determine which bits of the feature vector are effective for the task, we set a threshold δ^m for f^m so that each element in the vector can be marked as **1** or **0**. Finally, we have a learned binary mask vector $mask^m \in \mathbb{R}^{\tau^m \times d}$:

$$mask^m = \begin{cases} 1, & r_{ij}^m > \delta^m \\ 0, & r_{ij}^m \le \delta^m \end{cases}, i \in [1, \tau^m], \; j \in [1, d]. \quad (3)$$

Then, the filtered feature u_{mask}^m can be obtained by dot product of input feature $\hat{u}^m$ and $mask^m$,

$$u_{mask}^m = \hat{u}^m \circ mask^m \in \mathbb{R}^{\tau^m \times d}. \quad (4)$$

The study [8] also reports that the mask ratio σ^m, which is the proportion of zero elements in the mask, is related to feature learning capability. A high mask ratio may improve feature learning capability, whereas a low mask ratio may not help feature learning. Therefore, in training stage we define a mask ratio threshold σ_{mask}^m for each modality.

If $\sigma^m < \sigma_{mask}^m$, an additional random masking will be applied on u_{mask}^m with a mask ratio $(\sigma_{mask}^m - \sigma^m)$.

Subsequently, to integrate the effective feature bites in u_{mask}^m, we feed it into the encoder $E^m(\cdot)$ to obtain the hidden representation $h^m \in \mathbb{R}^{\tau^t \times d'}$,

$$h^m = E^m(u_{mask}^m). \quad (5)$$

To ensure the effectiveness of h^m, we design two constraints. The first one is to reconstruct u_{mask}^m by a decoder $D^m(\cdot)$, which ensures h^m to capture details of each modal features. The reconstruction loss is the *cosine similarity loss*,

$$\mathcal{L}_{recon}^m = 1 - \cos\left(D^m(h^m), u_{mask}^m\right). \quad (6)$$

The second constraint is applied in conjunction with the $mask^m$ learning. We design a sentiment prediction subtask for the filtered features. Specifically, the hidden representations h^t, h^a and h^v of the three modalities are concatenated and fed into the sentiment prediction network P, as shown in Fig. 2. The subtask is constrained by the *mean squared error loss (MSE)*,

$$\mathcal{L}_{sub\_task} = \frac{1}{N}\sum_{i=1}^{N} \| y_i - P([h_i^t, h_i^a, h_i^v]) \|_2^2, \tag{7}$$

where y_i is the ground truth, N is the number of samples and $\| \cdot \|_2^2$ is the squared L_2-norm.

2.4   Cross-modal Attention Fusion Module

Due to the significant "heterogeneity gap" [12] among different modal representations, it is necessary to unify the multimodal representations to enable the model to fully capture the consistent and complementary features between modalities. To this end, we employ cross-modal transformer to minimize the "heterogeneity gap" between modalities, while leveraging multi-head attention to effectively enhance each feature representation.

Specifically, in the $c \in \{1, ..., l^c\}$ attention head, any two modal representations h^α and h^β are embedded to obtain $Q_c^\alpha = h^\alpha W_c^{Q^\alpha}$, $K_c^\beta = h^\beta W_c^{K^\beta}$, $V_c^\beta = h^\beta W_c^{V^\beta}$, where $\alpha, \beta \in \{t, a, v\}, \alpha \neq \beta$, and $\{W_c^{Q^\alpha}, W_c^{K^\beta}, W_c^{V^\beta}\} \in \mathbb{R}^{d \times d'}$ are head-specific weights to linearly project the matrices into local spaces. The cross-modal interaction is defined as

$$head_c^{\beta \to \alpha} = softmax(\frac{Q_c^\alpha (K_c^\beta)^{\mathrm{T}}}{\sqrt{d'}})V_c^\beta. \tag{8}$$

Following the prior work [23], we also apply the residual connections and position-wise feed-forward sublayers to obtain the feature $Y^{\beta\alpha}$, which is calculated after cross-modal attention. Subsequently, the features are fused at the frame level, which is formulized as

$$z^{\beta\alpha} = (softmax(Y^{\beta\alpha}W^{\beta\alpha} + b^{\beta\alpha}))^{\mathrm{T}}Y^{\beta\alpha}, \tag{9}$$

$$Z = [z^{ta}, z^{at}, z^{tv}, z^{vt}, z^{av}, z^{va}], \tag{10}$$

where $W^{\beta\alpha} \in \mathbb{R}^{d' \times 1}$ is a weight for fusion, $z^{\beta\alpha} \in \{z^{ta}, z^{at}, z^{tv}, z^{vt}, z^{av}, z^{va}\} \in \mathbb{R}^{1 \times d'}$ is the utterance-level feature, and $Z \in \mathbb{R}^{6 \times d'}$ represents the concatenated feature that is used for sentiment prediction.

To optimize the model to do the MSA task, we use the *mean squared error loss (MSE)*,

$$\mathcal{L}_{pred} = \frac{1}{N}\sum_{i=1}^{N} \| y_i - \hat{y}_i \|_2^2, \tag{11}$$

where y_i is the ground truth and $\hat{y}_i$ is prediction.

Finally, the overall loss $\mathcal{L}$ is

$$\mathcal{L} = \mathcal{L}_{pred} + \lambda_1 \mathcal{L}_{recon} + \lambda_2 \mathcal{L}_{sub\_task}, \tag{12}$$

where λ_1 and λ_2 are hyperparameters that control the contributions of reconstruction loss and subtask loss.

3 Experiment

3.1 Datasets

We conduct experiments on two benchmark datasets for the MSA task, including CMU-MOSI [27] and CMU-MOSEI [2], which provide word-unaligned multimodal signals (text, visual and audio) for each utterance. The detailed statistics of two datasets are listed in Table 1.

Table 1. The statistics of the CMU-MOSI and CMU-MOSEI datasets.

Dataset	Training Set	Valid Set	Test Set	All
CMU-MOSI	1284	229	686	2199
CMU-MOSEI	16326	1871	4659	22856

CMU-MOSI: The CMU-MOSI dataset contains 2199 utterance video segments sliced from 93 videos in which 89 distinct narrators are sharing opinions on interesting topics. Each segment is manually annotated with a sentiment value ranged from -3 to $+3$, indicating the polarity (by positive/negative) and relative strength (by absolute value) of expressed sentiment.

CMU-MOSEI: The CMU-MOSEI dataset upgrades CMU-MOSI by expanding the size of the dataset. It contains 22856 annotated video segments (utterances), from 5000 videos, 1000 distinct speakers and 250 different topics.

3.2 Evaluation Metrics

Following the previous works [6], evaluations are conducted in two forms: classification and regression. For classification, we report binary classification accuracy (Acc-2), Weighted F1 score (F1-Score) and seven-class classification accuracy (Acc-7). Specifically, for CMU-MOSI and CMU-MOSEI datasets, we calculate Acc-2 and F1-Score in two ways: negative/non-negative (include zero) and negative/positive (exclude zero). For regression, we report Mean Absolute Error (MAE) and Pearson Correlation (Corr). Except for MAE, higher values denote better performance for all metrics.

Table 2. Performance comparison between MetaSelection without and with Learnable Masked AutoEncoder on three subsets.

Test Subset	Baseline					MetaSelection (ours)				
	MAE($\downarrow$)	Corr($\uparrow$)	Acc-2($\uparrow$)	F1-Score($\uparrow$)	Acc-7($\uparrow$)	MAE($\downarrow$)	Corr($\uparrow$)	Acc-2($\uparrow$)	F1-Score($\uparrow$)	Acc-7($\uparrow$)
One Agree	0.776	0.499	77.49	78.71	46.19	**0.683**	**0.562**	**84.20**	**85.38**	**52.54**
Two Agree	0.826	0.626	63.14	62.08	32.86	**0.745**	**0.708**	**71.56**	**71.39**	**40.92**
All Agree	0.601	0.823	93.65	94.08	49.22	**0.573**	**0.839**	**95.09**	**96.52**	**52.10**

3.3 Experimental Settings

The visual and audio features of the above two datasets come from UniMSE [11]. The BERT-base-uncased pre-trained model [14] is employed as the feature extractor for textual utterances. For each sample, the feature dimensions of visual (d^v), audio (d^a) and text (d^t) are 64, 64 and 768, respectively. All parameters in MetaSelection are optimized by Adam [15] with an initial learning rate of $1e-5$. And the hyperparameters are set as : $l^v = l^a = 1$, $l^t = 12$, $d = 256$, $\delta^a = \delta^v = 0.1$, $\delta^t = 0$, $\sigma^v_{mask} = \sigma^a_{mask} = 0.8$, $\sigma^t_{mask} = 0.3$, $d^{'} = 64$, $\lambda_1 = 0.1$ and $\lambda_2 = 0.05$. For unimodal transformers, the head numbers are all 8. For cross-modal transformer, the layer number is 2 and the head number is 4. All methods are trained and tested with one RTX 3090 GPU.

3.4 Evaluation on Samples with Conflicting Information

MetaSelection is proposed to address the issue of task-irrelevant or conflicting information across modalities of a sample. It is worth evaluating MetaSelection on such data. However, few of the existing datasets, including CMU-MOSI and CMU-MOSEI, provides the unimodal label for samples. Therefore, we select samples with conflicting information among modalities from the test set in MOSI. The selection is done by a recent SOTA method, Self-MM [26], which can robustly generate the unimodal label for each modality of a sample. The selected samples are grouped into three subsets: (a) *One Agree*: The label of only one modality is consistent with the ground truth (236 samples); (b) *Two Agree*: The labels of the two modalities are consistent with the ground truth (70 samples); (c) *All Agree*: The labels of all modalities are consistent with the ground truth (309 samples).

We treat the MetaSelection without Learnable Masked Autoencoder (LMAE) as the baseline, and compare it with MetaSelection on the three subsets. The results are listed in Table 2. One can see that LMAE does improve the model on all three subsets. Particularly, on the *One Agree* and *Two Agree* subsets, LMAE brings significant improvements, lowering MAE about 0.081–0.093, raising Corr about 0.063–0.082 and F1-Score about 6.57%–9.31%. On the other hand, on the *All Agree* subset, LMAE also improves the baseline on MAE, Corr and F1-Score by 0.028, 0.016 and 2.44%, respectively. It indicates two points: (1) LMAE significantly improves the model on samples with conflicting information among modalities; (2) LMAE can also refine the effective features from samples whose information of all modalities are consistent.

Table 3. Performance comparison between MetaSelection and seven SOTA methods on CMU-MOSI. In Acc-2 and F1-Score, the left of the "/" is calculated as "negative/non-negative" and the right is calculated as "negative/positive". The best results are in bold and the second-best results are underlined. $\Delta_{second-best}$ indicates the performance gain between the best and the second best. Results with † are from the original papers. Results with * are re-implemented using our features and settings for a fair comparison.

Model	CMU-MOSI				
	MAE($\downarrow$)	Corr($\uparrow$)	Acc-2($\uparrow$)	F1-Score($\uparrow$)	Acc-7($\uparrow$)
ICCN (2020) [21]†	0.862	0.714	–/83.07	–/83.02	39.01
MAG-BERT (2020) [20]†	0.712	0.796	84.2/86.1	84.1/86.0	–
MISA (2020) [7]†	0.783	0.761	81.8/83.4	81.7/83.6	42.3
BBFN (2021) [5]†	0.776	0.755	–/84.3	–/84.3	45.0
Self-MM (2021) [26]†	0.713	0.798	84.00/85.98	84.42/85.95	–
MMIM (2021) [6]†	0.700	0.800	84.14/86.06	84.00/85.98	46.65
DMD (2023) [16]†	–	–	–/83.5	–/83.5	41.9
MAG-BERT (2020) [20]*	0.751	0.777	82.34/84.12	82.27/84.11	41.90
MISA (2020) [7]*	0.749	0.785	81.20/82.93	81.19/82.98	43.73
Self-MM (2021) [26]*	<u>0.720</u>	<u>0.791</u>	83.18/85.00	83.09/84.97	46.21
MMIM (2021) [6]*	<u>0.720</u>	<u>0.791</u>	<u>83.24/ 85.06</u>	<u>83.18/ 85.06</u>	46.29
DMD (2023) [16]*	0.727	0.788	82.80/84.30	82.79/84.34	<u>46.36</u>
MetaSelection(ours)	**0.710**	**0.796**	**84.55/85.82**	**84.53/85.85**	**48.25**
$\Delta_{second-best}$	$\downarrow$0.010	$\uparrow$ 0.005	$\uparrow$ 1.31/$\uparrow$ 0.76	$\uparrow$1.35/$\uparrow$0.79	$\uparrow$1.89

3.5   Comparison with SOTA Methods

To demonstrate the effectiveness of MetaSelection on the MSA tasks, we compare it with seven SOTA methods, including modality-level selection methods (MMIM [6], Self-MM [26], ICCN [21]), feature-level selection methods (MISA [7], DMD [16]) and others (BBFN [5], MAG-BERT [20]). The comparison results are listed in Table 3 (CMU-MOSI) and Table 4 (CMU-MOSEI). We can see MetaSelection outperforms all competing methods on both datasets across all metrics (regression and classification combined). On the small-scale dataset, CMU-MOSI, modality-level methods outperform feature-level methods overall. The possible reason is that the conflicting and irrelevant information within modalities can be considered as noise. As is well-known, noise significantly affects model training on small datasets. Although feature-level methods decouple modal features into common and private spaces, they still inevitably learn the noise. In contrast, modality-level methods learn the entire modality containing noise by a low weight, which can reduce the impact of noise but also learn less effective information in the modality. MetaSelection not only eliminates conflicting features from modalities, but also preserves effective features, leading to an improved performance. CMU-MOSEI, being ten times larger than CMU-MOSI, contains more samples with consistent information across modalities. It pro-

Table 4. Performance comparison between MetaSelection and seven SOTA methods on CMU-MOSEI. The meanings of symbols † and * are the same as in Table 3.

Model	CMU-MOSEI				
	MAE($\downarrow$)	Corr($\uparrow$)	Acc-2($\uparrow$)	F1-Score($\uparrow$)	Acc-7($\uparrow$)
ICCN (2020) [21]†	0.565	0.713	–/84.18	–/84.15	51.58
MAG-BERT (2020) [20]†	–	–	84.7/–	84.5/–	–
MISA (2020) [7]†	0.555	0.756	83.6/85.5	83.8/85.3	52.2
BBFN (2021) [5]†	0.529	0.767	–/86.2	–/86.1	54.8
Self-MM (2021) [26]†	0.530	0.765	82.81/85.17	82.53/85.30	–
MMIM (2021) [6]†	0.526	0.772	82.24/85.97	82.66/85.94	54.24
DMD (2023) [16]†	–	–	–/84.8	–/84.7	54.6
MAG-BERT (2020) [20]*	0.549	0.756	<u>81.20</u>/ <u>85.14</u>	<u>81.71</u>/<u>85.11</u>	52.06
MISA (2020) [7]*	0.549	0.758	80.30/84.84	80.96/84.90	52.80
Self-MM (2021) [26]*	<u>0.535</u>	0.762	79.88/84.07	80.45/84.05	<u>53.42</u>
MMIM (2021) [6]*	0.557	0.739	79.82/84.32	80.48/84.36	51.29
DMD (2023) [16]*	0.540	<u>0.766</u>	80.83/85.03	81.41/85.04	52.35
MetaSelection(ours)	**0.528**	**0.769**	**82.66/85.79**	**82.83/85.70**	**54.02**
$\Delta_{second-best}$	$\downarrow$0.007	$\uparrow$ 0.003	$\uparrow$1.46/$\uparrow$0.65	$\uparrow$1.12/$\uparrow$0.59	$\uparrow$0.06

vides sufficient training data for all methods and enhance their generalization ability. Therefore, the performances of modality-level and feature-level methods are comparable. Our MetaSelection method still achieves the best performance on CMU-MOSEI. Although the improvement is not as pronounced as on CMU-MOSI, it demonstrates the robustness of MetaSelection on small-scale datasets.

3.6   Ablation Studies

We conducted ablation studies on the CMU-MOSI dataset to analyze the effectiveness of LMAE and Cross-Modal Attention module and four key components in LMAE. Table 5 shows all details. Firstly, one can see LMAE significantly boosts the performance of model since the feature selection can eliminate certain irrelevant or conflicting features. Meanwhile, Cross-Modal Attention Fusion contributes to a considerable performance gain because of feature fusion by modality interaction. Secondly, Feature Filter plays a crucial role in LMAE. Compared with the Random Mask used in [8], the learned Mask done by Feature Filter largely improves the performance across all metrics again. The reason could be that the Mask is learned to preserve effective features and eliminate irrelevant or conflicting features. Moreover, the subtask of sentiment prediction provides a supervision to the LMAE training, which also helps model. Finally, the threshold σ_{mask}^m ensures a sufficient mask ratio for model training, and further improve the robustness of model.

Table 5. Evaluations of LMAE and Cross-Modal Attention module and four key components in LMAE on the test set of CMU-MOSI.

Random Mask	Learnable Masked AutoEncoder				Cross-Modal Attention Fusion	MAE($\downarrow$)	Corr($\uparrow$)	Acc-2($\uparrow$)	F1-Score($\uparrow$)	Acc-7($\uparrow$)
	Feature Filter	threshold σ_{mask}^m	Auto-Encoder	subtask						
✗	✗	✗	✗	✗	✗	0.948	0.683	82.03/84.68	81.89/84.55	33.76
✗	✓	✓	✓	✓	✗	0.800	0.779	83.38/85.37	83.21/85.26	41.98
✗	✗	✗	✗	✗	✓	0.749	0.784	83.38/85.21	83.32/85.21	43.59
✗	✗	✗	✓	✗	✓	0.747	0.784	82.51/84.60	82.45/84.61	44.46
✓	✗	✗	✓	✗	✓	0.736	0.791	82.94/85.37	82.76/85.27	45.04
✗	✓	✗	✓	✗	✓	0.714	0.794	83.67/85.51	83.57/85.48	47.37
✗	✓	✗	✓	✓	✓	0.711	0.796	83.82/85.82	83.71/85.77	47.67
✗	✓	✓	✓	✓	✓	**0.710**	**0.796**	**84.55/85.82**	**84.53/85.85**	**48.25**

4   Conclusion

In this paper, we introduce a feature-level multimodal selection method, MetaSelection, for the MSA tasks. Unlike other feature-level selection methods, which decouple the features into common features and private features among modalities without considering the irrelevant and conflicting features involved in those private features, the LMAE in our MetaSelection can eliminate the irrelevant or conflicting features of each modality by a learned mask. To learn the mask, LMAE is jointly optimized by a prediction subtask and a reconstruction task during training. Afterward, the filtered features of modalities are fused by the cross-modal attention. Evaluations on samples with conflicting information across modalities and two benchmark datasets, CMU-MOSI and CMU-MOSEI, demonstrate the effectiveness of modal feature selection of our MetaSelection. In the future work, we will investigate more fine-grained feature relationships among modalities.

References

1. Andrew, G., Arora, R., Bilmes, J., Livescu, K.: Deep canonical correlation analysis. In: International Conference on Machine Learning, pp. 1247–1255 (2013)
2. Bagher Zadeh, A., Liang, P.P., Poria, S., Cambria, E., Morency, L.P.: Multimodal language analysis in the wild: CMU-MOSEI dataset and interpretable dynamic fusion graph. In: Annual Meeting of the Association for Computational Linguistics, vol. 1, pp. 2236–2246 (2018)
3. Bayoudh, K., Knani, R., Hamdaoui, F., Mtibaa, A.: A survey on deep multimodal learning for computer vision: advances, trends, applications, and datasets. Vis. Comput. **38**, 1–32 (2021)
4. Bousmalis, K., Trigeorgis, G., Silberman, N., Krishnan, D., Erhan, D.: Domain separation networks. Adv. Neural Inf. Process. Syst. **29** (2016)
5. Han, W., Chen, H., Gelbukh, A., Zadeh, A., Morency, L.P., Poria, S.: Bi-bimodal modality fusion for correlation-controlled multimodal sentiment analysis. In: International Conference on Multimodal Interaction, pp. 6–15 (2021)

6. Han, W., Chen, H., Poria, S.: Improving multimodal fusion with hierarchical mutual information maximization for multimodal sentiment analysis. In: Conference on Empirical Methods in Natural Language Processing, pp. 9180–9192 (2021)
7. Hazarika, D., Zimmermann, R., Poria, S.: MISA: modality-invariant and -specific representations for multimodal sentiment analysis. In: ACM International Conference on Multimedia, pp. 1122–1131 (2020)
8. He, K., Chen, X., Xie, S., Li, Y., Dollár, P., Girshick, R.: Masked autoencoders are scalable vision learners. In: IEEE/CVF Conference on Computer Vision and Pattern Recognition, pp. 16000–16009 (2022)
9. He, K., Zhang, X., Ren, S., Sun, J.: Deep residual learning for image recognition. In: IEEE Conference on Computer Vision and Pattern Recognition, pp. 770–778 (2016)
10. Hotelling, H.: Relations between two sets of variates. In: Breakthroughs in Statistics: Methodology and Distribution, pp. 162–190 (1992)
11. Hu, G., Lin, T.E., Zhao, Y., Lu, G., Wu, Y., Li, Y.: UniMSE: towards unified multimodal sentiment analysis and emotion recognition. In: Conference on Empirical Methods in Natural Language Processing, pp. 7837–7851 (2022)
12. Huang, X., Peng, Y.: Deep cross-media knowledge transfer. In: IEEE Conference on Computer Vision and Pattern Recognition, pp. 8837–8846 (2018)
13. Huang, Y., Du, C., Xue, Z., Chen, X., Zhao, H., Huang, L.: What makes multimodal learning better than single (provably). Adv. Neural. Inf. Process. Syst. 34, 10944–10956 (2021)
14. Kenton, J.D.M.W.C., Toutanova, L.K.: BERT: pre-training of deep bidirectional transformers for language understanding. In: North American Chapter of the Association for Computational Linguistics, vol. 1, pp. 4171–4186 (2019)
15. Kingma, D.P., Ba, J.: Adam: a method for stochastic optimization. In: International Conference on Learning Representations, pp. 1–8 (2015)
16. Li, Y., Wang, Y., Cui, Z.: Decoupled multimodal distilling for emotion recognition. In: IEEE/CVF Conference on Computer Vision and Pattern Recognition, pp. 6631–6640 (2023)
17. Majumder, N., Poria, S., Hazarika, D., Mihalcea, R., Gelbukh, A., Cambria, E.: DialogueRNN: an attentive RNN for emotion detection in conversations. In: AAAI Conference on Artificial Intelligence, vol. 33, pp. 6818–6825 (2019)
18. Mittal, T., Bhattacharya, U., Chandra, R., Bera, A., Manocha, D.: M3ER: multiplicative multimodal emotion recognition using facial, textual, and speech cues. In: AAAI Conference on Artificial Intelligence, vol. 34, pp. 1359–1367 (2020)
19. Poria, S., Hazarika, D., Majumder, N., Mihalcea, R.: Beneath the tip of the iceberg: current challenges and new directions in sentiment analysis research. IEEE Trans. Affect. Comput. (2020)
20. Rahman, W., et al.: Integrating multimodal information in large pretrained transformers. In: Annual Meeting of the Association for Computational Linguistics, pp. 2359–2369 (2020)
21. Sun, Z., Sarma, P., Sethares, W., Liang, Y.: Learning relationships between text, audio, and video via deep canonical correlation for multimodal language analysis. In: AAAI Conference on Artificial Intelligence, vol. 34, pp. 8992–8999 (2020)
22. Tishby, N., Zaslavsky, N.: Deep learning and the information bottleneck principle. In: IEEE Information Theory Workshop, pp. 1–5 (2015)
23. Vaswani, A., et al.: Attention is all you need. Adv. Neural Inf. Process. Syst. 30 (2017)
24. Wang, Y., et al.: A systematic review on affective computing: emotion models, databases, and recent advances. Inf. Fusion 83–84, 19–52 (2022)

25. Yang, D., Huang, S., Kuang, H., Du, Y., Zhang, L.: Disentangled representation learning for multimodal emotion recognition. In: ACM International Conference on Multimedia, pp. 1642–1651 (2022)
26. Yu, W., Xu, H., Yuan, Z., Wu, J.: Learning modality-specific representations with self-supervised multi-task learning for multimodal sentiment analysis. In: AAAI Conference on Artificial Intelligence, vol. 35, pp. 10790–10797 (2021)
27. Zadeh, A., Zellers, R., Pincus, E., Morency, L.P.: MOSI: multimodal corpus of sentiment intensity and subjectivity analysis in online opinion videos. arXiv preprint arXiv:1606.06259 (2016)
28. Zhang, Y., Chen, M., Shen, J., Wang, C.: Tailor versatile multi-modal learning for multi-label emotion recognition. In: AAAI Conference on Artificial Intelligence. vol. 36, pp. 9100–9108 (2022)
29. Zhou, Y., Liang, X., Zheng, S., Xuan, H., Kumada, T.: Adaptive mask co-optimization for modal dependence in multimodal learning. In: IEEE International Conference on Acoustics, Speech and Signal Processing, pp. 1–5. IEEE (2023)

Image Manipulation Localization Based on Multiscale Convolutional Attention

Runjie Liu[1], Guo Li[1], Wenchao Cui[1], Yirong Wu[2], Jian Zhang[3], and Shuifa Sun[3(✉)]

[1] School of Computer and Information Technology, China Three Corges University, Yichang, Hubei 443002, China
[2] Institute of Advanced Studies in Humanities and Social Sciences, Beijing Normal University, Beijing 100089, China
[3] School of Information Science and Technology, Hangzhou Normal University, Hangzhou 310036, China
watersun@hznu.edu.cn

Abstract. Current developments in image editing technology bring considerable challenges to the trustworthiness of multimedia data. Recent research has shown that convolutional attention mechanism can encode contextual information more effectively and efficiently than attentional mechanism, which has been validated in the field of semantic segmentation of images. However, convolutional attention-based networks focus only on the semantic information at the object level. The anomalous features introduced by tampering manipulation in the low-level information are ignored, resulting in poor localization of tampered regions. In this paper, leveraging convolutional attention mechanism, we propose an image tampering localization method based on multi-scale convolutional attention mechanism (IMLMCA), which fuses the low-level information to improve its capability to recognize the low-level anomalous information. To overcome the problems such as the unbalanced distribution of training positive and negative samples caused by the tampered regions, which usually occupy only a tiny part of the whole image, we introduce a mixed loss function consisting of focal-loss and Lovászloss to assist the method in learning the tampered features. The experimental results show that the proposed method achieves state-of-the-art performance on several publicly available tampered image datasets. For example, on CASIA dataset, the F1 score of the proposed method is 0.601, while the F1 score of the current best Transformer-based method, Objectformer, is 0.579. Codes are available at https://github.com/AlchemistLiu/IMLMCA.

Keywords: image tampering localization · convolutional neural network · multiscale convolutional attention · mixed loss function

1 Introduction

In recent years, the quality of generated multimedia contents has reached a high level, so that the boundary between real contents and synthetic contents has

Supported by The National Natural Science Foundation of China (U1703261; 61871258; 61972361).

Q. Liu et al. (Eds.): PRCV 2023, LNCS 14431, pp. 183–194, 2024.
https://doi.org/10.1007/978-981-99-8540-1_15

become increasingly blurred. Massive fake digital images have appeared, such as online rumors, insurance fraud, fake news, etc, which have a huge negative impact on human lives [1]. The purpose of image tampering localization is to locate the tampered regions in an image and identify the tampered image itself.

In image tampering localization tasks, it is required to pay attention to both the high-level semantic features and low-level features introduced by various tampering manipulation [2]. However there is no specific network framework designed to focus on both kinds of features. As shown in Fig. 1, a tampered image is generated by the copy-move manipulation, in which a tampered region originates from the original image. The data distribution in this region is similar to that of the original image. It is difficult for a general network model to extract anomalous information about tampering traces in such a tampered image. If tampered regions in a tampered image are generated by the splicing operation, it is more easily located in the tampered image since they come from different images.

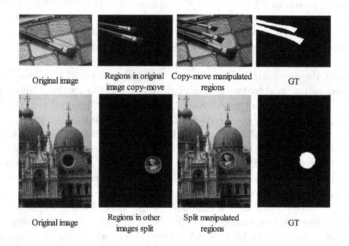

Fig. 1. Examples of tampered images generated by different tampering methods.

In image tampering localization tasks, the network model's ability of focusing on anomalous features in the tampered image is vital to performance improvement. Transformer methods based on self-attention mechanisms have recently achieved some advantages in image processing, including pixel-level image tampering localization. However, recent studies have shown that convolutional attention mechanisms can encode contextual information more effectively and efficiently than self-attention mechanisms, which has been validated in semantic segmentation of images [3]. In this study, a multi-scale convolutional attention image tampering localization method, named IMLMCA, is proposed. Based on convolutional attention framework, it utilizes low-level anomalous information to achieve high performance in locating tampered regions. The main contributions of this paper are as follows:

1) The encoder with a four-level multi-scale convolutional attention mechanism is used to simultaneously extract high-level semantic features and low-level minutiae features in image tampering localization tasks;

2) To address the issues of an unbalanced distribution of tampered image data and uneven sample localization difficulty, a mixed loss function integrating focal-loss and Lovászloss is designed to improve localization performance.

2   Related Work

2.1   Image Tampering Localization Based on Deep Learning

Zhou et al. [4] constructed a dual-stream Faster R-CNN to combine SRM features with RGB features for tampering localization. Bi et al. proposed RRU-Net [5] a ring residual network based on U-Net to solve the problem of gradient degradation in deep networks. Kwon et al. proposed CAT-Net [6], a dual-stream tampered image localization network to locate tampered regions based on HR-Net [7]. Chen et al. developed MVSS-Net [8], which combines the boundary features of tampered regions and noise features to learn more generalized tampering information using a multi-scale supervision approach. Wang et al. developed ObjectFormer [2], which captures subtle traces of manipulation that are not available in the RGB domain with visual Transformer.

2.2   Semantic Segmentation Based on Deep Learning Techniques

Currently, in semantic segmentation tasks, researchers focus on how to make the models better incorporate contextual information to improve the accuracy in object recognition. Chen et al. proposed to use dilated convolution in ASPP [9] to expand the convolution kernel field without the loss of resolution. Recently, Liu et al. proposed ConvNext [10], in which a large convolutional kernel is used to extract contextual information with reference to the structure of visual Transformer. It demonstrates that convolutional networks still have good feature extraction capabilities. Guo et al. [3]. proposed a multi-scale convolutional attention module to build the network encoder, in which using multiple convolutional layers with large convolutional kernels are used in parallel to capture global and contextual information, and attention operations are introduced to activate the interaction of multi-scale contextual information. This approach achieved the state-of-the-art performance in the semantic segmentation tasks, demonstrating that convolutional attention is a more efficient way to encode contextual information than self-attention.

3   Method

To address the anomalous distribution issue of feature information at different stages in tampered image localization tasks, we designed an encoder-decoder network model, IMLMCA, using the multi-scale convolutional attention module.

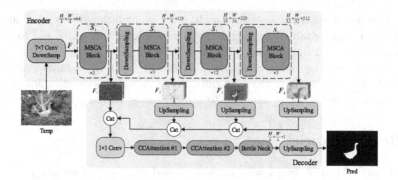

Fig. 2. Overall network structure diagram.

3.1 Overall Framework

As shown in Fig. 2, let the input of the model be an RGB image $\boldsymbol{Temp} \in \mathbb{R}^{H \times W \times 3}$. H and W are the height and width of the image, respectively, which are downsampled by the convolutional layer to generate features $\boldsymbol{F} \in \mathbb{R}^{\frac{H}{4} \times \frac{W}{4} \times 64}$. They are then fed into a tandem multi-scale convolutional attention (MSCA) module to extract different stages of features $\boldsymbol{F}_i(i = 1, 2, 3, 4)$ from low-level texture features to high-level abstract features. Except $\boldsymbol{F}_1$, the previous stage features are down-sampled to one-half of the original height and width at each stage. These features are resampled to the same size as $\boldsymbol{F}_1$ using bilinear interpolation. After they are concatenated in the channel dimension, the features are fed to the convolutional layer for feature mapping, and then the results are sent to the decoder.

The decoder uses the segmentation head proposed in CCNet [11]. The decoder mimics the structure of the self-attention module [12] by allowing a single feature from any position to perceive all other features at all positions to obtain complete contextual information. Also, the use of a convolutional layer instead of a fully connected layer avoids the computationally overloaded problem in the modules like CBAM [13]. The output is a single-channel binary map with the size of $(\frac{H}{4}, \frac{W}{4})$, which is scaled to the same height and width as the input image using a bilinear interpolation, and the final prediction $\boldsymbol{Pred}$ is obtained.

3.2 Multi-Scale Convolutional Attention Module (MSCA)

MSCA module takes the extracted multi-scale convolutional features to model the feature contextual information at different scales by matrix multiplication. Compared with transformer-based network, the convolutional attention module is more flexible in the input image size, which sufficiently fits the practical situation where the size of the tampered image is usually inconsistent.

The features $\boldsymbol{F}$ are input to the MSCA sub-module and processed by a depthwise convolutional layer with a convolutional kernel size of 5. Then, the multi-scale features are captured after a structure consisting of three asymmetric

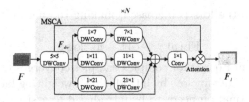

Fig. 3. Multi-scale Convolutional Attention Submodule.

convolutional layers with different sizes in parallel and added with the previously obtained features F_{dw}. Moreover, the features are mapped in channel dimension by a convolutional layer with a convolutional kernel size of 1. Finally, the results are multiplied with the previously extracted features to obtain the feature F_i containing contextual links. The process is shown in Eqs. (1) and (2).

$$F_{dw} = \text{DWConv}_{5\times5}(F) \tag{1}$$

$$F_i = F_{dw} \otimes \text{Conv}_{1\times1}(F_{dw} \oplus \sum_{i=0}^{3} \text{DWConv}_i(F_{dw})) \tag{2}$$

3.3 Mixed Loss Function

In image tampering localization tasks, tampered regions usually occupy only a tiny portion of the image. This leads to a highly unbalanced distribution of positive and negative samples represented by the tampered and untampered regions in the image. Generally, image tampering localization models tend to focus on the tampered regions generated by the splicing manipulations and do not focus well on the copy-move generated tampered regions, from which it can be inferred that the splicing tampered images are easy samples, compared with the copy-move tampered images. When a model with the cross-entropy loss is trained, loss value of negative samples dominates the model's gradient descent direction. The model's prediction confidence for easy samples keeps increasing and converges to 1, ignoring difficult samples. This leads to a devolution of the model, which can be used to recognize tampered images generated by splicing manipulations only. To solve the problems of uneven sample distribution and different sample difficulties, this paper uses a mixed loss with focal-loss [14] as the dominant term, and Lovászloss [15].

The equation of focal-loss is shown in (3), where p is the confidence level of the model outputs mapped between $[0, 1]$ after the sigmoid function, and α is the category weight to balance the positive and negative samples. The hyperparameter α is set to be less than 1, which can be used to reduces the weight of negative samples to decrease the effect of negative samples. λ denotes the weight of complex samples, and λ is set to be greater than 1 so that the relative loss of samples with high confidence is reduced and the contribution of complex samples to the loss is increased, making the network mainly focus on complex samples.

$$FL(p) = -\alpha(1-p)^\lambda \log p \tag{3}$$

The $1-p$ term in Eq. (3) for complex samples converges to a constant value and becomes a cross-entropy loss by introducing a weighting factor α. Lovászloss containing IoU information needs to be introduced to assist the network in predicting the boundary parts. The expression of Lovászloss is shown as Eqs. (4) and (5), where y^* is the image label, and $\widetilde{y}$ is the prediction results based on the confidence p. $J_c(y^*, \widetilde{y})$ denotes the ratio of the intersection between the label and the network prediction to the concatenation, and c is the category of the data.

$$\Delta J_c(y^*, \widetilde{y}) = 1 - J_c(y^*, \widetilde{y}) \tag{4}$$

$$J_c(y^*, \widetilde{y}) = \frac{|\{y^* = c\} \cap \{\widetilde{y} = c\}|}{|\{y^* = c\} \cup \{\widetilde{y} = c\}|} \tag{5}$$

The two losses are weighted and combined to become the loss function used by IMLMCA, as shown in Eq. (6).

$$L = \lambda_1 FL(p) + \lambda_2 \Delta J_c(y^*, \widetilde{y}) \tag{6}$$

4   Experiments

The evaluation metrics are the ROC area under the curve (AUC) scores and F1 scores at the pixel level.

4.1   Datasets

- CASIA [16]: CASIA dataset has two versions: CASIAv1 containing 921 tampered images and CASIAv2 containing 5123 tampered images. The images in the dataset are generated by splicing or copy-move manipulations. The images are also subjected to operations including blurring and filtering to mimic the tampered images of the natural scene.
- NIST16 [17]: It includes 584 tampered images. The tampered images were generated by splicing, copy-move and removal operations and post-processed to hide the visible tampering traces.
- IMD2020 [18]: It contains 35,000 authentic images taken by different cameras. The same number of tampered images are also generated using various methods.

For each dataset, the same approach as that in the literature [19,20] was used to obtain the training/testing set.

4.2   Implementation Details

Experiments were conducted on dual RTX2080 Ti GPUs. The batch size was 6, and the training iteration was 120 epochs. The hyperparameters α and λ of the focal-loss part were empirically set to 0.6 and 2, and the weights λ_1 and λ_2 in the mixed loss function were set to 0.5.

4.3   Baseline Models

We compared the image manipulation localization performance between our method and various baseline models. Those models are described as follows:

Table 1. Localization results of different approaches.

Methods	CASIA		NIST16		IMD2020		Params(M)
	AUC	F1	AUC	F1	AUC	F1	
RGB-N [4]	0.795	0.408	0.937	0.722	–	–	–
SPAN [19]	0.838	0.382	0.961	0.582	–	–	–
CAT-Net [6]	0.839	0.529	0.940	0.673	0.882	0.602	114.26
MVSS-Net [8]	0.862	0.446	0.974	0.654	0.872	0.553	146.81
TransForensics [21]	0.850	0.479	–	–	0.848	–	–
CFL-Net [22]	0.863	0.507	**0.997**	0.786	0.899	0.533	111.91
Objectformer [2]	0.882	0.579	0.996	**0.824**	0.821	–	–
Ours	**0.893**	**0.601**	0.993	0.821	**0.942**	**0.668**	**35.21**

- RGB-N [4]: Parallel RGB and noise streams are used to discover the inconsistency of tampered features and noise within the image.
- SPAN [19]: Using a spatial pyramid structure, the dependencies of image blocks are modeled using self-attention blocks.
- CAT-Net [6]: Image tampering localization is performed using image DCT information-assisted networks.
- MVSS-Net [8]: Tampered regions are located by attention mechanism using parallel RGB streams and noise streams combined with edge information.
- Transforensics [21]: The interaction between local image blocks at different scales is modeled using a self-attention encoder and a visual Transformer with a dense correction module.
- CFL-Net [22]: Based on a dual-stream network, a contrastive learning module is used to improve the accuracy of the network in identifying tampered regions.
- Objectformer [2]: Visual Transformer is used to model tampered image features with image DCT information-assisted networks.

Significant performance improvements can be observed from the results in Table 1. The tampered region localization results obtained by IMLMCA and some of the existing methods are visualized in Fig. 4. This indicates that IMLMCA can achieve better results in image tampering localization tasks. At the same time, IMLMCA has a lightweight advantage over the existing convolution-based image tampering localization networks, such as CAT-Net, MVSS-Net, and CFL-Net.

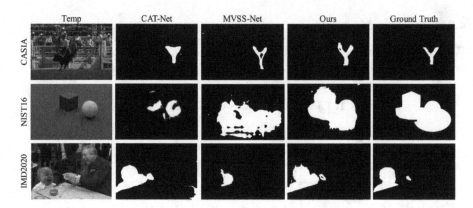

Fig. 4. Visualization of positioning results using different methods.

5   Results

5.1   Ablation Study

Res-50 is used as the network encoder and the segmentation head in CCNet is used as the decoder. The features extracted from the encoder at the second, third, and fourth stage are fed to the decoder, and cross entropy loss is chosen as the loss function. Except for the decoder, the other parts are replaced one by one with an encoder composed of MSCA modules. Ablation experiments are performed using all four-stage features, including the low-level feature F_1, in which a mixed loss function is used to validate the effectiveness of the proposed method. Ablation experiments are performed on CASIA dataset.

The quantitative results are shown in Table 2. Compared with the network that doesn't include a MSCA module, the proposed method improves the F1 score and AUC by 0.12 and 0.049, respectively, indicating that the encoder with MSCA module has a solid ability to extract anomalous information from the tampered regions. Compared with the network without the first-stage low-level feature F_1, the F1 score and AUC of the proposed method are improved by

Table 2. The results of ablation experiments on the CASIA dataset.

Encoder	Res-50	✓	✓	✓	✓				
	MSCA					✓	✓	✓	✓
Features Selection	F_2, F_3, F_4	✓	✓			✓	✓		
	F_1, F_2, F_3, F_4			✓	✓			✓	✓
Loss function	Cross entropy	✓		✓		✓		✓	
	Mixed loss		✓		✓		✓		✓
AUC		0.757	0.809	0.813	0.844	0.829	0.877	0.801	**0.893**
F1		0.399	0.437	0.475	0.481	0.522	0.577	0.441	**0.601**

0.024 and 0.016, respectively, demonstrating that introducing the low-level stage features can help the network locate the tampered regions better. The F1 and AUC of the network are decreased by 0.16 and 0.092 when a cross-entropy loss function is used, compared with IMLMCA.

When both the mixed loss function and the MSCA module are used, the network can locate the tampered regions in the images well. Otherwise, the features extracted by the network from small number of complex samples will affect the final localization accuracy. For tasks like image tampering localization, where the samples are un-evenly distributed and the data varies in difficulty, critical features extracted by the network need to be used in conjunction with a loss function that takes into account the data distribution issue.

5.2 Qualitative Analysis

In Fig. 5, the results by Grad-CAM visualization shows that with the introduction of the first-stage features, the network pays attention to the tampered regions in the images generated by the splicing manipulation, which contains related objects. The network has less impact on other kinds of splicing tampered images. The same enhancement effect is observed in the localization of copy-move tampered images. When the heat maps of the tampered images between the splicing and copy-move operations are compared, it is found that the first-stage low-level features are more helpful in improving the image tampering localization ability. This indicates that, in addition to the existing problems of data distribution, the low-level information extracted by the network encoder is the key to solving the original problem of inaccurate identification of complex samples in image tampering localization. IMLMCA has an advantage in solving these complex-sample identification problems.

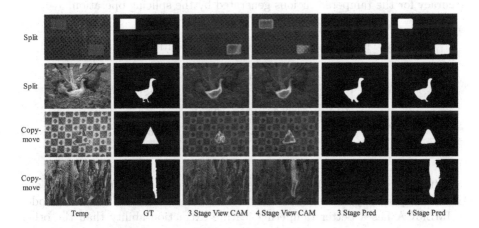

Fig. 5. The difference in the network's attention to tampered areas between using the first stage features and not using them, as well as the heat map and location prediction results.

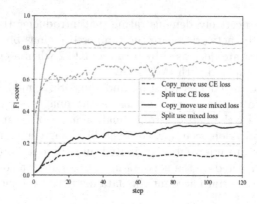

Fig. 6. When using the mixed loss and the cross-entropy loss, the trend of F1 score for network localization of tampered image regions generated by different tampering methods with increasing training steps.

Figure 6 demonstrates that the mixed loss function used by IMLMCA improves the network's ability to locate tampered regions by weighting the loss on complex samples. When the cross-entropy loss function is used, as the number of training epochs increases, the F1 score of the network for the localization of tampered images generated by the copy-move operation decreases. The F1 score rises for the tampered images generated by the splicing operation, indicating that the network can only identify the easy samples (splicing tampered images). After the mixed loss function is used, the F1 score of the network is still slightly increasing in the second half of the training (solid red line) when the network is used to locate the tampered regions generated by the copy-move operation. The solid green line indicates that the network maintains a higher recognition accuracy for the tampered regions generated by the splicing operation.

Table 3. The impact of different encoders on positioning performance.

Methods	AUC	Recall	Precision	F1
Res-base	0.844	0.493	0.471	0.481
HR-base	0.850	0.488	0.521	0.508
Ours	**0.893**	**0.557**	**0.652**	**0.601**

After the model encoder is replaced with the encoder part of ResNet and HRNet, respectively, the results are shown in Table 3. The MSCA based encoder in IMLMCA has a better tampering feature extraction ability than the other two approaches

6   Conclusions

We propose a convolutional attention mechanism-based image tampering localization network, IMLMCA, which demonstrates good extraction capability for high-level semantic features. It can also extract the low-level features beneficial to tampering localization. The number of network parameters are significantly reduced compared with the number of parameters in other convolutional tampering localization networks. The low-level information extracted by the network encoder is found to be more beneficial for solving complex problems. A mixed loss function that can solve the problems of unequal data distribution and different difficulty levels is used to guide the network to get better prediction results.

In further work, research will continue to improve the model's ability to localize tampered regions in images containing texture information.

References

1. Zheng, L., Zhang, Y., Thing, V.L.: A survey on image tampering and its detection in real-world photos. J. Vis. Commun. Image Represent. **58**, 380–399 (2019)
2. Wang, J., et al.: Objectformer for image manipulation detection and localization. In: Proceedings of the IEEE/CVF Conference on Computer Vision and Pattern Recognition, pp. 2364–2373 (2022)
3. Guo, M.H., Lu, C.Z., Hou, Q., Liu, Z.N., Cheng, M.M., Hu, S.M.: Segnext: rethinking convolutional attention design for semantic segmentation. In: Advances in Neural Information Processing Systems (2022)
4. Zhou, P., Han, X., Morariu, V.I., Davis, L.S.: Learning rich features for image manipulation detection. In: Proceedings of the IEEE Conference on Computer Vision and Pattern Recognition, pp. 1053–1061 (2018)
5. Bi, X., Wei, Y., Xiao, B., Li, W.: Rru-net: the ringed residual u-net for image splicing forgery detection. In: Proceedings of the IEEE/CVF Conference on Computer Vision and Pattern Recognition Workshops (2019)
6. Kwon, M.J., Yu, I.J., Nam, S.H., Lee, H.K.: Cat-net: compression artifact tracing network for detection and localization of image splicing. In: Proceedings of the IEEE/CVF Winter Conference on Applications of Computer Vision, pp. 375–384 (2021)
7. Sun, K., Xiao, B., Liu, D., Wang, J.: Deep high-resolution representation learning for human pose estimation. In: Proceedings of the IEEE/CVF Conference on Computer Vision and Pattern Recognition, pp. 5693–5703 (2019)
8. Chen, X., Dong, C., Ji, J., Cao, J., Li, X.: Image manipulation detection by multi-view multi-scale supervision. In: Proceedings of the IEEE/CVF International Conference on Computer Vision, pp. 14185–14193 (2021)
9. Chen, L.C., Zhu, Y., Papandreou, G., Schroff, F., Adam, H.: Encoder-decoder with atrous separable convolution for semantic image segmentation. In: Proceedings of the European Conference on Computer Vision, pp. 801–818 (2018)
10. Liu, Z., Mao, H., Wu, C.Y., Feichtenhofer, C., Darrell, T., Xie, S.: A convnet for the 2020s. In: Proceedings of the IEEE/CVF Conference on Computer Vision and Pattern Recognition, pp. 11976–11986 (2022)
11. Huang, Z., Wang, X., Huang, L., Huang, C., Wei, Y., Liu, W.: CCNET: criss-cross attention for semantic segmentation. In: Proceedings of the IEEE/CVF International Conference on Computer Vision, pp. 603–612 (2019)

12. Dosovitskiy, A., et al.: An image is worth 16x16 words: transformers for image recognition at scale. In: International Conference on Learning Representations (2021)
13. Woo, S., Park, J., Lee, J.Y., Kweon, I.S.: CBAM: convolutional block attention module. In: Proceedings of the European Conference on Computer Vision, pp. 3–19 (2018)
14. Lin, T.Y., Goyal, P., Girshick, R., He, K., Dollár, P.: Focal loss for dense object detection. In: Proceedings of the IEEE International Conference on Computer Vision, pp. 2980–2988 (2017)
15. Berman, M., Triki, A.R., Blaschko, M.B.: The lovász-softmax loss: a tractable surrogate for the optimization of the intersection-over-union measure in neural networks. In: Proceedings of the IEEE Conference on Computer Vision and Pattern Recognition, pp. 4413–4421 (2018)
16. Dong, J., Wang, W., Tan, T.: Casia image tampering detection evaluation database. In: 2013 IEEE China Summit and International Conference on Signal and Information Processing, pp. 422–426. IEEE (2013)
17. Guan, Y., et al.: Nist nimble 2016 datasets. https://mfc.nist.gov/
18. Novozamsky, A., Mahdian, B., Saic, S.: IMD 2020: a large-scale annotated dataset tailored for detecting manipulated images. In: Proceedings of the IEEE/CVF Winter Conference on Applications of Computer Vision Workshops, pp. 71–80 (2020)
19. Hu, X., Zhang, Z., Jiang, Z., Chaudhuri, S., Yang, Z., Nevatia, R.: SPAN: spatial pyramid attention network for image manipulation localization. In: Vedaldi, A., Bischof, H., Brox, T., Frahm, J.-M. (eds.) ECCV 2020. LNCS, vol. 12366, pp. 312–328. Springer, Cham (2020). https://doi.org/10.1007/978-3-030-58589-1_19
20. Liu, X., Liu, Y., Chen, J., Liu, X.: Pscc-net: progressive spatio-channel correlation network for image manipulation detection and localization. IEEE Trans. Circuits Syst. Video Technol. **32**(11), 7505–7517 (2022)
21. Hao, J., Zhang, Z., Yang, S., Xie, D., Pu, S.: Transforensics: image forgery localization with dense self-attention. In: Proceedings of the IEEE/CVF International Conference on Computer Vision, pp. 15055–15064 (2021)
22. Niloy, F.F., Bhaumik, K.K., Woo, S.S.: CFL-net: image forgery localization using contrastive learning. In: Proceedings of the IEEE/CVF Winter Conference on Applications of Computer Vision, pp. 4642–4651 (2023)

Bi-stream Multiscale Hamhead Networks with Contrastive Learning for Image Forgery Localization

Guo Li[1], Runjie Liu[1], Wenchao Cui[1], Yirong Wu[2], Ben Wang[3],
and Shuifa Sun[3(✉)]

[1] School of Computer and Information Technology, China Three Corges University,
Yichang, Hubei 443002, China
[2] Institute of Advanced Studies in Humanities and Social Sciences,
Beijing Normal University, Beijing 100089, China
[3] School of Information Science and Technology, Hangzhou Normal University,
Hangzhou 310036, China
watersun@hznu.edu.cn

Abstract. The purpose of image forgery localization task is to localize and segment forgery regions from a forgery image. Traditional forgery localization methods usually rely on specific manipulation traces, while contrastive learning-based methods focus on universal forgery traces. In this study, a general image tampering localization method based on contrastive learning, named Bi-Stream Multi-scale Hamhead Contrast Net (BMHC-Net), was proposed. At first, a multi-scale patch contrastive learning module was proposed, in which rich features of forgery images were projected to a feature space. The projection of rich features and real mask were divided into multi-scale patches, and contrastive learning was utilized to obtain the contrastive loss to locate the forgery regions of the image. Then, inspired by Hamberg network, Hamhead structure was utilized to construct a multi-stage feature fusion module that used attention mechanisms to focus on global information and efficiently fused multi-scale features. Experimental results demonstrate the effectiveness of the proposed method. For example, on IMD2020 dataset, BMHC-Net improved the overall performance in terms of F1-score by 0.1714, in which the contrastive loss module improved the system performance in terms of F1-score by 0.0418.

Keywords: deep learning · image forgery detection · matrix decomposition · supervised contrastive learning

1 Introduction

Due to the proliferation of digital technology and the improvement of image editing software, the difficulty of image editing has rapidly decreased [1]. As a

---

Supported by The National Natural Science Foundation of China (No. U1703261; No. 61871258).

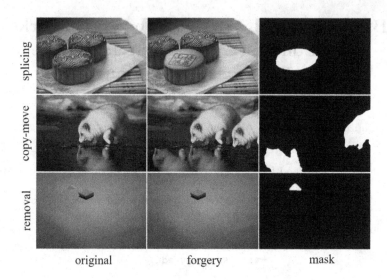

Fig. 1. Examples of forgery images.

result, many forgery images have appeared. With the widespread use of social media, the spread of various malicious forgery digital images has been accelerated. Due to their adverse effects, the development of image forgery detection and localization techniques becomes a must.

Existing deep learning-based methods for image forgery localization including ManTra-Net [2], and SPAN [3] can detect and localize various types of forgery regions manipulated by splicing, copy-move, or removal. These methods often rely on several traces left by specific forgery manipulations. However, in reality, forgery images, are often fabricated after multiple forgery manipulations and may be subjected to multiple post-processing, which may obscure forgery clues. Therefore, some methods designed explicitly for certain forgery traces are difficult to achieve the expected performance in practice. A more general algorithm that can detect multiple forgery manipulations is required.

In this study, a more general image forgery localization method, named BMHC-Net, Bi-Stream Multi-scale Hamhead Contrast Net (BMHC-Net) was proposed. This study makes the following key contributions:

1) Relying on supervised contrastive learning, a multi-scale patch contrastive learning module is proposed to project the rich features of forgery images to a feature space. The aim is to obtain the data distribution difference between the forgery and non-forgery regions through multi-scale patch contrastive learning, with the goal of to locateing the forgery regions of the image.

2) Inspired by Hamberg network [4], a multi-layer feature pyramid structure is constructed as multi-stage feature fusion module using Hamhead that focuses on global information. This structure selects and fuses multi-stage features to enhance the localization ability of the network for forgery regions.

2   Related Work

2.1   Image Forgery Detection and Localization

Early related work treated forgery detection as a classification problem, such as RGB-N [5]. In recent years, forgery localization methods produced results in the form of binary mask, which has become the mainstream research direction. Nowadays, various novel detection and localization methods are proposed. Liu et al. proposed PSCC-Net [6] fusing multi-scale features, which provides a new research perspective. Chen et al. proposed MVSS-Net [7], which utilizes two branches to extract edge information and noise information, respectively, and a dual attention mechanism is applied. Kwon et al. designed CAT-Net [8] that could learn DCT coefficient distribution to jointly learn the evidential features of compression artifacts in RGB and DCT domains. Objectformer [9] and TransForensics [10] adapted from Transformer and self-attention mechanism [11] introduce the self-attention mechanism into the image forgery localization field.

2.2   Contrastive Learning

The goal of contrastive learning is to learn the difference between diverse types of instances and the common features between similar instances so that the distance between instances in the same class instances is closer and the distance between instances in different class is opposite in the feature space. Contrastive learning has made considerable advances in unsupervised learning fields in recent years. Chen et al. proposed SimCLR [12], a simple framework for contrastive learning. Wang et al. compared pixel embeddings between different semantic classes in a supervised manner to assist segmentation [13]. Niloy et al. developed CFL-Net [14], in which contrastive learning loss is introduced into the field of image forgery localization in a general sense, achieving good results.

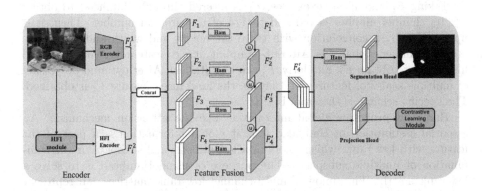

Fig. 2. The overall framework.

3   BMHC-Net

3.1   Overall Framework

The model proposed in this paper is shown in Fig. 2. The main structure is a dual-stream network consisting of a feature extractor, a muti-stage feature fusion module, a decoder. The backbone is ConvNext [15]. One stream is the image RGB stream, and the other is the high-frequency information stream. Precisely, an image I is inputted into the RGB encoder and the high-frequency encoder to extract the corresponding image forgery features, and multi-scale four-layer features with the same size are obtained and represented as F_i^n, $n = (1, 2)$, $i = (1, 2, 3, 4)$. $i = (1, 2, 3, 4)$ represents the four scales features and $n = (1, 2)$ represents two types of features from two encoderes, respectively. After feature fusion, the fused features, named F_i, will be sent to the segmentation head for forgery region prediction. Projection structure named P is generated after another mapping module and sent to enter the multi-scale patch contrastive learning module.

3.2   Multi-stage Feature Fusion Module

In the feature fusion stage, a feature pyramid is constructed with four layers of Hamhead to accomplish the dimensionality reduction and feature fusion at each stage. Hamhead utilizes matrix decomposition to replace the attention mechanism. It is more lightweight and can focus on global information at each layer to complete the aggregation of specific information, that is, forgery information. The whole process details of four layers are shown in Formula (1):

$$F_i' = \begin{cases} concat(Ham(F_i), UP(F_{i-1}')) & 1 < i \leq 4 \\ Ham(F_i) & i = 1 \end{cases} \tag{1}$$

Taking F_1 and F_2 as examples, F_1 is filtered through Hamhead to obtain F_1'. Its channel number is reduced to 1/4 of the original number. Then, it is up-sampled by a convolution with twice the height and width and 1/4 of the original number of channels. After up-sampling, it is concatenated with Hamhead filtered F_2 to obtain F_2'. Similarly, F_3' can be obtained. After pyramid processing of multiple scales of features in 4 layers, the final fused feature F_4' is obtained. The specific scheme is shown in Fig. 3.

Hamhead processes global information like the attention mechanism. An essential assumption of Hamhead is that global information is a part of the feature with low rank, which exists in a low-dimensional subspace or the combination of multiple subspaces. The purpose of using Hamhead is to solve a low-rank matrix. The union of multi-channel low-rank matrices can represent the forgery features or inherent features of the image. The idea is constructed by matrix decomposition, and its mathematical expression is described as follows:

$$F = \overline{F} + E = DC + E \tag{2}$$

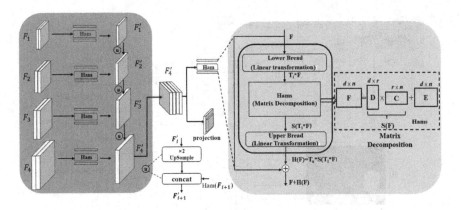

Fig. 3. Multi-scale feature fusion with Hamhead.

where F is the rich features before decomposition, D is the dictionary matrix, C is the corresponding coefficient matrix, $C = [c_1, ..., c_n] \in \mathbb{R}^{r \times n}$, $\overline{F} \in \mathbb{R}^{d \times n}$ is the low rank matrix recovered by matrix decomposition, and $E \in \mathbb{R}^{d \times n}$ is the residual matrix.

The objective function is shown in the Formula (3). Here, $\mathcal{L}$ represents the error between the reconstructed feature $\overline{F}$ and the original feature F. $\mathcal{R}_1$ and $\mathcal{R}_2$ respectively represent the regularization of the dictionary matrix D and the coefficient matrix C.

$$\min_{D,C} \mathcal{L}(F, DC) + \mathcal{R}_1(D) + \mathcal{R}_2(E) \tag{3}$$

The algorithm for solving the function is represented as $\mathcal{S}$, and the non-negative matrix factorization algorithm is selected. The upper and lower layers are linear transformations, which are represented as $\mathcal{T}_u$, and $\mathcal{T}_l$, respectively. Then the mathematical expression of the entire Hamhead structure is shown in Eq. (4), and its structure is shown on the right side of Fig. 3.

$$\mathcal{H}(F) = \mathcal{T}_u \mathcal{S}(\mathcal{T}_l F) \tag{4}$$

3.3   Multi-scale Patch Contrastive Learning Module

For the actual scenario of the forgery localization task, a more general algorithm is entailed to locate the forgery regions. The ultimate goal of the multi-scale patch contrastive learning module is to compare the forgery and non-forgery embeddings of the images from multiple scales to ensure that the feature distribution between the two types of regions is well separated in the feature space.

Since the shape of the embedded feature is H × W and the ground truth is known, each embedding's real label is known. The contrastive loss for single-pixel embedding can be introduced. However, the contrastive loss based on single-pixel does not consider the context information and has a considerable memory cost.

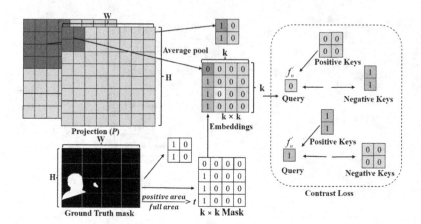

Fig. 4. Multi-scale patch contrastive learning loss module. For visualization purposes, projection in the figure is displayed as P with a shape of 256 × 8 × 8. P is divided into two different scales of 4 × 4 and 2 × 2. Taking 4 × 4 as an example, after pooling, we obtain a patch embedding with a size of 4 × 4, i.e., "k × k Embeddings with Label". The Mask is also divided into 4 × 4 patches. The contrastive loss of each embedding in "k × k Embeddings with Label" is calculated using Eq. (7).

To achieve a balance between context information and fine-grained traces, this study choose to divide P into multi-scale patches by average pooling. The scheme and specific description are shown in Fig. 4.

P is divided into patches evenly to obtain P', k can be 128, 64 and 32. Each patch in P' is embedded as f_n, $f_n \in \mathbb{R}^{256 \times h \times w}$, where $n \in (1, 2, ..., k^2)$, $h = H/k$, and $w = W/k$. The average value of all pixel-level embeddings in this patch is taken, then the patch-level embedding is obtained, so that each f_n becomes f'_n, $f'_n \in \mathbb{R}^{256 \times 1 \times 1}$. Similarly, the mask is divided into patches. After pooling, the pixel value of the forgery region is set to 1, and the pixel value of non-forgery region is set to 0.

After the average pooling, the mask is divided into patches, and the n_{th} patch is denoted as m_n, where $m_n \in \mathbb{R}^{1 \times 1}$. The value depends on the relative number of positive and negative examples (1, 0) of original pixels in this region. An adjustable threshold (denoted as t) is set to determine the value of m_n so that judgment is strict at a fine granularity and loose at a coarse granularity, thus achieving balance. The relationship between the value of m_n and the threshold is shown in the Formula (6), where z_j represents the key value corresponding to the j_{th} pixel in this region.

$$m_n = \begin{cases} 1 & \sum_{j=1}^{k^2} \frac{z_j}{k^2} \geq t \\ 0 & else \end{cases} \tag{5}$$

When k is taken as 128, 64, and 32, and the corresponding threshold t is taken as (1/2, 1/4, 1/4), the loss of the n_{th} patch embedding is given by formula (6):

$$\mathcal{L}_n = \frac{1}{|A_n|} \sum_{g^+ \in A_n} -\log \frac{exp(f'_n \cdot g^+/\tau)}{exp(f'_n \cdot g^+/\tau) + \sum_{g^- \in A_n} exp(f'_n \cdot g^-/\tau)} \tag{6}$$

where A_n also represents the set of all other patch level embeddings with the same label property as f'_n, and g^+ represents the overall property of the patch embedding, i.e., the patch f'_n. Similarly, g^- is the overall property of all negative patch embeddings whose labels differ from f'_n. The final contrastive loss is obtained by weighted averaging of the four partitions and all embeddings, where a_j is the weight for different coarse-grained partitioning methods. The larger the value of k, the more accurate the localization, and the greater its weight. Therefore, the contrastive learning loss is defined as:

$$\mathcal{L}_{con} = \sum_{j=1} \frac{a_j}{k_j^2} \sum_{n \in k_j^2} \mathcal{L}_n \tag{7}$$

The final loss function is:

$$\mathcal{L}_{total} = \mathcal{L}_{BCE} + \beta \times \mathcal{L}_{con} \tag{8}$$

where $\mathcal{L}_{BCE}$ is the binary cross-entropy loss, and β is the weight of the contrastive loss.

4   Experiment and Analysis

4.1   Experiment Setting

Three standard image forgery detection datasets CASIA [16], NIST16 [17] and IMD2020 [18] are used as experimental datasets to verify the performance of this method. For each dataset, the samples are randomly divided into training sets, verification sets, and test sets according to the ratio of 8:1:1. To simulate the diversity of data in the actual scene, random flipping and clipping methods were used. In this paper, AUC and F1-score are used as evaluation metrics to measure the performance of the method.

- CASIA [16]: CASIAv1 contains 921 forgery images, and CASIAv2 contains 5123 forgery images. The images in the dataset are generated by splicing or copy-move manipulations.
- NIST16 [17]: 584 forgery images are included. The forgery images were generated by splicing, copy-move, and removal manipulations.
- IMD2020 [18]: This dataset contains 2000 forgery images generated from multiple forgery manipulations.

The experiment was conducted on an Intel e5-2680 V3 CPU and single RTX3080Ti. The batch size was 5.

4.2 Baseline Models

The proposed method is compared with other methods on three public forgery image datasets, and the results are shown in Table 1. As shown in Table 1, the proposed method achieves better localization performance. It should be pointed out that on CASIA dataset, Objectformer [8] and TransForensics methods have F1-score, which are 0.02 and 0.07 higher than the proposed method, respectively. Objectformer is a forgery localization method based on the Transformer, while TransForensics is an image forgery localization method based on a self-attention mechanism. However, considering the performance of the proposed method on three datasets as a whole, it is more balanced and has better performance than Objectformer and TransForensics methods. Baseline methods are described as follows for comparison.

- RGB-N [5]: Bi-streams are used to discover the inconsistency of forgery features and noise within an image.
- SPAN [3]: A spatial pyramid structure and a self-attention mechanism are used.
- CAT-Net [8]: Image DCT and RGB information are used.
- MVSS-Net [7]: Attention mechanism is applied using RGB and noise streams with edge information.
- TransForensics [10]: Image blocks at different scales are modeled using a self-attention encoder.
- CFL-Net [14]: A typical contrastive learning module is used to improve performance.
- Objectformer [9]: With image DCT information, Visual Transformer is used to model forgery image features.

Figure 5 shows the prediction results of some methods on three datasets. A tampered image is randomly selected from the dataset as an example. The white part represents the region predicted to be tampered with, and the black

Table 1. Localization results of different approaches.

Methods	CASIA		NIST16		IMD2020	
	AUC	F1	AUC	F1	AUC	F1
RGB-N [5]	0.795	0.408	0.937	0.722	–	–
SPAN [3]	0.838	0.382	0.961	0.582	–	–
CAT-Net [8]	0.839	0.529	0.940	0.673	0.882	0.602
MVSS-Net [7]	0.862	0.446	0.974	0.654	0.872	0.553
TransForensics [10]	0.850	**0.627**	–	–	0.847	–
CFL-Net [14]	0.863	0.507	**0.997**	0.786	0.899	0.533
Objectformer [9]	0.882	0.579	0.996	0.824	0.821	–
Ours	**0.903**	0.552	0.996	**0.833**	**0.931**	**0.612**

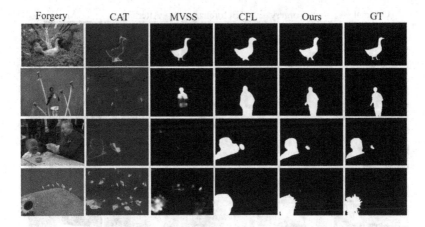

Fig. 5. Examples of forgery image, including normal types.

part represents the non-tampered region. It can be seen from Fig. 5 that other methods, such as CAT-Net and MVSS-Net, will show different localization performance on different images, which indicates that these methods are sensitive to certain forgery manipulations and do not have good generality. CFL-Net shows good localization performance on three datasets, but there are more false alarms. BMHC-Net proposed in this study utilizes contrastive learning and multi-stage Hamhead. Compared with other methods, it has better generality while maintaining higher localization performance.

4.3   Ablation Study

This experiment will examine the effect of each module on the localization performance of the proposed method. IMD2020 is used in the experiment, and the experimental results are listed in Table 2. Instead of Hamhead, ASPP module [19] are used for feature fusion and contrastive loss isn't used for ablation experiments. It can be seen from the experimental results that the overall performance is improved when the relevant modules are added to the network. This shows that the multiple modules have a positive effect on the overall performance of BMHC-Net.

In this study, a multi-scale patch level contrastive loss module is designed, and its performance at different scales is tested to verify the feasibility of BMHC-Net. The pixel-level contrastive learning Flops is set as 1, other Flops can be calculated, and the specific relationship is shown in Table 3. It can be found from Table 3 that the cost of calculation mainly comes from the more accurate and detailed division of k, and the detailed division is the main reason affecting the accuracy of the algorithm.

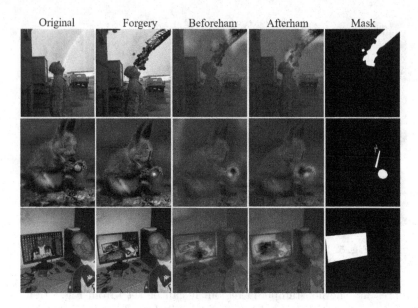

Fig. 6. Hamhead feature visualization.

4.4   Muti-level Hamhead

Based on the idea of disentangling linear and nonlinear features [20] the F is considered in terms of "linear and nonlinear" and divided into two parts: linear features (features represented by linear combinations of multiple low-dimensional subspaces) and nonlinear features, i.e., F contains features of both forgery and non-forgery. Finally, F contains linear intrinsic image features, nonlinear intrinsic image features, linear forgery features, and nonlinear forgery features, abbreviated as (LT, NT, LF, NF).

Without Hamhead, if the proposed distinguishes the four elements based on forgery as a judgment criterion, which may be a complex task. The introduction of Hamhead provides such an idea: it is difficult to divide (LT, NT, LF, NF), but it is relatively simple to divide (LT, LF) task.

A fundamental issue exists in the mechanism of muti-level Hamhead: The linear part of the sampling needs to be representative. The more decomposition structures, the more linear parts of the feature, and it is easier for the decoder

Table 2. The effect of each module

Hamhead	Contrastive loss	AUC	F1
×	×	0.886	0.473
√	×	0.918	0.538
×	√	0.898	0.485
√	√	**0.931**	**0.612**

Table 3. Multi-scale patch contrastive

k	Flops	AUC	F1
64	0.0039	0.928	0.589
32	0.00024	0.913	0.568
128,64,32 (Avg)	0.06665	0.929	0.608
128,64,32 (bias)	0.06665	**0.931**	**0.612**

Table 4. Multi-level repeat Hamhead.

Overlapping number	AUC	F1
0	0.931	0.612
1	0.918	0.570
2	0.905	0.556
3	0.892	0.541
4	0.893	0.533

to learn the difference of forgery and non-forgery regions. However, the feature richness will be weakened, and the representamethodtion will be untypical. The multi-level redundant Hamhead will inevitably makes F closer to the forgery extreme case of linear low rank, that is, rank 1, and the mechanism will ultimately lose its effect.

In this study, the ablation experiment on multi-level Hamhead is set after the final feature F_4'. The specific results are shown in Table 4. Heat map is used to illustrate the role of Hamhead in the proposed method in Fig. 6. This Hamhead is derived from the filtering part of the deepest feature in Fig. 3, where this Hamhead is marked in red. It can be observed from Fig. 6 that the values of unprocessed feature heat map are more scattered in heat map, while the values of thethe processed are more concentrated, indicating that the proposed method with Hamhead is more likely to focus on forgery regions.

5   Conclusion

In this study, we propose a more general method which utilizes matrix decomposition to decompose rich features to focus on global information. Contrastive loss is used to learn the difference of data distribution between forgery and non-forgery regions in the feature space. The experimental results show that on public CASIA, NIST16, and IMD2020 datasets, the proposed method in this study demonstrates a superior localization performance and achieves good generality.

References

1. Zheng, L., Zhang, Y., Thing, V.L.: A survey on image tampering and its detection in real-world photos. J. Vis. Commun. Image Represent. **58**, 380–399 (2019)
2. Wu, Y., AbdAlmageed, W., Natarajan, P.: Mantra-net: manipulation tracing network for detection and localization of image forgeries with anomalous features. In: Proceedings of the IEEE/CVF Conference on Computer Vision and Pattern Recognition, pp. 9543–9552 (2019)
3. Hu, X., Zhang, Z., Jiang, Z., Chaudhuri, S., Yang, Z., Nevatia, R.: SPAN: spatial pyramid attention network for image manipulation localization. In: Vedaldi, A., Bischof, H., Brox, T., Frahm, J.-M. (eds.) ECCV 2020. LNCS, vol. 12366, pp. 312–328. Springer, Cham (2020). https://doi.org/10.1007/978-3-030-58589-1_19

4. Geng, Z., Guo, M.H., Chen, H., Li, X., Wei, K., Lin, Z.: Is attention better than matrix decomposition? arXiv preprint arXiv:2109.04553 (2021)
5. Zhou, P., Han, X., Morariu, V.I., Davis, L.S.: Learning rich features for image manipulation detection. In: Proceedings of the IEEE Conference on Computer Vision and Pattern Recognition, pp. 1053–1061 (2018)
6. Liu, X., Liu, Y., Chen, J., Liu, X.: PSCC-net: progressive spatio-channel correlation network for image manipulation detection and localization. IEEE Trans. Circuits Syst. Video Technol. **32**(11), 7505–7517 (2022)
7. Chen, X., Dong, C., Ji, J., Cao, J., Li, X.: Image manipulation detection by multi-view multi-scale supervision. In: Proceedings of the IEEE/CVF International Conference on Computer Vision, pp. 14185–14193 (2021)
8. Kwon, M.J., Yu, I.J., Nam, S.H., Lee, H.K.: Cat-net: compression artifact tracing network for detection and localization of image splicing. In: Proceedings of the IEEE/CVF Winter Conference on Applications of Computer Vision, pp. 375–384 (2021)
9. Wang, J., et al.: Objectformer for image manipulation detection and localization. In: Proceedings of the IEEE/CVF Conference on Computer Vision and Pattern Recognition, pp. 2364–2373 (2022)
10. Hao, J., Zhang, Z., Yang, S., Xie, D., Pu, S.: Transforensics: image forgery localization with dense self-attention. In: Proceedings of the IEEE/CVF International Conference on Computer Vision, pp. 15055–15064 (2021)
11. Vaswani, A., et al.: Attention is all you need. Adv. Neural Inf. Process. Syst. **30** (2017)
12. Chen, T., Kornblith, S., Norouzi, M., Hinton, G.: A simple framework for contrastive learning of visual representations. In: International Conference on Machine Learning, pp. 1597–1607. PMLR (2020)
13. Wang, W., Zhou, T., Yu, F., Dai, J., Konukoglu, E., Van Gool, L.: Exploring cross-image pixel contrast for semantic segmentation. In: Proceedings of the IEEE/CVF International Conference on Computer Vision, pp. 7303–7313 (2021)
14. Niloy, F.F., Bhaumik, K.K., Woo, S.S.: CFL-Net: image forgery localization using contrastive learning. In: Proceedings of the IEEE/CVF Winter Conference on Applications of Computer Vision, pp. 4642–4651 (2023)
15. Liu, Z., Mao, H., Wu, C.Y., Feichtenhofer, C., Darrell, T., Xie, S.: A convnet for the 2020s. In: Proceedings of the IEEE/CVF Conference on Computer Vision and Pattern Recognition, pp. 11976–11986 (2022)
16. Dong, J., Wang, W., Tan, T.: Casia image tampering detection evaluation database. In: 2013 IEEE China Summit and International Conference on Signal and Information Processing, pp. 422–426. IEEE (2013)
17. Guan, H., et al.: MFC datasets: large-scale benchmark datasets for media forensic challenge evaluation. In: 2019 IEEE Winter Applications of Computer Vision Workshops (WACVW), pp. 63–72. IEEE (2019)
18. Novozamsky, A., Mahdian, B., Saic, S.: IMD 2020: a large-scale annotated dataset tailored for detecting manipulated images. In: Proceedings of the IEEE/CVF Winter Conference on Applications of Computer Vision Workshops, pp. 71–80 (2020)
19. Chen, L.C., Papandreou, G., Kokkinos, I., Murphy, K., Yuille, A.L.: Deeplab: semantic image segmentation with deep convolutional nets, atrous convolution, and fully connected crfs. IEEE Trans. Pattern Anal. Mach. Intell. **40**(4), 834–848 (2017)
20. He, T., Li, Z., Gong, Y., Yao, Y., Nie, X., Yin, Y.: Exploring linear feature disentanglement for neural networks. In: 2022 IEEE International Conference on Multimedia and Expo (ICME), pp. 1–6. IEEE (2022)

Fuse Tune: Hierarchical Decoder Towards Efficient Transfer Learning

Jianwen Cao, Tianhao Gong, and Yaohua Liu[✉]

Nanjing University of Information Science and Technology, Nanjing 210000, China
yaoh.liu@nuist.edu.cn

Abstract. This paper shows that integrating intermediate features brings superior efficiency to transfer learning without hindering performance. The implementation of our proposed Fuse Tune is flexible and simple: we preserve and concatenate features in the pre-trained backbone during forward propagation. A hierarchical feature decoder with shallow attentions is then designed to further dig the rich relations across features. Since the deep buried features are entirely exposed, back propagation in the cumbersome backbone is no longer required, which greatly improves the training efficiency. We make the observation that Fuse Tune performs especially well in self-supervised and label-scarce scenarios, proving its adaptive and robust representation transfer ability. Experiments demonstrate that Fuse Tune could achieve 184% temporal efficiency and 169% memory efficiency lead compared to Prompt Tune in ImageNet. Our code is open sourced in https://github.com/JeavanCode/FuseTune.

Keywords: Transfer Learning · Vision Representation · Feature Decoder · Computation and memory efficiency

1 Introduction

Due to their robust and generalized feature-extraction capabilities, pre-trained models with billions of parameters, such as Beit [1], Dino [2], and their improved variants [3, 4], have been widely adopted in various computer vision downstream tasks, including classification, segmentation, detection, generation, and so forth. It is worth noting that pre-trained models are typically categorized into supervised and self-supervised paradigm.

Supervised Learned (SL) pretraining requires a large amount of labeled data (such as ImageNet21K and JFT300M) to satisfy the data hungry backbones like ViT, thereby obtaining powerful representations. While Self-Supervised Learned (SSL) pretraining has recently emerged for its strong representation and label-free property, which is suitable for downstream transferring and scalable learning [5]. The core goal of pretraining is to improve the performance and efficiency of downstream tasks rather than training from

---

This work is partially supported by the Natural Science Foundation of Jiangsu Province under Grant BK20210642. The first author is an undergraduate student.

scratch. However, disparities between pre-trained and downstream tasks exist, requiring carefully designed transfer learning methods to narrow these gaps.

Previous transfer learning methods are usually computationally heavy, especially when the size of the pre-trained backbone grows. Therefore, how to efficiently transfer strong pre-trained models to target downstream tasks has become a research focus in the Big Model Era. Linear probe and Fine Tune are two most well-known and widely adopted transfer learning approaches. Linear probe refers to training a linear task-specific layer with the frozen pre-trained backbone. Due to its limited number of trainable parameters, Linear probe usually yields restricted performance, especially when pre-trained and downstream tasks share little in common. While Fine Tune refers to the strategy that all parameters in the pre-trained backbone are trainable for downstream tasks. Compared to Linear probe, Fine Tune usually has better performance. However, the performance improvement is achieved at the cost of significantly sacrificing computational efficiency.

Recently, various modern transfer learning methods have been developed, such as Prompt Tune [6], Adapter [7], Spot Tune [8], and etc. Most of these methods attempt to make transfer learning more parameter efficient while maintaining or even improving the performance of Fine Tune. Besides parameter efficiency, temporal efficiency is also of great importance in efficient transfer learning.

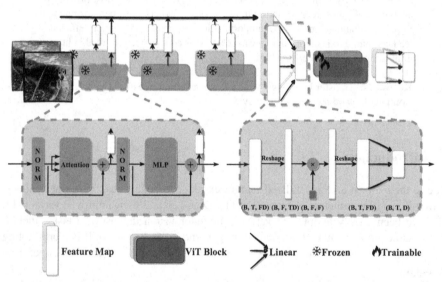

Fig. 1. The overall structure of our proposed Fuse Tune with optional Cross Fuse (Sect. 3.4). We first preserve the features extracted from each block of the frozen pre-trained backbone model and concentrate them together. After being reduced into a standard size of input for ViT, the processed features are finally transformed with a trainable Transformation module. The first enlarged part depicts the inner structure of the ViT block, while the second magnified part details the procedure of how the feature is aggregated into a standard size with batch matrix multiplication (marked as a red square). Notations B, T, F, and D are stated in Sect. 3.1.

Temporal efficiency is a more complex issue. It is not only determined by the number of trainable parameters but also by their specific placements within the network. Parameters close to the inputs require the entire process of back propagation, while parameters close to labels cost few resources to update. And that is exactly why when back propagation is involved in cumbersome backbones, considerable amount of computational cost is required, even when the number of trainable parameters is mineral like Visual Prompt Tuning (VPT) [6] and LoRA [9]. The observation therefore motivates us to explore an approach that circumvents the need for the back propagation procedure in the backbone.

Spurred by this idea, we propose an efficient yet powerful Fuse Tune strategy for transfer learning. Fuse Tune requires only 1/4 trainable parameters compared to Fine Tune yet yields comparable or even superior performance in various downstream tasks. It must be acknowledged that Fuse Tune is superior in temporal and memory efficiency instead of parameter efficiency when compared to parameter efficient methods like Prompt Tune and LoRA. We aim to design a temporal and memory efficient hierarchical decoder to get rid of the back propagation in the pre-trained backbone. Experiments show this design significantly accelerates training and consumes fewer computational resources. Our key contributions are summarized as follows:

- **Applying the feature decoder structure to transfer learning.** Decoders are mainly used to convert the abstract features extracted from encoders into a comprehensible form for pretext tasks. A natural question arises: is it possible to generalize decoders for the knowledge transferring in various downstream scenarios, instead of merely being limited to pretext tasks? Driven by this idea, we design a hierarchical feature decoder, which is a new paradigm for transfer learning.
- **Exploring transferability for large-scale target datasets.** Substantial distributional shifts are evident when comparing small-scale and large-scale datasets. A method that works well on small datasets may not be suitable for large-scale datasets [10]. Previous transfer learning methods mainly evaluate their transferred representation on VTAB [11], a benchmark that only contains tasks of small datasets, but overlook the transferability for large-scale datasets. In this paper, we extend experiments to ImageNet1K, a significantly larger and more intricate dataset, to ascertain the generalizability and robustness of our approach for downstream transferring.
- **Achieving state-of-the-art memory and temporal efficiency.** Learning transferable features often consumes large computation and memory resources due to the extra training on the backbone network. In this paper, we prove that only a small training requirement on an injected decoder could also demonstrate remarkable performance for various downstream tasks. The exclusion of any back propagation in the cumbersome backbone network enhances the efficiency of memory and time utilization. The advantage becomes more obvious when the backbone network grows larger, which is in line with the current trend of big models.

The rest of this paper is organized as follows. We first review some recent results on transfer learning, as well as supervised and self-supervised pre-trained methods in Sect. 2. Then, we detail our Fuse Tune Approach in Sect. 3. Numerical results are provided in Sect. 4, except for the standard image classification to demonstrate the effectiveness of our approach, we also perform robustness and ablation experiments. A summary is finally given in Sect. 5.

2 Related Work

2.1 Transfer Learning

Transfer learning refers to applying a model trained on a data-abundant source domain to a data-scarce target domain [12]. Source domain and target domain can be defined as $D_s = \{X_s, P_s(X_s)\}$ and $D_t = \{X_t, P_t(X_t)\}$ respectively, where X denotes the data samples and $P(X)$ is the label prediction. Transfer learning aims to transfer knowledge in $P_s(X_s)$ to boost the training performance of $P_t(X_t)$.

Recent works imply that transfer learning is effective for providing a restricted parameter subset and inducing a prior to the model [13]. Experiments show insignificant modification of weights in Fine Tune, which brings potentials for efficiently transfer knowledge between tasks [14].

2.2 Existing Transfer Methods

Recently, Transformer-based models have demonstrated powerful ability in capturing complex semantic information buried deep in the data sequence, we therefore focus on the transformer-based transfer learning in the following discussion.

Generally speaking, efficient transformer-based transfer learning could be roughly divided into two main categories: sifting the feature distribution like Adapter and Prompt Tune, and simulating the weights update for pre-trained backbone like LoRA.

Adapter [7] is first proposed for transfer learning in Natural Language Processing (NLP) and soon extended to Computer Vision (CV). To be specific, Adapter keeps the backbone frozen and adds a two-layer residual fully connected module to the end of each attention and MLP in the transformer block. Besides, the normalization layers in the backbone are also trainable. Experiments show that this simple application reaches comparable performance with Fine Tune.

Prompts are additional tokens that remind the pre-trained model of the specific downstream tasks. VPT [6] is a quintessential example applying prompt strategy. VPT prepends several randomly initialized tokens with the image embeddings as the input of each attention block. The pre-trained backbone is frozen during transferring and only the prompts are updated, thus making the number of trainable parameters negligible. However, since the additional tokens are prepended from the very first attention block to the last, full back propagation is inevitable. Moreover, these prepended tokens further exacerbate the already heavy computation and memory burden, which is even heavier than that of Fine Tune (2.2×, reported by the original paper).

Adapter and Prompt Tune modify features to transfer knowledge, while LoRA [9] modifies weights to simulate weights update. Specifically, if a weight matrix W is updated by ΔW after converging, LoRA assumes that this modification matrix of shape $D \times K$ could be approximated by a low-rank decomposition $A \times B$, where A and B are of shape $D \times R$ and $R \times K$ respectively. This greatly reduces the number of parameters to be learned due to R is typically a small number (e.g., 4). As the trainable parameters increase, the rank of LoRA will eventually converge to Fine Tune. Since LoRA simulates weights update, the ΔW trained for a specific task can be directly added to the original weights, which means computational overhead free during inference. However, the back propagation in the backbone is still inevitably involved during the training, leading to a heavy computation burden.

Unlike these works, our proposed Fuse Tune preserves the pre-trained feature information and excludes any back propagation or modification requirements for the backbone. This unprecedented characteristic makes Fuse Tune computational efficient, especially when the pre-trained backbone is cumbersome.

2.3 Self-supervised Pretraining in Vision

There are two main pretext tasks for self-supervised pretraining: Masked Image Modeling (MIM) [2–5] and Instance Discrimination (ID) [6, 7]. To be specific, MIM methods take masked images as input, aiming to reconstruct the original or semantic information of the image, focusing on the internal structural relationship within a sample. While contrastive learning methods based on ID tasks aim at figuring out discriminative differences across samples. In this paper, we employ two state-of-the-art representative models Beit2 and Dino2 as self-supervised pre-trained backbones for testing transfer learning ability of baselines. Beit2 is pre-trained in ImageNet1K with auxiliary information provided by Open CLIP [15]. While Dino2 is pre-trained in merged LVD142M and distilled in ImageNet1K. Both of these two models have learned highly discriminative features from large quantities of data.

Although learning transferable representations from SL pre-trained models has attracted intensive interests, the transferability of aforementioned methods for SSL pre-trained models has not been fully investigated. However, SSL is turning out to become more and more important [16, 17], especially in this era of Big Model. This paper attempts to explore the transferability of different transfer learning methods for both SL and SSL pre-trained models.

3 Approach

In order to enhance the performance of a pre-trained model in a target domain, previous transfer learning methods typically opt to update the parameters or features of a pre-trained backbone through back propagation. The process demands expensive computational and memory resources, particularly when dealing with excessively large models. To alleviate this problem and further utilize the characteristics of the features, we propose Fuse Tune, aiming to design a lightweight plug and play decoder to fuse features for transferring. Fuse Tune only requires tuning the parameters on the decoder, which excludes any back propagation on the backbone network and highly boosts the training efficiency.

3.1 Feature Selection

The overall structure is shown in Fig. 1. Fuse Tune only perform transformations on pre-trained features, the real focus is on digging internal and external relations across features. Fuse Tune could be taken as a hierarchical feature decoder which perform attention across features, hidden dimensions and tokens. Considering a ViT backbone with N blocks, when a sample is sent to the backbone, it would be split into T tokens with D hidden dimensions. We select F intermediate features by default, each of which lies after the attention module and MLP projection (i.e., $F = 2N$ by default). The features are preserved during the forward propagation and concatenated together along hidden dimension as shown in Fig. 1. To be specific, the shape of each feature in ViT is $B \times T \times D$, where B denotes the batch size. After concatenating, the shape of the concatenated features become $B \times T \times FD$. The exact number of features selected would be discussed in the ablation study section.

In the following sections, block number, token number, selected features' number, and hidden dimension are denoted as N, T, F and D respectively.

3.2 Feature Aggregation and Compression

In this subsection, we need to effectively and efficiently compress the concatenated feature of shape $B \times T \times FD$ into a standard feature size of ViT (i.e., $B \times T \times D$). Traditional options like linear projection and PCA [18] are either computationally heavy or poor in performance. To be specific, a valine linear projection would experimentally and theoretically cause information loss due to potential non-linear relationships between features. While multi-layer perceptron would cost unacceptable computation. There is a similar dilemma for PCA, the computational complexity of PCA is

$$Complexity = TDF \times min\{T, D, F\} + (DF)^3, \tag{1}$$

which cubically explodes as hidden dimension grows. Therefore, a parameter and computation efficient projector with strong performance is of great necessity.

Since this paper is devoted to develop an efficient transfer learning method, we select Linear projection for further improvement. To address the aforementioned information loss problem with the already efficient Linear projection, we further enhance the performance by proposing a novel feature aggregation method.

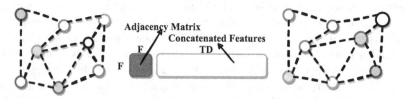

Fig. 2. Explanation for Feature decoder. Denote each feature as a node (with dimension TD) and their relations as weights for edges (elements in the adjacency matrix). By automatically learning the weights through back propagation, we can investigate the relations between the extracted features and further aggregate the features.

Deep features are transformed from shallow ones, the features obtained from the deep and shallow layers are closely connected. Drawing on this idea, the extracted features can be represented as connected nodes on a graph. We then obtain an adjacency matrix $A \in R^{F \times F}$. After randomly initializing this adjacency matrix and setting its elements as learnable parameters, batch matrix multiplication is performed with the reshaped feature map of size $B \times F \times TD$. Finally, we obtain aggregated features of size

$$(B, F, TD) = (B, F, F) \times (B, F, TD) \tag{2}$$

In this proposed feature aggregation, only F^2 (576 by default) additional parameters are required. It shows a significant improvement in performance without additional computation. This option is similar with the external attention mechanism [19] across features. Detailed diagrams are shown in the amplified part in Fig. 1 and Fig. 2.

3.3 Feature Transformation

So far, we have fused both low-level and high-level features, focusing on the connections across features from different layers. However, the internal relationship within a feature is yet to be thoroughly leveraged. To address the problem, we utilize a feature transformation for distribution adaption. Experiments prove that even a single attention block is sufficient due to previous efforts. We also observe a significant drop in performance without the feature transformation, which proves the necessity of this implementation.

3.4 Cross Fuse (Optional)

Once the features are decoded, they are ready to be sent to a task head for various downstream applications. Since the frozen backbone requires few computational resources during transferring, it is possible to fuse multiple pre-trained models for more robust representations. Experiments show that Cross Fuse improves downstream performance by a large margin, even when a simple linear projection is used to fuse different backbone's decoded features. It should be noted that Fine Tune and other transfer leaning methods require back propagation in the backbone. For this very reason, fusing multiple models is computationally heavy in these methods.

4 Experiments

4.1 Benchmarks and Baselines

To make a fair and comprehensive comparison with other state-of-the-art transfer learning methods, we select transformer models pre-trained on both supervised (ViT) and self-supervised (Beit2, Dino2) scenarios. We evaluate Fine Tune, Prompt Tune, Adapter, and the proposed Fuse Tune in ImageNet1K and various downstream tasks selected from VTAB.

There are three groups of datasets in VTAB, named VTAB-Nature, VTAB-Specialized and VTAB-Structured. The images in the VTAB-Nature are common nature images collected in everyday life like Cifar100 and Caltech101. The images in the VTAB-Specialized dataset are images from specialized fields, like satellite overhead views (Euro SAT) and pathological images (Patch Camelyon [20]). While VTAB-Structured involves structured images, such as predicting the distance to an object in a 3D scene, counting the number of objects in an image, and detecting the orientation of an object. To evaluate the transfer learning ability in scale for nature images, we select ImageNet1K rather than Cifar100 in VTAB-Nature. While for VTAB-Specialized and VTAB-Structured, we select the first dataset in each of them (i.e., Patch Camelyon and Clevr Count [21]). Hyperparameters for baselines (shown in Table 4) are controlled to be consistent.

We also make robustness comparisons when few labeled samples are available. Finally, an ablation study is carried to justify the sufficiency and necessity of our proposed designs individually.

4.2 Image Classification

Table 1. Fuse Tune classification results compared to other transfer learning methods. Fuse means fusing single model's features and Cross Fuse refers to fusing multiple models together. The Temporary efficiency and memory efficiency of Cross Fuse is compared to the single model fusing for Dino2. Dino2 pre-trained model is of patch size 14 while Beit2 and ViT are of patch size 16. We conduct 30 epochs transfer for ImageNet and 10 epochs for other benchmarks.

	ImageNet1K	Camelyon	Clevr	Average	Time	Memory
ImageNet21K Pre-trained ViT-B						
Fine Tune	80.62	87.08	**94.37**	**87.36**	1.6×	1.34×
Prompt	80.69	86.53	90.98	86.06	2.84×	2.69×
Adapter	**80.73**	86.33	87.91	84.99	1.17×	1×
Fuse	78.42	**87.16**	87.66	84.41	1×	1×
Dino2 Pre-trained ViT-B						
Fine Tune	84.79	89.50	99.72	91.34	1.61×	1.40×
Prompt	84.85	90.05	99.45	91.45	2.40×	2.23 ×
Adapter	83.59	90.23	99.59	91.13	1.19×	1.01 ×
Fuse	**85.75**	**90.31**	**99.81**	**91.96**	1×	1 ×
Beit2 Pre-trained ViT-B						
Fine Tune	83.38	90.79	97.75	90.64	1.61×	1.34 ×
Prompt	81.35	**92.07**	**98.48**	90.63	2.40×	2.69 ×
Adapter	80.27	90.23	97.65	89.38	1.19×	1×
Fuse	**83.45**	91.15	98.30	**90.96**	1×	1×
Dino2 and Beit2 Cross Fuse						
	86.26	**92.09**	**99.83**	**92.73**	1.7×	1.47×

From Table 1, Fuse Tune achieves the state-of-the-art performance especially in SSL pretraining and meanwhile consumes minimum computation resources. Even when binding Dino2 and Beit2 together (Cross Fuse) to obtain more stable features, Fuse Tune still consumes similar computational resources to finetuning single model. An astonishing 86.26 % accuracy on the ImageNet1K validation set is achieved by Cross Fuse, outperforming Fine Tune by at most 2.88 %. Temporary and Memory efficiency comparison is conducted on ImageNet1K benchmark.

Since the VPT paper does not offer settings for ImageNet transferring, we tried to prepend tokens ranging from 10, 50, 100 and 200, the best performance is achieved by VPT Deep with 200 prompt tokens. Sine the computational complexity of the attention mechanism grows quadratically with the number of tokens, more prompt tokens would lead to unacceptable costs. We do not observe significant improvement in performance with even more prompt tokens.

4.3 Robust Study

For downstream tasks, training data with labels could be scarce in plenty of scenarios. Hence, the ability of leveraging few labels while transferring is crucial for downstream tasks. In this section, we compare the transfer learning capabilities of Fuse Tune and other methods within a sparse sample space.

Table 2. 1% transferring (13 training examples for each class in ImageNet1K). All models are transferred with a 10-epoch schedule.

	ImageNet1K (1%)
ImageNet21K Pre-trained ViT-B	
Full	67.42
Prompt	75.20
Adapter	**76.37**
Fuse	71.18
Dino2 Pre-trained ViT-B	
Full	69.57
Prompt	73.72
Adapter	73.66
Fuse	**74.54**
Beit2 Pre-trained ViT-B	
Full	67.47
Prompt	66.30
Adapter	63.61
Fuse	**68.59**

We observe that our proposed Fuse Tune underperforms the existing approaches like Adapter and Prompt Tune in supervised ViT-B scenarios. It is understandable since ViT-B is a SL pre-trained model. Fuse Tune critically relies on the generalization ability of the pre-trained backbone since it keeps the backbone gradient frozen for downstream tasks.

Previous studies have indicated that the SL pre-trained models are inferior in feature generalization compared to the SSL pre-trained ones [15]. Considering the ubiquitous feature-abundant but label-scarce scenarios, the transferability of SSL pre-training models is much more important in nowadays [16]. Experiments in Table 2 show that in SSL pre-trained scenarios, Fuse Tune is superior to other existing methods. Specifically, in comparison to the conventional Fine Tune approach, our proposed Fuse Tune yielded a remarkable increase in accuracy by 4.97% and 1.12% for SSL pre-trained backbones Dino2 and Beit2, respectively.

4.4 Ablation Study

The ablation study shows the optimal settings which balance performance and efficiency. According to Table 3(a), while no feature transformation is utilized, the performance is significantly poorer than that with a single-block attention by 2.92%. Performance keeps growing by 0.51% if more blocks in the feature transformation are adopted, however, the number of trainable parameters also grows to 2.24× accordingly. While studying the influence of the first configuration, the rest configurations are set as the first column by default.

Table 3. 10 epoch pilot experiment for ablation study on ImageNet1K. The default settings in Fuse Tune (1 attention block with 24 intermediate features, 2 aggregation layers, and light argumentation) are marked in gray.

Blocks	Perform	Parameters
0	81.74	4.3M
1	84.66	11.4M
3	85.17	25.55M

(a)   Feature Transformation Depth

Feature	Aggre.	Perform	Params.
6	✗	84.66	11.4M
6	2	84.57	11.4M
12	✗	84.27	14.95M
24	✗	84.61	22.05M
24	2	85.54	22.05M

(b)   Extracted Feature number

Aggre.	Perform	Params.	Time
✗	84.61	22.05M	1×
1	84.98	22.05M	1×
2	85.54	22.05M	1.2×
4	83.53	22.05M	3.1×

(c)   Feature Graph Aggregation

Regulation	Perform
Light	85.54
Heavy	83.52

(d)   Regulation Strategy

Table 3(b) proves how many intermediate features should be extracted exactly. We observe that only fusing the last 6 layers' features outperforms fusing the entire 24 features without feature aggregation. It is understandable because only a linear projection is adopted for compressing all the features. With the help of feature aggregation, the intermediate features of 24 layers far exceed those of 6 layers (by 1% with negligible additional parameters). We also find out that the feature aggregation hardly has any help for the 6-layer fused features. It can be inferenced that the role of feature aggregation is to assist the model in better understanding the relationship between features and finally submit a more suitable representation. While valine linear projection is enough when there are few features.

Table 4. Hyperparameter settings for Fuse Tune.

Hyperparameters	Fuse Tune
Base learning rate	1e-4
Fine Tune epoch	10 (30 for ImageNet)
Warmup epoch	1 (3 for ImageNet)
Batch size	256
AdamW ε	1e-8
AdamW β	(0.9, 0.999)
Minimal learning rate	1e-6
Learning rate schedule	Cosine
Weight decay	0.0001 (0.05 for ImageNet)
Label smoothing ε	0.1
Rand augment [22]	9/0.5
Stochastic depth [23]	0.1
Cutmix [24] and Mixup [25]	×

Table 3(c) further explores the relationship between the number of aggregation layers and performance. We find out that the lack of layer aggregation results in lower performance than utilizing, and the most suitable number of layers was 2. Excessive depth of layers led to a decrease in performance and a significant increase in time consumption (3.1×). In Table 3(d), heavy regulation refers to training with Mixup [25] and Cutmix [24] while light refers to training without them.

Table 4 shows detailed training recipe for Fuse Tune. We follow recipes proposed in the original papers for reproducing the baselines except for Dino2 Fine Tune. Dino2 focuses on general representations and the authors do not perform Fine Tune for any specific downstream task, so we follow the reproduction of Dino Fine Tune in Beit.

5 Conclusion

In this paper, we propose Fuse Tune, a hierarchical decoder for further promoting the efficiency of transfer learning. So, there is only a training requirement for the injected feature decoder on top of the backbone network. The exclusion of back propagation in cumbersome backbone greatly speeds up the current transfer learning methods like Fine Tune and Prompt Tune. Experiments show that this speedup can be up to 2.84 times, and will theoretically continue to grow with an increasement of the model size. We wish our work could give a remarkable insight to researchers who care about transfer efficiency, and eventually benefit the development of the transfer learning community.

References

1. Bao, H., Dong, L., Piao, S., Wei, F.: Beit: bert pre-training of image transformers. arXiv preprint arXiv:2106.08254 (2021)
2. Caron, M., et al.: Emerging properties in self-supervised vision transformers. In: Proceedings of the IEEE/CVF International Conference on Computer Vision, pp. 9650–9660 (2021)
3. Peng, Z., Dong, L., Bao, H., Ye, Q., Wei, F.: Beit v2: masked image modeling with vector-quantized visual tokenizers. arXiv preprint arXiv:2208.06366 (2022)
4. Oquab, M., et al.: DINOv2: learning robust visual features without supervision. arXiv preprint arXiv:2304.07193 (2023)
5. Han, X., et al.: Pre-trained models: past, present and future. AI Open **2**, 225–250 (2021)
6. Jia, M., et al: Visual prompt tuning. In: Computer Vision–ECCV 2022: 17th European Conference, Tel Aviv, Israel, 23–27 October 2022, Proceedings, Part XXXIII, pp. 709–727 (2022)
7. Houlsby, N., et al.: Parameter-efficient transfer learning for NLP. In: International Conference on Machine Learning, pp. 2790–2799 (2019)
8. Guo, Y., Shi, H., Kumar, A., Grauman, K., Rosing, T., Feris, R.: Spottune: transfer learning through adaptive fine-tuning. In: Proceedings of the IEEE/CVF Conference on Computer Vision and Pattern Recognition, pp. 4805–4814 (2019)
9. Hu, E.J., et al.: Lora: low-rank adaptation of large language models. arXiv preprint arXiv: 2106.09685 (2021)
10. Malinin, A., et al.: Shifts: a dataset of real distributional shift across multiple large-scale tasks. arXiv preprint arXiv:2107.07455 (2021)
11. Zhai, X., et al.: A large-scale study of representation learning with the visual task adaptation benchmark. arXiv preprint arXiv:1910.04867 (2019)
12. Farahani, A., Pourshojae, B., Rasheed, K., Arabnia, H.R.: A concise review of transfer learning. In: 2020 International Conference on Computational Science and Computational Intelligence (CSCI), pp. 344–351 (2020)
13. Mao, H.H.: A survey on self-supervised pre-training for sequential transfer learning in neural networks. arXiv preprint arXiv:2007.00800 (2020)
14. Sanh, V., Wolf, T., Rush, A.: Movement pruning: adaptive sparsity by fine-tuning. Adv. Neural. Inf. Process. Syst. **33**, 20378–20389 (2020)
15. Cherti, M., et al.: Reproducible scaling laws for contrastive language-image learning. arXiv preprint arXiv:2212.07143 (2022)
16. Zhong, Y., Tang, H., Chen, J., Peng, J., Wang, Y.X.: Is self-supervised learning more robust than supervised learning? arXiv preprint arXiv:2206.05259 (2022)
17. Balestriero, R., et al.: A cookbook of self-supervised learning. arXiv preprint arXiv:2304. 12210 (2023)
18. Maćkiewicz, A., Ratajczak, W.: Principal components analysis (PCA). Comput. Geosci. **19**(3), 303–342 (1993)
19. Guo, M.H., Liu, Z.N., Mu, T.J., Hu, S.M.: Beyond self-attention: external attention using two linear layers for visual tasks. IEEE Trans. Pattern Anal. Mach. Intell. **45**, 5436–5447 (2022)
20. Veeling, B.S., Linmans, J., Winkens, J., Cohen, T., Welling, M.: Rotation equivariant CNNs for digital pathology. In: Frangi, A.F., Schnabel, J.A., Davatzikos, C., Alberola-López, C., Fichtinger, G. (eds.) MICCAI 2018. LNCS, vol. 11071, pp. 210–218. Springer, Cham (2018). https://doi.org/10.1007/978-3-030-00934-2_24
21. Johnson, J., Hariharan, B., Van Der Maaten, L., Fei Fei, L., Zitnick, C.L., Girshick, R.: Clevr: A diagnostic dataset for compositional language and elementary visual reasoning. In: Proceedings of the IEEE Conference on Computer Vision and Pattern Recognition, pp. 2901–2910 (2017)

22. Cubuk, E.D., Zoph, B., Shlens, J., Le, Q.V.: Randaugment: practical automated data augmentation with a reduced search space. In: Proceedings of the IEEE/CVF Conference on Computer Vision and Pattern Recognition Workshops, pp. 702–703 (2020)
23. Huang, G., Sun, Yu., Liu, Zhuang, Sedra, Daniel, Weinberger, Kilian Q.: Deep networks with stochastic depth. In: Leibe, B., Matas, J., Sebe, N., Welling, M. (eds.) ECCV 2016. LNCS, vol. 9908, pp. 646–661. Springer, Cham (2016). https://doi.org/10.1007/978-3-319-46493-0_39
24. Yun, S., Han, D., Oh, S.J., Chun, S., Choe, J., Yoo, Y.: Cutmix: regularization strategy to train strong classifiers with localizable features. In: Proceedings of the IEEE/CVF International Conference on Computer Vision, pp. 6023–6032 (2019)
25. Zhang, H., Cisse, M., Dauphin, Y.N., Lopez-Paz, D.: mixup: beyond empirical risk minimization. arXiv preprint arXiv:1710.09412 (2017)

Industrial-SAM with Interactive Adapter

Guannan Jiang[(✉)]

CATL, Ningde, China
JiangGN@catl.com

Abstract. Image segmentation is a fundamental task in computer vision that aims to partition an image into multiple regions or objects. The recently proposed Segment Anything Model (SAM) is a powerful and promptable model for image segmentation that can produce high-quality object masks from input prompts such as points or boxes. However, recent studies have found that SAM, which is trained on natural images, performs poorly on non-natural images, such as industrial images, where the objects and backgrounds may have different characteristics and distributions. To address this problem, we propose Industrial-SAM, a SAM adapted for industrial vision applications. Industrial-SAM leverages the pre-trained SAM as a backbone and fine-tunes it on domain-specific datasets of industrial images. To efficiently train Industrial-SAM and make it suitable for industrial vision, we design an Interactive Adapter. Interactive Adapter is a novel module that can incorporate human feedback during training to guide the model learning. It can also generate adaptive prompts for different images based on their content and difficulty. We conduct extensive experiments on the cap welding segmentation dataset and seal pin welding segmentation dataset. The experimental results show that Industrial-SAM can train excellent industrial vision models with the help of human feedback.

Keywords: Industrial Segmentation · Segment-anything Models · Efficient Training · Visual Adapter · Human Interactive Learning

1 Introduction

Image segmentation is a fundamental task in computer vision [5,7–10,13–15,20,24,28,32] that aims to partition an image into multiple regions or objects, each with a distinct semantic meaning. Image segmentation has many applications in various domains, such as medical imaging, self-driving cars, satellite imaging, and industrial vision. Industrial vision refers to the use of computer vision techniques to automate tasks in industrial settings, such as quality inspection, defect detection, and object recognition. Industrial vision poses unique challenges for image segmentation, as the objects and backgrounds may have different characteristics and distributions than natural images.

One of the recent advances in image segmentation is the Segment Anything Model (SAM) [17]. SAM is a powerful and promptable model that can produce high-quality object masks from input prompts such as points or boxes. SAM

Q. Liu et al. (Eds.): PRCV 2023, LNCS 14431, pp. 220–230, 2024.
https://doi.org/10.1007/978-981-99-8540-1_18

can also generate masks for all objects in an image without any prompts. SAM was trained on a large-scale dataset of 11 million images and 1.1 billion masks, covering a wide range of object categories and scenes. SAM is considered the first foundational model for computer vision, as it can transfer zero-shot to new image distributions and tasks.

However, recent studies [4,27,34] have found that SAM, which is trained on natural images, performs poorly on non-natural images, such as industrial vision. This is because SAM may not be able to generalize well to the different appearance and geometry of industrial objects and backgrounds. Moreover, SAM may not be able to handle the noise, occlusion, and illumination variations that are common in industrial images. Therefore, there is a need to adapt SAM for industrial vision applications.

In this paper, we propose Industrial-SAM, a SAM adapted for industrial vision. Industrial-SAM leverages the pre-trained SAM as a backbone and fine-tunes it on domain-specific datasets of industrial images. To efficiently train Industrial-SAM and make it suitable for industrial vision, we design an Interactive Adapter. Interactive Adapter is a novel module that can incorporate human feedback during training to guide the model learning. It can also generate adaptive prompts for different images based on their content and difficulty.

We conduct extensive experiments on the cap welding segmentation dataset and seal pin welding segmentation dataset, which is a challenging benchmark for industrial vision segmentation. The experimental results show that Industrial-SAM can train excellent industrial vision models with the help of human feedback.

The main contributions of this paper are:

- We propose Industrial-SAM, a SAM adapted for industrial vision applications.
- We design Interactive Adapter, an interactive adapter that can incorporate human feedback during training and generate adaptive prompts for different images.
- We evaluate Industrial-SAM on the cap welding segmentation dataset and seal pin welding segmentation dataset and show its superior performance.

The rest of this paper is organized as follows: Sect. 2 reviews related work on image segmentation and promptable models. Section 3 describes the proposed Industrial-SAM in detail. Section 4 presents the experimental setup and results. Section 5 concludes the paper.

2   Related Work

2.1   Image Segmentation

Image segmentation is a well-studied problem in computer vision that has many applications in various domains. Image segmentation methods can be broadly classified into two categories: supervised and unsupervised. Supervised methods can be further divided into semantic segmentation and instance segmentation. Semantic segmentation assigns a class label to each pixel in an image, while instance segmentation separates different objects of the same class.

Supervised methods usually rely on deep neural networks to learn features and classifiers from data. Some of the popular network architectures for image segmentation are Fully Convolutional Networks [23], U-Net [29], Mask R-CNN [8], and DeepLab [2]. These methods achieve impressive results on various benchmarks, such as PASCAL VOC, MS COCO, and Cityscapes. However, supervised methods have some limitations, such as requiring large amounts of annotated data, being prone to overfitting, and being unable to generalize well to new domains or tasks.

Unsupervised methods do not require annotated data for training, but instead rely on clustering techniques or generative models to group pixels into segments. Some of the classical clustering techniques for image segmentation are k-means [1], mean-shift [6], normalized cuts, and graph-based methods. These methods are simple and fast, but they often produce noisy or inconsistent segments, and they do not capture high-level semantic information.

2.2 Promptable Segmentation Models

Promptable models are a new paradigm for image segmentation that aim to overcome the limitations of both supervised and unsupervised methods. Promptable models are trained on large-scale datasets with diverse images and masks, and they can produce high-quality segments from input prompts such as points or boxes. Promptable models can also generate segments for all objects in an image without any prompts. Promptable models are flexible and adaptable, as they can transfer zero-shot to new domains or tasks without requiring additional data or fine-tuning.

The first promptable model for image segmentation is the Segment Anything Model (SAM) [17]. SAM is a powerful and promptable model that can produce high-quality object masks from input prompts such as points or boxes. SAM can also generate masks for all objects in an image without any prompts. SAM was trained on a large-scale dataset of 11 million images and 1.1 billion masks, covering a wide range of object categories and scenes. SAM is considered the first foundational model for computer vision, as it can transfer zero-shot to new image distributions and tasks.

However, SAM has some limitations when applied to industrial vision applications. SAM was trained on natural images, which may have different characteristics and distributions than industrial images. Moreover, SAM may not be able to handle the noise, occlusion, and illumination variations that are common in industrial images. Therefore, there is a need to adapt SAM for industrial vision applications.

In this paper, we propose Industrial-SAM, a SAM adapted for industrial vision applications. Industrial-SAM leverages the pre-trained SAM as a backbone and fine-tunes it on a domain-specific dataset of industrial images. To efficiently train Industrial-SAM and make it suitable for industrial vision, we design an

Interactive Adapter. Interactive Adapter can also generate adaptive prompts for different images based on their content and difficulty.

2.3 Parameter-Efficient Fine-Tuning

Parameter-efficient fine-tuning (PEFT) [3,11,12,16,18,19,22,25,30,31,36–39] is an important research direction for large-scale pre-trained models. The idea of PEFT is to fine-tune only a small fraction of lightweight modules in a large-scale pre-trained model, instead of the whole model, to obtain good performance on downstream tasks. PEFT was first developed and used in natural language processing (NLP) [11,12,18,19,22,26,30,37], and has been recently applied to computer vision (CV) [3,11,31,36,38,39]. The two main types of PEFT for large-scale vision models are adapter-based methods [3,11,31] and prompt tuning methods [38,39]. Adapter-based methods put small MLP networks into the vision model to customize it for downstream tasks. Prompt tuning methods append a few trainable tokens to the input sequence of vision Transformer to minimize the difference between pre-training and downstream data distributions. LoRA [12] learns low-rank parameters for the frozen weights of multi-head attentions [33]. Zhang et al. [36] propose a prompt search algorithm to automatically combine adapter, prompt tuning and LoRA together. Recently, Lian et al. [21] and Luo et al. [25] propose re-parameterized methods for visual adaptation, which can save additional costs during inference.

In this paper, we propose interactive adapter, a novel adapter method that incorporates human feedback during training to guide the model learning. Interactive adapter is inspired by interactive learning, which is a paradigm that allows humans to provide feedback or guidance to a learning agent during its learning process. Interactive learning can improve the efficiency and effectiveness of learning by reducing the data requirements, correcting errors, and providing explanations.

Interactive adapter combines the advantages of parameter-based adapter and prompt-based adapter. Interactive adapter inserts small modules between the layers of a pre-trained model and trains them on the target domain or task with human feedback. Interactive adapter can also generate adaptive prompts for different images based on their content and difficulty. Interactive adapter can leverage the power of deep neural networks and natural language prompts, while benefiting from human feedback to improve its performance and interpretability.

3 Method

3.1 Industrial-SAM

The overall framework of the proposed Industrial-SAM is illustrated in Fig. 1. Industrial-SAM is an image segmentation method based on human-computer interaction, which uses the self-attention mechanism to improve the accuracy and efficiency of segmentation. Industrial-SAM consists of three steps for training.

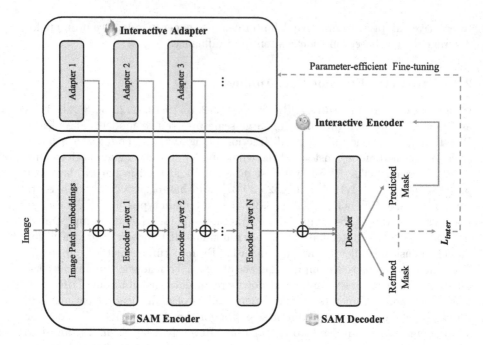

Fig. 1. Framework of **Industrial-SAM**. Industrial-SAM consists of three steps for training. First, the fixed SAM encoder takes the input image and generates the image features. Next, the fixed SAM decoder utilizes the image features and a box prompt that covers the entire image to predict the corresponding mask. Second, the predicted mask undergoes human feedback, where positive and negative points are clicked to correct the predicted mask and produce the refined mask. Third, the predicted mask and the refined mask are used to compute the interactive loss. The loss is then used to update the parameters of the interactive adapter. During inference, the segmentation results can be obtained either by a box prompt that covers the whole image or by clicking maps.

Predict Masks with Fixed SAM Encoder and Decoder. In this step, we use a fixed SAM encoder to receive the input image and generate the image features. The SAM encoder can capture the global and local dependencies among the pixels in the image. The image features are then fed into a fixed SAM decoder, which uses the image features and a box prompt that covers the entire image to predict the corresponding mask. The box prompt is a binary map that indicates the region of interest for segmentation. The SAM decoder also employs the SAM to refine the mask prediction based on the image features and the box prompt. The purpose of this step is to let the SAM encoder and decoder predict the global and local information of the image, which facilitates the subsequent human-computer interaction.

Human-Computer Interaction Feedback. In this step, we use human clicks to correct the predicted mask and generate the refined mask. The positive clicks indicate the regions that need to be preserved, while the negative clicks indicate the regions that need to be removed. The purpose of this step is to let humans participate in the segmentation process, providing more accurate and detailed annotation information, which improves the quality of segmentation.

Updating of the Interactive Adapter. In this step, we use the predicted mask and the refined mask to compute the interacting loss. The interacting loss measures the difference between the predicted mask and the refined mask. Then, we use the interacting loss to update the parameters of the interacting adapter. Specifically, the interacting loss is defined as:

$$\mathcal{L}_{inter}(\boldsymbol{m}_p, \boldsymbol{m}_r) = \mathcal{L}_{L1}(\boldsymbol{m}_p, \boldsymbol{m}_r) + \mathcal{L}_{dice}(\boldsymbol{m}_p, \boldsymbol{m}_r), \tag{1}$$

where $\mathcal{L}_{L1}(\cdot, \cdot)$ denotes the smooth L1 loss, $\mathcal{L}_{dice}(\cdot, \cdot)$ is the dice loss, $\boldsymbol{m}_p, \boldsymbol{m}_r$ denote the predicted and refined masks, respectively. The interactive adapter is a small network that connects after the output layer of the SAM encoder, which adjusts the predicted mask to fit the human-computer interaction feedback. The purpose of this step is to let the interactive adapter learn the patterns of human-computer interaction feedback, which improves the flexibility and adaptability of segmentation.

Finally, in inference stage, we can obtain segmentation results by two ways. One is using a box prompt that covers the whole image, which can quickly get a coarse segmentation result. The other is using clicking maps, which can generate more precise and detailed segmentation results according to user clicks.

3.2   Interactive Encoder

The designed interacting encoder follows the same design of the prompt encoder in SAM. The prompt encoder is responsible for encoding the prompts that are given by the user to guide the segmentation process. These prompts can be of different modalities, such as points, boxes, or texts. PointEmbedding module embeds each point prompt with an embedding vector, which is learned from an embedding layer. There are four types of point prompts: positive point, negative point, box corner point, and padding point. Each type has its own embedding layer. The padding point is used to pad the point prompts to a fixed length. NotAPointEmbedding module embeds each pixel that is not a point prompt with a special embedding vector, which is learned from an embedding layer. This module helps to distinguish between point prompts and non-point prompts. MaskDownscaling module downscales and encodes the mask prompts that are given by the user. The mask prompts are binary maps that indicate the region of interest for segmentation. The mask downscaling module consists of several convolutional layers, layer normalization layers, and activation layers. The output of this module has the same spatial size as the image embedding. NoMaskEmbedding module embeds each pixel that has no mask prompt with a special embedding vector, which is learned from an embedding layer. This module helps to distinguish between mask prompts and non-mask prompts. The output of the prompt encoder is a tensor that concatenates the embeddings of different

Fig. 2. Examples of cap welding and seal pin welding image segmentation.

types of prompts. This tensor has the same shape as the image embedding, and it is fed into the mask decoder along with the image embedding to predict the segmentation mask.

3.3   Interactive Adapter

The Interacting Adapter is a novel module that is added to the SAM model to improve its performance on challenging segmentation tasks. The adapter consists of two parts: a feature extractor and a feature fusion module. The feature extractor is a convolutional neural network that extracts high-level semantic features from the input image. The feature fusion module is a self-attention mechanism that learns to fuse the features from the feature extractor and the SAM model. The adapter structure can be easily plugged into any existing SAM model without changing its architecture or parameters. The adapter structure helps the SAM model adapt to segment industrial images.

4   Experiments

4.1   Experimental Settings

Datasets. Cap welding and seal pin welding image segmentation datasets. Seal pin welding and cap welding are two types of welding methods that are used to join metal parts together. Seal pin welding uses a small metal pin that is inserted into a hole in the metal parts and then heated to melt and fuse them together. Cap welding uses a metal cap that is placed over the joint of the metal parts and then heated to melt and bond them together. As shown in Fig. 2, the goal of seal pin welding and cap welding image segmentation is to separate the regions of the image that correspond to the welds from the regions that correspond to the metal parts. The output of the image segmentation is a binary mask that indicates which pixels belong to the welds and which pixels belong to the metal parts. Specifically, the image segmentation dataset of seal nail welding and top cover welding is a collection of images that are used to train and test AI models for detecting defects in lithium battery production. Seal nail welding is a process of sealing the battery shell with a steel nail after injecting electrolyte. Top cover welding is a process of welding the cap to the bare battery cell. Both processes are important for ensuring the quality and safety of the battery. The

Table 1. Results of cap welding segmentation dataset.

Method	Backbone	Train	Test	mIoU
OCRNet [35]	HRNet	supervised	-	76.34
U-Net [29]	U-Net	supervised	-	77.67
SAM [17]	ViT-B	supervised	Box	34.39
SAM [17]	ViT-B	supervised	1 click	58.50
SAM [17]	ViT-B	supervised	5 click	69.95
SAM [17]	ViT-B	supervised	10 click	75.90
Industrial-SAM, *ours*	ViT-B	human feedback (Box)	Box	50.50
Industrial-SAM, *ours*	ViT-B	human feedback (1 click)	Box	**80.31**
Industrial-SAM, *ours*	ViT-B	human feedback (5 click)	Box	79.79
Industrial-SAM, *ours*	ViT-B	human feedback (10 click)	Box	79.08

Table 2. Results of seal pin welding segmentation dataset.

Method	Backbone	Train	Test	mIoU
OCRNet [35]	HRNet	supervised	-	77.59
U-Net [29]	U-Net	supervised	-	79.87
SAM [17]	ViT-B	supervised	Box	45.93
SAM [17]	ViT-B	supervised	1 click	46.01
SAM [17]	ViT-B	supervised	5 click	43.17
SAM [17]	ViT-B	supervised	10 click	42.43
Industrial-SAM, *ours*	ViT-B	human feedback (Box)	Box	49.86
Industrial-SAM, *ours*	ViT-B	human feedback (1 click)	Box	69.33
Industrial-SAM, *ours*	ViT-B	human feedback (5 click)	Box	81.35
Industrial-SAM, *ours*	ViT-B	human feedback (10 click)	Box	**81.41**

dataset contains 311 images for seal nail welding training, 60 images for seal nail testing, 1601 images for top cover welding training, and 201 images for top cover welding testing. The images capture various types of defects such as bubbles, cuts, scratches, dirt, pits, bursts, bumps, skew, pinholes, etc.

4.2 Main Results

Results on Cap Welding Segmentation. Table 1 compares different methods for cap welding segmentation. The methods are evaluated by their mean intersection over union (mIoU), which measures how well the predicted regions match the ground truth regions. The methods use different backbones, training modes and test modes. The backbone is the base model that extracts features from the images. The training mode is how the model is trained, either by supervised learning or by human feedback. The test mode is how the model is tested, either by using bounding boxes or clicks to indicate the regions of interest.

The table shows that OCRNet and U-Net are two supervised methods that achieve high mIoU scores, but they require full annotations for training. SAM is a semi-supervised method that uses bounding boxes or clicks as weak annotations for training, but it has lower mIoU scores than the supervised methods. The proposed Industrial-SAM uses human feedback as a form of weak annotation for training. It outperforms SAM and even surpasses U-Net when using one click as human feedback. Industrial-SAM also achieves comparable results to OCRNet when using five or ten clicks as human feedback. This demonstrates that Industrial-SAM can effectively leverage human feedback to improve cap welding segmentation performance with minimal annotation cost.

Results on Seal Pin Welding Segmentation. Table 2 presents a comparison of different methods for segmenting seal pins in welding images, which is a task that involves detecting the regions of interest in metal pin images. The methods are assessed by their mean intersection over union (mIoU), a metric that quantifies the degree of overlap between the predicted and the ground truth regions. The methods differ in their backbones, training modes and test modes. The backbone refers to the base model that extracts features from the images. The training mode indicates how the model is trained, either by supervised learning or by human feedback. The test mode specifies how the model is tested, either by using bounding boxes or clicks to provide the regions of interest.

The table reveals that OCRNet and U-Net are two supervised methods that attain high mIoU scores, but they necessitate full annotations for training. SAM is a semi-supervised method that employs bounding boxes or clicks as weak annotations for training, but it yields lower mIoU scores than the supervised methods. The proposed Industrial-SAM utilizes human feedback as a form of weak annotation for training. It outshines SAM and even exceeds OCRNet and U-Net when using five or ten clicks as human feedback. Industrial-SAM also achieves results comparable to U-Net when using one click as human feedback. This illustrates that Industrial-SAM can effectively exploit human feedback to enhance seal pin welding segmentation performance with minimal annotation cost.

5   Conclusion

In this paper, we have presented Industrial-SAM, a SAM adapted for industrial vision applications. Industrial-SAM can leverage the pre-trained SAM as a backbone and fine-tune it on domain-specific datasets of industrial images with the help of human feedback. We have introduced Interactive Adapter, a novel module that can incorporate human feedback during training and generate adaptive prompts for different images. We have shown that Industrial-SAM can achieve state-of-the-art results on the cap welding segmentation dataset and seal pin welding segmentation dataset. Industrial-SAM is a powerful and promptable model for industrial vision segmentation that can benefit from human-machine collaboration.

References

1. Ahmed, M., Seraj, R., Islam, S.M.S.: The k-means algorithm: a comprehensive survey and performance evaluation. Electronics **9**(8), 1295 (2020)
2. Chen, L.C., Papandreou, G., Kokkinos, I., Murphy, K., Yuille, A.L.: DeepLab: semantic image segmentation with Deep Convolutional Nets, atrous convolution, and fully connected CRFs. IEEE Trans. Pattern Anal. Mach. Intell. **40**(4), 834–848 (2017)
3. Chen, S., et al.: AdaptFormer: adapting vision transformers for scalable visual recognition. CoRR abs/2205.13535 (2022)
4. Chen, T., et al.: SAM fails to segment anything? – SAM-adapter: adapting SAM in underperformed scenes: camouflage, shadow, and more. arXiv preprint arXiv:2304.09148 (2023)
5. Cheng, B., et al.: Panoptic-DeepLab: a simple, strong, and fast baseline for bottom-up panoptic segmentation. In: Proceedings of the IEEE/CVF Conference on Computer Vision and Pattern Recognition (2020)
6. Comaniciu, D., Meer, P.: Mean shift: a robust approach toward feature space analysis. IEEE Trans. Pattern Anal. Mach. Intell. **24**(5), 603–619 (2002)
7. de Geus, D., Meletis, P., Dubbelman, G.: Fast panoptic segmentation network. IEEE Robot. Autom. Lett. **5**, 1742–1749 (2020)
8. He, K., Gkioxari, G., Dollár, P., Girshick, R.: Mask R-CNN. In: Proceedings of the IEEE/CVF International Conference on Computer Vision (2017)
9. Hong, W., Guo, Q., Zhang, W., Chen, J., Chu, W.: LPSNet: a lightweight solution for fast panoptic segmentation. In: Proceedings of the IEEE/CVF Conference on Computer Vision and Pattern Recognition (2021)
10. Hou, R., et al.: Real-time panoptic segmentation from dense detections. In: Proceedings of the IEEE/CVF Conference on Computer Vision and Pattern Recognition (2020)
11. Houlsby, N., et al.: Parameter-efficient transfer learning for NLP. In: ICML (2019)
12. Hu, E.J., et al.: LoRA: low-rank adaptation of large language models. In: ICLR (2022)
13. Hu, J., et al.: ISTR: end-to-end instance segmentation with transformers. arXiv preprint arXiv:2105.00637 (2021)
14. Hu, J., et al.: Pseudo-label alignment for semi-supervised instance segmentation. arXiv preprint arXiv:2308.05359 (2023)
15. Hu, J., Huang, L., Ren, T., Zhang, S., Ji, R., Cao, L.: You only segment once: towards real-time panoptic segmentation. In: Proceedings of the IEEE/CVF Conference on Computer Vision and Pattern Recognition, pp. 17819–17829 (2023)
16. Jia, M., et al.: Visual prompt tuning. In: Avidan, S., Brostow, G., Cissé, M., Farinella, G.M., Hassner, T. (eds.) Computer Vision, ECCV 2022. LNCS, vol. 13693, pp. 709–727. Springer, Cham (2022). https://doi.org/10.1007/978-3-031-19827-4_41
17. Kirillov, A., et al.: Segment anything. arXiv preprint arXiv:2304.02643 (2023)
18. Lester, B., Al-Rfou, R., Constant, N.: The power of scale for parameter-efficient prompt tuning. arXiv preprint arXiv:2104.08691 (2021)
19. Li, X.L., Liang, P.: Prefix-tuning: optimizing continuous prompts for generation. In: ACL/IJCNLP (2021)
20. Li, Y., et al.: Fully convolutional networks for panoptic segmentation. In: Proceedings of the IEEE/CVF Conference on Computer Vision and Pattern Recognition (2021)

21. Lian, D., Zhou, D., Feng, J., Wang, X.: Scaling & shifting your features: a new baseline for efficient model tuning. In: Advances in Neural Information Processing Systems (NeurIPS) (2022)

22. Liu, P., Yuan, W., Fu, J., Jiang, Z., Hayashi, H., Neubig, G.: Pre-train, prompt, and predict: a systematic survey of prompting methods in natural language processing. arXiv preprint arXiv:2107.13586 (2021)

23. Long, J., Shelhamer, E., Darrell, T.: Fully convolutional networks for semantic segmentation. In: Proceedings of the IEEE Conference on Computer Vision and Pattern Recognition, pp. 3431–3440 (2015)

24. Lu, Y., Chen, Z., Chen, Z., Hu, J., Cao, L., Zhang, S.: CANDY: category-kernelized dynamic convolution for instance segmentation. In: 2023 IEEE International Conference on Acoustics, Speech and Signal Processing (ICASSP), ICASSP 2023, pp. 1–5. IEEE (2023)

25. Luo, G., et al.: Towards efficient visual adaption via structural re-parameterization. arXiv preprint arXiv:2302.08106 (2023)

26. Luo, G., Zhou, Y., Ren, T., Chen, S., Sun, X., Ji, R.: Cheap and quick: efficient vision-language instruction tuning for large language models. arXiv preprint arXiv:2305.15023 (2023)

27. Ma, J., Wang, B.: Segment anything in medical images. arXiv preprint arXiv:2304.12306 (2023)

28. Paszke, A., Chaurasia, A., Kim, S., Culurciello, E.: ENet: a deep neural network architecture for real-time semantic segmentation. arXiv preprint arXiv:1606.02147 (2016)

29. Ronneberger, O., Fischer, P., Brox, T.: U-Net: convolutional networks for biomedical image segmentation. In: Navab, N., Hornegger, J., Wells, W.M., Frangi, A.F. (eds.) MICCAI 2015, Part III. LNCS, vol. 9351, pp. 234–241. Springer, Cham (2015). https://doi.org/10.1007/978-3-319-24574-4_28

30. Shin, T., Razeghi, Y., Logan IV, R.L., Wallace, E., Singh, S.: AutoPrompt: eliciting knowledge from language models with automatically generated prompts. arXiv preprint arXiv:2010.15980 (2020)

31. Sung, Y.L., Cho, J., Bansal, M.: VL-adapter: parameter-efficient transfer learning for vision-and-language tasks. In: Proceedings of the IEEE/CVF Conference on Computer Vision and Pattern Recognition, pp. 5227–5237 (2022)

32. Tian, Z., Zhang, B., Chen, H., Shen, C.: Instance and panoptic segmentation using conditional convolutions. IEEE Trans. Pattern Anal. Mach. Intell. **45**, 669–680 (2023)

33. Vaswani, A., et al.: Attention is all you need. In: NIPS (2017)

34. Wu, J., et al.: Medical SAM adapter: adapting segment anything model for medical image segmentation. arXiv preprint arXiv:2304.12620 (2023)

35. Yuan, Y., Chen, X., Chen, X., Wang, J.: Segmentation transformer: object-contextual representations for semantic segmentation. arXiv preprint arXiv:1909.11065 (2019)

36. Zhang, Y., Zhou, K., Liu, Z.: Neural prompt search. CoRR abs/2206.04673 (2022)

37. Zhong, Z., Friedman, D., Chen, D.: Factual probing is [mask]: learning vs. learning to recall. arXiv preprint arXiv:2104.05240 (2021)

38. Zhou, K., Yang, J., Loy, C.C., Liu, Z.: Learning to prompt for vision-language models. CoRR abs/2109.01134 (2021)

39. Zhou, K., Yang, J., Loy, C.C., Liu, Z.: Conditional prompt learning for vision-language models. In: CVPR (2022)

Mining Temporal Inconsistency with 3D Face Model for Deepfake Video Detection

Ziyi Cheng[1], Chen Chen[2], Yichao Zhou[1], and Xiyuan Hu[1(✉)]

[1] School of Computer Science and Engineering, Nanjing University of Science and Technology, Nanjing 210094, China
{ziyicheng,yczhou,huxy}@njust.edu.cn
[2] Institute of Automation, Chinese Academy of Sciences, Beijing 100190, China
chen.chen@ia.ac.cn

Abstract. Recently, the abuse of face swapping technique including Deepfakes has garnered widespread attention. Facial manipulation techniques arise serious problem, including information confusion and erosion of trust. Previous methods for detecting deepfakes have primarily focused on identifying forged traces in independent frames or from two-dimensional face. As a result, the challenge of detecting forged traces from video sequence and higher dimensions has emerged as a critical research area. In this paper, we approach face swapping detection from a novel perspective by converting face frames into three-dimensional space. We argue that the Euclidean distance between landmarks in 2D faces and 3D facial reprojections, along with the shape and texture feature of the 3D face model, serve as clues to discern authenticity. Our framework incorporates a Recurrent Neural Network (RNN) to effectively exploit temporal features from videos. Extensive experiments conducted on several datasets demonstrate the effectiveness of our method. Particularly in cross-manipulation experiments, our approach outperforms state-of-the-art competitors, highlighting its potential as a robust solution for detecting deepfakes.

Keywords: Deepfake video detection · 3D morphable model · Temporal features

1 Introduction

The development of deep learning has advanced the update and iteration of face forgery technology. Deepfakes refer to synthesized or manipulated images or videos of a victim's portrait, generated using deep learning models. In particular, face swapping methods enable the source face to be swapped with a target face. The fake face produced by various manipulation techniques is difficult for human eyes to distinguish. Therefore, research on detection methods related to face swapping is essential to maintain social stability and information security.

State-of-the-art face manipulation techniques make it extremely challenging to uncover subtle forgery clues using only image analysis. Consequently, researchers have shifted their focus towards extracting forged evidence from

Q. Liu et al. (Eds.): PRCV 2023, LNCS 14431, pp. 231–243, 2024.
https://doi.org/10.1007/978-981-99-8540-1_19

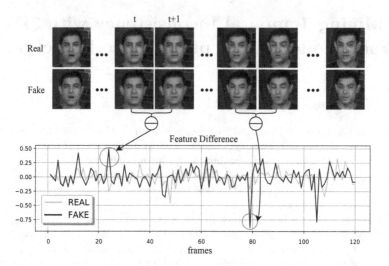

Fig. 1. The texture feature of 3D face model analysis for real and fake video. ⊖ represents the feature difference of two consecutive frames. Although the fake video is too realistic to distinguish, the feature in the forged video exhibits significant fluctuations in the figure (the place marked by the circle).

sources other than the original RGB images, including noise analysis, local regions and frequency information. Despite these efforts, existing approaches still rely on specific manipulation methods, limiting their ability to generalize well in unknown scenarios where images are captured by different devices and environments. As a result, emerging forgery methods continue to undermine the performance of existing face-swapping detection techniques. To address this issue, it is vital to expand beyond the information provided by two-dimensional RGB images and identify additional forgery traces for detection.

We make a key observation that although manipulated face videos display high fidelity in single frames, their temporal features still exhibit significant fluctuations. This inconsistency is an inherent drawback of Deepfake techniques because forged videos are generated on a frame-by-frame basis without considering the temporal continuity of the 3D face model. The shape and texture information of a real face in 3D space should exhibit temporal consistency, while fake videos manipulated using various methods do not demonstrate this characteristic. As illustrated in Fig. 1, features in 3D space, such as facial texture information, highlight the temporal differences between real and fake videos.

In this paper, we propose expanding the two-dimensional face image into a three-dimensional space to obtain more information for mining forgery clues. To better capture these temporal artifacts, we extract features, including distance between landmarks in 2D faces and 3D facial reprojections, as well as the shape and texture of the 3D face model. We design a Recurrent Neural Network (RNN) to extract deep temporal features from the feature sequences.

In summary, this paper makes the following three main contributions:

- We propose a novel face-swapping detection framework that converts two-dimensional face images into three-dimensional space. By expanding the information available in two-dimensional images, we uncover more forged clues to aid in completing the classification task.
- Our approach models temporal inconsistency, including the shape and texture information of the 3D face model. This insight into higher dimensional space enables the network to better distinguish manipulated videos with unknown operations.
- Our method achieves the state-of-the-art performance in terms of generalizing to different manipulation techniques.

2   Related Work

Thus far, most deepfake detection efforts have focused on methods based on independent frames. While these methods have achieved good results in the field of deepfake detection by leveraging the advantages of CNNs, they typically overlook the temporal information contained within face videos.

Video-Level Detection. The notion that videos contain more information than images has inspired researchers to explore the temporal features for forgery face detection. Several approaches, grounded in human physiological behavior, have yielded noteworthy results. Li et al. [14] utilized the frequency of human eye blinking in videos as a distinguishing feature to assess authenticity. Amerini et al. [2] estimated the optical flow field to take advantage of the difference in cross-frame motion. There approaches predominantly rely on CNNs to identify forged indicators within two-dimensional images, thereby encountering issues akin to those faced by frame-level detectors.

3D Face Reconstruction. In existing literature, 3D Morphable Models (3DMM) have proven to be pivotal in 3D facial modeling. With 3DMM, reconstruction can be executed through a composite analysis approach that leverages image intensity along with other features. Recently, numerous approaches have been proposed that utilize CNNs for efficient 3D face reconstruction. Deng et al. [5] presented Deep3D, a 3D face reconstruction method that leverages weakly-supervised learning. The authors investigate the shape aggregation problem using a CNN and introduce an exceptionally efficient and resilient network for achieving image-based 3D reconstruction.

3   Proposed Approach

Our proposed deepfake video detection framework comprises three components (as illustrated in Fig. 2): video preprocessing module, temporal feature embedding module and classification module. This framework leverages temporal inconsistency with 3D face model to detect forged videos.

3.1 Video Preprocessing

The video preprocessing module comprises a face detection and facial landmark detection procedure. In addition to the Region of Interest (ROI) of the face, the original video also contains a large background region. Since forgery techniques primarily target the face region, we employ a face detector to preserve the ROI of the face. We detect 68 facial landmarks that can represent the outline of the face on the preserved face frames. These landmarks are used to calculate the distance from the reprojection landmarks of the corresponding 3D face model in the subsequent feature embedding process.

This approach contrasts with methods that preprocess positive and negative samples in pairs for contrastive learning, such as IID [11], where pairing is not necessary. Consequently, the preprocessing requirements are less stringent when some forged videos cannot be matched with their corresponding original videos.

3.2 Temporal Inconsistency Modeling

The R-Net (ResNet-50) [10] takes the preprocessed face frames as input to produce coefficients for 3D face reconstruction, based on the Deep3D method. A 3DMM consists of two components: a shape model representing the object's geometry and an albedo model describing the object's texture or color. These models are constructed individually using Principal Component Analysis (PCA). Employing a 3DMM enables representation of face shape S and texture T through an affine model:

$$S = S(\alpha, \beta) = \overline{S} + B_s \alpha + B_{exp} \beta \tag{1}$$

$$T = T(\delta) = \overline{T} + B_t \delta \tag{2}$$

In this equation, $\overline{S}$ and $\overline{T}$ denote the average face shape and texture, while B_s, B_{exp} and B_t represent the PCA bases of shape, expression, and texture respectively. α, β and δ are the corresponding coefficient vectors used for generating a 3D face model. We utilize the BFM for $\overline{S}$, B_s, $\overline{T}$, and B_t, and employ the expression bases B_{exp} from [8]. By selecting a subset of the bases, we obtain $\alpha \in \mathbb{R}^{80}$, $\beta \in \mathbb{R}^{64}$ and $\delta \in \mathbb{R}^{80}$.

We can detect a set of 68 facial landmarks from the preprocessed frame, denoted as $P = \{P_i | P_i = (x_i, y_i), i = 1, ..., 68\}$. Once a 3D face model has been obtained, we then apply the face alignment method [3] to detect the 68 landmarks of the model. We project the corresponding vertices of the reconstructed shape onto the image and derive a set $P' = \{P_i' | P_i' = (x_i', y_i'), i = 1, ..., 68\}$. The Euclidean distance D between P and P' captures both the outline of the 2D face and the geometric features of the model, which can be expressed as:

$$D = \left\{ D_i | D_i = \sqrt{(x_i - x_i')^2 + (y_i - y_i')^2}, i = 1, ..., 68 \right\} \tag{3}$$

In frame f, the shape and texture features of the face in 3D space can be represented by vector x^f:

$$x^f = \left[\alpha^f, \delta^f, D^f \right] \in \mathbb{R}^{228}, f = 1, ..., F \tag{4}$$

where F is the length of input vector sequence. Then, the feature vector sequence of a video can be expressed as $X = [x^1, ..., x^F]$. To utilize the temporal information of videos, the difference of feature vector sequences between consecutive frames can be calculated as:

$$I_f = [\alpha^{f+1} - \alpha^f, \delta^{f+1} - \delta^f, D^f] \in \mathbb{R}^{228}, f = 1, ..., F - 1 \qquad (5)$$

The input of the classification procedure can be represented by $I = [I_1, ..., I_{F-1}]$.

3.3   Overall Architecture

The overall architecture of our proposed approach is illustrated in Fig. 2. We divide the video into successive frames and pass them through the video preprocessing module. A face detector is used to crop the ROI of the frame, and a landmarks detector is employed to obtain the landmark coordinates in each frame.

The cropped face images are inputted into the R-Net to obtain the regressed coefficients $\alpha \in \mathbb{R}^{80}$, $\beta \in \mathbb{R}^{64}$ and $\delta \in \mathbb{R}^{80}$. Based on the BFM model, the corresponding 3D face model is reconstructed. The features of each frame encompass the shape and texture information of the corresponding 3D face model. Subsequently, we embed the features $x^f = [\alpha^f, \delta^f, D^f]$ into the vector sequence X.

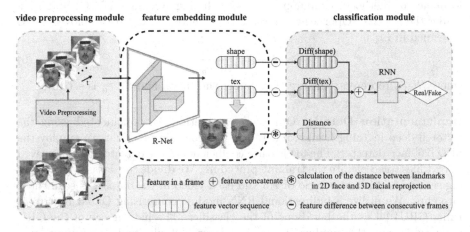

Fig. 2. Overview of proposed framework. 1) video preprocessing module: The video is divided into successive frames and processed with landmark extraction. 2) feature embedding module: The cropped face images are then fed into the R-Net to obtain the 3D face feature vector based on BFM model. 3) classification module: temporal inconsistency features are computed by difference operations on consecutive feature vectors and then fed into the RNN for further enhancement to determine the video's authenticity.

Next, we calculate the difference between feature vectors of consecutive frames and concatenate them to derive I_f. We then feed the concatenated vector sequence I into RNN to effectively mine temporal inconsistency as potential forgery clues. Additionally, fully-connected layers are integrated into the RNN's output to facilitate the classification process.

4   Experiments

4.1   Experimental Settings

Datasets. We conduct all experiments using three popular datasets, namely FaceForensics++ (FF++) [19], FaceShifter [12] and Celeb-DF [15].

FF++ consists of videos with varying quality based on two compression ratios (high quality (HQ) and low quality (LQ)). Four manipulation methods are employed, including identity swapping methods DeepFakes (DF), FaceSwap (FS) and expression swapping methods Face2Face (F2F), and NeuralTexture (NT). This typical recent dataset has been widely adopted.

FaceShifter is a new forgery dataset obtained by applying the FaceShifter (FST) manipulation method to the original video of FF++. This dataset is more realistic and challenging to differentiate between real and fake videos.

Celeb-DF contains real and DeepFakes synthesized videos with similar visual quality to those circulated online.

Evaluation Metrics. We use video-level authenticity to evaluate the performance of methods. Common metrics, including Accuracy (ACC) and Area Under the Receiver Operating Characteristic Curve (AUC), are employed. AUC is defined in Eq. 6:

$$AUC = \frac{Size_u}{Size_a} \tag{6}$$

where $Size_u$ represents the area size under the curve, and $Size_a$ is set to 1.

Implementation Details. Our proposed method is implemented using the Pytorch deep learning framework. In the preprocessing module, we employ the SFD [26] face detector. The 3D feature coefficients of the detected face are saved as text files to accelerate the training procedure. We divide the dataset into training set and test set at a ratio of 7:3 and each video in the dataset is segmented into clips. We utilize a bidirectional RNN module comprising Gated Recurrent Unit (GRU) [4] as the backbone of the classifier. The hidden size of the GRU is set to 64, and the number of recurrent layers defaults to 1. A dropout layer with a drop rate $dr_1 = 0.1$ is inserted between the input and RNN module, and another three dropout layers with $dr_2 = 0.5$ are used to separate the remaining layers. These settings are partially based on [22]. We adopt Adam optimizer with an initial learning rate of $lr = 0.0001$ and use the Lambda learning scheduler to linearly decay the learning rate to zero. The standard binary cross-entropy loss is utilized as our objective during training. This classification model will be trained for up to 400 epochs, and the batch size is set to 48.

4.2 Temporal Feature Visualization

We observe that there are noticeable differences in the normalized distance of some landmarks between real and fake video, as demonstrated in Fig. 3. Furthermore, the shape and texture features of the 3D face model in the forged video exhibit significant fluctuations. We visualize the features mentioned above, which include the distance between landmarks in the 2D face and the 3D facial reprojection, as well as the shape and texture features of the 3D face model.

4.3 Results and Comparisons

Intra-dataset Evaluation. To demonstrate the advantage of our approach, we conduct a comprehensive evaluation of our method and compare it with several state-of-the-art approaches on two benchmarks. Training and testing on the same datasets aim to explore the model's capacity for capturing deepfake cues in videos. The comparison results are derived from [25] and presented in Table 1. In addition, we compare our method with previous detection methods on the four subsets of FF++ (HQ). The results are shown in Table 2. The symbol † denotes re-implementation results from [7].

Cross-Dataset Evaluation. We assess the transferability of our method, given that it is trained on FF++ (HQ) but tested on Celeb-DF. Table 3 shows the comparison results with some recent methods. Results for some other methods are obtained from [15]. The experimental results demonstrate that our method exhibits competitive generalization ability across datasets.

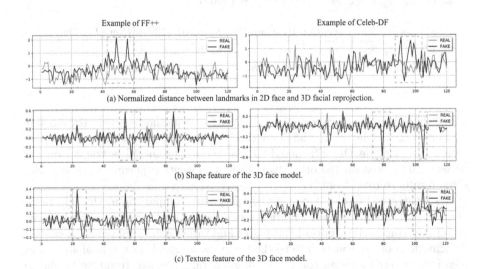

(a) Normalized distance between landmarks in 2D face and 3D facial reprojection.

(b) Shape feature of the 3D face model.

(c) Texture feature of the 3D face model.

Fig. 3. The left column is an example from FF++, and the right column from Celeb-DF. The temporal inconsistency of the features is shown in the green box. (Color figure online)

Table 1. AUC scores (%) comparisons with state-of-the-art approaches on two popular benchmarks. The results in red indicate the best performance, while those in blue represent the second-best performance.

Methods	FF++	Celeb-DF	Year
Capsule [18]	96.6	57.5	ICASSP 2019
TwoBranch [17]	93.2	76.6	ECCV 2020
TD-3DCNN [24]	72.2	88.8	IJCAI 2021
Lips [9]	97.1	82.4	CVPR 2021
ICT [6]	98.6	94.4	CVPR 2022
Ours	**97.3**	**90.4**	2023

Table 2. Comparison of ACC scores (%) on the FF++ (HQ) dataset with models trained on sub-datasets. The results in red indicate the best performance, while those in blue represent the second-best performance.

Methods	FF++			
	DeepFakes	Face2Face	FaceSwap	NeuralTextures
C3D [23]	**92.9**	**88.6**	91.8	89.6
FaceNetLSTM [20]	89.0	87.0	90.0	-
ADDNet-3d† [28]	92.1	83.9	**92.5**	78.2
Ours	99.7	92.0	98.1	**86.3**

Table 3. Cross-dataset evaluation by AUC scores (%). The best results in red, while those in blue represent the second-best performance.

Methods	MesoInception [1]	Capsule [18]	SMIL [13]	LRNet [22]	Ours
FF++HQ	83.0	96.6	96.8	97.3	97.3
Celeb-DF	53.6	**57.5**	56.3	56.3	58.0

Cross-Manipulation Evaluation. We verify our method's generalization ability across manipulation methods on the FF++ (HQ) dataset through experiments that include DF, FS, and FST methods. The model is trained on one dataset and tested on the other two. Table 4 presents experiment results, with some results sourced from [11].

As illustrated in Table 4, our method surpasses comparison methods in terms of mean AUC on unseen manipulation types. Some face swapping methods reconstruct and track a 3D model of the source and target actor, applying the tracked deformations of the source face to the target face model. In the final step, they blend the altered face on top of the original target video, resulting in clutter and inconsistency in the 3D information for fake videos. The shape and texture feature of the real face in 3D space should be temporally consistent, whereas fake videos manipulated through various methods exhibit the opposite characteristic. Thus, we mine the inconsistency of 3D face features over time to detect face forgery, leading to better generalization ability for diverse manipulation methods. These improvements primarily stem from the temporal cues extracted from the 3D space of the detected face based on BFM. The results presented in Table 4 effectively demonstrate our method's robust generalization ability across face swapping manipulation methods, extending the face information in frame images to 3D space and fully leveraging the temporal characteristics of the video.

4.4   Ablation Analysis

Influence of Input Length. We conduct ablation experiments to determine the optimal input length for our model and the results are displayed in Fig. 4. The input length refers to the number of consecutive frames from a video used as a single sample. The related experimental results indicate that an appropriate input length can enhance the model's performance and the efficiency of the training process. For low-quality video data, an input length of 60 yields the best results, which is also the model input length chosen for our other experiments.

Effect of Different Input Feature Combinations. To verify our hypothesis that forgery clues exist in the inconsistency of 3D face features over time, we

Table 4. Cross-manipulation evaluation in terms of AUC scores (%). Diagonal results indicate intra-testing performance. DF, FS and FST denote the DeepFakes, FaceSwap and FaceShifter datasets, respectively. The results in red indicate the best performance, while those in blue represent the second-best performance.

Train	Methods	DF	FS	FST	Mean↑
DF	MAT [27]	99.92	40.61	45.39	61.97
	GFF [16]	99.87	47.21	51.93	66.34
	DCL [21]	99.98	61.01	68.45	76.48
	IID [11]	99.51	63.83	73.49	**78.94**
	Ours	99.94	68.70	80.44	83.03
FS	MAT	64.13	99.67	57.37	73.72
	GFF	70.21	99.85	61.29	77.12
	DCL	74.80	99.90	64.86	79.85
	IID	75.39	99.73	66.18	**80.43**
	Ours	82.15	99.69	60.68	80.84
FST	MAT	58.15	55.03	99.16	70.78
	GFF	61.48	56.17	99.41	72.35
	DCL	63.98	58.43	99.49	73.97
	IID	65.42	59.50	99.50	**74.81**
	Ours	74.16	65.57	98.45	79.39

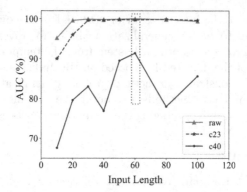

Fig. 4. AUC scores (%) on FF++ with varying input lengths. Models are trained on datasets with three different compression ratios to explore the effect of input length.

Table 5. AUC scores (%) on FF++ with different input feature combinations. The inputs consist of four components: the distance between landmarks in 2D face and 3D facial reprojection (dist), shape, expression and texture information from a 3D face model (shape, exp, tex). The best results are highlighted.

input	dist	shape	exp	tex	AUC
In-a	✓	✓		✓	**98.4**
In-b	✓				97.3
In-c		✓			91.4
In-d			✓		85.4
In-e				✓	93.1
In-f		✓		✓	95.9

conduct an ablation study on the effect of different input feature combinations on FST dataset. FST is more realistic and challenging to differentiate between real and fake instances. The results are displayed in Table 5. We evaluate six input combinations: dist+shape+tex, dist, shape, exp, tex, shape+tex, where dist, shape, exp and tex are short for the distance between landmarks in 2D face and 3D facial reprojection, shape, expression and texture information from a 3D face model, respectively.

Several noteworthy results can be observed in Table 5. First, the performance of In-d indicates that relying solely on expression related information does not effectively mine forgery clues in the video. Second, by comparing In-b, In-c, In-d and In-e, we find that the temporal consistency of distance plays a significant role in forgery detection, mainly because the distance captures both the outline of the 2D face and the geometric feature of the 3D face model. Third, we discover that the best results can be achieved by combining distance, shape, and texture feature as input.

5   Conclusion

In this paper, we propose a novel perspective for face-swapping detection by converting 2D face images to a 3D space. Our method leverages the shape and texture feature of the 3D face model to expose more forgery traces. Additionally, we compute the differences between feature vector sequences of consecutive frames to effectively mine temporal inconsistency. Extensive experiments conducted demonstrate the effectiveness of our method. Particularly in cross-manipulation experiments, our approach surpasses state-of-the-art competitors, showcasing its potential as a robust solution for detecting deepfakes.

Acknowledgements. This work was supported by the National Key R&D Program of China (2021YFF0602101) and National Natural Science Foundation of China (62172227).

References

1. Afchar, D., Nozick, V., Yamagishi, J., Echizen, I.: MesoNet: a compact facial video forgery detection network. In: 2018 IEEE International Workshop on Information Forensics and Security (WIFS), pp. 1–7. IEEE (2018)
2. Amerini, I., Galteri, L., Caldelli, R., Del Bimbo, A.: DeepFake video detection through optical flow based CNN. In: Proceedings of the IEEE/CVF International Conference on Computer Vision Workshops (2019)
3. Bulat, A., Tzimiropoulos, G.: How far are we from solving the 2D & 3D face alignment problem? (and a dataset of 230,000 3D facial landmarks). In: Proceedings of the IEEE International Conference on Computer Vision, pp. 1021–1030 (2017)
4. Cho, K., et al.: Learning phrase representations using RNN encoder-decoder for statistical machine translation. arXiv preprint arXiv:1406.1078 (2014)
5. Deng, Y., Yang, J., Xu, S., Chen, D., Jia, Y., Tong, X.: Accurate 3D face reconstruction with weakly-supervised learning: from single image to image set. In: Proceedings of the IEEE/CVF Conference on Computer Vision and Pattern Recognition Workshops (2019)
6. Dong, X., et al.: Protecting celebrities from DeepFake with identity consistency transformer. In: Proceedings of the IEEE/CVF Conference on Computer Vision and Pattern Recognition, pp. 9468–9478 (2022)
7. Gu, Z., et al.: Spatiotemporal inconsistency learning for DeepFake video detection. In: Proceedings of the 29th ACM International Conference on Multimedia, pp. 3473–3481 (2021)
8. Guo, Y., Cai, J., Jiang, B., Zheng, J., et al.: CNN-based real-time dense face reconstruction with inverse-rendered photo-realistic face images. IEEE Trans. Pattern Anal. Mach. Intell. **41**(6), 1294–1307 (2018)
9. Haliassos, A., Vougioukas, K., Petridis, S., Pantic, M.: Lips don't lie: a generalisable and robust approach to face forgery detection. In: Proceedings of the IEEE/CVF Conference on Computer Vision and Pattern Recognition, pp. 5039–5049 (2021)
10. He, K., Zhang, X., Ren, S., Sun, J.: Deep residual learning for image recognition. In: Proceedings of the IEEE Conference on Computer Vision and Pattern Recognition, pp. 770–778 (2016)

11. Huang, B., et al.: Implicit identity driven DeepFake face swapping detection. In: Proceedings of the IEEE/CVF Conference on Computer Vision and Pattern Recognition, pp. 4490–4499 (2023)
12. Li, L., Bao, J., Yang, H., Chen, D., Wen, F.: FaceShifter: towards high fidelity and occlusion aware face swapping. arXiv preprint arXiv:1912.13457 (2019)
13. Li, X., et al.: Sharp multiple instance learning for DeepFake video detection. In: Proceedings of the 28th ACM International Conference on Multimedia, pp. 1864–1872 (2020)
14. Li, Y., Chang, M.C., Lyu, S.: In Ictu Oculi: exposing AI created fake videos by detecting eye blinking. In: 2018 IEEE International Workshop on Information Forensics and Security (WIFS), pp. 1–7. IEEE (2018)
15. Li, Y., Yang, X., Sun, P., Qi, H., Lyu, S.: Celeb-DF: a large-scale challenging dataset for DeepFake forensics. In: Proceedings of the IEEE/CVF Conference on Computer Vision and Pattern Recognition, pp. 3207–3216 (2020)
16. Luo, Y., Zhang, Y., Yan, J., Liu, W.: Generalizing face forgery detection with high-frequency features. In: Proceedings of the IEEE/CVF Conference on Computer Vision and Pattern Recognition, pp. 16317–16326 (2021)
17. Masi, I., Killekar, A., Mascarenhas, R.M., Gurudatt, S.P., AbdAlmageed, W.: Two-branch recurrent network for isolating deepfakes in videos. In: Vedaldi, A., Bischof, H., Brox, T., Frahm, J.-M. (eds.) ECCV 2020, Part VII. LNCS, vol. 12352, pp. 667–684. Springer, Cham (2020). https://doi.org/10.1007/978-3-030-58571-6_39
18. Nguyen, H.H., Yamagishi, J., Echizen, I.: Capsule-forensics: using capsule networks to detect forged images and videos. In: 2019 IEEE International Conference on Acoustics, Speech and Signal Processing (ICASSP), ICASSP 2019, pp. 2307–2311. IEEE (2019)
19. Rossler, A., Cozzolino, D., Verdoliva, L., Riess, C., Thies, J., Nießner, M.: Face-Forensics++: learning to detect manipulated facial images. In: Proceedings of the IEEE/CVF International Conference on Computer Vision, pp. 1–11 (2019)
20. Sohrawardi, S.J., et al.: Poster: towards robust open-world detection of deepfakes. In: Proceedings of the 2019 ACM SIGSAC Conference on Computer and Communications Security, pp. 2613–2615 (2019)
21. Sun, K., Yao, T., Chen, S., Ding, S., Li, J., Ji, R.: Dual contrastive learning for general face forgery detection. In: Proceedings of the AAAI Conference on Artificial Intelligence, vol. 36, pp. 2316–2324 (2022)
22. Sun, Z., Han, Y., Hua, Z., Ruan, N., Jia, W.: Improving the efficiency and robustness of deepfakes detection through precise geometric features. In: Proceedings of the IEEE/CVF Conference on Computer Vision and Pattern Recognition, pp. 3609–3618 (2021)
23. Tran, D., Bourdev, L., Fergus, R., Torresani, L., Paluri, M.: Learning spatiotemporal features with 3D convolutional networks. In: Proceedings of the IEEE International Conference on Computer Vision, pp. 4489–4497 (2015)
24. Zhang, D., Li, C., Lin, F., Zeng, D., Ge, S.: Detecting deepfake videos with temporal dropout 3DCNN. In: IJCAI, pp. 1288–1294 (2021)
25. Zhang, D., Lin, F., Hua, Y., Wang, P., Zeng, D., Ge, S.: Deepfake video detection with spatiotemporal dropout transformer. In: Proceedings of the 30th ACM International Conference on Multimedia, pp. 5833–5841 (2022)
26. Zhang, S., Zhu, X., Lei, Z., Shi, H., Wang, X., Li, S.Z.: S3FD: single shot scale-invariant face detector. In: Proceedings of the IEEE International Conference on Computer Vision, pp. 192–201 (2017)

27. Zhao, H., Zhou, W., Chen, D., Wei, T., Zhang, W., Yu, N.: Multi-attentional deepfake detection. In: Proceedings of the IEEE/CVF Conference on Computer Vision and Pattern Recognition, pp. 2185–2194 (2021)
28. Zi, B., Chang, M., Chen, J., Ma, X., Jiang, Y.G.: WildDeepfake: a challenging real-world dataset for deepfake detection. In: Proceedings of the 28th ACM International Conference on Multimedia, pp. 2382–2390 (2020)

DT-TransUNet: A Dual-Task Model for Deepfake Detection and Segmentation

Junshuai Zheng[ID], Yichao Zhou, Xiyuan Hu[✉][ID], and Zhenmin Tang[ID]

Nanjing University of Science and Technology, Nanjing, China
`huxy@njust.edu.cn`

Abstract. The swift advancement of deepfake technology has given rise to concerns regarding its potential misuse, necessitating the detection of fake images and acquisition of supportive evidence. In response to this need, we present a novel dual-task network model called DT-TransUNet, which concurrently performs segmentation for both deepfake detection and deepfake segmentation. Additionally, we propose a new Multi-Scale Spatial Frequency Feature (MSSFF) module that employs the Stationary Wavelet Transform (SWT) to extract multi-scale high-frequency components and enhance these features using a texture activation function. When evaluated on multiple datasets, DT-TransUNet surpasses comparable methods in performance and visual segmentation quality, thereby validating the effectiveness and capability of the MSSFF module for deepfake detection and segmentation tasks.

Keywords: Dual-task Model · Deepfake Detection · Deepfake Segmentation · Stationary Wavelet Transform

1 Introduction

With the development of technologies such as Variational AutoEncoders (VAE) [9] and Generative Adversarial Networks (GAN) [7,14], the deepfake images and videos are generated highly realistic. Deepfake technology has yielded innovative applications in movies and short video, but it also brings some risks. Deepfake news has triggered a crisis of trust in society, and fake pornographic images have also brought trouble to people's life. In response to these problems, deepfake detection has emerged and become a hot topic in recent years.

In order to train and evaluate deepfake detection methods, researchers have developed several large-scale datasets, including FaceForensics++ (FF++), Deep Fake Detection (DFD), Celeb-DF, and Deepfake Detection Challenge (DFDC). Through the comparison and analysis of real and fake images, a large amount of research has been proposed about feature extraction methods and detection models. In summary, it is found that most studies regard the deepfake

This work was supported by the National Natural Science Foundation of China (62172227) and National Key R&D Program of China (2021YFF0602101).

detection as an image binary classification task. For an input image or video, deep learning models automatically extract features and classify it. While this end-to-end method can achieve high accuracy, its interpretability is weak and not conducive to understanding. The Face X-ray [17] method not only provides fake detection result but also segments the manipulated area, it offers a new direction for deepfake detection.

Segmentation of manipulated area not only offers a more visually intuitive representation, but also provides a more explainable result for the detection model. These methods generally include deepfake image classification task and manipulated area segmentation task, referred to as deepfake detection and segmentation. Though various features [18,22,25] and methods [8,23,28] are proposed for deepfake detection, there are less researches [21,27] on the deepfake segmentation.

Based on the above situation, this paper proposes a new mutli-task network model named DT-TransUNet. In this model, a new multi-scale spatial frequency feature module is designed to get high frequency feature. By fusing RGB image and frequency feature as inputs, the DT-TransUNet generates more accurate deepfake detection and segmentation results. In summary, the contributions of this paper are as follows:

(1) A dual-task model framework is proposed. DT-TransUNet is a dual-task model, which includes deepfake detection task and deepfake segmentation task. By sharing parameters from both tasks, it transfers knowledge and improves the accuracy between two tasks.
(2) A Multi-Scale Spatial Frequency Feature (MSSFF) module is proposed. This module employs the Stationary Wavelet Transform (SWT) method to extract multi-scale high-frequency components. Additionally, a texture activation function is designed to enhance the multi-scale spatial frequency feature.
(3) Experimental results show that our method outperforms similar methods, and has better segmentation visual effects in segmentation tasks. Finally, the MSSFF module is combined with other methods, the effectiveness and generality of this module are demonstrated for deepfake detection and segmentation tasks.

2 Related Work

As for deepfake detection, some scholars have investigated the differences of features between the real and fake image. These features include head pose [27], physiological signals [15,22], frequency characteristics, image features, etc. Li achieved deepfake detection by counting the difference in the number of eye blinks in real and fake videos, and using an LSTM network to learn blink features [18]. For frequency, Durall converted images into power spectra and extracted high-frequency information at specific frequencies for forgery detection [10]. Ameini proposed a deepfake detection method based on optical flow [1]. Guarnera used convolutional traces to detect forged images [13]. Li discovered facial warping in forged images and used it as a basis for detecting forgeries

[19]. Cozzolino employed 3D facial models and physical identity extraction to detect deepfake videos [6]. Yang distinguished deepfake images by identifying potential texture differences in images [26]. The introduction of these research methods has provided new insights and technical means for the development of deepfake detection.

According to the deepfake detection models, some scholars have made significant contributions [4,20,23,28]. Deng applied Efficient V2 to the task of deepfake detection [8]. Gafnguly proposed ViXnet [11], a fusion model of Vision Transformer and Xception. Coccomin integrated EfficientNet and Vision Transformer, achieving impressive results [5]. Furthermore, scholars such as Chen [3], Saif [24], and Chamot [2] have proposed new deepfake detection methods in the direction of space-time convolution, continuously advancing the development of deepfake detection technology.

For the research of deepfake detection and segmentation, there are only few studies now. In 2019, Nguyen proposed a dual-task approach [21]. Eff-YNet [25] adopted the Unet structure to combine EfficentNet and UNet. The Face X-ray method proposed by Li could also detect the manipulated area. Khalid [16] combined Swing Transformer and Unet network, and proposed a SWYNT network for deepfake segmentation. Guan proposed a method based on Xception collaborative learning recently [12], which is still in the preprint stage. These studies have played a positive role in promoting the development of deepfake detection technology, and provided new ideas and technical means for deepfake segmentation.

3   The Proposed Approach

3.1   Overall Framework

DT-TransUNet leverages multi-scale spatial frequency image and RGB image as model inputs, to explore forgery traces in both image spatial and frequency space. Simultaneously, the model outputs deepfake detection results and fake mask predictions through dual-task learning (Fig. 1).

DT-TransUNet adopts a model architecture similar to UNet, including six parts: MSSFF module, encoder, Transformer module, decoder, segmentation header and classification header. The MSSFF module is introduced in Sect. 3.2. The encoder module extracts features through four convolutional layers and a MaxPool layer, and each convolutional layer uses a BN layer and a ReLU function after. Transformer Block uses Vision Transformer as the backbone network, there are 6 transformer layers, each layer uses 8 multi-head attention heads, and the size of each patch is $7*7$. The decoder module mainly includes 1 deconvolution layer and 2 convolution layers, and the BN layer and SiLU function are connected after each convolution layer. The segmentation head uses the softmax function to generate the probability of segmentation images. The classification head adopts an adaptive global average pooling layer and 2 fully connected layers, and uses the softmax function to generate classification probabilities.

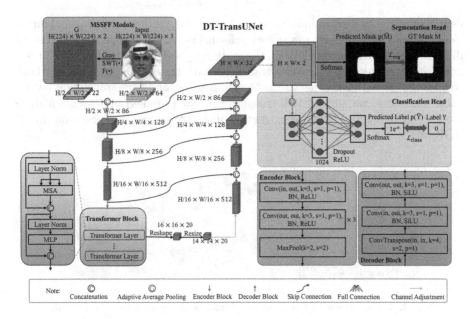

Fig. 1. The overview of our proposed DT-TransUNet.

The input image is an RGB image with a size of 224 × 224. First, the MSSFF module generates multi-scale spatial frequency feature with the same size. The RGB image is processed through the encoder module to obtain shallow spatial features with 64 channels. The multi-scale spatial frequency feature is also passed through a encoder module to acquire shallow frequency features with 22 channels. Shallow spatial and frequency features are combined by concatenation. The fused feature tensor is down-sampled three times by encoder block. Then, the features are fed into the Transformer module for global information modeling. The feature vector output by the Transformer is reshaped into a feature tensor with a size of 16 × 16 and 20 channels, and its resolution is adjusted to 14 × 14 using bilinear interpolation. Subsequently, the number of channels is modified to 512 through a 1 × 1 convolution operation, followed by four decoder blocks and 1 dimension adjustment block. During up-sampling, the resolution is doubled while the channel count is progressively reduced. Simultaneously, the same size encoder features and decoder features are connected by skip connections. The last feature tensor is then input into the segmentation head module and classification head to get results.

3.2 Multi-scale Spatial Frequency Feature Module

In deepfake images, there is a noticeable difference in high-frequency features between the face-swapped region and the background area. This difference can be extracted using frequency transform methods. Wavelet transform is a common method for extracting image spatial-frequency features. The high-frequency detail components contain information in horizontal, vertical, and diagonal directions, which can represent the salient information of the image. In contrast to discrete wavelet transform (DWT), the stationary wavelet transform (SWT) is a non-decimated version of wavelet transform with translation invariance, whose coefficients are with the same size as the input image and have translation invariance. Therefore, we adopt SWT transform in our proposed multi-scale feature module, as shown in Fig. 2, for extracting spatial-frequency features from input RGB image.

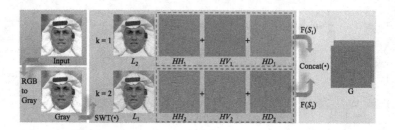

Fig. 2. The overview of our proposed Multi-Scale Spatial Feature Module.

The MSSFF module mainly consists of 2 steps:
(1) SWT Decomposition. For a given input image I, we first convert the RGB color image into gray. In order to obtain different scales high frequency information, the grayscale image I_g is converted to k layers by SWT, the formula is as follows:

$$\begin{cases} [L_1, HH_1, HV_1, HD_1] = SWTk(I_g) & k = 1 \\ [L_k, HH_k, HV_k, HD_k] = SWTk(L_{k-1}) & k > 1 \end{cases} \quad (1)$$

In the formula, $SWT_k(\cdot)$ represents the k-level SWT decomposition, with k being the number of decomposition layers. After one level of decomposition, four coefficient matrices are outputted. L_k is the approximation coefficient under the kth level decomposition, while HH_k, HV_k and HD_k represent the high-frequency coefficients in horizontal, vertical, and diagonal directions. By inputting L_k into $SWT_k(\cdot)$, the next level of decomposition can be obtained. To simplify subsequent calculations, there is only consider 2 layers high-frequency, so k is set to be 2.

(2) Texture Activation Function. The high-frequency wavelet coefficients after SWT contain a large number of numbers less than 0, which is not conducive to subsequent image processing and learning. In this regard, we use a texture enhancement method to enhance the high-frequency coefficients, thereby increasing the intensity and visualization of the texture. The formula is as follows:

$$\begin{aligned} S_1 = HH_1 + HV_1 + HD_1 \\ S_2 = HH_2 + HV_2 + HD_2 \end{aligned} \quad (2)$$

$$F(S_k) = \begin{cases} 0 & S_k(i,j) \leq 0 \\ \min((\frac{S_k(i,j)}{\max(S_k(i,j))}) * 255, 255) & S_k(i,j) > 0 \end{cases} \quad (3)$$

S_k represents the sum of high-frequency components in the low k layer, and $S_k(i, j)$ represents the element values in the i-th row and jth column of S_k.

$$G = Concat(F(S_1), F(S_2)) \tag{4}$$

Where $Concat(\cdot)$ stands for concatenating two feature map together. The feature map $G \in \mathbb{R}^{H \times W \times 2}$ is called a multi-scale spatial frequency feature. It can be seen from Fig. 2 that the high-frequency sub-map mainly includes high-frequency information such as edge information and noise, and there is an obvious stitching texture at the edge of the fake face stitching. The multi-scale spatial frequency feature can complement the information of image spatial features, which is helpful for the deepfake detection and segmentation.

3.3 Loss Function

The cross-entropy loss function is then used to measure the difference between the predicted probability $p(\hat{y})$ and the true label y. The cross-entropy function is defined as follows:

$$\mathcal{L}_{class} = -[y \log p(\hat{y}) + (1 - y) \log(1 - p(\hat{y}))] \tag{5}$$

In this case, the real image's y is set to 1, the deepfake image's y is set to 0.

Similarly, we input the features from the decoder into the segmentation head and use the softmax function to obtain the probability $p(\widehat{M})$ of the predicted forgery mask M. Given the ground truth mask $M \in \{0, 1\}^{h \times w}$, we employ the cross-entropy loss function as the segmentation loss to measure the difference between the masks $\widehat{M} \in \{0, 1\}^{h \times w}$ and M :

$$\mathcal{L}_{seg} = \sum_{i,j} -[Mij \log p(\widehat{M}ij) + (1 - Mij) \log(1 - p(\widehat{M}ij))] \tag{6}$$

In the equation, M_{ij} represents the pixel of mask. We set the forged pixel to 1 and the real pixel to 0.

Finally, the total loss is defined as:

$$\mathcal{L}_{total} = \mathcal{L}_{class} + \lambda \mathcal{L}_{seg} \tag{7}$$

The total loss is adjusted by the weight λ. In our experiment, the λ value is set to 1.

4   Experiments

4.1   Datasets

In our tasks, the objectives are to detect the deepfake image and segment the manipulated area. Datasets containing operation masks are required, so we selected the FaceForensics++ (FF++) dataset and the Deep Fake Detection

(DFD) dataset for our experiments. There are four sub-datasets (DF, FS, F2F, and NT) in FF++, we also tested the our model on those sub-datasets.

Due to the unbalanced proportion of fake and real videos in FF++ and DFD dataset, we adopt different image sampling strategies. For FF++ dataset, We randomly select 20 frames from each fake video and 80 frames from real videos. In the DF, F2F, FS and NT dataset, 50 frames from each video randomly selected. For the DFD dataset, we first randomly select the same number of fake videos as real ones, then randomly choose 50 frames from each video.

For the extracted image dataset, we first use the Retina face detector to identify the face area in the image and record its coordinates. Next, we crop the image to 1.3 times the size of the face area and use affine transformation to convert the face area into a 224×224 pixel image. Simultaneously, for the operation mask images of deepfake images, we perform concurrent cropping and warp affine transformation based on the same position of face area. This ensure consistency between the face image and operation mask. Finally, we divide the image dataset into training, validation, and test sets using an 8:1:1 ratio.

4.2   Mask Ground Truth Generation

Due to variations in the presentation of forged masks provided by different datasets, we propose a forgery mask generation strategy to address this. In Fig. 3, we demonstrate the mask generation results using this method.

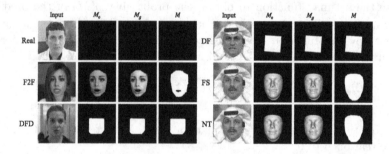

Fig. 3. The specific steps and results of mask generation.

First, we convert the original mask image M_o provided by the dataset into a grayscale image M_g. Then, we compare the pixel value of the grayscale image M_g with a preset threshold φ. If the pixel value is greater than the threshold φ, the pixel value is set to 1. Conversely, the pixel value is set to 0. Here, we set the threshold φ to be 10. Then, we perform dilation and erosion operations to generate the ground truth mask M.

4.3   Implementation Details

When training the network, we adopt the Adam optimizer, set the learning rate to 2e−4 and the weight decay to 1e−5. The batch size is 96. We train for 40 epochs together, with the first 6 epochs utilizing the Warmup strategy to gradually increase the learning rate. Simultaneously, we employ an adaptive learning rate adjustment strategy, reducing the learning rate to half its original value when the classification accuracy of the validation set fails to improve for two epochs.

4.4   Results on Deepfake Detection and Segmentation

In this section, we evaluate the proposed DT-TransUNet model on the FF++ and DFD datasets. We investigate the model's performance based on dual tasks: deepfake detection and deepfake segmentation. The evaluation metrics selected are commonly used classification evaluation metrics ACC and AUC, as well as segmentation evaluation metrics DICE and IOU.

There are few studies on dual-task models at present, and some research works only give classification accuracy values. So we implement Eff-YNet as a comparative experiment. For a fair comparison, we conduct experiments on Eff-YNet under the same experimental settings. In all tables, † represents the method implemented by myself.

Table 1. The results of the Eff-YNet and DT-TransUNet methods in FF++ and DFD datasets.

Method	FF++				DFD			
	DICE	IOU	ACC	AUC	DICE	IOU	ACC	AUC
Eff-YNet [25]†	83.25	81.53	86.57	86.61	82.04	78.96	88.13	88.16
Ours	**88.22**	**86.36**	**95.16**	**95.15**	**91.54**	**88.29**	**96.90**	**96.85**

We first validate our model on the FF++ and DFD datasets, the results are shown in Table 1. As for the FF++ dataset, our model achieves a 95.16% (ACC) and a 95.15% AUC for the deepfake detection task, which are 8.59% and 8.54% higher than the Eff-YNet method, respectively. At the same time, our method outperforms the Eff-YNet method in the deepfake segmentation task. About DFD dataset, our method also surpasses the Eff-YNet method, with 8.77% higher ACC and 8.69% higher AUC, 9.50% higher Dice and 9.33% higher IOU.

To further validate the effectiveness of our method, we conduct experiments on the four sub-datasets within the FF++ dataset: DeepFake, FaceSwap, Face2Face, and NeuralTextures. Additionally, we gather experimental results from other dual-task methods and present them in Tables 2 and 3 for comparison.

According to the data in Table 2, we can see that the Y-shaped AE method has the highest accuracy ACC in the DeepFake dataset, which is 0.05% higher than our method. But our method is 9.80% higher of IOU than the Y-shaped AE method. Compared with the CFL method, we outperform the ACC and

Table 2. The results of 6 methods under the DeepFake and FaceSwap datasets.

Method	DeepFakes				FaceSwap			
	DICE	IOU	ACC	AUC	DICE	IOU	ACC	AUC
Y-shaped AE [12]	-	85.46	**99.90**	-	-	84.60	95.81	-
Xception-att [12]	-	84.42	97.48	-	-	83.31	95.95	-
CFL (Xception) [12]	-	93.36	97.83	-	-	**94.57**	98.71	-
SWYNT [16]	-	-	97.12	97.00	-	-	**99.01**	**99.00**
Eff-YNet [25]†	92.61	90.49	95.22	95.22	90.09	88.63	92.07	92.10
Ours	**97.41**	**95.26**	99.85	**99.85**	**95.46**	93.91	98.15	98.17

IOU metrics by 2.02% and 1.90%, respectively. At the same time, our method is 2.73% and 2.85% higher than the SWYNT method in terms of ACC and AUC, respectively. As for FaceSwap dataset, the SWYNT method has the highest ACC and AUC metrics, while our method is only 0.86% below the optimal metrics.

Table 3. The results of 3 methods in the Face2Face and NeuralTextures datasets.

Method	Face2Face				NeuralTextures			
	DICE	IOU	ACC	AUC	DICE	IOU	ACC	AUC
Eff-YNet [25]†	94.38	93.10	96.15	96.13	82.66	81.24	85.28	85.28
SWYNT [16]	-	-	95.73	97.00	-	-	79.90	83.00
Ours	**96.11**	**94.82**	**98.77**	**98.78**	**88.01**	**86.31**	**93.28**	**93.29**

Through Table 3, we find that the four evaluation indicators of DT-TransUNet on the Face2Face and NeuralTextures datasets are higher than those of Eff-YNet and SWYNT.

Figure 4 displays the prediction forgery masks of the two methods for deepfake segmentation. For a real image, the mask should be entirely black, while the manipulated area should be white for a deepfake image. As seen in the figure, the segmentation results of Eff-YNet are unsatisfactory, with issues such as unclear outlines and inaccurate positioning. In contrast, the shape of mask predicted by DT-TransUNet are more closed to the ground truth, resulting in better deepfake segmentation. DT-TransUNet provides visual data support for the deepfake classification task.

4.5 Ablation Analysis

To evaluate the impact of the proposed MSSFF module on deepfake detection and segmentation, we design two sets of comparative experiments. First, we remove the MSSFF module from the DT-TransUNet model and compare

Fig. 4. The segmentation masks of the Eff-YNet and DT-TransUNet methods.

the experimental metrics between those two methods. Second, we combine the MSSFF module with the Eff-YNet method for experimental comparison. Both sets of experiments are performed on the DeepFake dataset. Table 4 presents the experimental results under various conditions, where (W/O) represents the method without the MSSFF module.

Table 4. The results of ablation experiment.

Method	DICE	IOU	ACC	AUC
Eff-YNet [25][†]	92.61	90.49	95.22	95.22
Eff-YNet [25][†]+MSSFF	**95.63(+3.02)**	**93.50(+3.07)**	**98.06(+2.84)**	**98.08(+2.86)**
DT-TransUNet (W/O)	96.54	94.42	98.97	98.96
DT-TransUNet	**97.41(+0.87)**	**95.26(+0.84)**	**99.85(+0.88)**	**99.85(+0.89)**

The experimental results show that the methods incorporating the MSSFF module perform better than the baseline methods. By adding the MSSFF module, the average metrics improve by approximately 3% and 0.87% for the Eff-YNet model and DT-TransUNet model, respectively. This experiment demonstrates the effectiveness of the multi-scale spatial frequency feature module for deepfake tasks, and validates the module's portability and versatility.

5 Conclusion

In this study, we introduced a novel multi-scale frequency feature module called MSSFF. MSSFF module leveraged the Stationary Wavelet Transform (SWT) to extract reliable high-frequency information across multiple scales, which enhancing feature learning capabilities for deepfake detection and segmentation task. Concurrently, we proposed a sophisticated dual-task deepfake detection and segmentation model named DT-TransUNet. This model not only classified deepfake images but also predicted the manipulated areas within them. Through experimental evaluations conducted on the FaceForensics++ dataset and various other datasets, our approach outperformed cutting-edge methods such as SWYNT and CLF. Thus, the effectiveness and versatility of the MSSFF module were demonstrated, highlighting its potential applicability in a broad range of scenarios.

References

1. Amerini, I., Galteri, L., Caldelli, R., Del Bimbo, A.: Deepfake video detection through optical flow based CNN. In: Proceedings of the IEEE/CVF International Conference on Computer Vision Workshops (2019)
2. Chamot, F., Geradts, Z., Haasdijk, E.: Deepfake forensics: cross-manipulation robustness of feedforward-and recurrent convolutional forgery detection methods. Forensic Sci. Int. Digit. Invest. **40**, 301374 (2022)
3. Chen, B., Li, T., Ding, W.: Detecting deepfake videos based on spatiotemporal attention and convolutional LSTM. Inf. Sci. **601**, 58–70 (2022)
4. Chollet, F.: Xception: deep learning with depthwise separable convolutions. In: Proceedings of the IEEE Conference on Computer Vision and Pattern Recognition, pp. 1251–1258 (2017)
5. Coccomini, D.A., Messina, N., Gennaro, C., Falchi, F.: Combining EfficientNet and vision transformers for video deepfake detection. In: Sclaroff, S., Distante, C., Leo, M., Farinella, G.M., Tombari, F. (eds.) Image Analysis and Processing, ICIAP 2022, Part III. LNCS, vol. 13233, pp. 219–229. Springer, Cham (2022). https://doi.org/10.1007/978-3-031-06433-3_19
6. Cozzolino, D., Rössler, A., Thies, J., Nießner, M., Verdoliva, L.: ID-Reveal: identity-aware deepfake video detection. In: Proceedings of the IEEE/CVF International Conference on Computer Vision, pp. 15108–15117 (2021)
7. Creswell, A., White, T., Dumoulin, V., Arulkumaran, K., Sengupta, B., Bharath, A.A.: Generative adversarial networks: an overview. IEEE Sig. Process. Mag. **35**(1), 53–65 (2018)
8. Deng, L., Suo, H., Li, D.: Deepfake video detection based on EfficientNet-V2 network. Comput. Intell. Neurosci. **2022**, 1–13 (2022)
9. Doersch, C.: Tutorial on variational autoencoders. arXiv preprint arXiv:1606.05908 (2016)
10. Durall, R., Keuper, M., Pfreundt, F.J., Keuper, J.: Unmasking deepfakes with simple features. arXiv preprint arXiv:1911.00686 (2019)
11. Ganguly, S., Ganguly, A., Mohiuddin, S., Malakar, S., Sarkar, R.: ViXNet: vision transformer with Xception network for deepfakes based video and image forgery detection. Exp. Syst. Appl. **210**, 118423 (2022)
12. Guan, W., Wang, W., Dong, J., Peng, B., Tan, T.: Collaborative feature learning for fine-grained facial forgery detection and segmentation. arXiv preprint arXiv:2304.08078 (2023)
13. Guarnera, L., Giudice, O., Battiato, S.: Fighting deepfake by exposing the convolutional traces on images. IEEE Access **8**, 165085–165098 (2020)

14. He, Z., Zuo, W., Kan, M., Shan, S., Chen, X.: AttGAN: facial attribute editing by only changing what you want. IEEE Trans. Image Process. **28**(11), 5464–5478 (2019)
15. Hernandez-Ortega, J., Tolosana, R., Fierrez, J., Morales, A.: DeepFakesON-Phys: deepfakes detection based on heart rate estimation. arXiv preprint arXiv:2010.00400 (2020)
16. Khalid, F., Akbar, M.H., Gul, S.: SWYNT: swin y-net transformers for deepfake detection. In: 2023 International Conference on Robotics and Automation in Industry (ICRAI), pp. 1–6. IEEE (2023)
17. Li, L., et al.: Face X-ray for more general face forgery detection. In: Proceedings of the IEEE/CVF Conference on Computer Vision and Pattern Recognition, pp. 5001–5010 (2020)
18. Li, Y., Chang, M.C., Lyu, S.: In Ictu Oculi: exposing AI created fake videos by detecting eye blinking. In: 2018 IEEE International Workshop on Information Forensics and Security (WIFS), pp. 1–7. IEEE (2018)
19. Li, Y., Lyu, S.: Exposing deepfake videos by detecting face warping artifacts. arXiv preprint arXiv:1811.00656 (2018)
20. Masi, I., Killekar, A., Mascarenhas, R.M., Gurudatt, S.P., AbdAlmageed, W.: Two-branch recurrent network for isolating deepfakes in videos. In: Vedaldi, A., Bischof, H., Brox, T., Frahm, J.-M. (eds.) ECCV 2020, Part VII. LNCS, vol. 12352, pp. 667–684. Springer, Cham (2020). https://doi.org/10.1007/978-3-030-58571-6_39
21. Nguyen, H.H., Fang, F., Yamagishi, J., Echizen, I.: Multi-task learning for detecting and segmenting manipulated facial images and videos. In: 2019 IEEE 10th International Conference on Biometrics Theory, Applications and Systems (BTAS), pp. 1–8. IEEE (2019)
22. Qi, H., et al.: DeepRhythm: exposing deepfakes with attentional visual heartbeat rhythms. In: Proceedings of the 28th ACM International Conference on Multimedia, pp. 4318–4327 (2020)
23. Rössler, A., Cozzolino, D., Verdoliva, L., Riess, C., Thies, J., Nießner, M.: Face-Forensics: a large-scale video dataset for forgery detection in human faces. arXiv preprint arXiv:1803.09179 (2018)
24. Saif, S., Tehseen, S., Ali, S.S., Kausar, S., Jameel, A.: Generalized deepfake video detection through time-distribution and metric learning. IT Prof. **24**(2), 38–44 (2022)
25. Tjon, E., Moh, M., Moh, T.S.: Eff-YNet: a dual task network for deepfake detection and segmentation. In: 2021 15th International Conference on Ubiquitous Information Management and Communication (IMCOM), pp. 1–8. IEEE (2021)
26. Yang, J., Xiao, S., Li, A., Lan, G., Wang, H.: Detecting fake images by identifying potential texture difference. Futur. Gener. Comput. Syst. **125**, 127–135 (2021)
27. Yang, X., Li, Y., Lyu, S.: Exposing deep fakes using inconsistent head poses. In: 2019 IEEE International Conference on Acoustics, Speech and Signal Processing (ICASSP), ICASSP 2019, pp. 8261–8265. IEEE (2019)
28. Zhang, X., Karaman, S., Chang, S.F.: Detecting and simulating artifacts in GAN fake images. In: 2019 IEEE International Workshop on Information Forensics and Security (WIFS), pp. 1–6. IEEE (2019)

Camouflaged Object Detection via Global-Edge Context and Mixed-Scale Refinement

Qilun Li, Fengqin Yao, Xiandong Wang, and Shengke Wang[✉]

Ocean University of China, Qingdao, China
{yaofengqin8312,wangxiandong,neverme}@stu.ouc.edu.cn,
{yaofengqin8312,wangxiandong,neverme}@ouc.edu.cn

Abstract. Camouflage object detection (COD), trying to segment objects that blend perfectly with the surrounding environment, is challenging and complex in real-world scenarios. However, existing deep learning methods often fail to accurately identify the boundary detail branches and complete structure of camouflaged objects. To address these challenges, we propose a novel Global-edge Context and Mixed-scale Refinement Network (GCMRNet) to handle the challenging COD task. Specifically, we propose a Global-edge Context Module (GCM), to effectively obtain long-distance context dependencies, which provides rich global context information. In addition, we also propose the Hierarchical Mixed-scale Refinement Module (HMRM) to conduct information interaction and feature refinement between channels, which aggregates the rich multi-level features for accurate COD. Extensive experimental results on three challenging benchmark datasets demonstrate that our proposed method outperforms 21 state-of-art methods under widely used evaluation metrics.

Keywords: Camouflage Object Detection · Salient Object Detection

1 Introduction

Camouflage is a common survival skill in nature, which uses its own structural and physiological characteristics to reduce the risk of predator identification. Camouflage also widely exists in people's daily life. For example, natural creatures hide in the surrounding environment to avoid predators or hunters. Camouflage object detection (COD) has been an emerging field of computer vision research in recent years. It is a challenging image processing method that aims to detect camouflaged objects hidden in the background area of the image and distinguish them from the background.

In recent years, camouflage object detection has attracted more and more attention from researchers and facilitates many valuable real-life applications, such as search and rescue [7], species discovery, and medical image analysis [10]. The existing frameworks can be roughly divided into two forms: traditional feature-based handmade structures and deep learning-based structures.

Q. Liu et al. (Eds.): PRCV 2023, LNCS 14431, pp. 256–268, 2024.
https://doi.org/10.1007/978-981-99-8540-1_21

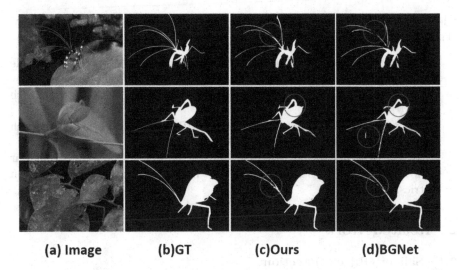

(a) Image   (b)GT   (c)Ours   (d)BGNet

Fig. 1. Visual examples of camouflaged object detection in some challenging scenarios. Compared with the recently proposed deep learning-based method BGNet, our method, can effectively segment the slender branch and complete structure, and produce accurate predictions with intact boundaries, as shown in red boxes. (Color figure online)

The traditional hand-made feature-based model mainly relies on hand-made features to describe the features of the object, including color, brightness, and angle. However, once the background environment changes, traditional methods of handcrafting features do not work well. Additionally, complex backgrounds are not a good fit for the traditional method of manually designing features, and model accuracy needs to be increased. To address these challenges, deep learning techniques have been adopted and shown great potential [13,44,50], more and more algorithms are introducing it into camouflaged object detection. As mentioned in [45], the edge prior is widely used as an effective auxiliary cue, which benefits the preservation of object structure, yet has been barely studied for COD. To our best knowledge, the BGNet [33] is the better to exploit the edge information to improve the performance of COD. However, it is noted that, despite the introduction of edge cues, BGNet [33] still loses some fine boundary-related details and thus weakens the performance of COD. As shown in Fig. 1, the BGNet [33] models lose the boundary detail branches and cannot effectively segment complete object.

To this end, we propose a novel COD model, called the Global-edge Context and Mixed-scale Refinement Network (GCMRNet). We designed two modules, including Global-edge Context Module (GCM) and Hierarchical Mixed-scale Refinement Module (HMRM). First, we designed the GCM module to provide rich contextual information in practical applications, which is critical for the accurate detection of camouflage objects. Second, we designed the HMRM module to conduct information interaction and feature refinement between channels, which aggregates the rich multi-level features for accurate COD. In summary, the main contributions of this paper are as follows:

- We propose a novel COD model, GCMRNet, which integrates mixed-scale features with the consideration of rich global context information.
- We propose an effective receptive field module, GCM, which provides rich context features for COD.
- We propose an effective fusion module, HMRM, which aggregates the rich multi-level features for accurate COD.
- Extensive experiments on three benchmark datasets demonstrate that our GCMRNet outperforms 21 state-of-the-art models in terms of four evaluation metrics.

2   Related Work

2.1   Salient Object Detection

The salient object detection [1,2,48] task is to detect the targets or regions of an image or video that are of most interest to the human eye based on visual saliency features, and then segment them at the pixel level. As demonstrated in [7,23,38], state-of-the-art SOD methods retrained on COD datasets are capable of generating high-quality predictive maps that may sometimes even outperform existing CNN-based COD algorithms.

EGNet [45] consists of three parts: edge feature extraction part, saliency target feature extraction part, and one-to-one guidance module. Edge features are used to help saliency target feature locate the target and make the boundary of the target more accurate. GCPANet [5] to effectively integrate low-level appearance features, high-level semantic features, and global context features through some progressive context-aware Feature Interweaved Aggregation (FIA) modules and generate the saliency map in a supervised way.

2.2   Camouflaged Object Detection

Camouflaged objects are ones that use their own shape, texture, or colors to 'perfectly' blend in with the surrounding background. The detection of camouflaged objects is difficult because of the high similarity between the foreground and background, making it difficult to separate the object from its surroundings. In the initial studies, traditional methods mostly used shape, texture, and color to distinguish camouflaged objects from the background [11,20,34,39]. However, these methods can only extract some low-level feature information. If the background becomes more complex, it is generally difficult to accurately extract the objects camouflaged.

Recently, an increasing number of CNN-based COD methods have been proposed and have become far more effective than traditional detection methods. They can identify more complex background information and achieve excellent performance. PFNet [25] initially locates potential target objects by exploring long-range semantic dependencies, and then focuses on the discovery and removal

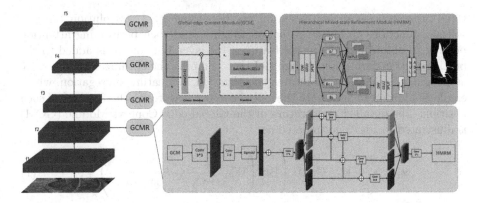

Fig. 2. The overall architecture of the proposed GCMRNet, which consists of two key components, i.e., Global-edge Context Module (GCM) and Hierarchical Mixed-scale Refinement Module (HMRM). See Sect. 3 for details

of distracting regions to progressively refine the segmentation results. Rank-Net [23] to simultaneously locate, segment, and rank camouflaged objects. The localization model is proposed to find the discriminative regions that make the camouflaged objects obvious. The segmentation model segments the full range of camouflaged objects. The ranking model infers the detectability of different camouflaged objects.

3   Method

3.1   Overall Architecture

As Fig. 2 shows, given a base architecture, the entire GCMR network aims to optimize the segment head at different layers in FPN, by developing a Global-edge Context Module (GCM), to effectively obtain long-distance context dependencies, which provides rich global context information. The Hierarchical Mixed-scale Refinement Module (HMRM) is designed to conduct information interaction and feature refinement between channels, which aggregates the rich multi-level features for accurate COD. The details of the components aforementioned are described in Sect. 3.2 and Sect. 3.3.

3.2   Global-Edge Context Module

Capturing long-range dependency, which aims to extract the global understanding of a visual scene, is proven to benefit a wide range of recognition tasks, such as image/video classification, object detection, and segmentation [14,15,42]. Therefore, we designed a Global-edge Context Module (GCM), to effectively obtain long-distance context dependencies, which provides rich global context information. As shown in Fig. 2, the simplified non-local block can be abstracted into three processes: (a) global attention pooling, using 1×1 convolution W_k and softmax functions to obtain attention weights, and then performing attention pooling to obtain global context features; (b) layer Batchnorm is added to the bottleneck transformation (before GeLU) to simplify the optimization and serve as a regularizer that can facilitate generalization. (c) feature aggregation, which employs addition to aggregate the global context features to the features of each position. The detailed architecture of the Global-edge Context Module (GCM) is illustrated in Fig. 2, formulated as

$$
\mathbf{z}_i = \mathbf{x}_i + W_{v2} \operatorname{GeLU} \left(\operatorname{BN} \left(W_{v1} \sum_{j=1}^{N_p} \frac{e^{W_k \mathbf{x}_j}}{\sum_{m=1}^{N_p} e^{W_k \mathbf{x}_m}} \mathbf{x}_j \right) \right),
\tag{1}
$$

where $\alpha_j = \frac{e^{W_k \mathbf{x}_j}}{\sum_m e^{W_k \mathbf{x}_m}}$ is the weight for global attention pooling, and $\delta(\cdot) = W_{v2} \operatorname{GeLU} (\operatorname{BN} (W_{v1}(\cdot)))$ denotes the bottleneck transform. Specifically, our GCM block consists of (a) global attention pooling for context modeling; (b) bottleneck transform to capture channel-wise dependencies; and (c) broadcast element-wise addition for feature fusion.

The global-edge context can facilitate a wide range of visual recognition tasks, and the flexibility of GCM blocks allows them to be inserted into network architectures for various computer vision problems.

3.3 Hierarchical Mixed-Scale Refinement Module

Similar to the multi-scale case, different channels also contain differentiated semantics. Therefore, it is necessary to explore the valuable clues contained in different channels. To this end, we designed HMRM for information interaction and feature refinement between channels, and enhanced the features from coarse-grained packet iteration to fine-grained channel modulation in the decoder, as shown in Fig. 2. Group-wise Iteration. We adopt 1×1 convolution to extend the channel number of feature map $\hat{f}_i$. The features are then divided into G groups $\{g_j\}_{j=1}^{G}$ along the channel dimension. Feature interaction between groups is carried out in an iterative manner. Specifically, the first group g1 is split into three feature sets $\{g_1'^k\}_{k=1}^{3}$ after a convolution block. Among them, the $g_1'^1$ is adopted for information exchange with the next group, and the other two are used for channel-wise modulation. It is noted that the output of group G with a similar input form to the previous groups only contains $g_{g'2}^{G}$ and $g_{g'2}^{G}$. Such an iterative mixing strategy strives to learn the critical clues from different channels and obtain a powerful feature representation. Channel-wise Modulation. The features $\left[\{g_j'^2\}_{j=1}^{G} \right]$ are concatenated and converted into the feature modulation vector α by a small convolutional network, which is employed to weight another concatenated feature $\left[\{g_j'^3\}_{j=1}^{G} \right]$. The weighted feature is then processed by a convolutional layer, which is defined as:

$$
\tilde{f}_i = \mathcal{A} \left(\hat{f}_i + \mathcal{N} \left(\mathcal{T} \left(\alpha \cdot \left[\{g_{g_j j}'^3\}_{j=1}^{G} \right] \right) \right) \right),
\tag{2}
$$

where A, N, and T represent the activation layer, the normalization layer, and the convolutional layer, respectively.

4 Experiments

4.1 Datasets

We conducted experiments on four publicly available camouflage object detection benchmark datasets: CAMO [18], CHAMELEON [31], COD10K [9], and NC4K [23]. In this paper, the training sets of CAMO, and COD10K as the model's training set, with a total of 4,040 images, and other data as the model's test set.

4.2 Implementation Details

The proposed GCMRNet is implemented with the deep learning framework PyTorch. As the settings in recent methods, the encoder is initialized with the parameters of ResNet-101 pretrained on ImageNet. We use kaiming-normal to initialize all the convolutional layers and linear layers. We utilize the Adam optimizer to train our mode. The learning rate and weight decay are set to $1e-4$ and 0.1, respectively. The entire model is trained for 25 epochs and takes about 2 h with a batch size of 16 in an end-to-end manner on an NVIDIA GeForce GPU (24 GB memory). During training and inference, the input image is simply resized to 416×416 and then fed into the network to obtain the predicted binary map. Random flipping is employed to augment the training data. This code can be available soon.

4.3 Evaluation Metrics

To present an in-depth evaluation of the performance of COD algorithms, we use five evaluation metrics that are widely used in image segmentation, including Emeasure [8], S-measure [6], F-measure [24], weighted F-measure [24] and Mean Absolute Error [28] denoted as E_ϕ, S_α, $F_{\max}$, F_β^w and $\mathcal{M}$, respectively.

4.4 Comparison with State-of-the-Arts

Since COD is an emerging field, there are not many methods for this topic, so we introduce some methods for salient object detection for comparison. Totally, we compare our GCMRNet with 21 state-of-the-art methods, including 10 SOD models, i.e., PoolNet [21], EGNet [45], SRCN [37], F^3Net [35], ITSD [47], CSNet [12], MINet [27], UCNet [43], PraNet [10] and BASNet [29], and 11 COD models, i.e., SINet [9], PFNet [26], S-MGL [41], R-MGL [41], LSR [23], UGTR [40], C^2FNet [32], JCSOD [19], C^2FNet [3], BGNet [33] and SegMaR [17]. The results of all these methods come from existing public data or are generated by models that are retrained based on the code released by the authors.

Table 1. Quantitative comparison with state-of-the-art methods for COD on three benchmarks using four widely used evaluation metrics (i.e., S_α, E_ϕ, F_β^w, and $\mathcal{M}$). "↑"/"↓" indicates that larger/smaller is better.

Method	Pub./Year	CAMO-Test				COD10K-Test				NC4K			
		S_α ↑	E_ϕ ↑	F_β^w ↑	$\mathcal{M}$ ↓	S_α ↑	E_ϕ ↑	F_β^w ↑	$\mathcal{M}$ ↓	S_α ↑	E_ϕ ↑	F_β^w ↑	$\mathcal{M}$ ↓
PoolNet [21]	CVPR'19	0.730	0.746	0.575	0.105	0.740	0.776	0.506	0.056	0.785	0.814	0.635	0.073
EGNet [45]	ICCV'19	0.732	0.800	0.604	0.109	0.736	0.810	0.517	0.061	0.777	0.841	0.639	0.075
SCRN [37]	ICCV'19	0.779	0.797	0.643	0.090	0.789	0.817	0.575	0.047	0.830	0.854	0.698	0.059
F³Net [35]	AAAI'20	0.711	0.741	0.564	0.109	0.739	0.795	0.544	0.051	0.780	0.824	0.656	0.070
ITSD [47]	CVPR'20	0.750	0.779	0.610	0.102	0.767	0.808	0.557	0.051	0.811	0.844	0.680	0.064
CSNet [12]	ECCV'20	0.771	0.795	0.642	0.092	0.778	0.809	0.569	0.047	0.750	0.773	0.603	0.088
MINet [27]	CVPR'20	0.748	0.791	0.637	0.090	0.770	0.832	0.608	0.042	0.812	0.862	0.720	0.056
UCNet [43]	CVPR'20	0.739	0.787	0.640	0.094	0.776	0.857	0.633	0.042	0.811	0.871	0.729	0.055
PraNet [10]	MICCAI'20	0.769	0.825	0.663	0.094	0.789	0.861	0.629	0.045	0.822	0.876	0.724	0.059
BASNet [29]	arxiv'21	0.749	0.796	0.646	0.096	0.802	0.855	0.677	0.038	0.817	0.859	0.732	0.058
SINet [9]	CVPR'20	0.745	0.804	0.644	0.092	0.776	0.864	0.631	0.043	0.808	0.871	0.723	0.058
PFNet [26]	CVPR'21	0.782	0.841	0.695	0.085	0.800	0.877	0.660	0.040	0.829	0.887	0.745	0.053
S-MGL [41]	CVPR'21	0.772	0.806	0.664	0.089	0.811	0.844	0.654	0.037	0.829	0.862	0.731	0.055
R-MGL [41]	CVPR'21	0.775	0.812	0.673	0.088	0.814	0.851	0.666	0.035	0.833	0.867	0.739	0.053
UGTR [40]	ICCV'21	0.784	0.821	0.683	0.086	0.817	0.852	0.665	0.036	0.839	0.874	0.746	0.052
LSR [23]	CVPR'21	0.787	0.838	0.696	0.080	0.804	0.880	0.673	0.037	0.840	0.895	0.766	0.048
C²FNet [32]	IJCAI'21	0.796	0.854	0.719	0.080	0.813	0.890	0.686	0.036	0.838	0.897	0.762	0.049
JCSOD [19]	CVPR'21	0.800	0.859	0.728	0.073	0.809	0.884	0.684	0.035	0.841	0.898	0.771	0.047
C²FNet [3]	TCSVT'22	0.800	0.869	0.730	0.077	0.811	0.891	0.691	0.036	0.850	0.899	0.776	0.045
BGNet [33]	IJCAI'22	0.812	0.870	0.749	0.073	0.831	0.901	0.722	0.033	0.851	0.907	0.788	0.044
SegMaR [17]	CVPR'22	0.815	0.872	0.742	0.071	0.833	0.895	0.724	0.033	0.856	0.908	0.786	0.043
Ours		**0.827**	**0.880**	**0.767**	**0.069**	**0.842**	**0.908**	**0.739**	**0.030**	**0.862**	**0.911**	**0.802**	**0.041**

Table 2. Quantitative comparison with state-of-the-art methods for COD on three benchmarks using four widely used evaluation metrics (i.e., S_α, $F_{\max}$, E_ϕ, and $\mathcal{M}$). "↑"/"↓" indicates that larger/smaller is better.

Method	Pub./Year	COD10K-Aquatic				COD10K-Terrestria				COD10K-Flying			
		S_α ↑	$F_{\max}$ ↑	E_ϕ ↑	$\mathcal{M}$ ↓	S_α ↑	$F_{\max}$ ↑	E_ϕ ↑	$\mathcal{M}$ ↓	S_α ↑	$F_{\max}$ ↑	E_ϕ ↑	$\mathcal{M}$ ↓
PiCANet [22]	CVPR'18	0.629	0.509	0.623	0.120	0.625	0.424	0.628	0.084	0.677	0.517	0.663	0.076
UNet++ [49]	MICCAI'18	0.599	0.437	0.659	0.121	0.593	0.366	0.637	0.081	0.659	0.467	0.708	0.068
BASNet [30]	CVPR'19	0.620	0.465	0.666	0.134	0.601	0.382	0.645	0.109	0.664	0.492	0.710	0.086
CPD [36]	CVPR'19	0.739	0.652	0.770	0.082	0.714	0.565	0.735	0.058	0.777	0.672	0.796	0.046
HTC [4]	CVPR'19	0.507	0.223	0.494	0.129	0.530	0.196	0.484	0.078	0.582	0.309	0.558	0.070
MSRCNN [16]	CVPR'19	0.614	0.465	0.685	0.107	0.611	0.418	0.671	0.070	0.674	0.523	0.742	0.058
PFANet [46]	CVPR'19	0.629	0.490	0.614	0.162	0.609	0.404	0.600	0.123	0.657	0.496	0.632	0.113
PoolNet [21]	CVPR'19	0.689	0.572	0.705	0.102	0.677	0.509	0.688	0.070	0.733	0.614	0.733	0.062
EGNet [45]	ICCV'19	0.725	0.630	0.775	0.080	0.704	0.550	0.748	0.054	0.768	0.651	0.803	0.044
SCRN [37]	ICCV'19	0.767	0.695	0.787	0.071	0.746	0.629	0.756	0.051	0.803	0.714	0.812	0.040
SINet [9]	CVPR'20	0.758	0.685	0.803	0.073	0.743	0.623	0.778	0.050	0.798	0.709	0.828	0.040
Ours		**0.831**	**0.796**	**0.896**	**0.044**	**0.817**	**0.739**	**0.889**	**0.031**	**0.870**	**0.827**	**0.932**	**0.020**

Quantitative Evaluation. Table 1 provides comparison results. On all datasets, the proposed model consistently and significantly outperforms all the COD methods in terms of the four metrics without using any post-processing techniques, where the best ones are highlighted in bold. In particular, while com-

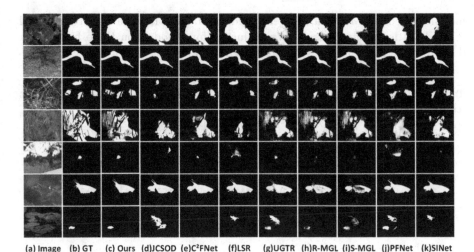

(a) Image   (b) GT   (c) Ours   (d)JCSOD   (e)C²FNet   (f)LSR   (g)UGTR   (h)R-MGL   (i)S-MGL   (j)PFNet   (k)SINet

Fig. 3. Visual comparison of the proposed model with eight state-of-the-art COD methods. Obviously, our method is capable of accurately segmenting various camouflaged objects with more clear boundaries.

pared with the second-best SegMaR, our method increases S_α by 1.16 %, E_ϕ by 0.89% and F_β^w by 2.47% on average. Compared with the third-best BGNet, our method increases S_α by 1.48 %, E_ϕ by 0.74% and F_β^w by 1.63% on average. In addition to the performance comparison on the three standard datasets, we also present the performance of the comparison model and the proposed model on the COD10K subset, as shown in Table 2, where the proposed model still achieves the best performance. Specifically, compared with the second-best SINet, our method increases S_α by 9.53 %, F_{max} by 17.12% and E_ϕ by 12.8% on average. Overall, our model achieves the best performance based on the available performance statistics and analysis, both on the subset and standard dataset.

Qualitative Evaluation. The Fig. 3 shows the visual comparison results of different methods on several typical examples. Our method achieves better results than these SOTA methods. We observed that our proposed GCMRNet can accurately capture camouflaged objects, reducing omissions compared to other methods. Simultaneously, as shown in Rows 3 and 6, our method is more precise than the other methods in detecting the boundary of the object. It deserves to be observed that our method is effective well in detecting small camouflaged objects, as showed in Rows 5 and 7. In a nutshell, our predictions have clearer and more complete object regions, as well as sharper contours.

4.5   Ablation Study

We designed ablation experiments to verify the effectiveness of our components, including GCM and HMRM. Model a is a basic model. Model b adds a GCM module based on Model a; Model c adds a HMRM module based on Model a; Model d is our final result. We evaluated four models on three benchmark datasets. The laboratory results are shown in Table 3.

Table 3. Quantitative evaluation for ablation studies on three datasets. The best results are highlighted in Bold. B: baseline.

Module	Method	CAMO-Test				COD10K-Test				NC4K			
		$S_\alpha\uparrow$	$E_\phi\uparrow$	$F_\beta^w\uparrow$	$\mathcal{M}\downarrow$	$S_\alpha\uparrow$	$E_\phi\uparrow$	$F_\beta^w\uparrow$	$\mathcal{M}\downarrow$	$S_\alpha\uparrow$	$E_\phi\uparrow$	$F_\beta^w\uparrow$	$\mathcal{M}\downarrow$
a	B	0.812	0.870	0.749	0.073	0.831	0.901	0.722	0.033	0.851	0.907	0.788	0.044
b	B+GCM	0.816	0.868	0.749	0.071	0.840	0.903	0.729	0.032	0.856	0.909	0.800	0.043
c	B+HMRM	0.819	0.870	0.754	**0.067**	0.838	0.901	0.732	0.031	0.858	0.906	0.796	0.044
d	Ours	0.827	**0.880**	**0.767**	0.069	0.842	**0.908**	**0.739**	**0.030**	0.862	**0.911**	**0.802**	**0.041**

The Effectiveness of GCM. By comparing Model a and Model b, we can clearly find that the evaluation index of Model b on three benchmark datasets has been greatly improved. By adding GCM modules, our model will perform better. It has been proven that GCM can obtain long-distance context dependences, which is conducive to obtaining global information so that segmenting visual scenes can be easier when global dependency information is captured.

The Effectiveness of HMRM. To verify the effectiveness of HMRM, we compare the experimental results of Model a and Model c. From the experimental results, we can see that if the HMRM module is removed, the performance of our model will degrade. HMRM can effectively perform information interaction and feature optimization between channels to obtain valuable semantic clues between different channels.

4.6   Discussion SOD and COD

It can be seen from the above experiments that our method shows outstanding performance in COD and SOD. Considering the differences between these two tasks, we are curious why our methods are consistent on these two seemingly different tasks. We attribute this to the generality and rationality of the designed structure. In fact, SOD and COD have obvious commonalities; That is, accurate segmentation must rely on capturing more global information and feature fusion. By establishing long-distance dependencies, our model can extract key and valuable cues from complex scenes and objects, which helps with precise positioning and smooth segmentation of objects. In addition, the proposed HMRM can effectively perform information interaction and feature optimization between channels, so as to obtain valuable semantic clues between different channels. All these components are built on the common needs of both tasks and provide a solid foundation for performance.

5   Conclusion

In this paper, we have proposed a novel GCMF-Net for the COD task. We propose a Global-edge Context Module (GCM), which exploits rich global context information from a backbone network. And we also propose the Hierarchical Mixed-scale Refinement Module (HMRM), which aggregates the rich multi-level features for accurate COD. Our model provides both global context and effective fusion, effectively satisfying the demands of accurate COD. Extensive experiments have validated the effectiveness of the method for COD and SOD tasks, outperforming the existing state-of-the-art methods. In the future, we will investigate compressing our model into a lightweight model suitable for mobile devices and improving its efficiency in real-time applications.

References

1. Liu, B., Zhou, Y., Liu, P., Sun, W., Li, S., Fang, X.: Saliency detection via double nuclear norm maximization and ensemble manifold regularization. Knowl. Based Syst. **183**, 104850 (2019)
2. Borji, A., Cheng, M.M., Hou, Q., Jiang, H., Li, J.: Salient object detection: a survey. Comput. Vis. Media **5**(2), 117–150 (2019)
3. Chen, G., Liu, S.J., Sun, Y.J., Ji, G.P., Wu, Y.F., Zhou, T.: Camouflaged object detection via context-aware cross-level fusion. IEEE Trans. Circ. Syst. Video Technol. **32**(10), 6981–6993 (2022)
4. Chen, K., et al.: Hybrid task cascade for instance segmentation. In: Proceedings of the IEEE/CVF Conference on Computer Vision and Pattern Recognition, pp. 4974–4983 (2019)
5. Chen, Z., Xu, Q., Cong, R., Huang, Q.: Global context-aware progressive aggregation network for salient object detection. In: Proceedings of the AAAI Conference on Artificial Intelligence, vol. 34, pp. 10599–10606 (2020)
6. Fan, D.P., Cheng, M.M., Liu, Y., Li, T., Borji, A.: Structure-measure: a new way to evaluate foreground maps. In: Proceedings of the IEEE International Conference on Computer Vision, pp. 4548–4557 (2017)
7. Fan, D.P., Ji, G.P., Cheng, M.M., Shao, L.: Concealed object detection. IEEE Trans. Pattern Anal. Mach. Intell. **44**(10), 6024–6042 (2021)
8. Fan, D.P., Ji, G.P., Qin, X., Cheng, M.M.: Cognitive vision inspired object segmentation metric and loss function. Sci. Sinica Informationis **51**, 1475 (2021)
9. Fan, D.P., Ji, G.P., Sun, G., Cheng, M.M., Shen, J., Shao, L.: Camouflaged object detection. In: Proceedings of the IEEE/CVF Conference on Computer Vision and Pattern Recognition, pp. 2777–2787 (2020)
10. Fan, D.-P., et al.: PraNet: parallel reverse attention network for polyp segmentation. In: Martel, A.L., et al. (eds.) MICCAI 2020. LNCS, vol. 12266, pp. 263–273. Springer, Cham (2020). https://doi.org/10.1007/978-3-030-59725-2_26
11. Feng, X., Guoying, C., Wei, S.: Camouflage texture evaluation using saliency map. In: Proceedings of the Fifth International Conference on Internet Multimedia Computing and Service, pp. 93–96 (2013)
12. Gao, S.-H., Tan, Y.-Q., Cheng, M.-M., Lu, C., Chen, Y., Yan, S.: Highly efficient salient object detection with 100k parameters. In: Vedaldi, A., Bischof, H., Brox, T., Frahm, J.-M. (eds.) ECCV 2020. LNCS, vol. 12351, pp. 702–721. Springer, Cham (2020). https://doi.org/10.1007/978-3-030-58539-6_42

13. Howard, A.G., et al.: MobileNets: efficient convolutional neural networks for mobile vision applications (2017)
14. Hu, H., Gu, J., Zhang, Z., Dai, J., Wei, Y.: Relation networks for object detection. In: Proceedings of the IEEE Conference on Computer Vision and Pattern Recognition, pp. 3588–3597 (2018)
15. Hu, J., Shen, L., Sun, G.: Squeeze-and-excitation networks. In: Proceedings of the IEEE Conference on Computer Vision and Pattern Recognition, pp. 7132–7141 (2018)
16. Huang, Z., Huang, L., Gong, Y., Huang, C., Wang, X.: Mask scoring R-CNN. In: Proceedings of the IEEE/CVF Conference on Computer Vision and Pattern Recognition, pp. 6409–6418 (2019)
17. Jia, Q., Yao, S., Liu, Y., Fan, X., Liu, R., Luo, Z.: Segment, magnify and reiterate: detecting camouflaged objects the hard way. In: Proceedings of the IEEE/CVF Conference on Computer Vision and Pattern Recognition, pp. 4713–4722 (2022)
18. Le, T.N., Nguyen, T.V., Nie, Z., Tran, M.T., Sugimoto, A.: Anabranch network for camouflaged object segmentation. Comput. Vis. Image Underst. **184**, 45–56 (2019)
19. Li, A., Zhang, J., Lv, Y., Liu, B., Zhang, T., Dai, Y.: Uncertainty-aware joint salient object and camouflaged object detection. In: Proceedings of the IEEE/CVF Conference on Computer Vision and Pattern Recognition, pp. 10071–10081 (2021)
20. Li, S., Florencio, D., Zhao, Y., Cook, C., Li, W.: Foreground detection in camouflaged scenes. In: 2017 IEEE International Conference on Image Processing (ICIP), pp. 4247–4251. IEEE (2017)
21. Liu, J.J., Hou, Q., Cheng, M.M., Feng, J., Jiang, J.: A simple pooling-based design for real-time salient object detection. In: Proceedings of the IEEE/CVF Conference on Computer Vision and Pattern Recognition, pp. 3917–3926 (2019)
22. Liu, N., Han, J., Yang, M.H.: PiCANet: learning pixel-wise contextual attention for saliency detection. In: Proceedings of the IEEE Conference on Computer Vision and Pattern Recognition, pp. 3089–3098 (2018)
23. Lv, Y., et al.: Simultaneously localize, segment and rank the camouflaged objects. In: Proceedings of the IEEE/CVF Conference on Computer Vision and Pattern Recognition, pp. 11591–11601 (2021)
24. Margolin, R., Zelnik-Manor, L., Tal, A.: How to evaluate foreground maps? In: Proceedings of the IEEE Conference on Computer Vision and Pattern Recognition, pp. 248–255 (2014)
25. Mei, H., Ji, G.P., Wei, Z., Yang, X., Wei, X., Fan, D.P.: Camouflaged object segmentation with distraction mining. In: Proceedings of the IEEE/CVF Conference on Computer Vision and Pattern Recognition (CVPR), pp. 8772–8781, June 2021
26. Mei, H., Ji, G.P., Wei, Z., Yang, X., Wei, X., Fan, D.P.: Camouflaged object segmentation with distraction mining. In: Proceedings of the IEEE/CVF Conference on Computer Vision and Pattern Recognition, pp. 8772–8781 (2021)
27. Pang, Y., Zhao, X., Zhang, L., Lu, H.: Multi-scale interactive network for salient object detection. In: Proceedings of the IEEE/CVF Conference on Computer Vision and Pattern Recognition, pp. 9413–9422 (2020)
28. Perazzi, F., Krähenbühl, P., Pritch, Y., Hornung, A.: Saliency filters: contrast based filtering for salient region detection. In: 2012 IEEE Conference on Computer Vision and Pattern Recognition, pp. 733–740. IEEE (2012)
29. Qin, X., et al.: Boundary-aware segmentation network for mobile and web applications. arXiv preprint arXiv:2101.04704 (2021)

30. Qin, X., Zhang, Z., Huang, C., Gao, C., Dehghan, M., Jagersand, M.: BASNet: boundary-aware salient object detection. In: Proceedings of the IEEE/CVF Conference on Computer Vision and Pattern Recognition, pp. 7479–7489 (2019)
31. Skurowski, P., Abdulameer, H., Błaszczyk, J., Depta, T., Kornacki, A., Kozieł, P.: Animal camouflage analysis: Chameleon database, vol. 2, no. 6, p. 7 (2018, unpublished)
32. Sun, Y., Chen, G., Zhou, T., Zhang, Y., Liu, N.: Context-aware cross-level fusion network for camouflaged object detection. arXiv preprint arXiv:2105.12555 (2021)
33. Sun, Y., Wang, S., Chen, C., Xiang, T.Z.: Boundary-guided camouflaged object detection. In: IJCAI, pp. 1335–1341 (2022)
34. Tankus, A., Yeshurun, Y.: Convexity-based visual camouflage breaking. Comput. Vis. Image Underst. **82**(3), 208–237 (2001)
35. Wei, J., Wang, S., Huang, Q.: F^3Net: fusion, feedback and focus for salient object detection. In: Proceedings of the AAAI Conference on Artificial Intelligence, vol. 34, pp. 12321–12328 (2020)
36. Wu, Z., Su, L., Huang, Q.: Cascaded partial decoder for fast and accurate salient object detection. In: Proceedings of the IEEE/CVF Conference on Computer Vision and Pattern Recognition, pp. 3907–3916 (2019)
37. Wu, Z., Su, L., Huang, Q.: Stacked cross refinement network for edge-aware salient object detection. In: Proceedings of the IEEE/CVF International Conference on Computer Vision, pp. 7264–7273 (2019)
38. Xu, X., Zhu, M., Yu, J., Chen, S., Hu, X., Yang, Y.: Boundary guidance network for camouflage object detection. Image Vis. Comput. **114**, 104283 (2021)
39. Xue, F., Yong, C., Xu, S., Dong, H., Luo, Y., Jia, W.: Camouflage performance analysis and evaluation framework based on features fusion. Multimedia Tools Appl. **75**(7), 4065–4082 (2016)
40. Yang, F., et al.: Uncertainty-guided transformer reasoning for camouflaged object detection. In: Proceedings of the IEEE/CVF International Conference on Computer Vision, pp. 4146–4155 (2021)
41. Zhai, Q., Li, X., Yang, F., Chen, C., Cheng, H., Fan, D.P.: Mutual graph learning for camouflaged object detection. In: Proceedings of the IEEE/CVF Conference on Computer Vision and Pattern Recognition, pp. 12997–13007 (2021)
42. Zhang, H., et al.: Context encoding for semantic segmentation. In: Proceedings of the IEEE Conference on Computer Vision and Pattern Recognition, pp. 7151–7160 (2018)
43. Zhang, J., et al.: UC-NET: Uncertainty inspired RGB-D saliency detection via conditional variational autoencoders. In: Proceedings of the IEEE/CVF Conference on Computer Vision and Pattern Recognition, pp. 8582–8591 (2020)
44. Zhang, X., Zhou, X., Lin, M., Sun, J.: ShuffleNet: an extremely efficient convolutional neural network for mobile devices. In: Proceedings of the IEEE Conference on Computer Vision and Pattern Recognition, pp. 6848–6856 (2018)
45. Zhao, J.X., Liu, J.J., Fan, D.P., Cao, Y., Yang, J., Cheng, M.M.: EGNet: edge guidance network for salient object detection. In: Proceedings of the IEEE/CVF International Conference on Computer Vision, pp. 8779–8788 (2019)
46. Zhao, T., Wu, X.: Pyramid feature attention network for saliency detection. In: Proceedings of the IEEE/CVF Conference on Computer Vision and Pattern Recognition, pp. 3085–3094 (2019)
47. Zhou, H., Xie, X., Lai, J.H., Chen, Z., Yang, L.: Interactive two-stream decoder for accurate and fast saliency detection. In: Proceedings of the IEEE/CVF Conference on Computer Vision and Pattern Recognition, pp. 9141–9150 (2020)

48. Zhou, T., Fan, D.P., Cheng, M.M., Shen, J., Shao, L.: RGB-D salient object detection: a survey. Comput. Vis. Media **7**(1), 37–69 (2021). https://doi.org/10.1007/s41095-020-0199-z
49. Zhou, Z., Rahman Siddiquee, M.M., Tajbakhsh, N., Liang, J.: UNet++: a nested U-Net architecture for medical image segmentation. In: Stoyanov, D., et al. (eds.) DLMIA/ML-CDS -2018. LNCS, vol. 11045, pp. 3–11. Springer, Cham (2018). https://doi.org/10.1007/978-3-030-00889-5_1
50. Zoph, B., Vasudevan, V., Shlens, J., Le, Q.V.: Learning transferable architectures for scalable image recognition. In: Proceedings of the IEEE Conference on Computer Vision and Pattern Recognition, pp. 8697–8710 (2018)

Multimedia Analysis and Reasoning

Enhancing CLIP-Based Text-Person Retrieval by Leveraging Negative Samples

Yumin Tian[1], Yuanbo Li[1], Di Wang[1(✉)], Xiao Liang[1], Ronghua Zhang[2], and Bo Wan[1]

[1] The Key Laboratory of Smart Human-Computer Interaction and Wearable Technology of Shaanxi Province, Xidian University, Xi'an, China
wangdi@xidian.edu.cn
[2] College of Information Science and Technology, Shihezi University, Shihezi, China

Abstract. Text-person retrieval (TPR) is a fine-grained cross-modal retrieval task that aims to find matching person images through detailed text descriptions. Various recent cross-modal pre-trained vision-language based models (e.g., CLIP [12]) have greatly improved the performance of text-image retrieval (TIR), but direct migration of these general models to the TPR has limited effectiveness: CLIP encodes images and text into global features separately, which makes it difficult to focus on fine-grained attributes of person, and due to the limited volume of TPR, fine-tuning CLIP may cause unstable parameter updates. In this paper, we offer a cost-effective approach to improve CLIP-based models by constructing negative samples and learning knowledge from them. Specifically, 1) replacing specific types of attribute words in the original text to generate hard negative text that do not match the semantics of the image, improving the model's understanding of fine-grained semantic knowledge; 2) a momentum contrastive learning framework for image intra-modal is introduced to scale up the data within a batch, thus providing enough negative samples and additional positive samples to learn person identity features, ensuring the gradient stability of the CLIP-based model. The proposed method achieves good results on three existing publicly available datasets.

Keywords: Text-person retrieval · CLIP · Negative mining

1 Introduction

Text-person retrieval is an extended task combining two tasks of cross-modal text-image retrieval and pedestrian re-identification. Compared with the former, TPR has a finer retrieval granularity, and compared with the latter, TPR needs to take into account the semantic gap between images and text. The goal of TPR is expect to find a person image matching the text description while also associating as many images belonging to the same identity by that image. This task has a wide range of applications, such as surveillance person tracking, smart

Q. Liu et al. (Eds.): PRCV 2023, LNCS 14431, pp. 271–283, 2024.
https://doi.org/10.1007/978-981-99-8540-1_22

transportation, and personal photo album creation, to help people find target people quickly in a huge amount of data.

With the appearances of large-scale vision-language pre-training models, great progress has been made in many image-text related downstream tasks. These pre-training models are jointly trained by large-scale text-image pairs with powerful cross-modal semantic understanding, and thus are capable of good semantic correspondence between images and text. Among these, CLIP [12] is the most widely used one. CLIP is trained by 30 billion image-text pairs, and the contained semantic knowledge is extremely rich, showing excellent performance on TIR. Some work has attempted to apply CLIP to TPR tasks, but simply migrating CLIP does not fully realised its potential. Specifically, as shown in Fig. 1, there are two challenges in migrating CLIP to TPR:

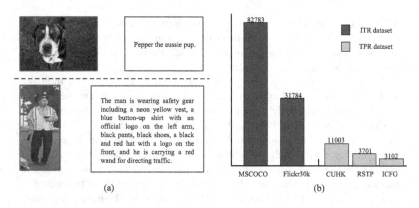

Fig. 1. Two challenges in migrating CLIP to the TPR. (a) Common sample used for CLIP pre-training usually contains only one or a few objects, while the sample in TPR contains multiple objects. (b) Generic TIR's dataset size far exceeds that of TPR.

Fine-Grained Concepts Learning. We note from the false positive samples in the visualization results from previous CLIP-based TPR works that the model is not very good at distinguishing some fine-grained information. Recent studies such as [3] have shown that the CLIP pre-training computes CLIP loss in a batch of random image-text pairs sampled from large-scale data, and the probability of a batch containing a set of the same objects is very low, so that when images and text contain just a common set of objects it is sufficient to be recognized as a matching pair, which leads to the problem of "object bias", i.e., fine-grained concepts like attributes, states, and relationships are not sufficiently represented and attended to, which results in the model not being able to distinguish well between similar image-text pairs. However, in TPR, a specific person is described through many terms that depict appearance characteristics, which will include multiple groups of objects (see Fig. 1.a), and there is a high probability that the same objects will be described between different person. The key to distinguish lies in one or several distinctive objects, hence requiring the model has a strong understanding of the fine-grained concepts.

Data Scarcity. Migrating CLIP prior knowledge to downstream tasks requires tuning the parameters of the CLIP pre-trained model using sufficient task-specific training data to maximize the performance of the task. However, compare to the generic TIR datasets, the volume of data in TPR is relatively small (see Fig. 1.b), and with the constrained batchsize, when use training data to fine-tune, CLIP pays more attention to the features and patterns of the new task and relatively ignores the features learned in the pre-training phase, which has a high probability to cause gradient oscillations, making it difficult for the model to adequately learn and represent the fine-grained attributes and distinguishing features between different identity person. Moreover, an identity in a batch usually has only one image, the probability of learning the generalized representation between different images of the same identity is low, the number of different identities is limited resulting in lack of diversity.

In order to solve the above problems, we propose two methods. First, generating hard negative text that differs only in individual attribute features compared with the original text, thus the model is forced to learn the correspondence between fine-grained semantic concepts in images and text, so that the model can make more accurate decisions by comprehensively combining semantic concepts in the context, thus alleviating the "object bias" problem of CLIP-based models. Second, introducing the momentum contrastive learning framework to store the images appearing in the previous batch as negative samples to expand the number of person identities in the current batch, enrich the diversity of contrastive learning, thus relieving the problem of data scarcity of TPR. Our contributions are summarized as follows:

Propose an approach to generate hard negative samples, which enables the model to better understand fine-grained semantic concepts;

Introduce the momentum contrastive learning framework for image modality to expand the training set scale, enriching diversity and improving training stability;

Achieve excellent performance on three existing public datasets and improve by more than 5% compared to using CLIP directly.

2   Related Works

Text-Based Person Retrieval. The task of text-person retrieval was first introduced by [11], they have made publicly available a dataset called CUHK-PEDES, which is the most widely used and largest TPR dataset. Earlier works such as [16] improve on it for image and text feature representation, cross-modal alignment methods, and metric functions. The cross-modal projection matching loss (CMPM) and cross-modal projection classification loss (CMPC) in [18] are widely used in TPR for cross-modal feature alignment and identity classification. Later, some works like [4,19] started to use BERT to obtain multi-level textual representation. NAFS [5] proposed a new stepped CNN for extracting full-scale image features by horizontally divide into a fixed number of horizontal strips, while using a locally constrained BERT for multiple scales fusion, and used visual nearest neighbors for reranking to further improve the performance. ACSA [9] implementing an asymmetric cross-attention module that implements attentional mechanism traction from different scales. These methods have achieved certain success, but their performance is limited by the lack of large-scale cross-modal pre-training and vision-language alignment.

Vision-Language Pre-training Models. With the appearance of multimodal pre-training models, the new paradigm of text-image tasks training is "pre-training model + task-specific fine-tuning", and pre-training models are widely migrated to multimodal downstream tasks, which have achieved good results on some benchmark datasets due to the rich prior knowledge. ALBEF [10] innovatively proposes to add modality fusion encoders after cross-modality contrastive to enable deeper interaction between image and text. Recently, some works have attempted to migrate prior knowledge from pre-trained models to TPR. TextReID [7] first tried to introduce CLIP, which uses a momentum contrastive learning framework to transfer the knowledge learned from large-scale image-text pairs. CFine [1] proposed a CLIP-driven cross-grain feature refinement and fine-grained correspondence discovery framework for deep interactions. ETP-VLP [15] introducing multiple integrity description constraints, and proposes a dynamic attribute hinting strategy to construct uniform attribute descriptions.

Negative Samples Mining. Negative samples mining was first introduced by the face recognition, similar to the person retrieval, requires accurately distinguish between different image of the same person and image of different people, and by mining more diverse negative samples. [13] proposes triple loss, which defines images closest to the anchor but belonging to different people as the most similar negative samples, and controls the distance between positive samples and the most similar negative samples by boundary parameters. In [6], in order to alleviate the problem of imbalance between positive and negative samples, hard negative samples mining is introduced to effectively select the useful samples with higher misclassification probability. In MoCo [8], the negative samples from the previous batch are encoded using the momentum encoder and stored into the queue, providing a large and consistent dictionary to enrich diversity. In [3], a rule-based data enhancement strategy for negative samples generation is proposed to enable VL models to better understand structured concepts.

3   Method

Our method is based on CLIP, and in this section we present the proposed method in detail, the overview of our method is shown in Fig. 2.

3.1   Feature Extraction

Text Feature Extraction. The input text is first sub-tokenized, and the start token and end token are added to the head and tail of the token sequence, and fill it to the maximum length L_1, then input to the text encoder to learn the semantic and contextual relations in the text. The end token t_{cls} of the text is used as the global representation of the text, and the other token feature vectors $\{t_1, t_2, \cdots, t_{L_1}\}$ used as the overall representation.

Image Feature Extraction. The input image, of size $H * W$, is divided into fixed size non-overlapping L_2 patches, each of size $P * P$, and these patches are spanned into a one-dimensional vector, where $L_2 = H * W/P^2$, a CLS token is added to the beginning of the patches sequence, and then it is fed into the image encoder to establish the dependencies between patches, and the encoded representation of the CLS token v_{cls} is used as the global representation of the image, and the other patch feature vectors $\{v_1, v_2, \cdots, v_{L_2}\}$ is used as the overall representation of the image.

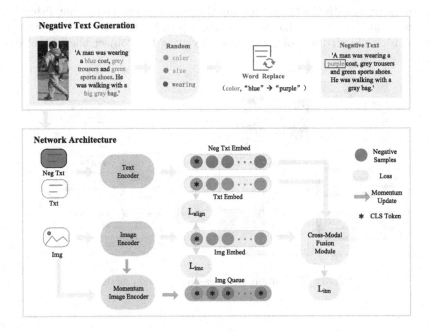

Fig. 2. Overview of the proposed method. It consists of an image encoder, a text encoder, an image momentum encoder and a cross-modal fusion module. The optimization objectives are cross-modal alignment loss L_{align}, image intra-modal contrastive loss L_{imc}, and image-text matching loss L_{itm}.

3.2 Fine-Grained Semantic Concept Learning

Negative Text Generation. We propose a low-cost data-driven strategy to enhance the sensitivity and comprehension of CLIP for fine-grained semantic concepts by replacing attribute words in each text in the training batch to generate hard negative text with the opposite semantics of the original text. Specifically, a specific person can be identified by three types of attributes: color, size, and wearing, so we consider generating the corresponding negative text from these three aspects. First, the lists associated with these three types is collected, which can be done by a simple Internet collection as well as traversing the dataset. Then, for each text description in the training set, randomly select a replacement type and the words of the text are compared with the list of that type, and when the sentence contains a word from the list, we randomly replace it with a different word from the same list. For example, when choosing a replacement word of color type, the word **"red"** in **"red-haired female"** is replaced with **"orange"** to get **"orange-haired female"**. This simple method of negative samples generation ensures that the composition of the text is approximately the same, but the semantics of the text has changed, it constitutes hard negative sample with the images forcing the model to learn to distinguish minor differences in the text, so that it is no longer possible to determine a person by the inclusion of individual objects, thus alleviating the "object bias" problem of CLIP.

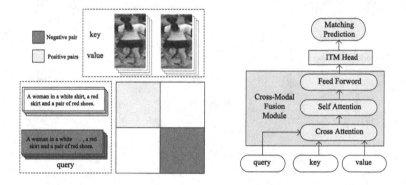

Fig. 3. Illustration for cross-modal fusion module.

Cross-Modal Fusion Module is shown in Fig. 3. The original image-text pair is positive pair, and the generated text and image constitute negative pair. To enhance the ability to discriminate difficult negative image-text pair, the cross-modal fusion module is designed to make the generated text and images deeply interactive. The text embedding is linearly mapped to query, and the image embedding is linearly mapped to key and value. The query, key, and value are fused by the cross-modal fusion module, and then the output into the ITM head to predict whether the image-text pair matches. The output of cross-modal fusion module p_{itm} can be calculated using the following equation:

$$MHA(q, k, v) = softmax(q * k^T / \sqrt{d_k}) * v \tag{1}$$

$$c_h = MHA(query, key, value) \tag{2}$$

$$p_{itm} = Linear(FFN(LN(MHA(Linear(c_h), Linear(c_h), Linear(c_h))))) \tag{3}$$

where $MHA(\cdot)$ is the multi-headed attention function, d_k is the dimension of the key vector, $softmax(\cdot)$ normalizes the vector, $Linear(\cdot)$ is a linear mapping function, $LN(\cdot)$ is the layer normalization calculation function, and $FFN(\cdot)$ is the feed-forward neural network component. We force the model to learn from hard negative pairs by calculating the ITM loss proposed in ALBEF [10]:

$$L_{itm} = E_{(I,T) \sim D} H(y_{itm}, p_{itm}) \tag{4}$$

where y_{itm} is a 2-dimensional one-hot vector representing the ground-truth label.

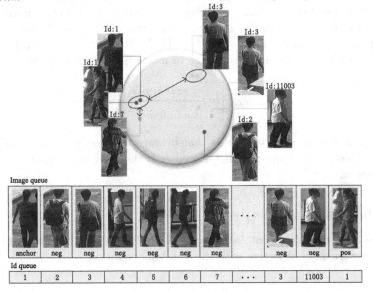

Fig. 4. Illustration of identity diversity from queue.

3.3 Image Modal Momentum Contrastive Learning

Inspired by the dynamic queue mechanism established in MoCo [8] that is independent of batch size, we try to introduce the contrastive learning framework within the image modality to enrich the diversity of identities in a batch, shown in Fig. 4. We consider the images belonging to the same person identity as different views, the views are positively paired with each other and the others are considered as negative. The number of identities in a batch can be expanded by using a dynamic queue to store a large number of negative images, encouraging the model to learn semantic associative representations within image modalities, enhancing the ability to discriminate between different identity and the learning of generalized identity representations belonging to the same person identity. Since the texts corresponding to different person images of the same identity may be described from different perspectives which are not identity universal, momentum encoders and queues are not used on the text side.

The image momentum encoder θ_k has the same architecture as the image encoder θ_q, but instead of gradient update, the parameters are updated by moving average according to the following formula, where $m \in [0,1)$ is the momentum parameter and k represents the current batch.

$$\theta_k = m\theta_{k-1} + (1-m)\theta_q \tag{5}$$

Input the image to the image encoder and the momentum image encoder, get the global features v_{cls} and v'_{cls} respectively, store v'_{cls} into the image queue Q while the identity id of the image is also pushed into the id queue. Define the similarity calculation as $sim(I, I') = (I)^T I'$. For each image in a batch, we calculate the softmax-normalized image to image queue similarity as:

$$L_{imc} = -E_{p\sim(I,Q)}\left[log\frac{exp(sim(v_{cls}, v_{cls}^+)/\tau)}{\sum_{k=1}^{K} exp(sim(v_{cls}, Q)/\tau)}\right] \tag{6}$$

where v_{cls}^+ is the image in the queue that belongs to the same identity as v_{cls}, τ is a learnable temperature parameter, and K is the number of negative samples in the queue.

3.4   Optimization Objective

We introduce the alignment loss proposed in ViTAA [16] as the cross-modal contrastive learning objective with the following equation:

$$L_{align} = \frac{1}{N}\sum_{i=1}^{N}\{log[1 + exp(-\tau_p(sim_i^+ - \alpha))] + log[1 + exp(\tau_n(sim_i^- - \beta))]\} \tag{7}$$

where S^+ means positive pair and S^- means negative pair, α denotes the lower bound for positive similarity and β denotes the upper bound for negative similarity, τ_p and τ_n denote the temperature parameters that adjust the slope of gradient.

The total optimization objective is the sum of the three losses:

$$L = L_{align} + L_{itm} + L_{imc} \tag{8}$$

It is worth mentioning that we find that L_{imc} can replace the id loss in traditional person retrieval. The traditional id loss requires designing the identity projection vector for each dataset with a different number of identities matrix, while using a queue can also achieve the purpose of distinguishing identities and is universal for datasets.

4 Experiments

4.1 Datasets and Configuration

Datasets. CUHK-PEDES [11] is the first TPR dataset with two textual descriptions per image for a total of 13,003 identities. ICFG-PEDES [2] with one textual descriptions per image for a total of 4102 identities. RSTPReid [19] with two textual descriptions per image for a total of 4101 identities.

Evaluation Metrics. We use the widely adopted top-k (k = 1, 5, 10) accuracy as the evaluation metrics, which ranks all gallery images based on their similarity relative to the text query. We also use mAP to evaluate the performance, mAP evaluates the model by combining both the index and the number of correct results. The higher top-k and mAP indicate better performance.

Implementation Details. We use CLIP as backbone. In training, the image queue length is set to 16384, and the momentum update parameters are 0.995, using the Adam optimizer with an initial learning rate of 1e−5, and all experiments were performed on an RTX3090 24 GB GPU. We use Rerank proposed in [5] to further improve the retrieval performance during inference. We take CLIP network structure with CMPM and CMPC losses as baseline. When inferencing, the cross-modal fusion module is removed to improve efficiency.

4.2 Comparisons with State-of-the-Art Methods

We compared the results of our proposed method with other state-of-the-art methods on CUHK-PEDES, ICFG-PEDES and RSTPReid. The results are shown in Table 1 and Table 2. Rerank means to using visual neighbors for re-ranking after initial ranking. The same rerank method is used in NAFS [5] and ASCA [9] to get better performance. Compared to baseline, there is significant improvement on all three datasets. We are substantially ahead of previous approaches without rerank. After using rerank, our method is substantially ahead of all methods, because our method learns rich identity distinguishability information during negative samples learning, which allows the model to more accurately associate visual neighbors. Specifically, the best performing CFine [17] and ETP-VLP [15] are CLIP-based models, compared to them, our performance is boosted significantly on the three datasets, this demonstrates that our method can effectively enhance CLIP-based models in TPR.

Table 1. Comparison of state-of-the-art methods on CUHK-PEDES.

Methods	Ref	Rank1	Rank5	Rank10	mAP
GNA-RNN [11]	CVPR17	19.05	-	53.64	-
ViTAA [16]	ECCV20	55.97	75.84	83.52	-
DSSL [19]	MM21	59.98	80.41	87.56	-
SSAN [2]	arXiv21	61.37	80.15	86.73	-
NAFS [5]	arXiv21	61.50	81.19	87.51	-
TextReID [7]	BMVC21	64.08	81.73	88.19	60.08
AXM-Net [4]	AAAI22	64.44	80.52	86.77	57.38
IVT [14]	ECCVW22	65.59	83.11	89.21	-
ASCA [9]	TMM22	68.67	85.61	90.66	-
CFine [17]	arXiv22	69.57	85.93	91.15	-
ETP-VLP [15]	arXiv23	70.16	86.10	90.98	66.32
baseline		66.38	84.19	89.91	60.83
Ours		**73.03**	**87.94**	**92.67**	**66.63**
Ours+Rerank		**77.63**	**91.61**	**94.99**	**71.95**

Table 2. Comparison of state-of-the-art methods on ICFG-PEDES and RSTPReid.

Methods	ICFG-PEDES				RSTPReid			
	R1	R5	R10	mAP	R1	R5	R10	mAP
SSAN [2]	54.23	72.63	79.53	-	-	-	-	-
DSSL [19]	-	-	-	-	32.43	55.08	63.19	-
IVT [14]	56.04	73.60	80.22	-	46.70	70.00	78.80	-
Cfine [17]	60.83	76.55	82.42	-	50.55	72.50	81.60	-
ETP-VLP [15]	60.64	75.97	81.76	42.78	50.65	72.45	81.20	43.11
baseline	59.76	76.33	82.71	37.98	49.40	64.10	71.40	36.81
ours	**65.02**	**79.88**	**84.6**	42.53	**56.30**	**72.80**	80.00	**45.82**
ours+Rerank	80.49	91.28	94.3	53.32	65.60	81.90	87.50	54.26

4.3   Ablation Study

We analyzed the proposed method on the three datasets, the results are shown in Table 3. We use CLIP network structure with CMPM and CMPC losses as baseline, the base performance are shown in No. 1. In No. 2, we replace the loss in baseline with align loss L_{align} as a cross-modal learning loss, and using our proposed method for fine-grained semantic concept learning method (T-Neg). In No. 3, we using our proposed image intra-modal momentum contrastive learning method (I-Neg). It can be seen that each component of our proposed method on the three datasets has a very significant improvement compare to baseline in Rank1 and mAP. In No. 4, when we use all components together, the performance can be further improved to reach the best.

Table 3. Ablation results of the proposed method evaluated on the three datasets.

No. 1	baseline	L_{align}	T-Neg	I-Neg	CUHK-PEDES		ICFG-PEDES		RSTPReid	
					Rank1	mAP	Rank1	mAP	Rank1	mAP
1	✓				66.70	60.71	59.76	37.98	49.20	37.14
2	✓	✓	✓		+5.35	+2.33	+4.28	+2.58	+6.30	+4.23
3	✓	✓		✓	+5.39	+5.58	+3.89	+4.44	+5.00	+7.75
4	✓	✓	✓	✓	+6.33	+5.92	+5.26	+4.55	+7.10	+8.68

4.4 Qualitative Results

Figure 5 shows some visual retrieval results, the first row is the top10 retrieval result of baseline, while the second row is our proposed method, and the red boxes are the correct person images retrieved. It can be seen that ours are more accurate and comprehensive.

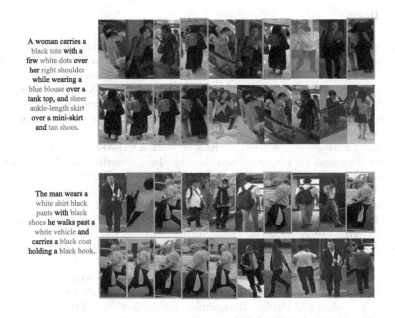

A woman carries a black tote with a few white dots over her right shoulder while wearing a blue blouse over a tank top, and sheer ankle-length skirt over a mini-skirt and tan shoes.

The man wears a white shirt black pants with black shoes he walks past a white vehicle and carries a black coat holding a black book.

Fig. 5. The qualitative results compares the top-10 retrieval results from the baseline and our proposed method.

5   Conclusion

In this paper, we propose a method to improve CLIP-based TPR models using negative samples. We propose a way to learn cross-modal fine-grained semantic concepts by generating hard negative samples, thus can target a specific person based on the comprehensive fine-grained information. While introducing a momentum-contrastive learning framework to fully leverage the information from negative images, hence enhancing batch diversity and capturing consistent features across different views of the same identity. Experiments on three benchmark datasets demonstrate the effectiveness of our proposed method. The method serves as a reference for improving the migration of cross-modal pretrained models to fine-grained downstream tasks with limited data.

Acknowledgements. This work was supported in part by the National Natural Science Foundation of China under Grants 62072354, 61972302, 62276203 and 62072355, in part by the Foundation of National Key Laboratory of Human Factors Engineering under Grant 6142222210101, in part by the Fundamental Research Funds for the Central Universities under Grants QTZX23084, QTZX23105, and QTZX23108, and in part by the Fund Project of XJPCC under Grants 2023CB005, 2022CB002-08, and 2022CA007.

References

1. Chung, J., Gulcehre, C., Cho, K., Bengio, Y.: Empirical evaluation of gated recurrent neural networks on sequence modeling. arXiv preprint arXiv:1412.3555 (2014)
2. Ding, Z., Ding, C., Shao, Z., Tao, D.: Semantically self-aligned network for text-to-image part-aware person re-identification. arXiv preprint arXiv:2107.12666 (2021)
3. Doveh, S., et al.: Teaching structured vision & language concepts to vision & language models. In: Proceedings of the IEEE/CVF Conference on Computer Vision and Pattern Recognition, pp. 2657–2668 (2023)
4. Farooq, A., Awais, M., Kittler, J., Khalid, S.S.: AXM-Net: implicit cross-modal feature alignment for person re-identification. In: Proceedings of the AAAI Conference on Artificial Intelligence, vol. 36, pp. 4477–4485 (2022)
5. Gao, C., et al.: Contextual non-local alignment over full-scale representation for text-based person search. arXiv preprint arXiv:2101.03036 (2021)
6. Girshick, R.: Fast R-CNN. In: Proceedings of the IEEE International Conference on Computer Vision, pp. 1440–1448 (2015)
7. Han, X., He, S., Zhang, L., Xiang, T.: Text-based person search with limited data. arXiv preprint arXiv:2110.10807 (2021)
8. He, K., Fan, H., Wu, Y., Xie, S., Girshick, R.: Momentum contrast for unsupervised visual representation learning. In: Proceedings of the IEEE/CVF Conference on Computer Vision and Pattern Recognition, pp. 9729–9738 (2020)
9. Ji, Z., Hu, J., Liu, D., Wu, L.Y., Zhao, Y.: Asymmetric cross-scale alignment for text-based person search. IEEE Trans. Multimedia **25**, 7699–7709 (2023). https://doi.org/10.1109/TMM.2022.3225754
10. Li, J., Selvaraju, R., Gotmare, A., Joty, S., Xiong, C., Hoi, S.C.H.: Align before fuse: vision and language representation learning with momentum distillation. Adv. Neural. Inf. Process. Syst. **34**, 9694–9705 (2021)

11. Li, S., Xiao, T., Li, H., Zhou, B., Yue, D., Wang, X.: Person search with natural language description. In: Proceedings of the IEEE Conference on Computer Vision and Pattern Recognition, pp. 1970–1979 (2017)
12. Radford, A., et al.: Learning transferable visual models from natural language supervision. In: International Conference on Machine Learning, pp. 8748–8763. PMLR (2021)
13. Schroff, F., Kalenichenko, D., Philbin, J.: FaceNet: a unified embedding for face recognition and clustering. In: Proceedings of the IEEE Conference on Computer Vision and Pattern Recognition, pp. 815–823 (2015)
14. Shu, X., et al.: See finer, see more: implicit modality alignment for text-based person retrieval. In: Karlinsky, L., Michaeli, T., Nishino, K. (eds.) Computer Vision, ECCV 2022 Workshops, Part V. LNCS, vol. 13805, pp. 624–641. Springer, Cham (2023). https://doi.org/10.1007/978-3-031-25072-9_42
15. Wang, G., Yu, F., Li, J., Jia, Q., Ding, S.: Exploiting the textual potential from vision-language pre-training for text-based person search. arXiv preprint arXiv:2303.04497 (2023)
16. Wang, Z., Fang, Z., Wang, J., Yang, Y.: *ViTAA*: visual-textual attributes alignment in person search by natural language. In: Vedaldi, A., Bischof, H., Brox, T., Frahm, J.-M. (eds.) ECCV 2020. LNCS, vol. 12357, pp. 402–420. Springer, Cham (2020). https://doi.org/10.1007/978-3-030-58610-2_24
17. Yan, S., Dong, N., Zhang, L., Tang, J.: Clip-driven fine-grained text-image person re-identification. arXiv preprint arXiv:2210.10276 (2022)
18. Zhang, Y., Lu, H.: Deep cross-modal projection learning for image-text matching. In: Ferrari, V., Hebert, M., Sminchisescu, C., Weiss, Y. (eds.) ECCV 2018. LNCS, vol. 11205, pp. 707–723. Springer, Cham (2018). https://doi.org/10.1007/978-3-030-01246-5_42
19. Zhu, A., et al.: DSSL: deep surroundings-person separation learning for text-based person retrieval. In: Proceedings of the 29th ACM International Conference on Multimedia, pp. 209–217 (2021)

Global Selection and Local Attention Network for Referring Image Segmentation

Haixin Ding, Shengchuan Zhang$^{(\boxtimes)}$, and Liujuan Cao

Key Laboratory of Multimedia Trusted Perception and Efficient Computing, Ministry of Education of China, Xiamen University, Xiamen 361005, People's Republic of China
zsc_2016@xmu.edu.cm

Abstract. Referring image segmentation (RIS) aims to segment the target object based on a natural language expression. The challenge lies in comprehending both the image and the referring expression simultaneously, while establishing the alignment between these two modalities. Recently, the visual-language large-scale pre-trained model CLIP can well align the modalities. However, the alignment in these models is based on the global image. And RIS requires aligning global text features with local visual features, rather than global visual features. To this end, features extracted by CLIP can not be directly applied to RIS. In this paper, we propose a novel framework called *Global Selection and Local Attention Network* (GLNet), which builds upon CLIP. GLNet comprises two modules: *Global Selection and Fusion Module* (GSFM) and *Local Attention Module* (LAM). GSFM utilizes text information to adaptively select and fuse visual features from low-level and middle-level. LAM leverages attention mechanisms on both local visual features and local text features to establish relationships between objects and text. Extensive experiments demonstrate the exceptional performance of our proposed method in referring image segmentation. On RefCOCO+, GLNet achieves significant performance gains of +2.38%, +2.78%, and +2.50% on the three splits compared to SADLR.

Keywords: Referring image segmentation · vision-language · global-local · image segmentation

1 Introduction

Referring image segmentation (RIS) [1–6] aims at segmenting the target object described by a natural language expression in an image. RIS not only needs

This work was supported by National Key R&D Program of China (No. 2022ZD0118202), the National Science Fund for Distinguished Young Scholars (No. 62025603), the National Natural Science Foundation of China (No. U21B2037, No. U22B2051, No. 62176222, No. 62176223, No. 62176226, No. 62072386, No. 62072387, No. 62072389, No. 62002305 and No. 62272401), and the Natural Science Foundation of Fujian Province of China (No. 2021J01002, No. 2022J06001).

Q. Liu et al. (Eds.): PRCV 2023, LNCS 14431, pp. 284–295, 2024.
https://doi.org/10.1007/978-981-99-8540-1_23

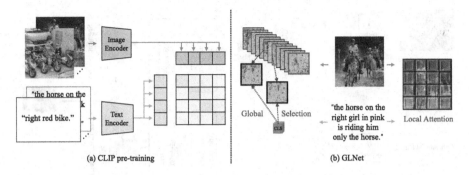

Fig. 1. (a) CLIP leverages large-scale image-text pairs to pre-train the model using contrastive learning. (b) The way Global Selection and Local Attention Network (GLNet) capture information related to the target in the local features. (Color figure online)

accurately to locate the target object described in the natural language expression but also to segment out a fine mask. For accurate object localization, it is crucial to align features from images and text. Recently, as shown in Fig. 1(a), CLIP [7] has made great progress in learning the aligned representation between global visual features and text features by using large-scale image-text pairs. However, this alignment is not suitable for RIS, which requires aligning local visual features with global text features.

Therefore, as illustrated in Fig. 1(b), we propose Global Selection and Local Attention Network (GLNet) as an extension of CLIP to improve RIS. We adopt two perspectives to leverage the knowledge of CLIP for RIS, *i.e.* Global Selection and Local Attention. Firstly, we use the global text feature to adaptively select and fuse it with low-level and mid-level visual features from the vision transformer. Visual features from different layers may focus on different objects in the image. For instance, in Fig. 1(b), the highest-level visual features (red block) capture all objects, while the low-level and mid-level features (black blocks) focus on the object related to the referring expression. Secondly, we employ attention mechanisms on local visual features and text features. This enables GLNet to learn relationships among objects in the image and the correspondence between natural language expressions and image objects. Compared with SADLR [8], the GLNet achieves a more notable performance of +2.38%, +2.78%, and +2.50% IoU gains on three splits in RefCOCO.

In summary, our contributions to this study are three-fold:

- We propose a novel framework called Global Selection and Local Attention Network (GLNet) for Referring Image Segmentation (RIS). GLNet effectively leverages the knowledge obtained from the aligned image and text feature spaces in CLIP, considering both the Global Selection and Local Attention perspectives.
- To better utilize the visual features from the low-level and middle-level of the vision transformer, we develop the Global Selection and Fusion Mod-

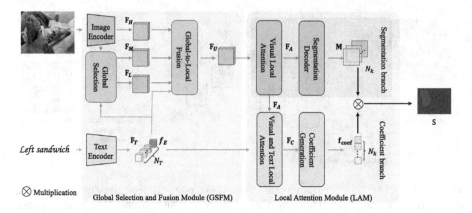

Fig. 2. The overview of the proposed Global selection and Local attention Network (GLNet). GLNet consists of Global Selection and Fusion Module (GSFM) and Local Attention Module (LAM).

ule (GSFM), which adaptively selects and fuses these features based on global text features. Furthermore, to establish alignment relationships among objects in an image and between objects and text, we design the Local Attention Module (LAM), which employs an attention mechanism.

2 Related Works

2.1 Referring Image Segmentation

Referring Image Segmentation (RIS) aims to segment the target object in an image based on a natural language expression. Early approaches, such as concatenation-and-convolution methods [1,2,9], employ CNN and RNN to extract visual and text features, respectively. These features are then concatenated to predict the segmentation mask. With the increasing popularity of attention mechanisms in computer vision, various methods [10–16] develop to establish relationships between text features and visual features using different attention modules. To capture long-range dependencies between visual and text features, several methods [17–21] propose multi-modal fusion decoders based on the transformer [22]. Recently, iterative approaches [8,23] are proposed to facilitate progressive interaction between the two modalities.

3 Method

3.1 Overview

As shown in Fig. 2, the Global Selection and Local Attention Network (GLNet) comprises the Global Selection and Fusion Module (GSFM) and the Local Attention Module (LAM). GSFM utilizes the global text features f_E to adaptively fuse

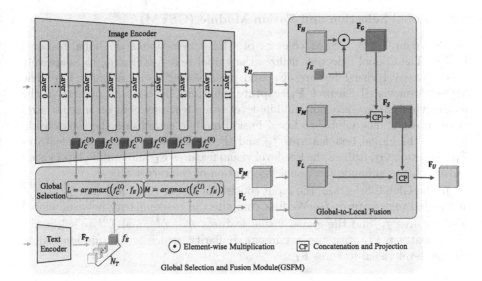

Fig. 3. Global Selection and Fusion Module (GSFM). GSFM consists of Global Selection and Global-to-Local Fusion.

with the low-level visual features $\mathbf{F_L}$ and the middle-level visual features $\mathbf{F_M}$. LAM employs attention mechanisms to fuse the fusion local visual features $\mathbf{F_U}$ and local text features $\mathbf{F_T}$.

In section **Features Extraction**, we provide the details of feature extraction. Subsequently, in section **Global Selection and Fusion Module (GSFM)** and **Local Attention Module (LAM)** discuss the specific details of GSFM and LAM, respectively.

3.2 Features Extraction

Text Encoder. For the given natural language expression, we extract local text features $\mathbf{F_T} \in \mathbb{R}^{N_T \times C}$ using a modified Transformer [22] architecture described in [24]. Here, N_T represents the number of text tokens. The expression sequence is enclosed with [SOS] and [EOS] tokens to indicate the start and end of sequences. The activations of the [EOS] token are considered as the global text features $f_{\mathbf{E}} \in \mathbb{R}^{1 \times C}$ for the entire natural language expression.

Image Encoder. Following the design of vision transformer (ViT) [25], the image $\mathbf{I} \in \mathbb{R}^{H_I \times W_I \times 3}$ is patched and projected denoted as $\mathbf{I_P} \in \mathbb{R}^{HW \times C}$, where $(H, W) = (H_I/P, W_I/P)$ and P indicates the resolution of each image patch. Then, I_P is fed into ViT which employs 12 transformer layers. And the output of each layer is defined as $F_V^{(i)} \in \mathbb{R}^{H \times W \times C}, i = 0, ..., 11$. The visual features of the final layer are recorded as the highest-level visual features $\mathbf{F_H}$. Each layer's visual features contain a learnable embedding in the form of a class token. The class token is recorded as the global visual features $f_{\mathbf{C}}^{(i)} \in \mathbb{R}^C$.

3.3   Global Selection and Fusion Module (GSFM)

As shown in Fig. 3, GSFM consists of Global Selection and Global-to-Local Fusion. The Global Selection utilizes the global text features f_E to adaptively select low-level visual features $\mathbf{F_L}$ and middle-level visual features $\mathbf{F_M}$. The highest-level visual features $\mathbf{F_H}$ capture information about all objects in the image, while the low-level and middle-level visual features focus on specific parts of the objects. The Global-to-Local Fusion module establishes the relationship between the global text features f_E and the highest-level local image features $\mathbf{F_H}$. Additionally, it fuses the low-level visual features $\mathbf{F_L}$ and middle-level visual features $\mathbf{f_M}$ to enhance the visual information of the target object.

In the Global Selection, the input consists of global visual features $f_C^{(3)}$ to $f_C^{(8)}$ and global text features f_E. We first compute the *cos* similarity between global text features f_E and the global visual features $f_C^{(3)}$, $f_C^{(4)}$, and $f_C^{(5)}$ to select the visual features $\mathbf{F_V^l}$ with the maximum similarity, which is then designated as the low-level visual features $\mathbf{F_L}$:

$$l = argmax(f_E \cdot f_C^{(i)}), \quad i = 3, 4, 5, \ \mathbf{F_L} = \mathbf{F_V^{(l)}}. \tag{1}$$

The middle-level visual features are obtained by computing the *cos* similarity of global text features f_E with the global visual features $f_C^{(6)}$, $f_C^{(7)}$, and $f_C^{(8)}$:

$$m = argmax(f_E \cdot f_C^{(j)}), \quad j = 6, 7, 8, \ \mathbf{F_M} = \mathbf{F_V^{(m)}}. \tag{2}$$

In the Global-to-Local Fusion, the global text features f_E and highest-level local visual features $\mathbf{F_H}$ are initially fused by element-wise multiplication:

$$f_G = f_H \odot f_E. \tag{3}$$

Here, f_G represents the element of the Global-to-Local fusion features $\mathbf{F_G}$, and f_H represents the element of the highest-level visual features $\mathbf{F_H}$.

Next, the Global-to-Local fusion features $\mathbf{F_G}$ are fused with the low-level visual features $\mathbf{F_L}$ and middle-level visual features $\mathbf{F_M}$ through concatenation and projection:

$$\begin{aligned} \mathbf{F_S} &= W_S[\mathbf{F_G}, \mathbf{F_M}], \\ \mathbf{F_U} &= W_U[\mathbf{F_S}, \mathbf{F_L}]. \end{aligned} \tag{4}$$

In these equations, W_S and W_U represent projection matrices, $[,]$ denotes concatenation, and $\mathbf{F_U}$ represents the fused local visual features.

3.4   Local Attention Module (LAM)

As shown in Fig. 4, LAM consists of two branches: the Segmentation branch and the Coefficient branch. The Segmentation branch includes Visual Local Attention and Segmentation Decoder, which employ attention mechanisms to establish relationships among objects in the image and generate response masks $\mathbf{M}$ for

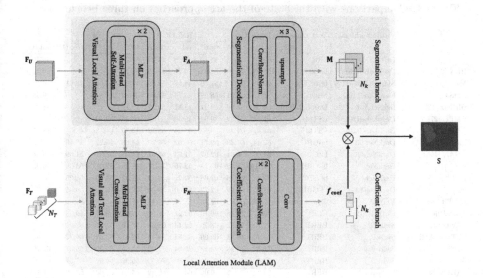

Fig. 4. Local Attention Module (LAM). LAM consists of a Segmentation branch and a Coefficient branch. The Segmentation branch contains Visual Local Attention and Segmentation Decoder. The Coefficient branch contains Visual and Text local attention and Coefficient Generation.

these objects. The Coefficient branch consists of Visual and Text Local Attention and Coefficient Generation, which identify correspondences between local text features and local visual features to generate coefficients $f_{\mathbf{coef}}$ for different response masks $\mathbf{M}$.

In the Segmentation branch, the Fusion local visual features $\mathbf{F_U}$ are fed into Visual Local Attention to obtain attention local visual features $\mathbf{F_A}$:

$$\begin{aligned} \mathbf{F'_A} &= \mathrm{MHSA}(\mathbf{F_U}) + \mathbf{F_U}, \\ \mathbf{F_A} &= \mathrm{MLP}(\mathbf{F'_A}) + \mathbf{F'_A}. \end{aligned} \tag{5}$$

Here, $MHSA$ denotes Multi-Head Self-Attention, and MLP represents Multi-Layer Perceptron. Specifically, a head in $MHSA$ is calculated as the following:

$$MHSA^{(h)}(\mathbf{F_U}) = softmax(\frac{\mathbf{F_U}\mathbf{F_U}^T}{\sqrt{d_{\mathbf{F_U}}}})\mathbf{F_U}. \tag{6}$$

The response masks $\mathbf{M}$ are generated using the attention visual features $\mathbf{F_A}$ in the Segmentation Decoder. The Segmentation Decoder comprises three stacked 3×3 convolution layers and one 1×1 convolution layer, followed by the Sigmoid activation function. Upsampling layers are inserted between the layers to control the output size if desired.

In Coefficient branch, the attention local visual features $\mathbf{F_A}$ and local text features $\mathbf{F_T}$ are fed into Visual and Text Local Attention to obtain Coefficient features $\mathbf{F_K}$:

Table 1. Comparisons with the state-of-the-art approaches on three benchmarks

Method	Visual Backbone	Text Encoder	RefCOCO			RefCOCO+			G-Ref		
			val	testA	testB	val	testA	testB	val (U)	test (U)	val (G)
DMN [9]	DPN92	SRU	49.78	54.83	45.13	38.88	44.22	32.29	-	-	36.76
RRN [2]	DeepLab-R101	bi-LSTM	55.33	57.26	53.93	39.75	42.15	36.11	-	-	36.45
MAttNet [26]	MaskRCNN-R101	bi-LSTM	56.51	62.37	51.70	46.67	52.39	40.08	47.64	48.61	-
CMSA [11]	DeepLab-R101	None	58.32	60.61	55.09	43.76	47.60	37.89	-	-	39.98
BRINet [13]	ResNet-101	DeepLab-R101	60.98	62.99	59.21	48.17	52.32	42.11	-	-	48.04
CMPC [27]	DeepLab-R101	LSTM	61.36	64.53	39.98	49.56	53.44	43.23	59.64	-	-
LSCM [5]	DeepLab-R101	LSTM	61.47	64.99	59.55	49.34	53.12	43.50	-	-	48.05
MCN [12]	DarkNet-53	bi-GRU	62.44	64.20	59.71	50.62	54.99	44.69	49.22	49.40	-
CGAN [4]	DarkNet-53	bi-GRU	64.86	68.04	62.07	51.03	55.51	44.06	51.01	51.69	46.54
ISFP [15]	DarkNet-53	bi-GRU	65.19	68.45	62.73	52.70	56.77	46.39	52.67	53.00	50.08
LTS [28]	DarkNet-53	bi-GRU	65.43	67.76	63.08	54.21	58.32	48.02	54.40	54.25	-
VLT [17]	DarkNet-53	bi-GRU	65.65	68.29	62.73	55.50	59.20	49.36	52.99	56.65	49.76
ReSTR [20]	ViT-B	Transformer	67.22	69.30	64.45	55.78	60.44	48.27	-	-	54.48
CRIS [19]	CLIP-R101	CLIP	70.47	73.18	66.10	62.27	68.08	53.68	59.87	60.36	-
LAVT [21]	Swin-B	BERT	72.73	75.82	68.79	62.14	68.38	55.10	61.24	62.09	60.50
VLT (TPAMI) [29]	Swin-B	BERT	72.96	75.96	69.60	63.53	68.43	56.92	63.49	66.22	62.80
GRSE [16]	Swin-B	BERT	73.82	76.48	70.18	66.04	71.02	57.65	65.00	65.97	62.70
M3Att [23]	Swin-B	BERT	73.60	76.23	70.36	65.34	70.50	56.98	64.92	67.37	63.90
SADLR [8]	Swin-B	BERT	74.24	76.25	70.06	64.28	69.09	55.19	63.60	63.56	61.16
GLNet (ours)	CLIP-ViTB	CLIP	**74.70**	**77.35**	**70.41**	**66.66**	**71.87**	**57.69**	**65.51**	**65.31**	**63.46**

$$\mathbf{F'_K} = \mathrm{MHCA}(\mathbf{F_A}, \mathbf{F_T}) + \mathbf{F_A},$$
$$\mathbf{F_K} = \mathrm{MLP}(\mathbf{F'_K}) + \mathbf{F'_K}. \tag{7}$$

Here, $MHCA$ denotes Multi-Head Cross-Attention:

$$MHCA^{(h)}(\mathbf{F_A}, \mathbf{F_T}) = softmax(\frac{\mathbf{F_A F_T}^T}{\sqrt{d_{\mathbf{F_T}}}})\mathbf{F_T}. \tag{8}$$

The coefficients $f_{\mathbf{coef}}$ are generated by the Coefficient Generation which comprises two stacked 3×3 convolution layers and one 1×1 convolution layer, followed by the Tanh activation function.

Finally, the segmentation mask $\mathbf{S}$ is obtained by weighting and summing the coefficients $f_{\mathbf{coef}}$ used for the response mask $\mathbf{M}$:

$$\mathbf{S} = f_{\mathbf{coef}} \otimes \mathbf{M}. \tag{9}$$

To guide the training process, a Binary Cross Entropy loss is applied to the segmentation mask $\mathbf{S}$.

4 Experiments

4.1 Experiment Settings

We use ViT-B [25] as the image encoder and Transformer [22] as the text encoder. Both the image and text encoders are initialized with CLIP [7]. Input images are resized to 416×416. The number N_K of response masks $\mathbf{M}$ and coefficients

f_{coef} is 32. The maximum length for the input natural language expression is set to 17 for RefCOCO and RefCOCO+, and 22 for G-Ref, including the [SOS] and [EOS] tokens. We use the Adam [30] optimizer to train the network for 30 epochs with an initial learning rate of 2.5×10^{-5}, and we decay the learning rate in the 15th, 20th, and 25th epochs with a decay rate of 0.1. We train the model with a batch size of 16 on a single Tesla A100 GPU.

4.2   Datasets

We conduct extensive experiments on three benchmarks: RefCOCO [31], Ref-COCO+ [31], and G-Ref [32].

RefCOCO consists of 19,994 images from MSCOCO [33], and it is divided into four sets: train, validation, testA, and testB, with 120,624, 10,834, 5,657, and 5,095 samples, respectively. The target objects in testA are people, and the categories of testB are objects.

RefCOCO+ has the source with RefCOCO. It comprises 19,994 images, which are split into train (120,191), validation (10,758), testA (5,726), and testB (4,889) subsets. RefCOCO+ is more challenging than RefCOCO because it prohibits the use of location words, allowing only words that describe physical characteristics to be used.

G-Ref is collected from Amazon Mechanical Turk. It contains 26,711 images with 54,822 annotated objects and 104,560 annotated texts. The G-Ref dataset has two different partitions, one by UMD [34] and the other by Google [32].

4.3   Comparisons with State-of-the-Art Approaches

As shown in Table 1, our proposed GLNet is compared with state-of-the-art approaches on three benchmarks. On RefCOCO, GLNet outperforms SADLR [8] in terms of IoU performance, with improvements of +0.46%, +1.10%, and +0.35% on the three splits. On the more challenging dataset RefCOCO+, GLNet achieves significant performance gains of +2.38%, +2.78%, and +2.50% on the three splits compared to SADLR [8]. On the most difficult dataset G-Ref, GLNet surpasses SADLR [8] on both the validation and test subsets from the UMD partition, with margins of +1.91% and +1.75% respectively. GLNet also achieves a gain of +2.30% over SADLR [8] on the validation set of the Google partition. These experiments state the effectiveness of the proposed GLNet in RIS.

4.4   Ablation Studies

Table 2 presents the results of the ablation studies conducted on RefCOCO. Compared to directly using the highest-level visual features $\mathbf{F_H}$ (where $H = 11$) to segment the mask, the GSFM adaptively selects and fuses the low-level visual features $\mathbf{F_L}$ (where $L = l$) and middle-level visual features $\mathbf{F_M}$ (where $M = m$). This integration of GSFM brings improvements of +1.46%, +0.89%, and +0.95% on the three splits. Additionally, the segmentation branch in LAM

Table 2. Ablation study on the RefCOCO validation, testA and testB datasets. L: The low-level visual features $\mathbf{F_L}$ are from the L-th layer in ViT. M: The middle-level visual features $\mathbf{F_M}$ are from the M-th layer in ViT. H: The highest-level visual features $\mathbf{F_H}$ are from the 11-th layer in ViT. - : Corresponding visual features are not used. Seg: Segmentation branch. Coef: Coefficient branch.

	GSFM			LAM		IoU	prec@0.5	prec@0.6	prec@0.7	prec@0.8	prec@0.9
	L	M	H	Seg	Coef						
val	-	-	11	-	-	70.46	80.93	76.02	69.13	55.00	20.86
	9	10	11	-	-	71.60	82.31	78.09	71.41	58.33	22.60
	5	8	11	-	-	71.78	82.32	78.05	71.77	59.03	23.95
	1	m	11	-	-	71.92	82.46	77.85	71.90	59.25	24.91
	1	m	11	Seg	-	74.13	85.37	81.41	75.94	62.94	26.23
	1	m	11	Seg	Coef	**74.70**	**86.08**	**82.61**	**77.19**	**64.50**	**26.58**
testA	-	-	11	-	-	73.93	85.59	81.72	76.06	61.48	21.90
	9	10	11	-	-	74.33	86.08	82.39	76.02	62.77	24.02
	5	8	11	-	-	74.79	86.30	82.95	77.35	64.96	24.02
	1	m	11	-	-	74.82	85.89	83.06	76.91	64.15	25.40
	1	m	11	Seg	-	76.79	89.18	86.51	80.90	67.59	26.99
	1	m	11	Seg	Coef	**77.35**	**90.06**	**87.53**	**82.37**	**69.52**	**26.76**
testB	-	-	11	-	-	67.22	75.50	69.24	61.76	49.14	23.33
	9	10	11	-	-	67.62	76.15	70.02	63.35	51.59	24.78
	5	8	11	-	-	67.58	75.89	70.40	63.49	52.65	25.84
	1	m	11	-	-	68.17	76.05	71.06	64.63	52.95	26.96
	1	m	11	Seg	-	70.23	79.58	74.50	**68.47**	55.81	**29.10**
	1	m	11	Seg	Coef	**70.41**	**79.68**	**75.01**	68.38	**56.78**	28.85

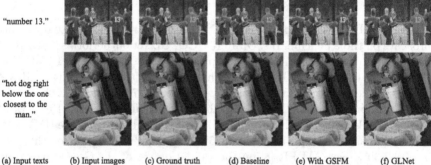

"number 13."

"hot dog right below the one closest to the man."

(a) Input texts   (b) Input images   (c) Ground truth   (d) Baseline   (e) With GSFM   (f) GLNet

Fig. 5. Visualization of the components of GLNet. (a) Input texts. (b) Input images. (c) Ground truth. (d) Baseline: using the highest-level visual features $\mathbf{F_H}$ of ViT to generate the segmentation mask. (e) With GSFM: using GSFM to select and fuse features to generate the segmentation mask. (f) GLNet.

contributes performance gains of +2.21%, +1.97%, and +2.06% on the three splits, respectively. The introduction of the coefficient branch further improves performance on the three splits by +0.57%, +0.56%, and +0.18%, respectively. In summary, every proposed component has its contribution.

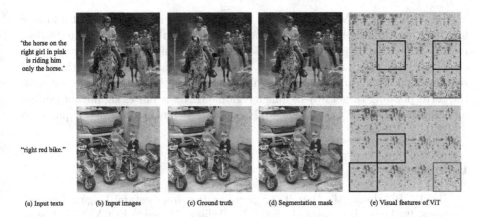

Fig. 6. The visualization selected visual features of GSFM. (a) Input texts. (b) Inputs images. (c) Ground truth. (d) Segmentation mask. (e) The visual features of ViT. The black blanks represent the visual features selected by GSFM. The red blanks represent the highest-level visual features.

4.5 Visualization and Qualitative Analysis

The Components of GLNet. As depicted in Fig. 5, the Baseline (d) utilizes only the highest-level visual features F_H from ViT, resulting in incorrect object prediction and incomplete segmentation masks. When GSFM is employed (e), the correct target object can be located, demonstrating that the visual features selected and fused by GSFM better enable target object localization. GLNet (f) can accurately locate the target object and generate more precise segmentation masks. These findings highlight that LAM focuses on the local visual features most relevant to the natural language expression.

Visualization of Selected Visual Features in GSFM. Figure 6 illustrates the visualization of selected visual features in GSFM. In column (e), the highest-level visual features in the red block pay attention to all objects in the image. However, some of these objects are not relevant to the natural language expression. Meanwhile, the selected low-level visual features and middle-level visual features in the black blocks focus on specific local regions that are more related to the natural language expression, rather than all objects. This observation demonstrates the effectiveness of our proposed GSFM and the necessity of combining all masks adaptively.

5   Conclusion

This paper introduces a novel framework called Global Selection and Local Attention Network (GLNet) for Referring Image Segmentation (RIS). Our approach leverages the knowledge acquired in CLIP and incorporates two key modules: the Global Selection and Fusion Module (GSFM) and the Local Attention

Module (LAM). GSFM utilizes global text features to adaptively select and fuse low-level and middle-level visual features, thereby enhancing the visual information related to target objects. LAM employs attention mechanisms to focus on local text features and local visual features, facilitating the establishment of relationships among objects and aligning the natural language expression with the corresponding target object image region. GLNet achieves significant performance gains of +2.38%, +2.78%, and +2.50% on the three splits compared to SADLR.

References

1. Hu, R., Rohrbach, M., Darrell, T.: Segmentation from natural language expressions. In: Leibe, B., Matas, J., Sebe, N., Welling, M. (eds.) ECCV 2016. LNCS, vol. 9905, pp. 108–124. Springer, Cham (2016). https://doi.org/10.1007/978-3-319-46448-0_7
2. Li, R., et al.: Referring image segmentation via recurrent refinement networks. In: 2018 CVPR, pp. 5745–5753 (2018)
3. Chen, D.-J., Jia, S., Lo, Y.-C., Chen, H.-T., Liu, T.-L.: See-through-text grouping for referring image segmentation. In: 2019 ICCV, pp. 7453–7462 (2019)
4. Luo, G., et al.: Cascade grouped attention network for referring expression segmentation. In: Proceedings of the 28th ACM International Conference on Multimedia (2020)
5. Hui, T., et al.: Linguistic structure guided context modeling for referring image segmentation arXiv arXiv:2010.00515 (2020)
6. Jiao, Y., et al.: Two-stage visual cues enhancement network for referring image segmentation. In: Proceedings of the 29th ACM International Conference on Multimedia (2021)
7. Radford, A., et al.: Learning transferable visual models from natural language supervision. In: International Conference on Machine Learning (2021)
8. Yang, Z., et al.: Semantics-aware dynamic localization and refinement for referring image segmentation. arXiv arXiv:2303.06345 (2023)
9. Margffoy-Tuay, E., Pérez, J., Botero, E., Arbeláez, P.: Dynamic multimodal instance segmentation guided by natural language queries. arXiv arXiv:1807.02257 (2018)
10. Shi, H., Li, H., Meng, F., Wu, Q.: Key-word-aware network for referring expression image segmentation. In: Ferrari, V., Hebert, M., Sminchisescu, C., Weiss, Y. (eds.) ECCV 2018. LNCS, vol. 11210, pp. 38–54. Springer, Cham (2018). https://doi.org/10.1007/978-3-030-01231-1_3
11. Ye, L., Rochan, M., Liu, Z., Wang, Y.: Cross-modal self-attention network for referring image segmentation. In: 2019 CVPR, pp. 10494–10503 (2019)
12. Luo, G., et al.: Multi-task collaborative network for joint referring expression comprehension and segmentation. In: 2020 CVPR, pp. 10031–10040 (2020)
13. Hu, Z., Feng, G., Sun, J., Zhang, L., Lu, H.: Bi-directional relationship inferring network for referring image segmentation. In: 2020 CVPR, pp. 4423–4432 (2020)
14. Feng, G., Hu, Z., Zhang, L., Lu, H.: Encoder fusion network with co-attention embedding for referring image segmentation. In: 2021 CVPR, pp. 15501–15510 (2021)
15. Liu, C., Jiang, X., Ding, H.: Instance-specific feature propagation for referring segmentation. IEEE Trans. Multimedia 25, 3657–3667 (2023)

16. Liu, C., Ding, H., Jiang, X.: GRES: generalized referring expression segmentation (2023)
17. Ding, H., Liu, C., Wang, S., Jiang, X.: Vision-language transformer and query generation for referring segmentation. In: 2021 ICCV, pp. 16301–16310 (2021)
18. Kamath, A., Singh, M., LeCun, Y., Misra, I., Synnaeve, G., Carion, N.: MDETR - modulated detection for end-to-end multi-modal understanding. In: 2021 ICCV, pp. 1760–1770 (2021)
19. Wang, Z., et al.: CRIS: clip-driven referring image segmentation (2021)
20. Kim, N.H., Kim, D., Lan, C., Zeng, W., Kwak, S.: ReSTR: convolution-free referring image segmentation using transformers. arXiv arXiv:2203.16768 (2022)
21. Yang, Z., et al.: LAVT: language-aware vision transformer for referring image segmentation. arXiv arXiv:2112.02244 (2022)
22. Vaswani, A., et al.: Attention is all you need. arXiv arXiv:1706.03762 (2023)
23. Liu, C., Ding, H., Zhang, Y., Jiang, X.: Multi-modal mutual attention and iterative interaction for referring image segmentation. IEEE Trans. Image Process. **32**, 3054–3065 (2023)
24. Radford, A., Wu, J., Child, R., Luan, D., Amodei, D., Sutskever, I.: Language models are unsupervised multitask learners (2019)
25. Dosovitskiy, A., et al.: An image is worth 16 × 16 words: transformers for image recognition at scale. arXiv arXiv:2010.11929 (2020)
26. Yu, L., et al.: MAttNet: modular attention network for referring expression comprehension. In: 2018 CVPR, pp. 1307–1315 (2018)
27. Huang, S., et al.: Referring image segmentation via cross-modal progressive comprehension. In: 2020 CVPR, pp. 10485–10494 (2020)
28. Jing, Y., Kong, T., Wang, W., Wang, L., Li, L., Tan, T.: Locate then segment: a strong pipeline for referring image segmentation. In: 2021 CVPR, pp. 9853–9862 (2021)
29. Ding, H., Liu, C., Wang, S., Jiang, X.: VLT: vision-language transformer and query generation for referring segmentation. IEEE Trans. Pattern Anal. Mach. Intell. **45**(6), 7900–7916 (2023)
30. D. P. Kingma, J. Ba, Adam: A method for stochastic optimization, CoRR abs/1412.6980
31. Yu, L., Poirson, P., Yang, S., Berg, A.C., Berg, T.L.: Modeling context in referring expressions. arXiv arXiv:1608.00272 (2016)
32. Mao, J., Huang, J., Toshev, A., Camburu, O.-M., Yuille, A. L., Murphy, K.P.: Generation and comprehension of unambiguous object descriptions. In: 2016 CVPR, pp. 11–20 (2016)
33. Lin, T.-Y., et al.: Microsoft COCO: common objects in context. In: Fleet, D., Pajdla, T., Schiele, B., Tuytelaars, T. (eds.) ECCV 2014. LNCS, vol. 8693, pp. 740–755. Springer, Cham (2014). https://doi.org/10.1007/978-3-319-10602-1_48
34. Kazemzadeh, S., Ordonez, V., Matten, M., Berg, T.L.: ReferItGame: referring to objects in photographs of natural scenes. In: EMNLP (2014)
35. Liu, C., Lin, Z.L., Shen, X., Yang, J., Lu, X., Yuille, A.L.: Recurrent multimodal interaction for referring image segmentation. In: 2017 ICCV, pp. 1280–1289 (2017)
36. He, K., Gkioxari, G., Dollár, P., Girshick, R.B.: Mask R-CNN. In: 2017 ICCV, pp. 2980–2988 (2017)
37. Nagaraja, V.K., Morariu, V.I., Davis, L.S.: Modeling context between objects for referring expression understanding. In: Leibe, B., Matas, J., Sebe, N., Welling, M. (eds.) ECCV 2016. LNCS, vol. 9908, pp. 792–807. Springer, Cham (2016). https://doi.org/10.1007/978-3-319-46493-0_48

MTQ-Caps: A Multi-task Capsule Network for Blind Image Quality Assessment

Yijie Wei[1], Bincheng Wang[1], Fangfang Liang[1,2], and Bo Liu[1,2(✉)]

[1] College of Information Science and Technology, Hebei Agricultural University,
Baoding 071000, Hebei, China
boliu@hebau.edu.cn
[2] Hebei Key Laboratory of Agricultural Big Data, Hebei Agricultural University,
Baoding, Hebei, China

Abstract. Blind image quality assessment (BIQA) is a field that aims to predict the quality of images without the use of reference images. This field has garnered considerable attention due to its potential applications in areas such as visual understanding and computational imaging. The two primary challenges in BIQA are the differentiation of distortion types and the prediction of quality scores. However, previous approaches typically address only one of these tasks, neglecting the associations between them. In this paper, we propose a multi-task capsule network for BIQA (MTQ-Caps). The proposed MTQ-Caps includes two types of quality capsules routed from CNN-based features: distortion capsules, which correspond to specific distortion types, and content capsules, which encapsulate semantic information for quality score estimation. Consequently, MTQ-Caps utilizes both synthetic distortions and human subjective judgments in a multi-task learning approach. Experimental results demonstrate that our proposed method outperforms state-of-the-art methods on synthetic databases and achieves competitive performance on authentic databases, even without distortion descriptions.

Keywords: Blind image quality assessment · capsule network · multi-task

1 Introduction

Image quality assessment (IQA) is a mechanism that enables computers to evaluate the quality of an image in a manner similar to human perception. Automated image filtering based on intrinsic image quality can effectively reduce the accumulation of low-quality images, a common issue with the growth of social networks. It can also provide samples of varying difficulty for recognition tasks. The challenge of IQA lies in the fact that both objective and subjective perceptions contribute to the descriptions of image quality. For instance, image quality can be degraded by various distortions such as Gaussian blur, whitening, JPEG

Q. Liu et al. (Eds.): PRCV 2023, LNCS 14431, pp. 296–308, 2024.
https://doi.org/10.1007/978-981-99-8540-1_24

compression, etc. Factors such as unknown subjects and overexposure also influence people's judgment of image quality to a certain extent.

IQA can generally be divided into three categories: full-reference IQA (FR-IQA), reduced-reference IQA (RR-IQA), and no-reference IQA (NR-IQA) or blind IQA (BIQA). The main difference among these methods is their dependence on reference images. The introduction of a series of IQA databases has facilitated tasks from single quality prediction [4,13] to complex image quality analysis aided by semantic information [19,24]. Earlier databases, such as the LIVE database [18], contain only 29 reference images and five synthetic distortions. More recent databases, such as LIVE Challenge [2] and KonIQ-10K [6], are more complex and diverse in size and in the subjects and styles of included images. Over the past few years, the rapid development of deep learning in computer vision has brought about improved solutions to IQA [7,8,19,25]. These methods leverage the fitting ability of deep models but do not fully explore the relationship between multiple distortion types and opinion scores, resulting in quality predictions that lack interpretability. In many practical applications, specifying the cause of quality degradation is more important than a simple quality grading.

In this paper, we aim to unify distortion type identification and quality estimation into a multi-task learning framework. To enhance the model's ability to represent different distortion types, we extend the Capsule Networks (CapsNets) [17], resulting in the multi-task capsule network for BIQA (MTQ-Caps). In MTQ-Caps, we adopt the so-called distortion capsule to represent each distortion type, with the distortion confidence calculated based on the length of the corresponding capsule. Additionally, we introduce content capsules to encapsulate multi-scale semantic information related to quality assessment. By optimizing these two sets of capsules through backpropagation, the proposed framework is designed to incorporate both distortion categories and subjective quality scores. The contributions of our work are summarized as follows:

- We propose a multi-task deep learning approach for predicting image distortion types and assessing quality. Two branches of the model, one for distinguishing distortion categories and the other for regressing quality scores, share the same backbone network. We demonstrate that the performance of the proposed approach can be enhanced by merging the two tasks.
- Given that CapsNets can learn complex patterns through specific capsules, we devise two groups of capsules: distortion capsules, which serve distortion types, and content capsules, which contain quality-related semantic features at multiple scales. The prediction of distortion types and quality assessment scores can be calculated using the length of the corresponding capsules.
- We experimentally demonstrate the stability and generalization of the model for both authentic and synthetic IQA databases.

2   Related Work

2.1   Deep Learning for Blind Image Quality Assessment

In the past decades, traditional image quality assessment (IQA) tasks were typically addressed using two methods: hand-crafted feature-based BIQA [23] and statistical feature-based BIQA [15]. However, these conventional approaches require significant computing resources and entail lengthy processing times. Furthermore, while scene statistic features provide a global view of image quality, they are inadequate for measuring complex distortions present in authentically distorted images.

With the rapid development of deep learning, IQA based on deep learning has gained significant popularity. Kang et al. [7] proposed a method that utilizes a simple CNN to address BIQA. Zhang et al. [24] introduced a framework based on the concept of semantic obviousness to measure image quality. In order to make the objective evaluation results of the model more in line with the human subjective perception of image quality, Kim et al. [8] developed a method that learns visual sensitivity features of the human visual system (HVS) without any prior knowledge. Bosse et al. [1] proposed a unified framework that enables joint learning of local quality and local weights, which represent the relative importance of local quality to global quality estimates. Additionally, Lin et al. [11] presented a quality regression network that incorporates perceptual difference information into network learning. They also introduced an adversarial learning approach that utilizes error maps of distorted images to guide quality prediction and generate information-rich features.

To solve the problem of complex distortion due to the variety of real distorted image content subjects, Su et al. [19] employed an adaptive parametric generative hyper network that learns the parameters of the quality prediction network from features extracted by a multi-scale feature extractor, enhancing the model's adaptability to the data. Zhu et al. [27] applied meta-learning to image quality assessment, leveraging deep meta-learning to acquire meta-knowledge shared across different types of distortion. Similar to the model-independent training strategy in [26], the meta-learning approach is also applicable to multiple models. And VIPNet [21] obtained distorted information representations related to subjective perception rules through different quality perception modules.

When distortion types are known, this prior information can be used to model distortion types and assist in learning quality scores [4,13,25]. Ma et al. [13] proposed a multi-task learning method that utilizes a fusion layer to reuse the information obtained for distortion type prediction in mass fraction prediction. Zhang et al. [25] developed a model that incorporates a pre-trained image distortion type classification branch for feature extraction, along with the VGG-16 pre-trained on ImageNet. The model employs bilinear pooling for feature fusion. The introduction of distortion type information through the distortion classification branch enhances feature generalization, but the subsequent bilinear pooling increases computational workload. Golestaneh et al. [4] proposed a novel approach to feature fusion, which involves combining different features from the quality score prediction branch. This fusion concept is very similar to the combination of multi-scale feature maps mentioned in our method.

2.2 Capsule Networks

Since Sabour et al. [17] proposed the Capsule Networks (CapsNets), generating informative capsule representations through dynamic routing provides a new learning paradigm for recognition tasks. Kosiorek et al. [9] proposed a novel representation learning method based on unsupervised learning, which achieved competitive results in unsupervised object classification with the integration of capsules. Additionally, Mazzia et al. [14] proposed an efficient routing method using self-attention, which performs routing only once, resulting in a reduction in model parameters and computation time.

In summary, the architecture of CapsNets shows promise in capturing hierarchical relationships among features. The dynamic routing mechanism employed in CapsNets enables information transfer between different levels of abstraction, which is crucial for capturing the complex relationships among various quality features within an image. Building upon these insights, we propose a multi-task learning model based on the CapsNets. In our model, we incorporate distortion type information to assist in IQA.

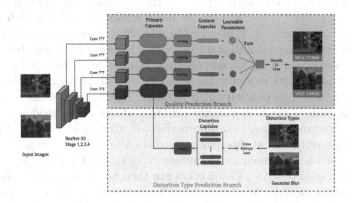

Fig. 1. The overall architecture of MTQ-Caps

3   Proposed Method

This section aims to present how we can extend the CapsNets to predict image quality scores. The architecture of our model is shown in Fig. 1. The proposed model consists of three parts: a backbone network to extract features, a multi-scale sub-capsule network for the quality prediction, and sub-capsule network to re-transform the feature maps into some distortion capsules.

3.1 Semantic Feature Extraction Architecture

In the feature extraction phase, we use a typical deep learning architecture for feature extraction. To enhance the adaptability of the model, feature maps at different scales are fed to our following networks.

Here, we employ the ResNet-50 [5] as the backbone network. Semantic features are extracted from three specific blocks, while global features originate from the last block. The parameters are initialized using a pre-trained model on ImageNet. The final average pooling layer and fully connected layer of ResNet-50 are omitted since they are unnecessary for semantic feature extraction. Consequently, we obtain local features from conv2_10, conv3_12, and conv4_18, and global features from conv5_9 using ResNet-50.

3.2 Sub-capsule Network for Quality Prediction

According to [14,17], most CapsNets are used for classification tasks, but for the regression task, IQA, we use carefully designed CapsNets to make it work on IQA.

First, we use the multi-scale features extracted by ResNet-50 as the input of the CapsNets. Because of the different scales of the input features, we have fine-tuned the convolutional layers used to generate the primary capsules. Here, we refer to the CapsNets with inputs for the 1st, 2nd, 3rd and 4th stages as C_1, C_2, C_3, C_4. For C_1, C_2, and C_3, we use a convolutional layer with a kernel of 7×7 and step size is 2. While for C_4, whose input contains more semantic information, we use a convolutional layer with a kernel of 3×3 and step size is 2. The primary capsules in C_2, C_3 and C_4 are uniformly designed as 8-dimensional, but in C_1, the primary capsules are setted into 32-dimensional. Such a design will produce more capsules with rich semantic information, while a relatively balanced number of capsules will increase the stability of our model.

For the dynamic routing algorithm performed after the generation of primary capsules, we specify that at this quality prediction branch, each capsule network will route out one 32-dimensional content capsule, for a total of four content capsules. The quality score of each stage can then be represented by the length of each content capsule. We can get the final image quality score prediction value by weighting the sum of the quality scores of each stage. The initialized weights used are incremented by C_1, C_2, C_3, C_4, i.e. C_4 has the highest weight relative to the other three. Given the backbone structure of ResNet-50, we know that the deeper the network, the stronger the semantic information contained in the features, so it is a more sensible choice to assign a higher weight to the quality score obtained by the content capsule in C_4. Also, for features that contain less semantic information but more local information, we would expect it to be more helpful for this work, so it would be wise to assign a relatively small initialization weight to it. For databases that don't contain distortion type supervisory information, our network will simply use this branch and not include the distortion type prediction branch as explained next. Since smooth l1 loss is used as our loss function, the optimization of our network will be defined as

$$\mathcal{L}_1 = \mathcal{L}_{smooth_{L_1}}(x) = \begin{cases} 0.5x^2 & \text{if } |x| < 1 \\ |x| - 0.5 & \text{otherwise} \end{cases} \tag{1}$$

where $x = \sum_{i=1}^{4} p_i \|c_i\|_2 - y$, the p_i means the i-th learnable parameter, c_i denotes the i-th capsule of 4 content capsules, and y is the groud-truth.

3.3   Sub-capsule Network for Distortion Type Prediction

Given that the CapsNets can perform well on classification task [9,14,17], we utilize it for the distortion type prediction, which will also ensure uniformity in the proposed network structure. This branch is only used for synthetically distorted databases because such databases contain distortion type information, so when encountering authentically distorted databases, we just use the quality prediction branch to obtain quality scores. In the network design of this branch, we do not use the reconstruction network [17] in the CapsNets, which would make our model more complex and would not be effective for image quality assessment.

For this distortion type prediction branch, primary capsules are the primary capsules of C_4 in the quality prediction branch, as shown in Fig. 1. Unlike the quality prediction branch, the dynamic routing algorithm here will route out the same number of distortion capsules as the distortion types in the databases. And the distortion capsules are designed as 32-dimensional. In this branch, we only use one capsule network for the following reasons: (i) Multi-scale features contain both global and local information, similar to the tendency of talent to make quality judgments, so we feed the multi-scale features to well-designed multiple CapsNets. (ii) The prediction of distortion types tends to make more use of global features, leaving less contribution from local features.

The Table 4 demonstrates that our introduction of a distortion type prediction branch is an effective improvement for image quality assessment. Since the classification of Distortion Type is a multi-classification task, So cross-entropy loss function is employed as our loss function, which defined as

$$\mathcal{L}_2 = \mathcal{L}_{cross\_entropy}(q, p) = -\frac{1}{N} \sum_{i=0}^{N-1} \sum_{k=0}^{K-1} q_{i,k} \log p_{i,k} \tag{2}$$

where N is the number of samples, K is the number of categories, $q_{i,k}$ denotes a variable (0 or 1), 1 if the category is the same as the category of sample i, 0 otherwise. $p_{i,k}$ denotes the predicted probability of belonging to category k for observation sample i.

For databases with available distortion type supervision information, we will introduce this branch, and the whole optimization is defined as

$$\mathcal{L} = \alpha * \mathcal{L}_1 + \beta * \mathcal{L}_2 \tag{3}$$

where α and β are artificially specified hyperparameters.

4   Experiments

4.1   Databases

To distinguish between the acquisition methods of distorted images, we used two types of databases to evaluate the performance of the model: synthetically distorted image databases and authentically distorted image databases. The details of these databases are shown in Table 1. Among them, the value of MOS is positively correlated with the level of image quality, while that of DMOS is opposite.

Table 1. Benchmark IQA databases for performance evaluation.

Database	LIVE [18]	CSIQ [10]	TID2013 [16]	KonIQ-10k [6]	LIVEC [2]
Content	29	30	25	10073	1162
MOS (DMOS) Range	[0, 100] (DMOS)	[0,1] (DMOS)	[0, 9]	[0, 100]	[0, 100]
Number of Distorted Images	779	866	3000	10073	1162
Distortion Type	Synthetic	Synthetic	Synthetic	Authentic	Authentic
Number of Distortion Types	5	6	24	N/A	N/A
Subjective Environment	Laboratory	Laboratory	Laboratory	Crowdsourcing	Crowdsourcing

4.2   Evaluation Criteria

In our experiments, we use Spearman's rank order correlation coefficient (SRCC) and Pearson's linear correlation coefficient (PLCC) to measure the prediction accuracy. The two criteria both range from -1 to 1 and higher values indicate better model performance. The SRCC evaluates the correlation between ground-truth and prediction by using the monotonic equation, and it defined as

$$SRCC = 1 - \frac{6\sum_{i=1}^{N} d_i^2}{N(N^2 - 1)} \tag{4}$$

where N represents the number of distorted images, and d_i is the difference between the ranks of i-th image in ground-truth and prediction.

$$PLCC = \frac{\sum_{i=1}^{N} (x_i - \bar{x}_i)(\hat{x}_i - \bar{\hat{x}}_i)}{\sqrt{\sum_{i=1}^{N} (x_i - \bar{x}_i)^2}\sqrt{\sum_{i=1}^{N} (\hat{x}_i - \bar{\hat{x}}_i)^2}} \tag{5}$$

where N represents the number of distorted images, x_i and $\hat{x}_i$ represent the ground truth and the predicted values of the i-th image respectively, $\bar{x}_i$ and $\bar{\hat{x}}$ denote the means of ground-truth and predicted scores.

4.3   Implementation Details

We used PyTorch to implement our model, trained and tested it on NVIDIA GeForce RTX 3090 GPUs. For the Eq. 2, α and β are setted 0.9 and 0.1. In our approach, we first normalized the subjective scores of images on all databases to [0, 1]. We used random sampling and a random horizontal flip of 25 patches for data augmentation. All training photos are randomly cropped to a size of 224×224 pixels for input to our model. We trained 16 epoches using the Adam optimizer with a weight decay of 1e−5. Mini-batch size is 64, the learning rate is 1e−5, and it decays by a factor of 10 every five epochs. For the testing strategy, a 25-patch data augmentation of 224×224 pixels in size will be randomly sampled and averaged at the end to obtain the test SRCC and PLCC.

For the quality prediction branch, we initialize the learnable parameters to [0.15, 0.15, 0.25, 0.45], which corresponds to the output of stage1, 2, 3, and 4 after the capsule network. For the distortion type prediction branch, its capsule network part shares the same primary capsule layer as the capsule network of the 4-th stage of the quality prediction branch, but reconnects a distortion capsule layer that outputs the number of distortion types.

For the division of the database, we used 80% images for training and the remaining 20% for testing. Also, for the synthetic image databases, in order to avoid overlap of content, we have taken the approach of dividing according to reference images. We run ten times of this random train-test splitting operation and the reported results are the mean SRCC and PLCC values obtained by these ten times.

4.4   Comparisons with the State-of-the-Art Methods

Evaluation on Distorted Databases. To demonstrate the effectiveness and generalization of the model on both types of databases, we compared some traditional methods and generic state-of-the-art BIQA methods based on deep learning. The tested SRCC and PLCC of our approach and state-of-the-art BIQA methods are listed in Table 2. The best results are marked in bold and the suboptimal approach is underlined.

We showed three synthetically distorted databases, including LIVE [18], CSIQ [10] and TID2013 [16], to evaluate the compared methods. It can be seen that our method on the synthetically distorted databases outperforms the other methods listed. This fully demonstrates that the subtask of distortion classification for artificially degraded images can assist the model to optimize the perception of overall image quality, thereby improving evaluation performance.

To verify the effectiveness of the proposed approach on authentically distorted databases, we employ two databases, i.e., LIVEC [2] and KonIQ-10k [6]. The results are shown in Table 2. Although there is no distortion type information, our model can also be adapted to such databases due to the feature representation of the capsule. On the LIVEC, our approach surpasses other models to achieve state-of-the-art, and is at least 2% higher than other models. Our approach also achieves competitive performance on the KonIQ-10k, achieving the second-best performance after TReS [3] and VIPNet [21]. The results show that our model doesn't rely much on distortion type information when faced with distorted images, and distortion type information is an improvement rather than a necessity for our model.

Table 2. Comparison results (SRCC and PLCC) on five IQA databases.

Methods	LIVE		CSIQ		TID2013		KonIQ-10k		LIVEC	
	SRCC	PLCC	SRCC	PLCC	SRCC	PLCC	SRCC	PLCC	SRCC	PLCC
BRISQUE [15]	0.939	0.942	0.746	0.829	0.573	0.651	0.665	0.681	0.608	0.629
ILNIQE [23]	0.902	0.865	0.806	0.808	0.521	0.648	0.432	0.508	0.507	0.523
WaDIQaM-NR [1]	0.954	0.963	0.852	0.844	0.761	0.787	0.797	0.805	0.671	0.680
PQR [22]	0.965	0.971	0.873	0.901	0.849	0.864	0.880	0.884	0.857	0.882
DBCNN [25]	0.968	0.971	0.946	<u>0.959</u>	0.816	0.865	0.875	0.884	0.851	0.869
MetaIQA [27]	0.960	0.959	0.899	0.908	0.856	0.868	0.887	0.856	0.802	0.835
HyperIQA [19]	0.962	0.966	0.923	0.942	0.840	0.858	0.906	0.917	<u>0.859</u>	<u>0.882</u>
AIGQA [12]	0.960	0.957	0.927	0.952	0.871	0.893	-	-	0.751	0.761
GraphIQA [20]	<u>0.979</u>	<u>0.980</u>	0.947	<u>0.959</u>	-	-	<u>0.911</u>	0.915	0.845	0.862
TReS [3]	0.969	0.968	0.922	0.942	0.863	0.883	**0.915**	**0.928**	0.846	0.877
VIPNet [21]	0.968	0.971	<u>0.963</u>	**0.966**	<u>0.891</u>	<u>0.902</u>	**0.915**	**0.928**	0.827	0.848
MTQ-Caps	**0.981**	**0.983**	**0.964**	**0.966**	**0.898**	**0.915**	**0.915**	<u>0.919</u>	**0.879**	**0.903**
std	±0.003	±0.003	±0.002	±0.002	±0.004	±0.004	±0.002	±0.002	±0.002	±0.002

Table 3. Cross-database SRCC comparison of BIQA methods.

Type	Training	Testing	DBCNN [25]	HyperIQA [19]	TReS [3]	**Ours**
Synthetic	LIVE	TID2013	0.524	0.551	<u>0.562</u>	**0.573**
		CSIQ	0.758	0.744	<u>0.761</u>	**0.779**
Synthetic	TID2013	LIVE	**0.891**	0.856	<u>0.888</u>	**0.891**
		CSIQ	<u>0.807</u>	0.692	0.708	**0.827**
Authentic	KonIQ-10k	LIVEC	0.755	0.785	<u>0.786</u>	**0.793**
	LIVEC	KonIQ-10k	0.754	**0.772**	0.733	<u>0.770</u>

Evaluation on Cross Databases. Conducting evaluations across different databases is a critical step in assessing the generality and stability of a model. Specifically, we designed cross-database experiments of different scales according to the types of databases, and compared them with several methods with superior effects. The comparison results are shown in the Table 3. It can be seen that our proposed method achieves the best performance in most settings. The results demonstrate that our method has excellent generalization and stability in the face of evaluation tasks with different scales and image subjects.

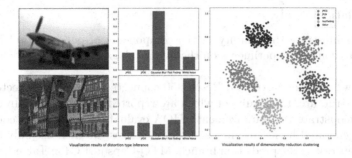

Fig. 2. Visualization results.

Visualization of Distortion Types. In our approach, we use the distortion-type predicted branch to extract hierarchical quality descriptions of image regions. The low-level capsules before this branch capture basic descriptions of different distortion properties, While the higher-level capsules after the routing process of this branch provide more semantic descriptions, which can better represent specific distortion types.

To verify the usefulness of this type of capsules, we have visualized the quality description capsule corresponding to the sample images in the LIVE database. Specifically, we use the model inference image, and use the normalized module lengths of the distortion capsules generated in the distortion-type predicted branch as the information representation of the distortion type. The visualization results are illustrated on the left side of Fig. 2. The modulus value of the distortion capsules represents the confidence that the distortion region in the image contains a certain type of distortion. To further verify the usefulness of distortion capsules for discriminating distortion types, we visualize the largest distortion capsules corresponding to sample images in the LIVE database, which contains five types of synthetic distortions. Since the features extracted from images with the same distortion type have similar distribution characteristics, we reasoned over 20% of the test images in the LIVE database. We used the t-SNE algorithm to reduce the dimensionality of the semantic capsules after the semantic branching process, transforming similar distortion-type features in the high-dimensional space to the low-dimensional space. As shown on the right side of Fig. 2, the closer points indicate distribution features that obey similar distribution, i.e., semantic features containing similar distortion types, which are distinguished by dimensionality reduction. They demonstrated that capsules representing the same type of distortion are well-clustered, while the boundaries between different distortion types are also distinct. This indicates that the high-level distortion capsules in the distortion-type predicted branch are able to effectively distinguish between different types of distortions.

4.5   Ablation Study

To further illustrate the validity of the components of our proposed method, we chose to conduct experiments on the five databases to complete the ablation study.

First, we use the pre-trained ResNet-50 connected to a fully connected layer as a base structure, the results of which are reported in the Table 4. In order to better demonstrate the improvement of IQA evaluation accuracy by combining capsules generated by feature maps of different sizes, we then use only the 4-th capsule network as the prediction branch, MTQ-Caps just C4 in Table 4. Clearly, the combination of content capsules containing feature information of different sizes improves the evaluation performance of MTQ-Caps on all databases. We also report the results obtained by using only the quality score prediction branch without distortion type information (i.e. MTQ-Caps w/o Distortion Type Prediction in Table 4) to illustrate the stability and generalization of our model. The SRCC and PLCC will continue to be raised when we employ the distortion type prediction branch, which further demonstrates the validity of the branch we have introduced. DBCNN [25] also makes use of distortion type information, but our method can still outperform it even without using the distortion type prediction branch. The architecture corresponding to the results in bold in the final row is the one we have used in Fig. 1. Here we introduce a total of four stages of capsules in the quality prediction branch to receive multi-scale features, which enables our model to achieve state-of-the-art on multiple databases containing different distortion types.

Table 4. The results (SRCC and PLCC) of ablation study.

Methods	LIVE		CSIQ		TID2013		LIVEC		KonIQ-10k	
	SRCC	PLCC	SRCC	PLCC	SRCC	PLCC	SRCC	PLCC	SRCC	PLCC
ResNet-50	0.923	0.947	0.915	0.920	0.782	0.794	0.827	0.852	0.851	0.873
MTQ-Caps just C4	0.943	0.950	0.936	0.938	0.845	0.860	0.836	855	0.862	0.880
MTQ-Caps w/o Distortion Type Prediction	<u>0.970</u>	<u>0.973</u>	<u>0.940</u>	<u>0.943</u>	<u>0.879</u>	<u>0.892</u>	-	-	-	-
MTQ-Caps	**0.981**	**0.983**	**0.964**	**0.966**	**0.898**	**0.915**	**0.879**	**0.903**	**0.911**	**0.918**

5   Conclusion

In this paper, we propose a multi-task quality capsule network (MTQ-Caps) for solving the BIQA task. The fusion of multi-scale features in the form of capsules enables our model to perform effectively on both synthetically and authentically distorted databases. Furthermore, for databases that contain distortion type information, we introduce the distortion type prediction capsule sub-network to enhance performance. The experimental results demonstrate the immense potential of our model and validate the beneficial effects of the capsule networks.

Acknowledgments. This work was supported by the National Natural Science Foundation of China (Nos. 61972132, 62106065), the S&T Program of Hebei (Nos. 20327404D, 21327404D), the Natural Science Foundation of Hebei Province, China (No. F2020204009), and the Research Project for Self-cultivating Talents of Hebei Agricultural University (No. PY201810).

References

1. Bosse, S., Maniry, D., Müller, K.R., Wiegand, T., Samek, W.: Deep neural networks for no-reference and full-reference image quality assessment. IEEE Trans. Image Process. **27**(1), 206–219 (2017)
2. Ghadiyaram, D., Bovik, A.C.: Massive online crowdsourced study of subjective and objective picture quality. IEEE Trans. Image Process. **25**(1), 372–387 (2015)
3. Golestaneh, S.A., Dadsetan, S., Kitani, K.M.: No-reference image quality assessment via transformers, relative ranking, and self-consistency. In: Proceedings of the IEEE/CVF Winter Conference on Applications of Computer Vision, pp. 1220–1230 (2022)
4. Golestaneh, S.A., Kitani, K.: No-reference image quality assessment via feature fusion and multi-task learning. arXiv preprint arXiv:2006.03783 (2020)
5. He, K., Zhang, X., Ren, S., Sun, J., et al.: Deep residual learning for image recognition. In: Proceedings of the IEEE Conference on Computer Vision and Pattern Recognition, pp. 770–778 (2016)
6. Hosu, V., Lin, H., Sziranyi, T., et al.: KonIQ-10k: an ecologically valid database for deep learning of blind image quality assessment. IEEE Trans. Image Process. **29**, 4041–4056 (2020)
7. Kang, L., Ye, P., Li, Y., Doermann, D.: Convolutional neural networks for no-reference image quality assessment. In: Proceedings of the IEEE Conference on Computer Vision and Pattern Recognition, pp. 1733–1740 (2014)
8. Kim, J., Lee, S.: Deep learning of human visual sensitivity in image quality assessment framework. In: Proceedings of the IEEE Conference on Computer Vision and Pattern Recognition, pp. 1676–1684 (2017)
9. Kosiorek, A., Sabour, S., Teh, Y.W., Hinton, G.E.: Stacked capsule autoencoders. In: Advances in Neural Information Processing Systems, vol. 32 (2019)
10. Larson, E.C., Chandler, D.M.: Most apparent distortion: full-reference image quality assessment and the role of strategy. J. Electron. Imaging **19**(1), 011006–011006 (2010)
11. Lin, K.Y., Wang, G.: Hallucinated-IQA: no-reference image quality assessment via adversarial learning. In: Proceedings of the IEEE Conference on Computer Vision and Pattern Recognition, pp. 732–741 (2018)
12. Ma, J., et al.: Blind image quality assessment with active inference. IEEE Trans. Image Process. **30**, 3650–3663 (2021)
13. Ma, K., Liu, W., Zhang, K., Duanmu, Z., et al.: End-to-end blind image quality assessment using deep neural networks. IEEE Trans. Image Process. **27**(3), 1202–1213 (2017)
14. Mazzia, V., Salvetti, F., Chiaberge, M.: Efficient-capsnet: capsule network with self-attention routing. Sci. Rep. **11**(1), 14634 (2021)
15. Mittal, A., Moorthy, A.K., Bovik, A.C.: No-reference image quality assessment in the spatial domain. IEEE Trans. Image Process. **21**(12), 4695–4708 (2012)
16. Ponomarenko, N., et al.: Image database TID2013: peculiarities, results and perspectives. Sig. Process. Image Commun. **30**, 57–77 (2015)

17. Sabour, S., Frosst, N., Hinton, G.E.: Dynamic routing between capsules. In: Advances in Neural Information Processing Systems, vol. 30 (2017)
18. Sheikh, H.R., Sabir, M.F., Bovik, A.C.: A statistical evaluation of recent full reference image quality assessment algorithms. IEEE Trans. Image Process. **15**(11), 3440–3451 (2006)
19. Su, S., Yan, Q., Zhu, Y., Zhang, C., Ge, X., Sun, J., et al.: Blindly assess image quality in the wild guided by a self-adaptive hyper network. In: Proceedings of the IEEE Conference on Computer Vision and Pattern Recognition, pp. 3667–3676 (2020)
20. Sun, S., Yu, T., Xu, J., Zhou, W., Chen, Z.: GraphIQA: learning distortion graph representations for blind image quality assessment. IEEE Trans. Multimedia (2022)
21. Wang, X., Xiong, J., Lin, W.: Visual interaction perceptual network for blind image quality assessment. IEEE Trans. Multimedia (2023)
22. Zeng, H., Zhang, L., Bovik, A.C.: A probabilistic quality representation approach to deep blind image quality prediction. arXiv preprint arXiv:1708.08190 (2017)
23. Zhang, L., Zhang, L., Bovik, A.C.: A feature-enriched completely blind image quality evaluator. IEEE Trans. Image Process. **24**(8), 2579–2591 (2015)
24. Zhang, P., Zhou, W., Wu, L., et al.: SOM: semantic obviousness metric for image quality assessment. In: Proceedings of the IEEE Conference on Computer Vision and Pattern Recognition, pp. 2394–2402 (2015)
25. Zhang, W., Ma, K., Yan, J., Deng, D., Wang, Z.: Blind image quality assessment using a deep bilinear convolutional neural network. IEEE Trans. Circuits Syst. Video Technol. **30**(1), 36–47 (2018)
26. Zhang, W., Ma, K., Zhai, G., Yang, X.: Learning to blindly assess image quality in the laboratory and wild. In: 2020 IEEE International Conference on Image Processing (ICIP), pp. 111–115. IEEE (2020)
27. Zhu, H., Li, L., Wu, J., Dong, W., Shi, G.: MetaIQA: deep meta-learning for no-reference image quality assessment. In: Proceedings of the IEEE/CVF Conference on Computer Vision and Pattern Recognition, pp. 14143–14152 (2020)

VCD: Visual Causality Discovery
for Cross-Modal Question Reasoning

Yang Liu[✉], Ying Tan, Jingzhou Luo, and Weixing Chen

Sun Yat-sen University, Guangzhou, China
liuy856@mail.sysu.edu.cn, {tany86,luojzh5}@mail2.sysu.edu.cn

Abstract. Existing visual question reasoning methods usually fail to explicitly discover the inherent causal mechanism and ignore jointly modeling cross-modal event temporality and causality. In this paper, we propose a visual question reasoning framework named Cross-Modal Question Reasoning (CMQR), to discover temporal causal structure and mitigate visual spurious correlation by causal intervention. To explicitly discover visual causal structure, the Visual Causality Discovery (VCD) architecture is proposed to find question-critical scene temporally and disentangle the visual spurious correlations by attention-based front-door causal intervention module named Local-Global Causal Attention Module (LGCAM). To align the fine-grained interactions between linguistic semantics and spatial-temporal representations, we build an Interactive Visual-Linguistic Transformer (IVLT) that builds the multi-modal co-occurrence interactions between visual and linguistic content. Extensive experiments on four datasets demonstrate the superiority of CMQR for discovering visual causal structures and achieving robust question reasoning. The supplementary file can be referred to https://github.com/ YangLiu9208/VCD/blob/main/0793_supp.pdf.

Keywords: Visual Question Answering · Visual-linguistic · Causal Inference

1 Introduction

Event understanding has become a prominent research topic in video analysis because videos have good potential to understand event temporality and causality. Since the expressivity of natural language can potentially describe a richer event space that facilitates the deeper event understanding, in this paper, we focus on event-level visual question reasoning task, which aims to fully understand richer multi-modal event space and answer the given question in a causality-aware way. To achieve event-level visual question reasoning, the model is required to achieve fine-grained understanding of video and language content involving various complex relations such as spatial-temporal visual relation, linguistic semantic relation, and visual-linguistic causal dependency. Most of the existing visual question reasoning methods [16] use recurrent neural networks

© The Author(s), under exclusive license to Springer Nature Singapore Pte Ltd. 2024
Q. Liu et al. (Eds.): PRCV 2023, LNCS 14431, pp. 309–322, 2024.
https://doi.org/10.1007/978-981-99-8540-1_25

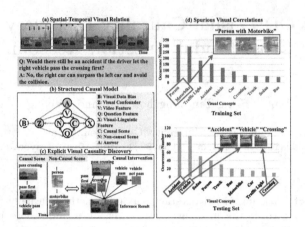

Fig. 1. An example of event-level counterfactual visual question reasoning task and its structured causal model (SCM). The counterfactual inference is to obtain the outcome of certain hypothesis that do not occur. The SCM shows how the confounder induces the spurious correlation. The green path is the unbiased visual question reasoning. The red path is the biased one. (Color figure online)

(RNNs) [31], attention mechanisms [33] or Graph Convolutional Networks [15] for relation reasoning between visual and linguistic modalities. Although achieving promising results, the current visual question reasoning methods suffer from the following two common limitations.

First, the existing visual question reasoning methods usually focus on relatively simple events where temporal understanding and causality discovery are simply not required to perform well, and ignore more complex and challenging events that require in-depth understanding of the causality, spatial-temporal dynamics, and linguistic relations. As shown in Fig. 1(a), the event-level counterfactual visual question reasoning task requires the outcome of certain hypothesis that does not occur in the given video. If we just simply correlate relevant visual contents, we cannot to get the right inference result without discovering the hidden spatial-temporal and causal dependencies. Second, most of the visual question reasoning models tend to capture the spurious visual correlations rather than the true causal structure, which leads to an unreliable reasoning process [21]. As shown in the SCM from Fig. 1(b), the concepts "person" and "motorbike" are dominant in training set (Fig. 1(c) and (d)) and thus the predictor may learn the spurious correlation between the "person" with the "motorbike" without looking at the collision region (causal scene C) to reason how actually the accident happens. Taking a causal look at VideoQA, we partition the visual scenes into two parts: 1) causal scene C, which holds the question-critical information, 2) non-causal scene N, which is irrelevant to the answer. Such biased dataset entails two causal effects: the visual bias B and non-causal scene N lead to the confounder Z, and then affects the visual feature V, causal scene C, question feature Q, visual-linguistic feature X, and the answer A.

To address the aforementioned limitations, we propose an event-level visual question reasoning framework named Cross-Modal Question Reasoning (CMQR). Experiments on SUTD-TrafficQA, TGIF-QA, MSVD-QA, and MSRVTT-QA datasets show the advantages of our CMQR over the state-of-the-art methods. The main contributions can be summarized as follows:

- We propose a causality-aware visual question reasoning framework named Cross-Modal Question Reasoning (CMQR), to discover cross-modal causal structures via causal interventions and achieve robust visual question reasoning and answering.
- We introduce the Visual Causality Discovery (VCD) architecture that learns to find the temporal causal scenes for a given question and mitigates the unobservable visual spurious correlations by an attention-based causal front-door intervention module named Local-Global Causal Attention Module (LGCAM).
- We construct an Interactive Visual-Linguistic Transformer (IVLT) to align and discover the multi-modal co-occurrence interactions between linguistic semantics and spatial-temporal visual concepts.

2   Related Works

2.1   Visual Question Reasoning

Compared with the image-based visual question reasoning [2], event-level visual question reasoning is much more challenging due to the existence of temporal dimension. To accomplish this task, the model needs to capture spatial-temporal and visual-linguistic relations. To explore relational reasoning, Xu et al. [38] proposed an attention mechanism to exploit the appearance and motion knowledge with the question as a guidance. Later on, some hierarchical attention and co-attention based methods [12,17] are proposed to learn appearance-motion and question-related multi-modal interactions. Le et al. [16] proposed hierarchical conditional relation network (HCRN) to construct sophisticated structures for representation and reasoning over videos. Lei et al. [17] employed sparse sampling to build a transformer-based model named CLIPBERT and achieve video-and-language understanding. However, previous works tend to implicitly capture the spurious visual-linguistic correlations, while we propose the Visual Causality Discovery (VCD) to explicitly uncover the visual causal structure.

2.2   Causal Inference in Visual Learning

Compared with conventional debiasing techniques [37], causal inference [21,28] shows its potential in mitigating the spurious correlations and disentangling the desired model effects for better generalization. Counterfactual and causal inference have attracted increasing attention in visual explanations [35], scene graph generation [32], image recognition [36], video analysis [24], and vision-language tasks [18,41]. However, most of the existing visual tasks are relatively simple. Although some recent works CVL [1], Counterfactual VQA [26], and CATT [41] focused on visual question reasoning tasks, they adopted structured causal model (SCM) to eliminate either the linguistic or visual bias without considering explicit cross-modal causality discovery. Differently, our CMQR aims for event-level visual question reasoning that requires fine-grained understanding of spatial-temporal and visual-linguistic causal dependency. Moreover, our Visual Causality Discovery (VCD) applies front-door causal interventions to explicitly find question-critic visual scene.

3 Methodology

3.1 Visual Representation Learning

The goal of event-level visual question reasoning is to deduce an answer $\tilde{a}$ from a video $\mathcal{V}$ with a given question q. The video $\mathcal{V}$ of L frames is divided into N equal clips. Each clip of C_i of length $T = \lfloor L/N \rfloor$ is presented by two types of visual features: frame-wise appearance feature vectors $F_i^a = \{f_{i,j}^a | f_{i,j}^a, j = 1, \ldots, T\}$ and motion feature vector at clip level f_i^m. In our experiments, the vision-language transformer with frozen parameters XCLIP [25] (other visual backbones are evaluated in Table 6) is used to extract the frame-level appearance features F^a and the clip-level motion features F^m. We use a linear layer to map F^a and F^m into the same d-dimensional feature space.

3.2 Linguistic Representation Learning

Each word of the question is respectively embedded into a vector of 300 dimension by Glove [29] word embedding, which is further mapped into a d-dimensional space using linear transformation. Then, we represent the corresponding question and answer semantics as $Q = \{q_1, q_2, \cdots, q_L\}$, $A = \{a_1, a_2, \cdots, a_{L_a}\}$, where L, L_a indicate the length of Q and A, respectively. To obtain contextual linguistic representations that aggregates dynamic long-range temporal dependencies from multiple time-steps, a Bert [3] model is employed to encode Q and the answer A, respectively. Finally, the updated representations for the question and answer candidates can be written as:

$$
\begin{aligned}
Q &= \{q_i | q_i \in \mathbb{R}^d\}_{i=1}^L \\
A &= \{a_i | a_i \in \mathbb{R}^d\}_{i=1}^{L_a}
\end{aligned}
\tag{1}
$$

3.3 Visual Causality Discovery

For visual-linguistic question reasoning, we employ Pearl's structural causal model (SCM) [28] to model the causal effect between video-question pairs (V, Q), causal scene C, non-causal scene N, and the answer A, as shown in Fig. 2(a). We hope to train a video question answering model to the learn the true causal effect $\{V, Q\} \rightarrow C \rightarrow X \rightarrow A$: the model should reason the answer A from video

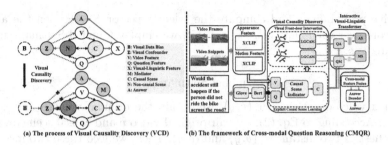

(a) The process of Visual Causality Discovery (VCD)   (b) The framework of Cross-modal Question Reasoning (CMQR)

Fig. 2. The process of Visual Causality Discovery (VCD) (a) and framework of Cross-modal Question Reasoning (b). The green path is the unbiased question reasoning. The red path is the biased one. The black path is the division of causal and non-causal visual scenes. (Color figure online)

feature V, causal-scene C and question feature Q instead of exploiting the non-causal scene N and spurious correlations induced by the confounders Z (i.e., the existence of non-causal scene and overexploiting the co-occurrence between visual concepts and answer). In our SCM, the non-interventional prediction can be expressed by Bayes rule:

$$P(A|V,Q) = \sum_z P(A|V,Q,z)P(z|V,Q) \qquad (2)$$

However, the above objective learns not only the main direct correlation from $\{V,Q\} \to X \to A$ but also the spurious one from the back-door path $\{V,Q\} \leftarrow Z \to A$. An intervention on $\{V,Q\}$ is denoted as $do(V,Q)$, which cuts off the link $\{V,Q\} \leftarrow Z$ to block the back-door path $\{V,Q\} \leftarrow Z \to A$ and eliminate the spurious correlation. In this way, $\{V,Q\}$ and A are deconfounded and the model can learn the true causal effect $\{V,Q\} \to C \to X \to A$.

Explicit Causal Scene Learning. Inspired by the fact that only part of the visual scenes are critical to answering the question, we split the video V into causal scene C and non-causal scene N (see the black path in Fig. 2(a)). Specifically, given the causal scene C and question Q, we assume that the answer A is determined, regardless the variations of the non-causal scene N: $A \perp N|C,Q$, where $\perp$ denotes the probabilistic independence. Thus, we build an explicit causal scene learning (ECSL) module to estimate C.

For a video-question pair (v, q), we encode video instance v as a sequence of K visual clips. The ECSL module aims to estimate the causal scene $\hat{c}$ according to the question q. Concretely, we first construct a cross-modal attention module to indicate the probability of each visual clip belongs to causal scene ($p_{\hat{c}} \in \mathbb{R}^K$):

$$p_{\hat{c}} = \text{softmax}(G_v^1(v) \cdot G_q^1(q)^\top) \qquad (3)$$

where G_v^1 and G_q^1 are fully connected layers to align cross-modal representations. However, the soft mask makes $\hat{c}$ overlap. To achieve a differentiable selection on attentive probabilities and compute the selector vector $S \in \mathbb{R}^K$ on the attention score over each clip (i.e., $p_{\hat{c},i}, i \in K$), we employ Gumbel-Softmax [8] and estimate $\hat{c}$ as:

$$\hat{c} = \text{Gumbel-Softmax}(p_{\hat{c},i}) \cdot v \tag{4}$$

For a video-question pair (v, q), we obtain the original video v and causal scene $\hat{c}$. According to Eq. (2), we pair original video v and causal scene $\hat{c}$ with q to synthesizes two instances: (v, q) and $(\hat{c}, q)$. Then, we feed these two instances into visual front-door causal intervention (VFCI) module to deconfound $\{V, Q\}$ and A.

Visual Front-Door Causal Intervention. In visual domains, it is hard to explicitly represent confounders due to complex data biases. Fortunately, the front-door adjustment give a feasible way to calculate $P(A|do(V), Q)$. In Fig. 2(a), an additional mediator M can be inserted between X and A to construct the front-door path $\{V, Q\} \to X \to M \to A$ for transmitting knowledge. For visual question reasoning, an attention-based model $P(A|V, Q) = \sum_m P(M = m|V, Q)P(A|M = m)$ will select a few regions from the original video V and causal scene C based on the question Q to predict the answer A, where m denotes the selected knowledge from M. Thus, the answer predictor can be represented by two parts: two feature extractors $V \to X \to M$, $C \to X \to M$, and an answer predictor $M \to A$. In the following, we take the visual interventional probability $P(A|do(V), Q)$ for original video V as an example (the $P(A|do(C), Q)$ for causal scene C is implemented in the same way):

$$
\begin{aligned}
&P(A|do(V), Q) \\
&= \sum_m P(M = m|do(V), Q)P(A|do(M = m)) \\
&= \sum_m P(M = m|V, Q) \sum_v P(V = v)P(A|V = v, M = m)
\end{aligned}
\tag{5}
$$

To implement visual front-door causal intervention Eq. (5) in a deep learning framework, we parameterize the $P(A|V, M)$ as a network $g(\cdot)$ followed by a softmax layer since most of visual-linguistic tasks are classification formulations, and then apply Normalized Weighted Geometric Mean (NWGM) [39] to reduce computational cost:

$$
\begin{aligned}
P(A|do(V), Q) &\approx \text{Softmax}[g(\hat{M}, \hat{V})] \\
&= \text{Softmax}\left[g(\sum_m P(M = m|f(V))m, \sum_v P(V = v|h(V))v)\right]
\end{aligned}
\tag{6}
$$

where $\hat{M}$ and $\hat{V}$ denote the estimations of M and V, $h(\cdot)$ and $f(\cdot)$ denote the network mappings. The derivation details from Eq. (5)–(6) is given in the Appendix 2.

Actually, $\hat{M}$ is essentially an in-sample sampling process where m denotes the selected knowledge from the current input sample V, $\hat{V}$ is essentially a cross-sample sampling process since it comes from the other samples. Therefore, both $\hat{M}$ and $\hat{V}$ can be calculated by attention networks [41].

Therefore, we propose a Local-Global Causal Attention Module (LGCAM) that jointly estimates $\hat{M}$ and $\hat{V}$ to increase the representation ability of the causality-aware visual features. $\hat{M}$ can be learned by local-local visual feature F_{LL}, $\hat{V}$ can be learned by local-global visual feature F_{LG}. Here, we take the computation of F_{LG} as the example to clarify our LGCAM. Specifically, we firstly calculate $F_L = f(V)$ and $F_G = h(V)$ and use them as the input, where $f(\cdot)$ denotes the visual feature extractor (frame-wise appearance feature or motion feature) followed by a query embedding function, and $h(\cdot)$ denotes the K-means based visual feature selector from the whole training samples followed by a query embedding function. Thus, F_L represents the visual feature of the current input sample (local visual feature) and F_G represents the global visual feature. The F_G is obtained by randomly sampling from the whole clustering dictionaries with the same size as F_L. The LGCAM takes F_L and F_G as the inputs and computes local-global visual feature F_{LG} by conditioning global visual feature F_G to the local visual feature F_L. The output of the LGCAM is denoted as F_{LG}:

$$
\begin{aligned}
&\textbf{Input} : Q = F_L, K = F_G, V = F_G \\
&\textbf{Local-Global} : H = [W_V V, W_Q Q \odot W_K K] \\
&\textbf{Activation Mapping} : H' = \text{GELU}(W_H H + b_H) \qquad (7)\\
&\textbf{Attention Weights} : \alpha = \text{Softmax}(W_{H'} H' + b_{H'}) \\
&\textbf{Output} : F_{LG} = \alpha \odot F_G
\end{aligned}
$$

where $[.,.]$ denotes concatenation operation, $\odot$ is the Hadamard product, W_Q, W_K, W_V, $W_{H'}$ denote the weights of linear layers, b_H and $b_{H'}$ denote the biases of linear layers. From Fig. 2(b), the visual front-door causal intervention module has two branches for appearance and motion features. Therefore, the F_{LG} has two variants, one for appearance branch F_{LG}^a, and the other for motion branch F_{LG}^m. The F_{LL} can be computed similarly as F_{LG} when setting $Q = K = V = F_L$. Finally, the F_{LG} and F_{LL} are concatenated $F_C = [F_{LG}, F_{LL}]$ for estimating $P(A|do(V), Q)$.

3.4 Interactive Visual-Linguistic Transformer

To align the fine-grained interactions between linguistic semantics and spatial-temporal representations, we build an Interactive Visual-Linguistic Transformer (IVLT) that contains four sub-modules, namely Question-Appearance (QA), Question-Motion (QM), Appearance-Semantics (AS) and Motion-Semantics (MS). The QA (QM) module consists of an R-layer Multi-modal Transformer Block (MTB) for multi-modal interaction between the question and the appearance (motion) features. Similarly, the AS (MS) uses the MTB to infer the appearance (motion) information given the questions.

For QA and QM modules, the input of MTB is question representation Q obtained from Sect. 3.2 and causality-aware visual representations F_C^a, F_C^m obtained from Sect. 3.3, respectively. To maintain the positional information of the video sequence, the appearance feature F_C^a and motion feature F_C^m are firstly added with the learned positional embeddings P^a and P^m, respectively. Thus, for $r = 1, 2, \ldots, R$ layers of the MTB, with the input $F_C^a = [F_C^a, P^a]$, $F_C^m = [F_C^m, P^m]$, Q^a, and Q^m, the multi-modal output for QA and QM are computed as:

$$\hat{Q}_r^a = U_r^a + \sigma^a(\text{LN}(U_r^a))$$
$$\hat{Q}_r^m = U_r^m + \sigma^m(\text{LN}(U_r^m))$$
$$U_r^a = \text{LN}(\hat{Q}_{r-1}^a) + \text{MMA}^a(\hat{Q}_{r-1}^a, F_C^a)$$
$$U_r^m = \text{LN}(\hat{Q}_{r-1}^m) + \text{MMA}^m(\hat{Q}_{r-1}^m, F_C^m)$$

$$(8)$$

where $\hat{Q}_0^a = Q^h, \hat{Q}_0^m = Q^h$, U_r^a and U_r^m are the intermediate feature at r-th layer of the MTB. $\text{LN}(\cdot)$ denotes the layer normalization operation and $\sigma^a(\cdot)$ and $\sigma^m(\cdot)$ denote the two-layer linear projections with GELU activation. $\text{MMA}(\cdot)$ is the Multi-head Multi-modal Attention layer. We denote the output semantics-aware appearance and motion features of QA and MA as $L^a = \hat{Q}^a = \hat{Q}_R^a$ and $L^m = \hat{Q}^m = \hat{Q}_R^m$, respectively.

Similar to Eq. (8), given the visual appearance and motion feature F_{LG}^a, F_{LG}^m and question semantics L^a, L^m, the multi-modal output for AS and MS are computed as the same way. The output visual clues of QA and MA are denoted as $F_s^a = \hat{L}_R^a$ and $F_s^m = \hat{L}_R^m$, respectively. Then, the output of the AS and MS are concatenated to make the final visual output $F = [F_s^a, F_s^m] \in \mathbb{R}^{2d}$. The output of the QA and QM are concatenated to make the final question semantics output $L = [L^a, L^m] \in \mathbb{R}^{2d}$.

3.5   Cross-Modal Feature Fusion and Training

Finally, we apply different answer decoders [16] to (v, q) and $(\hat{c}, q)$ and obtain original prediction and causal prediction losses:

$$\mathcal{L}_o = \text{XE}(\text{CMQR}(v, q), a)$$
$$\mathcal{L}_c = \text{XE}(\text{CMQR}(\hat{c}, q), a)$$

$$(9)$$

where XE denotes the cross-entropy loss, a is the ground-truth answer, CMQR denotes our proposed framework. Furthermore, to make the predictions of original and causal scene consistent, we apply KL-divergence between the predictions of (v, q) and $(\hat{c}, q)$:

$$\mathcal{L}_a = \text{KL}(\text{CMQR}(v, q), \text{CMQR}(\hat{c}, q))$$

$$(10)$$

Finally, the learning objective of our CMQR is:

$$\mathcal{L}_{\text{CMQR}} = \mathcal{L}_o + \lambda_c \mathcal{L}_c + \lambda_a \mathcal{L}_a$$

$$(11)$$

Table 1. Results on SUTD-TrafficQA dataset.

Method	Basic	Attri.	Intro.	Counter.	Fore.	Rev.	All
VQAC[†] [14]	34.02	49.43	<u>34.44</u>	39.74	38.55	49.73	36.00
MASN[†] [30]	33.83	<u>50.86</u>	34.23	41.06	41.57	<u>50.80</u>	36.03
DualVGR[†] [34]	33.91	50.57	33.40	<u>41.39</u>	41.57	50.62	36.07
HCRN [16]	-	-	-	-	-	-	36.49
HCRN[†] [16]	<u>34.17</u>	50.29	33.40	40.73	<u>44.58</u>	50.09	36.26
Eclipse [40]	-	-	-	-	-	-	<u>37.05</u>
CMQR (ours)	**36.10**	**52.59**	**38.38**	**46.03**	**48.80**	**58.05**	**38.63**

Table 2. Comparison with state-of-the-art methods on TGIF-QA.

Method	Action↑	Transition↑	FrameQA↑	Count↓
GMIN [6]	73.0	81.7	57.5	4.16
L-GCN [7]	74.3	81.1	56.3	3.95
HCRN [16]	75.0	81.4	55.9	3.82
HGA [11]	75.4	81.0	55.1	4.09
QueST [10]	75.9	81.0	59.7	4.19
Bridge [27]	75.9	<u>82.6</u>	57.5	**3.71**
QESAL [19]	76.1	82.0	57.8	3.95
ASTG [13]	76.3	82.1	<u>61.2</u>	<u>3.78</u>
CASSG [22]	77.6	**83.7**	58.7	3.83
HAIR [20]	77.8	82.3	60.2	3.88
CMQR (ours)	**78.1**	82.4	**62.3**	3.83

4   Experiments

In this paper, we evaluate our CMQR on four datasets: **SUTD-TrafficQA** [40], **TGIF-QA** [9], **MSVD-QA** [38], and **MSRVTT-QA** [38]. More details of these datasets are given in Appendix 4. For fair comparisons, we follow [16] to divide the videos into 8 clips for all datasets. The XCLIP [23] with ViT-L/14 pretrained on Kinetics-600 dataset is used to extract the appearance and motion features. For the question, we adopt the pre-trained 300-dimensional Glove [29] word embeddings to initialize the word features in the sentence. For parameter settings, we set the dimension d of hidden layer to 512. For the Multi-modal Transformer Block (MTB), the number of layers r is set to 3 for SUTD-TrafficQA, 8 for TGIF-QA, 5 for MSVD-QA, and 6 for MSRVTT-QA. The number of attentional heads H is set to 8. The dictionary is initialized by applying K-means over the whole visual features from the whole training set to get 512 clusters and is updated during end-to-end training. We train the model using the Adam optimizer with an initial learning rate 2e-4, a momentum 0.9, and a weight decay

0. The learning rate reduces by half when the loss stops decreasing after every 5 epochs. The batch size is set to 64. All experiments are terminated after 50 epochs. λ_c and λ_a are all set to 0.1.

4.1   Comparison with State-of-the-Art Methods

The results in Table 1 demonstrate that our CMQR achieves the best performance for six reasoning tasks including basic understanding, event forecasting, reverse reasoning, counterfactual inference, introspection and attribution analysis. Specifically, the CMQR improves the state-of-the-art method Eclipse [40] by 1.58%. Compared with the re-implemented methods $VQAC^\dagger$, $MASN^\dagger$, $DualVGR^\dagger$, $HCRN^\dagger$ and $IGV^\dagger$, our CMQR outperforms these methods for introspection and counterfactual inference tasks that require causal relational reasoning among the causal, logic, and spatial-temporal structures of the visual and linguistic content. These results show that our CMQR has strong ability in modeling multi-level interaction and causal relations between the language and spatial-temporal structure.

To evaluate the generalization ability of CMQR on other event-level datasets, we conduct experiments on TGIF-QA, MSVD-QA, and MSRVTT-QA datasets, as shown in Table 2, 3, 4. From Table 2, we can see that our CMQR achieves the best performance for *Action* and *FrameQA* tasks. Additionally, our CMQR also achieves relatively high performance for *Transition* and *Count* tasks. For the *Transition* task, the CMQR also outperforms nearly all comparison methods. For the *Count* task, we also achieve a competitive MSE value. From Table 3, our CMQR outperforms all the comparison methods by a significant margin. For *What*, *Who*, and *When* types, the CMQR outperforms all the comparison methods. It can be observed in Table 4 that our CMQR performs better than the best performing method IGV [18]. For *What*, *Who*, and *When* question types, the CMQR performs the best. Moreover, our CMQR can generalize well across different datasets and has good potential to model multi-level interaction and causal relations between the language and spatial-temporal structure.

4.2   Ablation Studies

We further conduct ablation experiments to verify the contributions of five essential components: 1) Explicit Causal Scene Learning (ECSL), 2) Visual Front-door Causal Intervention (VFCI), 3) Visual Causality Discovery (VCD), 4) Visual Causality Discovery (VCD), and Interactive Visual-Linguistic Transformer (IVLT). From Table 5, our CMQR achieves the best performance across all datasets and tasks. Without ECSL, the performance drops significantly due to the lack of the causal scene. This shows that the ECSL indeed find the causal scene that facilitate question reasoning. The performance of CMQR w/o ECSL, CMQR w/o VFCI are all lower than that of the CMQR. This validates that both the causal scene and visual front-door causal intervention are indispensable and contribute to discover the causal structures, and thus improve the model performance. The performance of CMQR w/o IVLT is higher than that of CMQR

Table 3. Comparison with SOTAs on MSVD.

Method	What	Who	How	When	Where	All
GMIN [6]	24.8	49.9	84.1	75.9	53.6	35.4
QueST [10]	24.5	52.9	79.1	72.4	50.0	36.1
HCRN [16]	-	-	-	-	-	36.1
CASSG [22]	24.9	52.7	84.4	74.1	53.6	36.5
QESAL [19]	25.8	51.7	83.0	72.4	50.0	36.6
Bridge [27]	-	-	-	-	-	37.2
HAIR [20]	-	-	-	-	-	37.5
VQAC [14]	26.9	53.6	-	-	-	37.8
MASN [30]	-	-	-	-	-	38.0
HRNAT [5]	-	-	-	-	-	38.2
ASTG [13]	26.3	55.3	82.4	72.4	50.0	38.2
DualVGR [34]	28.6	53.8	80.0	70.6	46.4	39.0
CMQR (Ours)	37.0	59.9	81.0	75.8	46.4	46.4

Table 4. Comparison with SOTAs on MSRVTT.

Method	What	Who	How	When	Where	All
HRNAT [5]	-	-	-	-	-	35.3
DualVGR [34]	29.4	45.5	79.7	76.6	36.4	35.5
HCRN [16]	-	-	-	-	-	35.6
VQAC [14]	29.1	46.5	-	-	-	35.7
CASSG [22]	29.8	46.3	84.9	75.2	35.6	36.1
GMIN [6]	30.2	45.4	84.1	74.9	43.2	36.1
QESAL [19]	30.7	46.0	82.4	76.1	41.6	36.7
Bridge [27]	-	-	-	-	-	36.9
HAIR [20]	-	-	-	-	-	36.9
ClipBert [17]	-	-	-	-	-	37.4
ASTG [13]	31.1	48.5	83.1	77.7	38.0	37.6
CMQR (ours)	32.2	50.2	82.3	78.4	38.0	38.9

Table 5. Ablation study on three datasets.

Datasets	CMQR w/o ECSL	CMQR w/o VFCI	CMQR w/o VCD	CMQR w/o IVLT	CMQR
SUTD	37.68	37.42	37.28	37.75	**38.63**
MSVD	44.9	44.7	43.6	44.8	**46.4**
MSRVTT	38.1	38.0	37.5	37.7	**38.9**

w/o VCD shows that visual and linguistic causal intervention modules contribute more than the IVLT due to the existence of cross-modal bias. With all components, our CMQR performs the best because all components contribute to our CMQR.

To validate the effectiveness of our causal module VCD in non-causal models, we apply the VCD to three state-of-the-art models Co-Mem [4], HGA [11] and HCRN [16]. As shown in Table 6, our VCD brings each backbone model a sharp gain across all benchmark datasets (+0.9%–6.5%), which evidences its model-agnostic property. To be noticed, for the causal and temporal questions (i.e., SUTD-TrafficQA), our VCD shows equivalent improvements on all four backbones (+1.05%–2.02%). These results validate that our VCD is effective in capturing the causality and reducing the spurious correlations across different models.

To conduct parameter analysis and validate the whether our CMQR could generalize to different visual appearance and motion features, we evaluate the performance of the CMQR using different visual backbones, please refer to Appendix 5. More visualization analysis are given in Appendix 6.

Table 6. The VCD module is applied to non-causal models.

Models	SUTD-TrafficQA	MSVD-QA	MSRVTT-QA
Co-Mem [4]	35.10	34.6	35.3
Co-Mem [4]+ VCD	**37.12** (+2.02)	**40.7** (+6.1)	**38.0** (+2.7)
HGA [11]	35.81	35.4	36.1
HGA [11]+ VCD	**37.23** (+1.42)	**41.9** (+6.5)	**38.2** (+2.1)
HCRN [16]	36.49	36.1	35.6
HCRN [16]+ VCD	**37.54** (+1.05)	**42.2** (+6.1)	**37.8** (+2.2)
Our Backbone	37.42	44.7	38.0
Our Backbone + VCD	**38.63** (+1.21)	**46.4** (+1.9)	**38.9** (+0.9)

5   Conclusion

In this paper, we propose an event-level visual question reasoning framework named Cross-Modal Question Reasoning (CMQR), to explicitly discover cross-modal causal structures. To explicitly discover visual causal structure, we propose a Visual Causality Discovery (VCD) architecture that learns to discover temporal question-critical scenes and mitigate the visual spurious correlations by front-door causal intervention. To align the fine-grained interactions between linguistic semantics and spatial-temporal visual concepts, we build an Interactive Visual-Linguistic Transformer (IVLT). Extensive experiments on four datasets well demonstrate the effectiveness of our CMQR for discovering visual causal structure and achieving robust event-level visual question reasoning.

References

1. Abbasnejad, E., Teney, D., Parvaneh, A., Shi, J., Hengel, A.V.D.: Counterfactual vision and language learning. In: Proceedings of the IEEE/CVF Conference on Computer Vision and Pattern Recognition, pp. 10044–10054 (2020)
2. Antol, S., et al.: VQA: visual question answering. In: Proceedings of the IEEE International Conference on Computer Vision, pp. 2425–2433 (2015)
3. Devlin, J., Chang, M.W., Lee, K., Toutanova, K.: Bert: pre-training of deep bidirectional transformers for language understanding. arXiv preprint arXiv:1810.04805 (2018)
4. Gao, J., Ge, R., Chen, K., Nevatia, R.: Motion-appearance co-memory networks for video question answering. In: Proceedings of the IEEE Conference on Computer Vision and Pattern Recognition, pp. 6576–6585 (2018)
5. Gao, L., Lei, Y., Zeng, P., Song, J., Wang, M., Shen, H.T.: Hierarchical representation network with auxiliary tasks for video captioning and video question answering. IEEE Trans. Image Process. **31**, 202–215 (2022)
6. Gu, M., Zhao, Z., Jin, W., Hong, R., Wu, F.: Graph-based multi-interaction network for video question answering. IEEE Trans. Image Process. **30**, 2758–2770 (2021)

7. Huang, D., Chen, P., Zeng, R., Du, Q., Tan, M., Gan, C.: Location-aware graph convolutional networks for video question answering. In: Proceedings of the AAAI Conference on Artificial Intelligence, vol. 34, pp. 11021–11028 (2020)

8. Jang, E., Gu, S., Poole, B.: Categorical reparameterization with gumbel-softmax. arXiv preprint arXiv:1611.01144 (2016)

9. Jang, Y., Song, Y., Yu, Y., Kim, Y., Kim, G.: TGIF-QA: toward spatio-temporal reasoning in visual question answering. In: Proceedings of the IEEE Conference on Computer Vision and Pattern Recognition, pp. 2758–2766 (2017)

10. Jiang, J., Chen, Z., Lin, H., Zhao, X., Gao, Y.: Divide and conquer: question-guided spatio-temporal contextual attention for video question answering. In: Proceedings of the AAAI Conference on Artificial Intelligence, vol. 34, pp. 11101–11108 (2020)

11. Jiang, P., Han, Y.: Reasoning with heterogeneous graph alignment for video question answering. In: Proceedings of the AAAI Conference on Artificial Intelligence, vol. 34, pp. 11109–11116 (2020)

12. JiayinCai, C., Shi, C., Li, L., Cheng, Y., Shan, Y.: Feature augmented memory with global attention network for VideoQA. In: IJCAI, pp. 998–1004 (2020)

13. Jin, W., Zhao, Z., Cao, X., Zhu, J., He, X., Zhuang, Y.: Adaptive spatio-temporal graph enhanced vision-language representation for video QA. IEEE Trans. Image Process. 30, 5477–5489 (2021)

14. Kim, N., Ha, S.J., Kang, J.W.: Video question answering using language-guided deep compressed-domain video feature. In: Proceedings of the IEEE/CVF International Conference on Computer Vision, pp. 1708–1717 (2021)

15. Kipf, T.N., Welling, M.: Semi-supervised classification with graph convolutional networks. arXiv preprint arXiv:1609.02907 (2016)

16. Le, T.M., Le, V., Venkatesh, S., Tran, T.: Hierarchical conditional relation networks for video question answering. In: Proceedings of the IEEE/CVF Conference on Computer Vision and Pattern Recognition, pp. 9972–9981 (2020)

17. Lei, J., et al.: Less is more: clipbert for video-and-language learning via sparse sampling. In: Proceedings of the IEEE/CVF Conference on Computer Vision and Pattern Recognition, pp. 7331–7341 (2021)

18. Li, Y., Wang, X., Xiao, J., Ji, W., Chua, T.S.: Invariant grounding for video question answering. In: Proceedings of the IEEE/CVF Conference on Computer Vision and Pattern Recognition, pp. 2928–2937 (2022)

19. Liu, F., Liu, J., Hong, R., Lu, H.: Question-guided erasing-based spatiotemporal attention learning for video question answering. IEEE Trans. Neural Netw. Learn. Syst. (2021)

20. Liu, F., Liu, J., Wang, W., Lu, H.: Hair: hierarchical visual-semantic relational reasoning for video question answering. In: Proceedings of the IEEE/CVF International Conference on Computer Vision, pp. 1698–1707 (2021)

21. Liu, Y., Wei, Y.S., Yan, H., Li, G.B., Lin, L.: Causal reasoning meets visual representation learning: a prospective study. Mach. Intell. Res. 19, 1–27 (2022)

22. Liu, Y., Zhang, X., Huang, F., Zhang, B., Li, Z.: Cross-attentional spatio-temporal semantic graph networks for video question answering. IEEE Trans. Image Process. 31, 1684–1696 (2022)

23. Liu, Z., et al.: Swin transformer: hierarchical vision transformer using shifted windows. In: Proceedings of the IEEE/CVF International Conference on Computer Vision, pp. 10012–10022 (2021)

24. Nan, G., et al.: Interventional video grounding with dual contrastive learning. In: Proceedings of the IEEE/CVF Conference on Computer Vision and Pattern Recognition, pp. 2765–2775 (2021)

25. Ni, B., et al.: Expanding language-image pretrained models for general video recognition. In: Avidan, S., Brostow, G., Cissé, M., Farinella, G.M., Hassner, T. (eds.) ECCV 2022, Part IV. LNCS, vol. 13664, pp. 1–18. Springer, Cham (2022). https://doi.org/10.1007/978-3-031-19772-7_1

26. Niu, Y., Tang, K., Zhang, H., Lu, Z., Hua, X.S., Wen, J.R.: Counterfactual VQA: a cause-effect look at language bias. In: Proceedings of the IEEE/CVF Conference on Computer Vision and Pattern Recognition, pp. 12700–12710 (2021)

27. Park, J., Lee, J., Sohn, K.: Bridge to answer: structure-aware graph interaction network for video question answering. In: Proceedings of the IEEE/CVF Conference on Computer Vision and Pattern Recognition, pp. 15526–15535 (2021)

28. Pearl, J., Glymour, M., Jewell, N.P.: Causal Inference in Statistics: A Primer. Wiley, Hoboken (2016)

29. Pennington, J., Socher, R., Manning, C.D.: Glove: global vectors for word representation. In: Proceedings of the 2014 Conference on Empirical Methods in Natural Language Processing (EMNLP), pp. 1532–1543 (2014)

30. Seo, A., Kang, G.C., Park, J., Zhang, B.T.: Attend what you need: motion-appearance synergistic networks for video question answering. In: Proceedings of the 59th Annual Meeting of the Association for Computational Linguistics and the 11th International Joint Conference on Natural Language Processing (Volume 1: Long Papers), pp. 6167–6177 (2021)

31. Sukhbaatar, S., Szlam, A., Weston, J., Fergus, R.: End-to-end memory networks. Adv. Neural. Inf. Process. Syst. **2015**, 2440–2448 (2015)

32. Tang, K., Niu, Y., Huang, J., Shi, J., Zhang, H.: Unbiased scene graph generation from biased training. In: Proceedings of the IEEE/CVF Conference on Computer Vision and Pattern Recognition, pp. 3716–3725 (2020)

33. Vaswani, A., et al.: Attention is all you need. In: Advances in Neural Information Processing Systems, pp. 5998–6008 (2017)

34. Wang, J., Bao, B., Xu, C.: DualVGR: a dual-visual graph reasoning unit for video question answering. IEEE Trans. Multimedia **24**, 3369–3380 (2021)

35. Wang, P., Vasconcelos, N.: Scout: self-aware discriminant counterfactual explanations. In: Proceedings of the IEEE/CVF Conference on Computer Vision and Pattern Recognition, pp. 8981–8990 (2020)

36. Wang, T., Zhou, C., Sun, Q., Zhang, H.: Causal attention for unbiased visual recognition. In: Proceedings of the IEEE/CVF International Conference on Computer Vision, pp. 3091–3100 (2021)

37. Wang, T., et al.: The devil is in classification: a simple framework for long-tail instance segmentation. In: Vedaldi, A., Bischof, H., Brox, T., Frahm, J.-M. (eds.) ECCV 2020. LNCS, vol. 12359, pp. 728–744. Springer, Cham (2020). https://doi.org/10.1007/978-3-030-58568-6_43

38. Xu, D., et al.: Video question answering via gradually refined attention over appearance and motion. In: Proceedings of the 25th ACM International Conference on Multimedia, pp. 1645–1653 (2017)

39. Xu, K., et al.: Show, attend and tell: neural image caption generation with visual attention. In: International Conference on Machine Learning, pp. 2048–2057 (2015)

40. Xu, L., Huang, H., Liu, J.: SUTD-TrafficQA: a question answering benchmark and an efficient network for video reasoning over traffic events. In: Proceedings of the IEEE/CVF Conference on Computer Vision and Pattern Recognition, pp. 9878–9888 (2021)

41. Yang, X., Zhang, H., Qi, G., Cai, J.: Causal attention for vision-language tasks. In: Proceedings of the IEEE/CVF Conference on Computer Vision and Pattern Recognition, pp. 9847–9857 (2021)

Multimodal Topic and Sentiment Recognition for Chinese Data Based on Pre-trained Encoders

Qian Chen, Siting Chen, Changli Wu, and Jun Peng[✉]

School of Informatics, Xiamen University, Xiamen 361005, Fujian, China
{chenq,chensiting,wuchangli,pengjun}@stu.xmu.edu.cn

Abstract. With the rapid development of mobile internet technology, massive amounts of multimodal data have emerged from major online platforms. However, defining multimodal topic categories is a more subjective task, and the lack of a public Chinese dataset hinders the progress of the task. And fine-grained multimodal sentiment classification is extremely challenging, especially for aspect-level classification. In this paper, we build a Chinese dataset based on Weibo for multimodal topic classification and propose a pre-trained encoder-based model for topic classification and fine-grained sentiment classification. The experimental results on Multi-ZOL and our proposed dataset show that the proposed model outperforms the benchmark model on the multi-modal topic classification task, achieving an accuracy of 0.99. On the fine-grained multimodal sentiment classification task, the proposed model achieves an accuracy of 0.5860 and a macro-F1 of 0.5791, which is competitive with SOTA.

Keywords: multimodal · topic classification · sentiment classification

1 Introduction

With the rapid development of mobile internet technology, social media platforms have become popular among the public due to their convenience, flexibility, and interactivity. As a result, the Internet has carried more and more information presented in different data formats, many of which appear in the form of image and text combinations or video and text combinations, giving birth to massive multimodal data. These multimodal data contain a large amount of valuable information about sudden events, hot topics, user emotions, user stances [14], *etc.*, attracting more and more attention from scholars and relevant industry practitioners.

In multimodal data on the Internet, text, images, videos, and voice data all contain corresponding emotional information, which seems different on the surface but complements each other. Traditional text or speech emotion recognition usually only uses a single modal information, which is not enough to recognize complex emotional information. Multimodal data can provide richer emotional

Q. Liu et al. (Eds.): PRCV 2023, LNCS 14431, pp. 323–334, 2024.
https://doi.org/10.1007/978-981-99-8540-1_26

information and can more clearly and accurately reflect user tendencies, preferences, and other emotions. The recognition and utilization of this emotional information are of great significance in public opinion monitoring [19], business decision-making [10], political elections [15], box office predictions [5], stock market predictions, and other areas.

In multimodal topic and sentiment classifications, there are two challenging problems. Firstly, how to extract valuable information from the images. Secondly, how to fuse image features with textual features for better representation. In this paper, we propose a multimodal topic and sentiment recognition model for Chinese data based on pre-trained encoders. It consists of a pre-trained XLNet [26] for encoding Chinese text and a pre-trained ResNet [22] for encoding images. Specifically, we use the output of the third-to-last layer of the pre-trained ResNet as image features. And the last 4 hidden-states of XLNET are concatenated as text features.

In addition, inspired by MCAN [29], we apply multiple self-attention and guided-attention layers to facilitate information interaction between multimodal features and enhance feature representation. Furthermore, we apply multimodal fusion based on multi-layer perceptron for attention reduction and classification.

Since there are no publicly available Chinese multimodal topic classification datasets on the Internet, which hinders the development of Chinese multimodal topic classification, we build a new dataset with 18167 pieces of Chinese multimodal data. This dataset will be released for future research on multimodal classification tasks in the Chinese language.

Our main contributions are as follows: 1) Propose a Chinese multimodal topic classification dataset based on Weibo data. 2) Propose a pre-trained encoder-based and cross-modal attention-based model for the multimodal topic and sentiment recognition. 3) Conduct experiments on the self-built Chinese topic classification dataset and Multi-ZOL dataset, achieving SOTA and near-SOTA performance.

2   Related Work

2.1   Topic Classification

There are very few articles that directly discuss topic classification tasks, and many scholars tend to focus on the detection and classification of event-related content [30] and the detection and tracking of popular topics [3]. We believe that such classification and topic classification are similar for any event or popular topic, it can be classified into a subjectively defined topic category.

The detection and classification of event-related content is initially proposed in the field of text retrieval. Early work in this field is completed by Agarwal and Rambow [1]. The authors detect noun entities and their relationships in text documents to infer events, such as interactive events or observed events.

Event classification from social media is a multimodal task that includes both text and images. One major challenge is the joint multimodal modeling of event categories. In early fusion, text and visual descriptors are concatenated with each other at the feature level [8]. In late fusion, separate models are generated for image and text information, and then combined at the decision level [18]. Wang et al. represent text and images with textual and visual word representations and fuse them by extracting joint latent topics using the multimodal extension of LDA [7,21]. [13] uses a novel annotation method,called Japanese Entity Labeling with Dynamic Annotation,to deepen the understanding of the effectiveness of images for multi-modal text classification.

2.2 Multimodal Sentiment Classification

In multimodal sentiment analysis, image-text multimodal sentiment classification is closely related to social media. Image-text multimodal can be further divided into coarse-grained and fine-grained sentiment classification based on granularity.

Yu et al. [28] innovatively used a pre-trained convolutional neural network to extract image features and a text convolutional neural network to extract text features, and fused these multimodal features by direct concatenation to train a logistic regression model. Xu et al. [23] extract corresponding scene and object features from images, and Xu et al. [24] propose a shared attention mechanism to model the relationship between images and text. Wang et al. [20] propose a multimodal fusion network based on graph neural networks, which is more intuitive than traditional methods in fusing corresponding information between images and text.

Other work mainly focuses on fine-grained image-text sentiment classification. Xu et al. [25] propose an aspect term-based interactive memory multimodal sentiment classification network. [12] provides a multimodal sentiment classification approach using visual distillation and attention network. ITMSC [2] can automatically adjust the contribution of images in the fusion representation through the exploitation of semantic descriptions of images and text similarity relations. [9] proposes a novel multimodal sentiment classification model based on gated attention mechanism. KAMT [31] designs an external knowledge enhanced multi-task representation learning network. Yu et al. [27] propose an aspect term-based multimodal ABSA sentiment classification model.

3   Method

Figure 1 shows the overall framework of the proposed multimodal topic and sentiment recognition model for Chinese data. As can be seen, it consists of a pre-trained XLNet for encoding Chinese text and a pre-trained ResNet for encoding images. In addition, multiple self-attention and guided-attention layers are applied to facilitate information interaction between multimodal features and enhance feature representation. And multimodal fusion based on MLP for attention reduction and classification (Fig. 2).

3.1 Encoders

Different from the original network, which uses Faster R-CNN pre-trained on COCO [17] to extract image features, we use the output of the third-to-last layer of the pre-trained ResNet101 [22] as image features.

"Jieba" and LSTM [11] are used to extract text features. Specifically, after tokenizing, each text is converted to a $n \times 300$ matrix through an embedding layer, where n is the length of the text. Finally, the word embeddings are passed through a single-layer LSTM network, and the final hidden-state is used as the text feature. In addition, we also use XLNet pre-trained on Chinese data as a text encoder. The last 4 hidden-states are concatenated as text features.

3.2 Attention Layer

As mentioned above, multiple Self-Attention (SA) and Guided-Attention (GA) layers are applied to enhance feature representations. SA takes single-modal feature X and outputs attended feature Z, while GA takes multi-modal features X and Y, and outputs attended feature Z for X guided by Y. Specifically, the single-modal feature X is transformed into query feature Q, key feature K, and value feature V, all with a dimension of d, and the SA feature f is obtained according to the following formula.

$$f = A(Q, K, V) = \text{softmax}\left(\frac{QK}{\sqrt{d}}\right) V \qquad (1)$$

To further improve the representation power of SA features, the Multi-head SA (MSA) is introduced. It consists of a multi-head attention layer and a point-wise feed-forward layer. Each head in MSA is SA and parallel to each other. When given a set of input features $X = [x_1; ...; x_m] \in R^{m \times d_x}$, the multi-head attention learns the pairwise relationships between pairs of instances $\langle x_i, x_j \rangle \in X$. It outputs attention-weighted features $Z \in R^{m \times d}$ by taking a weighted sum over all instances in X. The feed-forward layer further transforms them through two fully connected layers, which include a ReLU activation layer and a dropout layer. Additionally, the outputs of these two layers are optimized through residual and layer normalization [6]. Specifically, the MSA features F is obtained by calculating the SA features of each head:

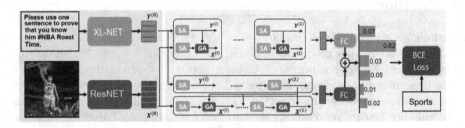

Fig. 1. The overall framework of the network.

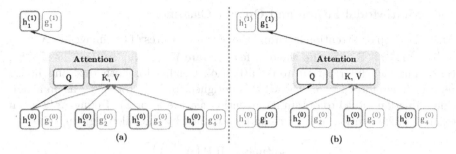

Fig. 2. Self-Attention (SA) unit and Guided-Attention (GA) unit.

$$F = MSA(Q, K, V) = [\text{head}_1, \text{head}_2, \ldots, \text{head}_h] W^O \tag{2}$$

$$\text{head}_j = SA\left(QW_j^Q, KW_j^K, VW_j^V\right) \tag{3}$$

Here, W_j^Q, W_j^K, W_j^V, W_j^O are projection weight matrices. Then, the original features X and F are transformed using fully connected layers to obtain MSA features F^α and F_Z^α through the following equation F^α.

$$F^\alpha = \text{Layer Norm}(X + F) \tag{4}$$

$$F_Z^\alpha = FC(\sigma(FC(F^\alpha))) \tag{5}$$

σ is the activation function, e.g., ReLU. Then, F_Z^α and F^α are transformed to final MSA output Z using layer normalization and a fully connected layer, as shown in Eq. 6.

$$Z = FC(\text{Layer Norm}((F^\alpha + F_Z^\alpha))) \tag{6}$$

GA takes multi-modal input features $X \in R^{m \times d_x}$ and $Y = [y_1; \ldots; y_n] \in R^{n \times d_y}$, where Y guides the attention learning of X. Note that X and Y can be used to represent features of different modalities. And GA models the pairwise relationships between each pair of instances $\langle x_i, y_j \rangle$ from X and Y, respectively. The implementation is similar to the SA.

3.3 Deep Cross-Attention Layer

Given the input image features X and text features Y mentioned above, deep Multi-modal Cross-Attention (MCA) learning can be achieved to process the input features. The deep cross-attention layer consists of L stacked MCA layers, which are denoted as $MCA^{(1)}, MCA^{(2)}, \ldots, MCA^{(L)}$. The two input features of $MCA^{(l)}$ are represented as $X^{(l-1)}$ and $Y^{(l-1)}$, and their output features are represented as $X^{(l)}$ and $Y^{(l)}$, which will be further fed into $MCA^{(l+1)}$ in a cyclic manner. For the first MCA layer, let its input features be $X^0 = X$ and $Y^0 = Y$.

$$\left[X^{(l)}, Y^{(l)}\right] = MCA^{(l)}\left(\left[X^{(l-1)}, Y^{(l-1)}\right]\right) \tag{7}$$

3.4   MultiModal Fusion and Output Classifier

After the deep co-attention learning, the output features of the image are $X^{(L)} = [x_1^{(L)}; ...; x_m^{(L)}] \in \mathbb{R}^{m \times d_x}$ and the text features are $Y^{(L)} = [y_1^{(L)}; ...; y_n^{(L)}] \in \mathbb{R}^{n \times d_y}$, containing rich information about attention weights for both text and image regions. Therefore, a two-layer MLP is designed for $Y^{(L)}$ (or $X^{(L)}$) as an attention reduction model to obtain its attention feature $\tilde{y}$ (or $\tilde{x}$). Taking $X^{(L)}$ as an example, its attention feature $\tilde{x}$ can be obtained by the following equations:

$$\alpha = \text{softmax}\left(\text{MLP}\left(X^{(L)}\right)\right)$$
$$\tilde{x} = \sum_{i=1}^{m} \alpha_i x_i^{(L)} \tag{8}$$

Here, $\alpha = [\alpha_1, \alpha_2, \ldots, \alpha_m] \in \mathbb{R}^m$ is the learned attention weights. Similarly, we can obtain the attention feature $\tilde{y}$ of $Y^{(L)}$. After obtaining $\tilde{y}$ and $\tilde{x}$, multimodal feature fusion can be performed using the following equation:

$$Z = \text{Layer Norm}\left(W_x^T \tilde{x} + W_y^T \tilde{y}\right) \tag{9}$$

Here, W_x and $W_y \in R^{d \times d_z}$ are two linear projection matrices, d_z is the shared dimension of the fused features. LayerNorm is used to stabilize training [66]. Finally, the fused feature vector z is subject to a non-linear transformation to obtain a vector $s \in R^N$, and a classifier is trained using a cross-entropy (CE) loss function to classify s.

4   Experiments

4.1   Datasets

We evaluate our approach in two datasets, including Multi-ZOL and our own-constructed Weibo topic classification dataset.

Multi-ZOL is collected from ZOL.com, which is the leading IT information and business web portal in China. It contains 12,587 comments on 114 brands and 1,318 types of mobile phones, among which 7,359 are single-modal comments and 5,288 are multimodal comments. Each sample contains a text, a set of images, and one to six aspects, i.e., cost performance, performance configuration, battery life, appearance and feel, shooting effect, and screen. Each aspect has an integer sentiment score ranging from 1 to 10, and there are a total of 28,469 aspects.

Weibo topic classification dataset is collected by us given that there is no publicly available Chinese dataset for multimodal topic classification. All data are from Weibo, a leading social media website in China. We crawle original Weibo posts within the past year that contain both text and images, based on keywords. The keywords include 31 hot topics, such as "Jay Chou", "Russia-Ukraine Conflict", "National Defense", etc. We manually classified these keywords into six categories, i.e., entertainment, games, military, politics, economy, and sports. The dataset includes 18,167 multimodal Weibo posts that contain both text and images. An example of data content is shown in Figure 3 The specific statistical information is shown in Table 1. Considering the quantity and quality of each keyword obtained through crawling, we divide the above-mentioned keywords according to Table 2.

Categories	Image	Text
Entertainment		Jay Chou's "Peninsula Iron Box" K Peninsula Iron Box @QQ music
Politics		12/3 European perspective is different from the United States, Europe is close to Ukraine and Russia, and Europe is all over the world
Sports		Please use one sentence to prove that you know him #NBA RoastTime.

Fig. 3. Example of Weibo topic classification dataset.

By analyzing the crawled raw data, this study found that Weibo data is complex and diverse. Due to the keyword-based crawling method, it is difficult to quickly clean the crawled raw data. The challenges are including:

1) Many data entries are not directly related to their assigned categories. 2) Weibo contains many text-based images, such as webpage screenshots and memo screenshots. Many images in Weibo have no actual content, such as solid color images or low-quality/damaged images. 3) There is a significant amount of content duplication and hot topic exploitation on Weibo, including low-quality content posted by marketing accounts and fake accounts.

To address the challenges mentioned above, and leveraging existing technologies and knowledge, we design and execute the following data preprocessing steps, resulting in a relatively clean Weibo topic classification dataset.

First, We write a Python script to automatically recognize text in images and remove images which are heavily covered by text. Specifically, we remove images with more than 10 text elements or text area coverage exceeding 10%. Second, removal of duplicate and similar data. Third, manual remove unrelated

Table 1. Dataset statistics of Weibo topic categories.

Attribute	Statistics
Weibo data	18167
Numer of categories	6
Average text length	103.86
Maximum text length	2994
Minimum text length	2

Table 2. Keywords and statistics for each category in self-constructed Weibo dataset.

Category	Count	Keywords
Economics	2866	GDP, stock market, exchange, fund, bitcoin, finance, Europe & energy, stocks, investment
Politics	2364	Ministry of Foreign Affairs, China-US, US sanctions, US systems
Military	2833	Russia-Ukraine conflict, bombing, air strike, national defense, military, Ukraine, weapons
Entertainment	6198	Dilraba, Xu Song, Yang Mi, Zhang Jie, Jay Chou
Sports	1370	NBA, Chinese national football team, table tennis, Tokyo Olympics
Games	2536	Honor of Kings, Genshin Impact

and content-less images. The statistics of the category quantities are shown in Table 2.

4.2   Implementation Details

Baseline Model. We compare our model with the research of Kumar [16]. The size of the input image is set to 224×224 and the dimension of extracted features is 2048. The maximum text length was set to 200 for the Weibo dataset and 320 for the Multi-ZOL dataset with the Aspec Term inserted at the beginning of the text. When extracting text features with LSTM, the vocabulary size is 65425, and the dimensions of word embedding and sentence features are both 100. The batch-size during training is 64, and the learning rate is 1e-4. For the pre-trained XLNet, the output text feature dimension is 768, batch-size during training is 32, the learning rate is 1e-5. The proportion of the training set, validation set, and test set for the Weibo dataset is 6:2:2, and 8:1:1 for Multi-ZOL.

Modified MCAN. We integrate the MCAN into the framework of the baseline model and adjust the training strategy accordingly. We use an image size of 224×224, the image feature dimension, maximum text length, text feature dimension, and multimodal fusion feature are set to 2048, 200, 512, and 1024, respectively. The potential dimension of the multi-head attention, number of heads, and potential dimension of each head is the same as Vaswani [4]. During training, we use the Adam optimizer with $\beta_1 = 0.9$ and $\beta_2 = 0.98$. The base learning rate is $min(2.5te^{-5}, 1e^{-4})$, where t is the current epoch size starting from 1. After 5 epochs of training, the learning rate begins to decay at a rate of multiplying by 1/5 every 2 epochs. All models are trained for 20 epochs with the same batch size. Experiments are done on four 2080TI GPUs

5   Results and Analysis on Weibo Dataset

In the topic classification experiment on Weibo topic classification dataset, we used the accuracy of the validation set and the test set as evaluation metrics. In addition to the baseline model and MCAN model, we also conducted topic classification experiments using only image or text data from the dataset. Table 3 shows the experimental results of this task.

Table 3. Experimental results of topic classification.

Multimodal FusionMethod	Feature Extraction Model	Validation Set Accuracy	Test Set Accuracy
Image Only	ResNet101	0.8808	0.8773
Text Only	LSTM	0.9083	0.8891
	XLNet	0.9821	0.9786
Baseline Model	LSTM+ResNet101	0.9293	0.9179
	XLNet+ResNet101	0.9912	0.9887
Collaborative Attention Network	LSTM+ResNet101	0.9482	0.9351
	XLNet+ResNet101	**0.9921**	**0.9909**

As Table 3 shows, The ResNet101-based model using only image data performed poorly because the semantic information conveyed by images in social media data is often much less than that conveyed by the text. In addition, in the absence of guidance from text, the entity information contained in images is often more complex, and difficult to obtain useful information for topic classification. Therefore, the performance of the LSTM-based model using only text data outperforms the ResNet101-based model using only image data.

However, due to the lack of attention mechanism and the failure to address performance degradation caused by long-term dependencies, the performance of the LSTM model was significantly lower than that of the XLNet model.

After incorporating both image and text modalities, the performance of the baseline model was better than that of the corresponding single-modality model. However, it can be seen that the combination of LSTM and ResNet101 still could not surpass the single-modality XLNet model, which reflects the huge advantage of language models such as XLNet in topic classification tasks.

The collaborative attention network achieved better performance than the baseline model by communicating semantic information between modalities. Its fusion is more detailed and robust compared to the relatively separate fusion of the baseline model. While due to the relative simplicity of the topic classification task itself, the advantage of the collaborative attention network was not significantly reflected.

6   Results and Analysis on Multi-ZOL Dataset

In the sentiment classification experiment, we use the accuracy of the test set and Macro-F1 as evaluation metrics. In addition to the baseline model and modified, we also conduct sentiment classification experiments using only text from the dataset. Table 4 shows the experimental results of this task.

Table 4. Experimental Results of Sentiment Classification.

Multimodal Fusion Method	Feature Extraction Model	Accuracy	Macro-F1
Text Only	LSTM	0.5592	0.5464
	XLNet	0.5643	0.5569
Baseline Model	LSTM+ResNet101	0.5678	0.5574
	XLNet+ResNet101	0.5761	0.5694
Collaborative Attention Network	LSTM+ResNet101	0.5801	0.5726
	XLNet+ResNet101	**0.5860**	**0.5797**
MIMN(SOTA) [25]	Bi-LSTM+CNN	0.6159	0.6051

Considering the need to incorporate Aspect Term information, the single-modality model using only images is not suitable for this task. Therefore, we conduct experiments using only text data with the LSTM model, but found that its performance was poor due to the lack of attention mechanism and failure to address performance degradation caused by long-term dependencies.

After incorporating both image and text modalities, the performance of the baseline model is better than that of the single-modality model. It can be observed that the combination of LSTM and ResNet101 slightly outperforms the single-modality XLNet model. This may be because in multimodal data of review types, images often convey distinct user emotions, which can complement the emotional orientation of the text.

The collaborative attention network achieves better performance than the baseline model by communicating semantic information between modalities. Its fusion is more detailed and robust compared to the relatively separate fusion of the baseline model. However, there is still a considerable gap compared to the MIMN model proposed by Xu, et al. [25] and others for this task. This is likely because the multimodal fusion model we use did not treat Aspect Term information and text context information equally. Aspect term is the lack of enough attention throughout the entire multimodal fusion process when we just simply concatenate them and input them directly into the network.

7   Conclusion

Multimodal learning involves multiple modalities, such as images and text. Compared with traditional single-modal learning, multimodal learning can obtain more in-depth information and interaction information between modalities, providing a powerful tool for analyzing complex and diverse Internet data. In this paper, we build a multimodal topic classification dataset based on Weibo data and transfer the MCAN to multimodal topic classification and sentiment classification. We achieve an accuracy of 0.99 for multi-modal topic classification. And for sentiment classification, we achieve an accuracy of 0.5860 and a macro-F1 of 0.5791, which is competitive with SOTA.

References

1. Agarwal, A., Rambow, O.: Automatic detection and classification of social events. In: Proceedings of the 2010 Conference on Empirical Methods in Natural Language Processing, pp. 1024–1034 (2010)
2. An, J., Zainon, N.W., Mohd, W., Hao, Z.: Improving targeted multimodal sentiment classification with semantic description of images. Comput. Mater. Continua **75**(3) (2023)
3. Asgari-Chenaghlu, M., Feizi-Derakhshi, M.R., Balafar, M.A., Motamed, C., et al.: Topicbert: a transformer transfer learning based memory-graph approach for multimodal streaming social media topic detection. arXiv preprint arXiv:2008.06877 (2020)
4. Vaswani, A.: Attention is all you need. In: Advances in Neural Information Processing Systems, vol. 30:i (2017)
5. Asur, S., Huberman, B.A.: Predicting the future with social media. In: 2010 IEEE/WIC/ACM International Conference on Web Intelligence and Intelligent Agent Technology, vol. 1, pp. 492–499. IEEE (2010)
6. Ba, J.L., Kiros, J.R., Hinton, G.E.: Layer normalization. arxiv e-prints. arXiv preprint arXiv:1607.06450 (2016)
7. Barnard, K., Duygulu, P., Forsyth, D., De Freitas, N., Blei, D.M., Jordan, M.I.: Matching words and pictures. J. Mach. Learn. Res. **3**, 1107–1135 (2003)
8. Brenner, M, Izquierdo, E.: Multimodal detection, retrieval and classification of social events in web photo collections. In: ICMR 2014 Workshop on Social Events in Web Multimedia (SEWM), pp. 5–10 (2014)
9. Yongping, D., Liu, Y., Peng, Z., Jin, X.: Gated attention fusion network for multimodal sentiment classification. Knowl.-Based Syst. **240**, 108107 (2022)
10. He, W., Wang, F.-K., Akula, V.: Managing extracted knowledge from big social media data for business decision making. J. Knowl. Manag. **21**(2), 275–294 (2017)
11. Hochreiter, S., Schmidhuber, J.: Long short-term memory. Neural Comput. **9**(8), 1735–1780 (1997)
12. Hou, S., Tuerhong, G., Wushouer, M.: Visdanet: visual distillation and attention network for multimodal sentiment classification. Sensors **23**(2), 661 (2023)
13. Ma, C., Shen, A., Yoshikawa, H., Iwakura, T., Beck, D., Baldwin, T.: On the effectiveness of images in multi-modal text classification: an annotation study. ACM Trans. Asian Low-Resour. Lang. Inf. Process. **22**(3), 1–19 (2023)
14. Mansour, S.: Social media analysis of user's responses to terrorism using sentiment analysis and text mining. Procedia Comput. Sci. **140**, 95–103 (2018)

15. O'Connor, B., Balasubramanyan, R., Routledge, B., Smith, N.: From tweets to polls: linking text sentiment to public opinion time series. In: Proceedings of the International AAAI Conference on Web and Social Media, vol. 4, pp. 122–129 (2010)

16. Kumar, R.: Multimodal-tweets-classification using CrisisMMD dataset (2020). https://github.com/ravindrakumar-iitkgp/MultimodalTweetsClassification

17. Ren, S., He, K., Girshick, R., Sun, J.: Faster R-CNN: towards real-time object detection with region proposal networks. In: Advances in Neural Information Processing Systems, vol. 28 (2015)

18. Sutanto, T., Nayak, R.: ADMRG@ mediaeval 2013 social event detection. In: MediaEval (2013)

19. Tong, Y., Sun, W.: Multimedia network public opinion supervision prediction algorithm based on big data. Complexity **2020**, 1–11 (2020)

20. Wang, K., Shen, W., Yang, Y., Quan, X., Wang, R.: Relational graph attention network for aspect-based sentiment analysis. arXiv preprint arXiv:2004.12362 (2020)

21. Wang, Z., Cui, P., Xie, L., Zhu, W., Rui, Y., Yang, S.: Bilateral correspondence model for words-and-pictures association in multimedia-rich microblogs. ACM Trans. Multimedia Comput. Commun. Appl. (TOMM) **10**(4), 1–21 (2014)

22. Zifeng, W., Shen, C., Van Den Hengel, A.: Wider or deeper: revisiting the resnet model for visual recognition. Pattern Recogn. **90**, 119–133 (2019)

23. Xu, N., Mao, W.: Multisentinet: a deep semantic network for multimodal sentiment analysis. In: Proceedings of the 2017 ACM on Conference on Information and Knowledge Management, pp. 2399–2402 (2017)

24. Xu, N., Mao, W., Chen, G.: A co-memory network for multimodal sentiment analysis. In: The 41st International ACM SIGIR Conference on Research & Development in Information Retrieval, pp. 929–932 (2018)

25. Nan, X., Mao, W., Chen, G.: Multi-interactive memory network for aspect based multimodal sentiment analysis. In: Proceedings of the AAAI Conference on Artificial Intelligence, vol. 33, pp. 371–378 (2019)

26. Yang, Z., et al.: XLNet: generalized autoregressive pretraining for language understanding. In: Advances in Neural Information Processing Systems, vol. 32 (2019)

27. Yu, J., Jiang, J.: Adapting BERT for target-oriented multimodal sentiment classification. In: IJCAI (2019)

28. Yuhai, Yu., Lin, H., Meng, J., Zhao, Z.: Visual and textual sentiment analysis of a microblog using deep convolutional neural networks. Algorithms **9**(2), 41 (2016)

29. Yu, Z., Yu, J., Cui, Y., Tao, D., Tian, Q.: Deep modular co-attention networks for visual question answering. In: Proceedings of the IEEE/CVF Conference on Computer Vision and Pattern Recognition, pp. 6281–6290 (2019)

30. Zeppelzauer, M., Schopfhauser, D.: Multimodal classification of events in social media. Image Vis. Comput. **53**, 45–56 (2016)

31. Zhang, Y., Tiwari, P., Rong, L., Chen, R., AlNajem, N.A., Hossain, M.S.: Affective interaction: attentive representation learning for multi-modal sentiment classification. ACM Trans. Multimedia Comput. Commun. Appl. **18**(3s), 1–23 (2022)

Multi-feature Fusion-Based Central Similarity Deep Supervised Hashing

Chao He[1,2,3], Hongxi Wei[1,2,3]([✉]), and Kai Lu[1,2,3]

[1] School of Computer Science, Inner Mongolia University, Hohhot, China
cswhx@imu.edu.cn
[2] Provincial Key Laboratory of Mongolian Information Processing Technology,
Hohhot, China
[3] National and Local Joint Engineering Research Center of Mongolian Information
Processing Technology, Hohhot, China

Abstract. The deep image hashing aims to map the input image into simply binary hash codes via deep neural networks. Nevertheless, previous deep supervised hashing methods merely focus on the high-level features of the image and neglect the low-level features of the image. Low-level features usually contain more detailed information. Therefore, we propose a multi-feature fusion-based central similarity deep supervised hashing method. Specifically, a cross-layer fusion module is designed to effectively fuse image features of high and low levels. On top of that, a channel attention module is introduced to filter out the useless information in the fused features. We perform comprehensive experiments on three widely-studied datasets: NUS-WIDE, MS-COCO and ImageNet. Experimental results indicate that our proposed method has superior performance compared to state-of-the-art deep supervised hashing methods.

Keywords: Fusion · Deep supervised hashing · Central similarity

1 Introduction

The goal of image retrieval is to search the most similar images to the query image in the gallery, which ideally have the same categorical attributes as the query image [1]. The common approach for solving this task is to encode the images into feature vectors with definite length and perform nearest neighbor search in the corresponding vector space. Nevertheless, when dealing with large-scale image data, those methods require extremely large memory load and computational cost. Hashing is one of the effective methods to tackle the above problems, with extremely fast retrieval speed and low memory usage [2,3]. It not only represents the images as compact binary codes, but also preserves the similarity between images [4].

In recent years, with the development of deep learning technology, many hashing methods based on deep learning have been proposed [5–10], which can well represent the nonlinear hash function for generating the hash code of the input data. On these basis, some methods have contributed improvements in other aspects. Deep Learning to Hash by Continuation (HashNet) [11] used to perform deep learning hashing by continuous method with convergence guarantees, which addressed the ill-posed gradient and data imbalance problems in the end-to-end framework of deep feature learning and binary hash encoding. On top of HashNet, Deep Cauchy Hashing for Hamming Space Retrieval (DCH) [12] proposed a pairwise cross-entropy loss and a quantization loss based on the Cauchy distribution, which penalized significantly on similar image pairs with Hamming distance larger than a given Hamming radius threshold and learned nearly lossless hash codes. Maximum-Margin Hamming Hashing (MMHH) [13] introduced a max-margin t-distribution loss, where the t-distribution concentrates more similar data points to be within the Hamming ball, and the margin characterizes the Hamming radius such that less penalization is applied to similar data points within the Hamming ball. The method enabled constant-time search through hash lookups in Hamming space retrieval. Subsequently, Deep Polarized Network for Supervised Learning of Accurate Binary Hashing Codes (DPN) [14] proposed a polarization loss which is the only differentiable loss term that simultaneously minimizes the inter-class and maximizes the intra-class Hamming distance.

Nonetheless, all the above methods merely capture pairwise or triplet data similarity from a local perspective to learn the hash function, which can hardly tackle the problems of imbalanced data and insufficient data distribution coverage. Therefore, Central Similarity Quantization for Efficient Image and Video Retrieval (CSQ) [15] proposed a new global similarity metric, termed as central similarity, which encourages hash codes of similar data pairs to approach a common center (i.e., hash center), and hash codes of dissimilar data pairs to converge to different centers. It provided two methods for generating hash centers: one is to directly construct hash centers with maximum mutual Hamming distance by utilizing the properties of Hadamard matrix, and the other is to generate hash centers by randomly sampling from Bernoulli distributions. Both approaches can generate hash centers that have sufficient Hamming distance from each other. Over the basis of CSQ, Deep Hashing With Hash Center Update For Efficient Image Retrieval (DCSH) [16] combined hash center update [17] and classification loss to further improve the performance on multi-label datasets. Yet most of previous approaches [5,7,8,10,15,16] only utilize the high-level features of image and neglect the low-level features of image. Therefore, this paper proposes a Multi-Feature Fusion-based central similarity Deep Supervised Hashing method (MFFDSH). Our contributions are mainly summarized as follows:

1) We design a Cross-layer Fusion Module (CFM), which introduces Receptive Field Blocks (RFB) [18] to effectively extract image features from different levels and fuse these features to obtain more enriched image feature information.

2) Channel Attention Module (CAM) [19] is introduced to focus on the more important information in the fused image features, which can eliminate the useless information effectively.

3) We conduct experiments on three widely used public datasets: NUS-WIDE [20], MS-COCO [21] and ImageNet [22]. The experimental results demonstrate that MFFDSH is superior to existing deep supervised hashing methods.

2 Related Work

2.1 Symmetric Deep Supervised Hashing in Image Retrieval

Due to the rapid development of deep learning, deep supervised hashing methods have achieved superior performance on image retrieval tasks. The majority of these methods are symmetric deep supervised hashing methods such as CNNH [10], DSH [8], DNNH [5], HashNet [11], DCH [12], etc., which exploit the similarity or dissimilarity between pairs of images to learn one deep hash function for query images and database (retrieval) images. Symmetric deep supervised hashing methods utilize only fewer training images and supervised information to obtain superior performance, which have been preferred by many researchers.

2.2 Asymmetric Deep Supervised Hashing in Image Retrieval

Symmetric deep supervised hashing methods are broadly used in image retrieval tasks and have achieved well performance. Nevertheless, the training of these hashing methods is typically time-consuming, which makes them hard to effectively utilize the supervised information for cases with large-scale database. Asymmetric Deep Supervised Hashing (ADSH) [23] is the first asymmetric strategy method for large-scale nearest neighbor search that treats query images and database (retrieval) images in an asymmetric way. More specifically, ADSH learns a deep hash function only for query points, while the hash codes for database points are directly learned. On this basis, many asymmetric hashing methods have been proposed such as ADMH [24], JLDSH [25], DHoASH [26], JLMT [27]. Each of these methods imposes some extent of improvement on the network architecture or loss function. Compared with traditional symmetric deep supervised hashing methods, asymmetric deep supervised hashing method can utilize the supervised information of database (retrieval) images to obtain superior performance which has better training efficiency. However, it is impractical for tens of millions of images to all have label information.

2.3 Central Similarity Deep Supervised Hashing in Image Retrieval

All of the above deep supervised hashing methods learn hash functions by capturing pairwise or triplet data similarities from a local perspective, which may harm the discriminability of the generated hash codes. CSQ [15] proposes a new

338   C. He et al.

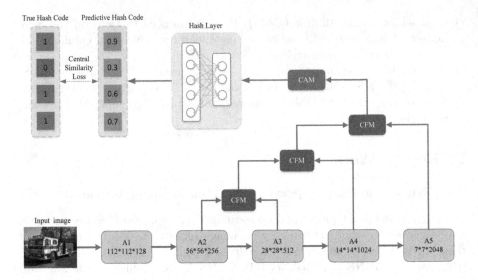

Fig. 1. The overall architecture of MFFDSH.

global similarity metric, called central similarity. Here, hash values of similar data points are encouraged to approach a common center, whereas pairs of dissimilar hash codes converge to distinct centers. DSCH [16] exploits the supervised information of the training images to further improve the performance. The central similarity deep supervised hashing methods can capture the global data distribution, which tackle the problem of data imbalance and generate high-quality hash functions efficiently. Nevertheless, these methods ignore the low-level features of the image, including edges, textures, etc.

3   MFFDSH

3.1   Model Architecture

Figure 1 demonstrates the overall architecture of MFFDSH. In the first place, MFFDSH adopts ResNet50 [28] to extract five groups of features at different levels, denoted as A_i $(i = 1, 2, 3, 4, 5)$. In the next step, the Cross-layer Fusion Module (CFM) is utilized to extend the receptive field and fuse the features at different levels so that more abundant image feature information can be obtained. Immediately after that, the Channel Attention Module (CAM) is exploited to focus on the more important fused features, and finally the hash representation of the image is obtained after the hash layer.

3.2   Cross-Layer Fusion Module (CFM)

As the high level of the network usually represents more semantic features, the spatial location characteristics are lost after multiple pooling operations. However, the low level of the network corresponds more to information such as details

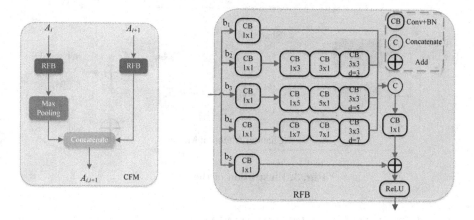

Fig. 2. Illustration of the CFM and RFB.

and spatial locations. Thus it is beneficial to fuse low-level features with high-level features [29]. In accordance with this observation, we utilize the next four levels (A2, A3, A4, A5) for feature fusion, where the former two levels have high-resolution features that contain abundant detail information and correspond to the low level of the network, while the latter two levels contain enriched semantic information and correspond to the high level of the network.

CFM can fuse different levels of image features effectively and primarily includes RFB module, max pooling and concatenation operation, whose structure is shown on the left side of Fig. 2. Among them, the RFB module mainly contains five branches, denoted as b_k ($k = 1, 2, 3, 4, 5$). In each branch, the number of channels is first reduced to 128 with a 1×1 convolutional layer. Next, there are three additional convolutional layers in branches b_2, b_3, and b_4, which are $1 \times (2k - 1)$ convolutional layer, $(2k - 1) \times 1$ convolution layer and 3×3 expansion convolution layer with expansion rate $(2k - 1)$. The RFB concatenates the outputs of the former four branches and then reduces the number of channels to 128 with 1×1 convolution operation. eventually, the dimensionality-reduced features are summed element by element with the output of the last branch and the entire module is fed to the ReLU activation function to obtain the final features. The specific structure of the RFB is illustrated on the right side of Fig. 2. After obtaining the features from the RFB output, we use the way of concatenating two adjacent levels to gain the final fused features. The specific fusion process of CFM can be described as follows:

$$A_{i,i+1} = Concat\left(MaxPool(\text{RFB}(A_i)), \text{RFB}(A_{i+1})\right), i \in 2, 3, 4 \quad (1)$$

$$A_{i+1} = A_{i,i+1}, i \in 2, 3, 4 \quad (2)$$

where $A_{i,i+1}$ represents the fused features.

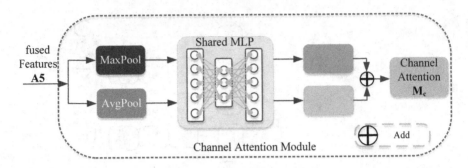

Fig. 3. Illustration of the CAM.

3.3   Channel Attention Module (CAM)

CAM is predominantly utilized to filter useless information and retain more important image features [19], whose structure is demonstrated in Fig. 3. The spatial information of the fused feature map is first aggregated by average-pooling and max-pooling operations to generate two different spatial context descriptors. Secondly, both descriptors are fed into a shared network which is composed of multi-layer perceptrons (MLP) and a hidden layer. After the shared network is applied to each descriptor, we merge the output feature vectors using element-wise summation. In short, the channel attention is computed as:

$$M_c(A_5) = \sigma\left(MLP\left(AvgPool\left(A_5\right)\right) + MLP\left(MaxPool\left(A_5\right)\right)\right) \tag{3}$$

where A_5 represents the final fused features and σ denotes the sigmoid function.

3.4   Loss Function

Given the generated centers $C = \{c_1, \ldots, c_q\}$ for training data X with q categories, we obtain the semantic hash centers $C' = \left\{c_1', c_2', \ldots, c_N'\right\}$ for single or multi-label data, where c_i' denotes the hash center of the data sample x_i. We refer to [15] for generating hash centers and adopt its loss function:

$$L_C = \frac{1}{K} \sum_i^N \sum_{k \in K} \left[c_{i,k}' \log h_{i,k} + \left(1 - c_{i,k}'\right) \log\left(1 - h_{i,k}\right)\right] \tag{4}$$

where h, N, K denote the hash code, the number of training sets, and the length of the hash code, respectively. Existing optimization methods cannot guarantee that the generated hash codes completely converge to the hash center, thus quantization loss L_Q is introduced to refine the generated hash codes h_i.

$$L_Q = \sum_i^N \sum_{k=1}^K \left(\log \cosh\left(|2h_{i,k} - 1| - 1\right)\right). \tag{5}$$

Finally, we have central similarity optimization problem:

$$\min_{\Theta} L_T = L_C + \lambda_1 L_Q \tag{6}$$

where θ is the set of all parameters for deep hash function learning, and λ_1 is the hyper-parameter.

4   Experiments

4.1   Datasets and Evaluation Metric

Datasets. We use three common datasets, which are ImageNet [22], NUS-WIDE [20] and MS-COCO [21]. On ImageNet, we follow the experimental setup in [11] and randomly select 100 categories. All the training images of these 100 categories are regarded as the retrieval set and the test images are treated as the query set. Finally, 130 images from each category were randomly selected as the training set. As ImageNet is a single-label dataset, we directly generate one hash center for each category. Both NUS-WIDE and MS-COCO are multi-label datasets with 21 and 80 categories respectively. Thus we generated the hash centers according to the method in [15].

Evaluation Metric. We adopt the Mean Average Precision (MAP) of different bits 16, 32, 64 to evaluate the quality of the retrieval images. Specifically, the MAP results are calculated based on the top 5000 returned samples from the NUS-WIDE and MS-COCO datasets, 1000 returned samples from the ImageNet dataset, respectively.

4.2   Implementation Details

We select the recent deep supervised hashing methods for comparison, which are CNNH [10], DNNH [5], DHN [9], HashNet [11], DCH [12], MMHH [13], CSQ [15], DCSH [16]. The batch size is set to 64, the learning rate is $1e^{-5}$, and the total number of training epochs is 60. To speed up the training, the Adam [30] optimizer is utilized. Similar to the work in [15], λ_1 is set to 0.25.

4.3   Experimental Results and Analysis

Table 1 demonstrates the experimental results on three benchmark datasets, where the MAP values of the top five methods are derived from [15]. For fair comparison, asymmetric deep supervised hashing methods are not listed here for the reason that none of the methods in the table utilize supervised information (label information) from the database (retrieval) images. From Table 1, we can observe that our MFFDSH achieves the best performance on both 32 bits and 64 bits hash codes, yet the performance on 16 bits hash codes is inferior to CSQ and DCSH, which may be owing to the limited representation capability of 16 bits hash codes. This indicates that the proposed MFFDSH model can extract more abundant image features and also illustrate that detailed information of images is beneficial.

Table 1. The corresponding results (MAP) on the three benchmark datasets.

Method	NUS-WIDE			MS-COCO			ImageNet		
	16 bits	32 bits	64 bits	16 bits	32 bits	64 bits	16 bits	32 bits	64 bits
CNNH [10]	0.655	0.659	0.647	0.599	0.617	0.620	0.315	0.473	0.596
DNNH [5]	0.703	0.738	0.754	0.644	0.651	0.647	0.353	0.522	0.610
DHN [9]	0.712	0.759	0.771	0.719	0.731	0.745	0.367	0.522	0.627
HashNet [11]	0.757	0.775	0.790	0.745	0.773	0.788	0.622	0.701	0.739
DCH [12]	0.773	0.795	0.818	0.759	0.801	0.825	0.652	0.737	0.758
MMHH [13]	-	-	-	0.735	0.783	0.807	-	-	-
CSQ [15]	**0.810**	0.825	0.839	0.796	0.838	0.861	**0.851**	0.865	0.873
DCSH [16]	-	0.841	-	**0.805**	0.847	0.861	-	-	-
MFFDSH (Ours)	0.808	**0.842**	**0.848**	0.793	**0.857**	**0.877**	0.849	**0.874**	**0.880**

Table 2. The comparative results of different configurations.

Method	NUS-WIDE			MS-COCO			ImageNet		
	16 bits	32 bits	64 bits	16 bits	32 bits	64 bits	16 bits	32 bits	64 bits
MFFDSH-AF	**0.810**	0.825	0.839	**0.796**	0.838	0.861	**0.851**	0.865	0.873
MFFDSH-A	0.806	0.840	0.845	0.791	0.851	0.877	0.845	0.872	0.880
MFFDSH	0.808	**0.842**	**0.848**	0.793	**0.857**	**0.877**	0.849	**0.874**	**0.880**

4.4 Ablation Study

To further analyze the overall design of our proposed method, a detailed ablation study is performed to illustrate the effectiveness of each component. Specifically, we investigated two variants of MFFDSH, denoted as MFFDSH-AF and MFFDSH-A. MFFDSH-A indicates the elimination of the CAM module and MFFDSH-AF indicates the elimination of both CAM and CFM modules. Table 2 demonstrates the experimental results of the ablation studies. As can be observed in Table 2, MFFDSH-A increases the performance, which indicates that the CFM module can effectively fuse the low-level and high-level image features; MFFDSH-AF can further increase the performance on the majority of hash code bits, which illustrates to some extent the effectiveness of the CAM module.

5 Conclusion

In order to tackle the problem that image detail information is ignored, we propose a multi-feature fusion-based central similarity deep supervised hashing method. Specifically, we design a cross-layer fusion module that can effectively fuse the image detail information at the low level with the image semantic information at the high level, followed by introducing a channel attention module that focuses on the more important information in the fused features, and finally outputs the hash code of the image through the hash layer. Extensive experiments have been conducted on three benchmark datasets, and the experimental results demonstrate that MFFDSH is superior to existing deep supervised hashing methods.

Acknowledgments. This study is supported by the Natural Science Foundation of Inner Mongolia Autonomous Region under Grant 2019ZD14, the Project for Science and Technology of Inner Mongolia Autonomous Region under Grant 2019GG281, and the Program for Young Talents of Science and Technology in Universities of Inner Mongolia Autonomous Region under Grant NJYT-20-A05.

References

1. Yandex, A.B., Lempitsky, V.: Aggregating local deep features for image retrieval. In: Proceedings of the IEEE International Conference on Computer Vision (ICCV), pp. 1269–1277. IEEE (2015)
2. Zhang, D., Wang, J., Cai, D., Lu, J.: Self-taught hashing for fast similarity search. In: Proceedings of the 33rd International ACM SIGIR Conference on Research and Development in Information Retrieval (SIGIR), pp. 18–25. ACM (2010)
3. Liu, W., Wang, J., Ji, R., Jiang, Y.G., Chang, S.F.: Supervised hashing with kernels. In: Proceedings of the IEEE Conference on Computer Vision and Pattern Recognition (CVPR), pp. 2074–2081. IEEE (2012)
4. Guo, Y., Ding, G., Liu, L., Han, J., Shao, L.: Learning to hash with optimized anchor embedding for scalable retrieval. IEEE Trans. Image Process. **26**(3), 1344–1354 (2017)
5. Lai, H., Pan, Y., Liu, Y., Yan, S.: Simultaneous feature learning and hash coding with deep neural networks. In: Proceedings of the IEEE Conference on Computer Vision and Pattern Recognition (CVPR), pp. 3270–3278. IEEE (2015)
6. Shen, F., Shen, C., Liu, W., Tao, S.H.: Supervised discrete hashing. In: Proceedings of the IEEE Conference on Computer Vision and Pattern Recognition (CVPR), pp. 37–45. IEEE (2015)
7. Li, W.J., Wang, S., Kang. W.C.: Feature learning based deep supervised hashing with pairwise labels. In: Proceedings of the Twenty-Fifth International Joint Conference on Artificial Intelligence (IJCAI), pp. 1711–1717. Morgan Kaufmann (2016)
8. Liu, H., Wang, R., Shan, S., Chen. X.: Deep supervised hashing for fast image retrieval. In: Proceedings of the IEEE Conference on Computer Vision and Pattern Recognition (CVPR), pp. 2064–2072. IEEE (2016)
9. Zhu, H., Long, M.S., Wang, J.M., Cao, Y.: Deep hashing network for efficient similarity retrieval. In: Proceedings of the thirty AAAI Conference on Artificial Intelligence (AAAI), pp. 2415–2421. AAAI (2016)
10. Xia, R.K., Pan, Y., Lai, H.J., Liu, C., Yan, S.C.: Supervised hashing for image retrieval via image representation learning. In: Proceedings of the Twenty-Eighth AAAI Conference on Artificial Intelligence (AAAI), pp. 2156–2162. AAAI (2014)
11. Cao, Z.J., Long, M.S., Wang, J. M., Yu, P.S.: HashNet: deep learning to hash by continuation. In: Proceedings of the IEEE International Conference on Computer Vision (ICCV), pp. 5609–5618. IEEE (2017)
12. Cao, Y., Long, M.S., Liu, B., Wang, J.M.: Deep cauchy hashing for hamming space retrieval. In: Proceedings of the IEEE Conference on Computer Vision and Pattern Recognition (CVPR), pp. 1229–1237. IEEE (2018)

13. Kang, R., Cao, Y., Long, M.S., Wang, J.M., Yu, P.S.: Maximum-margin hamming hashing. In: Proceedings of the IEEE International Conference on Computer Vision (ICCV), pp. 8251–8260. IEEE (2019)
14. Fan, L. X., Ng, K.W., Ju, C., Zhang, T., Chan, C.S.: Deep polarized network for supervised learning of accurate binary hashing codes. In: Proceedings of the International Joint Conference on Artificial Intelligence (IJCAI), pp. 825–831. Morgan Kaufmann (2020)
15. Yuan, L., et al.: Central similarity quantization for efficient image and video retrieval. In: Proceedings of the IEEE Conference on Computer Vision and Pattern Recognition (CVPR), pp. 3080–3089. IEEE (2020)
16. Jose, A., Filbert, D., Rohlfing, C., Ohm, J.R.: Deep hashing with hash center update for efficient image retrieval. In: Proceedings of the IEEE International Conference on Acoustics, Speech, and Signal Processing (ICASSP), pp. 4773–4777. IEEE (2022)
17. Hong, W.X., Chang, Y.T., Qin, H.F., Hung, W.C., Tsai, Y.H., Yang, M.H.: Image hashing via linear discriminant learning. In: Proceedings of the IEEE Winter Conference on Applications of Computer Vision (WACV), pp. 2520–2528. IEEE (2020)
18. Wu, Z., Su, L., Huang, Q.: Cascaded partial decoder for fast and accurate salient object detection. In: Proceedings of the 32nd International Conference on Computer Vision and Pattern Recognition (CVPR), pp. 3907–3916. IEEE (2019)
19. Woo, S., Park, J., Lee, J.-Y., Kweon, I.S.: CBAM: convolutional block attention module. In: Ferrari, V., Hebert, M., Sminchisescu, C., Weiss, Y. (eds.) ECCV 2018. LNCS, vol. 11211, pp. 3–19. Springer, Cham (2018). https://doi.org/10.1007/978-3-030-01234-2_1
20. Chua, T., Tang, J., Hong, R., Li, H., Luo, Z., Zheng, Y.: NUS-WIDE: a real-world web image database from national university of Singapore. In: Proceedings of the 8th ACM International Conference on Image and Video Retrieval (CIVR). ACM (2009)
21. Lin, T.-Y., et al.: Microsoft COCO: common objects in context. In: Fleet, D., Pajdla, T., Schiele, B., Tuytelaars, T. (eds.) ECCV 2014. LNCS, vol. 8693, pp. 740–755. Springer, Cham (2014). https://doi.org/10.1007/978-3-319-10602-1_48
22. Russakovsky, O., et al.: Imagenet large scale visual recognition challenge. Int. J. Comput. Vision 115(3), 211–252 (2015)
23. Jiang, Q.Y., Li, W.J.: Asymmetric deep supervised hashing. In: Proceedings of the Thirty-Second AAAI Conference on Artificial Intelligence (AAAI), pp. 3342–3349. AAAI (2018)
24. Ma, L., Li, H.L., Meng, F.M., Wu, Q.B., Ngan, K.N.: Discriminative deep metric learning for asymmetric discrete hashing. Neurocomputing 380, 115–124 (2020)
25. Gu, G.H., Liu, J.T., Li, Z.Y., Huo, W.H., Zhao, Y.: Joint learning based deep supervised hashing for large-scale image retrieval. Neurocomputing 385, 348–357 (2020)
26. Yang, Y.C., Zhang, J.X., Wang, Q., Liu, B.: Deep high-order asymmetric supervised hashing for image retrieval. In: Proceedings of the International Joint Conference on Neural Networks (IJCNN), pp. 1–7. IEEE (2020)
27. Wei, H.X., He, C.: Joint learning method based on transformer for image retrieval. In: Proceedings of the International Joint Conference on Neural Networks (IJCNN), pp. 1–7. IEEE (2022)
28. He, K.M., Zhang, X.Y., Ren, S.Q., Sun, J.: Deep residual learning for image recognition. In: Proceedings of the International Conference on Computer Vision and Pattern Recognition (CVPR), pp. 770–778. IEEE (2016)

29. Sindagi, V., Patel, V.M.: Multi-level bottom-top and top-bottom feature fusion for crowd counting. In: Proceedings of the IEEE International Conference on Computer Vision (ICCV), pp. 1002–1012. IEEE (2019)
30. Kingma, D.P., Ba, J.: Adam: a method for stochastic optimization. In: Proceedings of the International Conference on Learning Representations (ICLR), Ithaca (2015)

VVA: Video Values Analysis

Yachun Mi[1,2], Yan Shu[2], Honglei Xu[2], Shaohui Liu[1,2(✉)], and Feng Jiang[2]

[1] State Key Laboratory of Communication Content Cognition, People's Daily Online, Beijing 100733, China
[2] Harbin Institute of Technology, Harbin 150001, China
shliu@hit.edu.cn

Abstract. User-generated content videos have attracted increasingly attention due to its dominant role in social platforms. It is crucial to analyze values in videos because the extensive range of video content results in significant variations in the subjective quality of videos. However, the research literature on Video Values Analysis (VVA) is very scarce, which aims to evaluate the compatibility between video content and the social mainstream values. Meanwhile, existing video content analysis methods are mainly based on classification techniques, which can not adequate VVA due to their coarse-grained manners. To tackle this challenge, we propose a framework to generate more fine-grained scores for diverse videos, termed as Video Values Analysis Model (VVAM), which consists of a feature extractor based on R3D, a feature aggregation module based on Transformer and a regression head based on MLP. In addition, considered texts in videos can be key clues to improve VVA, we design a new pipeline, termed as Text-Guided Video Values Analysis Model (TG-VVAM), in which texts in videos are spotted by OCR tools and a cross-modal fusion module is used to combine the vision and text features. To further facilitate the VVA, we construct a large-scale dataset, termed as Video Values Analysis Dataset (VVAD), which contains 53,705 short videos of various types from main social platforms. Experiments demonstrate that our proposed VVAM and TG-VVAM achieves promising results in the VVAD.

Keywords: Video values analysis · Video values analysis model · Text-guided video values analysis model · Video values analysis dataset

1 Introduction

In recent years, user-generated content (UGC) videos have become a dominant form of content on platforms such as TikTok, YouTube and Facebook, which have attracted tens of millions and billions of users respectively [1,2]. Users' diverse intentions in creating videos often lead to the presence of vulgar, negative, or

Supported by State Key Laboratory of Communication Content Cognition, People's Daily Online under Grant 2020YFB1406902 and A12003.

Q. Liu et al. (Eds.): PRCV 2023, LNCS 14431, pp. 346–358, 2024.
https://doi.org/10.1007/978-981-99-8540-1_28

harmful content in certain videos, which spread information that goes against mainstream societal values. Therefore, it is significant to analyze values in videos.

Despite that, no previous solutions are able to tackle Video Values Analysis (VVA), as shown in Fig. 1, which aims to assess the alignment between video content and the mainstream values of society. From our insight, VVA faces two main challenges. Firstly, the current methods [5,6,10,24,27] for video content analysis primarily rely on classification techniques, which are insufficient for adequately addressing the fine-grained VVA task. This leads to the lack of a baseline as well as the corresponding evaluation criteria. Additionally, there are no reliable benchmarks for evaluation, hindering the research community from tackling this task.

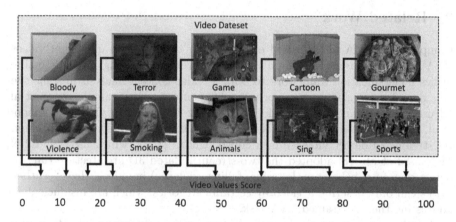

Fig. 1. Illustration of Video Values Analysis (VVA). Videos in the dataset are scored according to their contents, with higher scores representing videos that are more in line with mainstream values.

In this paper, inspired by the video quality assessment (VQA) task [16,28, 29] which takes a fine-grained manner to evaluate the quality of videos, we propose a framework to address the problem, termed as Video Values Analysis Model (VVAM). VVAM consists of a R3D [11] model for feature extraction, a Transformer [25] architecture for capturing dependencies from different video patch contexts and a regression head based on MLP. In addition, we observe texts in videos are crucial for VVA in some scenarios. Thus, based on VVAM, Text-Guided Video Values Analysis Model (TG-VVAM) is proposed. With the assistance of OCR tools [4], texts in videos can be spotted. Then a cross-modal fusion module is applied to integrates visual features of video and text features extracted by the Sentence-BERT [21].

As there lack large-scale dataset for Video Values Analysis, we propose a large-scale video values analysis dataset (VVAD) consisting of 53,705 short videos of diverse types from social platforms. Each video in the dataset is scored according to its content, with higher scores representing videos that are more in line with mainstream values.

Our main contributions are summarized as follows:

(1) This paper proposes video values analysis (VVA) for the first time. Instead of using the coarse-grained classification method, we employ the values scores to assess how well the various contents of the video match the mainstream social values.
(2) We introduce VVAM, a powerful baseline for VVA.
(3) Based on VVAM, we propose TG-VVAM, in which texts in videos are spotted to effectively enhance the VVA performance.
(4) We construct a large-scale dataset VVAD consisting of 53,705 short videos, each of which is scored according to its values orientation.

2   Related Work

Compared to VQA task focused on evaluating the objective quality of videos, VVA is largely absent in current research literature. Existing methods mainly utilize a classification framework to achieve video content analysis in some specific domains (e.g., violent or pornographic content). For example, Asad et al. [3] used VGG16 and LSTM to identify violent behaviors in videos. Mehmood et al. [19] used a two-stream CNN structure that consists of spatial and temporal streams to detect violence in a crowd. Unlike [3,19], which used 2D CNN, Mehmood et al. [18] adopted the two-stream Inflated 3D architecture (i3D) to detect abnormal behaviors (loitering, falling, violence) in videos. However, these works can only describe a video by certain categories, which are insufficient for addressing the fine-grained VVA task.

In addition to using single-frame features, Some works also introduce additional information to improve the accuracy of classification of pornographic content. Pu et al. [20] constructed a multimodal cross-fusion network (MCN) for multimodal feature fusion, which uses the appearance, motion, and audio features of the video as inputs to detect violent behaviors. Wang et al. [26] used spatial, audio, motion, and temporal features in video and fuse them through a cross-modal deep learning network. This results in a significant improvement in the accuracy of video classification. Mazinani and Ahmadi [17] designed a decision maker with a reinforcement learning structure that is not only capable of detecting pornographic frames in real time, but also adaptive. However, all of them neglect the texts in the videos, which can be a significant clue for analyzing values in the videos.

3   Video Values Analysis Dataset

According to the survey, the absence of studies on video values analysis leads to a lack of publicly available datasets. To conduct our study, we crawl 53,705 short video data from several video platforms (e.g., Shake, Twitter, Facebook, youtube, etc.). We crawl videos in three main categories (positive, negative and neutral) based on whether the video content is positive or not. The videos are then carefully categorized according to their specific content. Specifically, we identify 13 categories of short videos with positive values, 4 categories with neutral values, and 9 categories with negative values.

Table 1. Summary of Video Values Analysis Dataset.

Values	Category	Number	Format	Time Duration	Max length	Score range
Positive	Knowledge	1948	RGB	10–23 s	351	86.7–95.1
	Sports	1695	RGB	10–25 s	379	88.2–99.7
	gourmet	1856	RGB	8–17 s	269	82.5–92.6
	Technology	1733	RGB	11–19 s	287	79.3–90.6
	Art	1394	RGB	9–20 s	312	75.4–86.2
	Documentary	1523	RGB	10–16 s	250	66.6–80.4
	Fashion	1486	RGB	12–22 s	343	66.4–77.9
	Military parades	1120	RGB	11–24 s	370	86.7–95.4
	Wellness	1491	RGB	13–25 s	369	78.6–89.1
	Laugh	1076	RGB	8–16 s	254	71.2–83.8
	Sing	1629	RGB	10–23 s	358	75.9–84.4
	Entertainment	1548	RGB	9–20 s	301	68.2–79.3
	Cartoon	1293	RGB	15–25 s	382	60.8–77.5
Neutral	Everyday lives	5218	YUV	7–23 s	253	40.6–51.8
	News	5684	YUV	10–21 s	319	45.7–59.3
	History	5369	YUV	10–24 s	370	50.1–62.6
	Animals	5463	YUV	9–18 s	275	43.4–57.2
Negative	Bloody	1158	RGB	15–25 s	383	0.4–11.2
	Violence	1524	RGB	10–24 s	360	8.5–19.1
	Terror	981	RGB	10–19 s	287	6.3–20.7
	Explosion	1492	RGB	9–17 s	264	16.9–28.8
	Gunslinging	1417	RGB	11–20 s	313	10.7–18.4
	Smoking	1054	RGB	6–16 s	252	16.3–28.1
	Alcoholism	1347	RGB	8–17 s	258	20.7–35.9
	Game	1685	RGB	14–24 s	363	29.8–41.2
	Racism	1521	RGB	6–16 s	252	11.1–19.7

For each short video, we resize its resolution to 320*240 to facilitate our computation. Moreover, we devise a set of scoring criteria based on the positivity of the content and solicited 50 experts in computer vision to rate the content using these criteria. Since a video may contain more than one type of content, we score each video by taking into account the weight of each factor in the video. For instance, in documentaries, the content may be related to history, animals, etc., so the content score of the documentary is within a range. The details of the dataset are presented in Table 1.

4   Method

In this section, we will introduce the specific implementation of the proposed method in detail. The first subsection will introduce the Video Values Analysis Model (VVAM) method. The second subsection will introduce the Text-Guided Video Values Analysis Model (TG-VVAM) method.

4.1 Video Values Analysis Model (VVAM)

As shown in Fig. 2, the proposed model consists of two main modules, namely video feature extraction module and feature aggregation module.

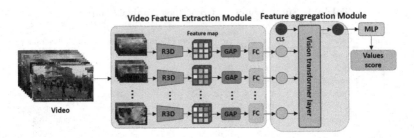

Fig. 2. Video Values Analysis Model (VVAM) architecture. The model consists of an R3D-based feature extractor, a Transformer-based feature aggregation module, and a regression head.

Video Feature Extraction Module: We apply the residual 3D convolutional neural network (R3D) [11] as a feature extractor for the video. The R3D model can be considered as a 3D version of ResNet [12]. The model is an extension of ResNet and 2D CNN, and its main features are the utilization of 3D convolution to capture the spatio-temporal features of video data, and the implementation of residual connections to speed up model training and enhance model performance. The input to the R3D network is a sequence of video frames that undergoes a series of 3D convolution, batch normalization and nonlinear activation operations to obtain a multidimensional tensor representing the features of the video.

The R3D pre-trained on the Kinetic [15] dataset was used as a feature extractor. Moreover, we take the output of the last pooling layer of the R3D network as the extracted features. The obtained feature maps are then subjected to a global average pooling (GAP) operation, and the pooled results are fed into the feature fusion module through a fully connected (FC) layer.

Feature Aggregation Module: Transformer encoder [25] is used to aggregate features. Specifically, we set the number of layers of the encoder to 12 and the number of heads of the multi-headed attention to 24. The dimensionality of each input and output token of Transformer encoder is 1024 dimensions. In addition, an additional cls-token is set and the output of the Transformer encoder corresponding to the cls-token is used as the extracted video feature. Moreover, we do not use a pre-trained model, but train our model from scratch on our dataset VVAD.

Regression Head: The MLP layer, which consists of two fully connected layers with 512 and 128 neurons in the hidden layer, transforms the output corresponding to the classification token (CLS) into a one-dimensional numerical score.

4.2   Text-Guided Video Values Analysis Model (TG-VVAM)

Based on VVAM, considered texts in videos can be key clues to improve VVA, we design a new pipeline, termed as Text-Guided Video Values Analysis Model (TG-VVAM), in which texts in videos are spotted by OCR tools and a cross-modal fusion module is used to combine the vision and text features. As shown in Fig. 3.

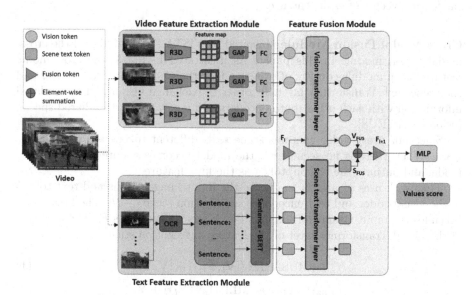

Fig. 3. Text-Guided Video Values Analysis Model (TG-VVAM) architecture. R3D model [11] is used to extract video image features. The OCR [4,22,23]and Sentence-BERT [21] models extract the video text features and the dual encoder is used to fuse the features.

The proposed model consists of four main modules. First, we select multiple consecutive video clips (not necessarily adjacent between clips) from the video, and we apply R3D, introduced in Sect. 4.1, to extract video features. Simultaneously, we use optical character recognition (OCR) techniques to extract the text information in the video frames. Moreover, we propose a novel multi-modal fusion mechanism that can fuse video features with the semantic features of textual cues in the video. Finally, the fused features are fed into the analysis head to obtain the values score of the video.

Text Feature Extraction Module: For the text information in the video screen, we use an existing OCR tool [4,22,23] to extract the text information. Before applying OCR for text extraction, we extract the video frames from the video with moderate spacing, and to ensure that the subtitles are captured as much as possible, we set the interval to 10. Then apply OCR to return all recognition results. Finally, the extracted text is de-duplicated, i.e. duplicate text or sentences are removed.

After getting the text information in the video, the next step is to extract the feature vectors of the text. We choose the Sentence-BERT (SBERT) [21] model as the text encoder. In our approach, we employ one of the sub-networks in the pre-trained SBERT as a text encoder, and simultaneously, perform a meanpooling operation on the token embedding output from the BERT component to obtain the feature vector of each sentence.

Cross-Modal Fusion Module: Inspired by ViSTA [7], we designed the cross-modal fusion module in this paper. As illustrated in Fig. 3, the cross-model feature fusion module consists of two encoders: the visual Transformer layer and the scene text Transformer layer. In order to exchange visual and text-related information with each other, both encoders add a new token, the shared special fusion token [FUS].

Since not every video contains scene text, different tokens are used in this work. When there is no scene text, the model becomes a pure visual encoder model and outputs the [vision] token as the final feature.

This work uses V_l and S_l to denote the input image tokens and text tokens of the visual coder and text encoder of aggregation stage $l - th$. The first fusion token for the input visual and scene text encoders is denoted as F_l. The workflow of the visual Transformer layer of the feature aggregation stage is:

$$\mathbf{Y}_l \leftarrow \text{MHSA}\left(\text{LN}\left([\mathbf{V}_l; \mathbf{F}_l]\right)\right) + [\mathbf{V}_l; \mathbf{F}_l] \tag{1}$$

$$[\mathbf{V}_{l+1}; \mathbf{V}_{\text{FUS}}] \leftarrow \text{MLP}\left(\text{LN}\left(\mathbf{Y}_l\right)\right) + \mathbf{Y}_l \tag{2}$$

where V_{FUS} is the feature corresponding to the visual part of the output of the fusion token. The workflow of the scene text Transformer layer in the aggregation stage is similar.

$$\mathbf{Y}_l \leftarrow \text{MHSA}\left(\text{LN}\left([\mathbf{V}_l; \mathbf{F}_l]\right)\right) + [\mathbf{V}_l; \mathbf{F}_l] \tag{3}$$

$$[\mathbf{V}_{l+1}; \mathbf{V}_{\text{FUS}}] \leftarrow \text{MLP}\left(\text{LN}\left(\mathbf{Y}_l\right)\right) + \mathbf{Y}_l \tag{4}$$

$$\mathbf{Y}_l \leftarrow \text{MHSA}\left(\text{LN}\left([\mathbf{S}_l; \mathbf{F}_l]\right)\right) + [\mathbf{S}_l; \mathbf{F}_l] \tag{5}$$

$$[\mathbf{S}_{l+1}; \mathbf{S}_{\text{FUS}}] \leftarrow \text{MLP}\left(\text{LN}\left(\mathbf{Y}_l\right)\right) + \mathbf{Y}_l \tag{6}$$

where S_{FUS} is the feature corresponding to the scene text part of the output of the fusion token. The input fusion features of the next layer are computed by their element-by-element summation, defined as follows:

$$\mathbf{F}_{l+1} = \mathbf{V}_{FUS} + \mathbf{S}_{FUS} \tag{7}$$

With the approach introduced above, visual features V and scene text features S are learned through separate Transformer layers. Since it is shared between the two encoders, the special fusion token [FUS] acts as a bridge between the two encoders. Due to the visual and scene text aggregation layers, the learning of image features and scene text features interact with each other through indirect fusion tokens. Unlike updating the shared token twice, this work directly sums the predicted fusion tokens from the visual and scene text Transformer layers to form the fusion token during the aggregation process. Ultimately, the fusion token gets the values score through the MLP module.

5 Experiments

In this section, We will first describe the experimental setup. Then, we compare multiple visual feature extraction models and fusion models to demonstrate the effectiveness of our framework VVAM. Finally, we perform an ablation study of the proposed TG-VVAM method to assess the effectiveness of each component.

5.1 Experimental Setup

Dataset. We conduct experiments on our proposed video values analysis dataset. Each category of these short video videos is divided into a training set, a validation set and a test set according to the ratio of 7:2:1.

Evaluation Metircs. We adopt the Spearman's rank correlation coefficient (SROCC) as the evaluation metric. The Spearman's rank correlation coefficient is used to measure the strength of the monotonic relationship between two continuous variables, which does not assume any distribution of the data and is robust to outliers. The formula for computing SROCC is given below. Let X and Y be sets containing N elements, and after sorting both X and Y in ascending or descending order, the position of an element is its rank.

$$\rho = \frac{\sum_{i=1}^{N}(x_i - \tilde{x})(y_i - \tilde{y})}{\sqrt{\sum_{i=1}^{N}(x_i - \tilde{x})^2 \sum_{i=1}^{N}(y_i - \tilde{y})^2}} = 1 - \frac{6\sum_{i=1}^{N} d_i^2}{N(N^2 - 1)} \tag{8}$$

where x_i and y_i are the ranks of the values taken for observation i, respectively, $\tilde{x}$ and $\tilde{y}$ the mean ranks of X and Y, respectively, N is the total number of observations, and $d_i = x_i - y_i$ denotes the number of rank differences of the dichotomous paired variables.

Implementation Details. We use PyTorch as the training platform and four NVIDIA GeForce RTX 3090 cards to train the model. The mean square error (MSE) is used as the loss function of the model. The initial learning rate is set to $1 \times 10 - 2$, and the cosine annealing strategy is used to adjust the learning rate. Uniformly, we use the Adam optimizer and train the model with 500 epochs.

5.2 Comparison Experiments on VVAM

In order to determine the basic architecture of VVAM, we conduct preliminary experiments on the visual extraction part and the feature fusion part of the model,as shown in Table 2. The table shows the average SROCC performance of

Table 2. A Comparison of Classical Visual Feature Extraction and Feature Fusion Methods.

Feature Extraction	Feature Fusion	Sing	Knowledge	Everyday lives	News	Bloody	Violence	Average
ResNet50	MLP	0.785	0.783	0.733	0.741	0.792	0.798	0.784
	GRU	0.826	0.819	0.757	0.801	0.839	0.833	0.825
	LSTM	0.832	0.840	0.791	0.816	0.850	0.845	0.839
	Transformer	0.855	0.852	0.831	0.837	0.862	0.864	0.855
C3D	MLP	0.773	0.778	0.742	0.748	0.788	0.792	0.780
	GRU	0.824	0.832	0.803	0.806	0.850	0.848	0.836
	LSTM	0.849	0.852	0.837	0.846	0.872	0.868	0.850
	Transformer	0.869	0.876	0.854	0.858	0.893	0.881	0.871
R3D	MLP	0.790	0.794	0.757	0.766	0.813	0.804	0.797
	GRU	0.858	0.850	0.839	0.845	0.878	0.871	0.856
	LSTM	0.877	0.871	0.851	0.856	0.884	0.889	0.873
	Transformer	**0.899**	**0.901**	**0.887**	**0.879**	**0.912**	**0.916**	**0.902**

different combinations of visual and fusion models over the entire dataset, and shows the results for the six categories of data in the dataset.

Comparison on Visual Module. We compare the performance of different visual feature extraction methods including: (1) ResNet-50 [12], pre-trained on ImageNet [9] and using the feature maps of its res5c layer as the extracted features. (2) C3D [24], pre-trained on Sports-1M [14] and using the output of fc6 layer as the extracted features. (3) R3D [11], pre-trained on Kinetics [15] and using the output of the last pooling layer as the extracted features. The results are shown in Table 2. We observe that the average SROCC values for Resnet50, C3D and R3D are 0.826, 0.834 and 0.857. This indicates that 3D convolutional neural networks are more suitable for processing video data compared to 2D convolution. In addition, the R3D model has better performance than the C3D model in the 3D CNN.

Comparison on Fusion Module. We compare the performance of different feature fusion methods including: (1) Multi-layer perceptron (MLP). (2) Gate Recurrent Unit (GRU) [8]. (3) Long short-term memory (LSTM) [13]. (4) Transformer [25]. The results are shown in Table 2. We observe that the average SROCC values for MLP, GRU, LSTM and Transformer are 0.787, 0.839, 0.854 and 0.876. This shows that Transformer has better data fusion capability relative to traditional MLP and RNN (GRU and LSTM).

With the two comparison experiments above, we decided to use R3D as the extraction module for visual features and considered using Transformer for data fusion. To verify the validity of our proposed fusion model, we continue with the ablation experiments.

5.3 Ablation Studies

For the feature vectors obtained from the video feature extraction module and the text feature extraction module, we use the two commonly used methods of feature splice: (1) add. (2) concatenate. To ensure that the visual features and text features can be added element by element, the obtained visual features and text features are processed to the same length using a fully connected layer. For the cross-modal feature fusion module, a Transformer Encoder is used here to model the feature vector, and the output corresponding to the extra set token is used as the extracted fused features, which are then passed through an MLP module to obtain the score.

As can be seen from Table 3, our proposed cross-modal feature fusion method shows significant performance improvement relative to the traditional Transformer Encoder-based method, with 3.0% and 2.1% improvement relative to add and concatenate, respectively.

To verify the effect of each module of the TG-VVAM on the model, we perform the following ablation experiments. The results are shown in Table 4. We observe that the results of using visual features alone are always much better than those of using text features alone, regardless of whether the cross-modal fusion module is used or not, while adding text features on top of visual features, the model results only gain a small improvement. Therefore, we can conclude that the visual information of the video plays the most important role for the model, while the textual information of the video scene only assists the visual

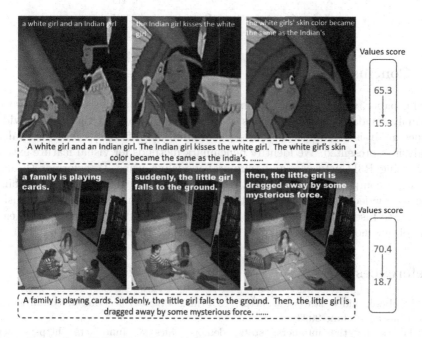

Fig. 4. Examples of VVAM and TG-VVAM performance on two negative videos. The two rows of images represent the two videos, and the dashed boxes represent the text of the scenes extracted from the videos. In the values score box on the right, red represents the VVAM results and green represents the TG-VVAM results. (Color figure online)

information to further improve the model. As shown in the two specific examples in Fig. 4, fusing text features does allow the model to be improved.

Table 3. Ablation Study on Cross-Modal Feature Fusion

Feature Extraction	Feature Splice	Fusion Model	SROCC
R3D + OCR	add	one encoder	0.908
	concatenate	one encoder	0.916
	ours	**dual fusion encoder**	**0.935**

Table 4. Ablation Study on Each Module of TG-VVAM

R3D	OCR+SBERT	Fusion Model	SROCC
✓	✗	✗	0.613
✓	✓	✗	0.736
✓	✗	✓	0.902
✗	✓	✗	0.373
✗	✓	✓	0.411
✓	✓	✓	0.935

6   Conclusion

In this paper, we present for the first time video values analysis, a new approach to analyze video content by means of regression. To conduct the study of video values analysis, a large-scale dataset containing 53,705 videos for video values analysis is presented. We identify benchmark methods for visual feature extraction using R3D networks and temporal modeling using Transformer through extensive comparison experiments. Further, we propose TG-VVAM in combination with video scene text information, which uses a dual encoder feature fusion module to fuse visual features and scene text features. The effectiveness of each part of our model is demonstrated by ablation experiments.

References

1. Facebook video statistics. https://99firms.com/blog/facebook-video-statistics/. Accessed 22 June 2023
2. Tiktok by the numbers: stats, demographics & fun facts. https://www.omnicoreagency.com/tiktok-statistics/. Accessed 22 June 2023
3. Asad, M., Yang, J., He, J., Shamsolmoali, P., He, X.: Multi-frame feature-fusion-based model for violence detection. Vis. Comput. **37**, 1415–1431 (2021)
4. Baidu: video-to-text-OCR. https://github.com/HenryLulu/video-to-text-ocr-demo. Accessed 15 June 2023

5. Carreira, J., Zisserman, A.: Quo vadis, action recognition? A new model and the kinetics dataset. In: Proceedings of the IEEE Conference on Computer Vision and Pattern Recognition, pp. 6299–6308 (2017)
6. Chen, Y., Kalantidis, Y., Li, J., Yan, S., Feng, J.: Multi-fiber networks for video recognition. In: Proceedings of the European Conference on Computer Vision (ECCV), pp. 352–367 (2018)
7. Cheng, M., et al.: ViSTA: vision and scene text aggregation for cross-modal retrieval. In: Proceedings of the IEEE/CVF Conference on Computer Vision and Pattern Recognition, pp. 5184–5193 (2022)
8. Chung, J., Gulcehre, C., Cho, K., Bengio, Y.: Empirical evaluation of gated recurrent neural networks on sequence modeling. arXiv preprint arXiv:1412.3555 (2014)
9. Deng, J., Dong, W., Socher, R., Li, L.J., Li, K., Fei-Fei, L.: Imagenet: a large-scale hierarchical image database. In: 2009 IEEE Conference on Computer Vision and Pattern Recognition, pp. 248–255. IEEE (2009)
10. Feichtenhofer, C.: X3D: expanding architectures for efficient video recognition. In: Proceedings of the IEEE/CVF Conference on Computer Vision and Pattern Recognition, pp. 203–213 (2020)
11. Hara, K., Kataoka, H., Satoh, Y.: Learning spatio-temporal features with 3D residual networks for action recognition. In: Proceedings of the IEEE International Conference on Computer Vision Workshops, pp. 3154–3160 (2017)
12. He, K., Zhang, X., Ren, S., Sun, J.: Deep residual learning for image recognition. In: Proceedings of the IEEE Conference on Computer Vision and Pattern Recognition, pp. 770–778 (2016)
13. Hochreiter, S., Schmidhuber, J.: Long short-term memory. Neural Comput. $9(8)$, 1735–1780 (1997)
14. Karpathy, A., Toderici, G., Shetty, S., Leung, T., Sukthankar, R., Fei-Fei, L.: Large-scale video classification with convolutional neural networks. In: Proceedings of the IEEE Conference on Computer Vision and Pattern Recognition, pp. 1725–1732 (2014)
15. Kay, W., et al.: The kinetics human action video dataset. arXiv preprint arXiv:1705.06950 (2017)
16. Li, D., Jiang, T., Jiang, M.: Quality assessment of in-the-wild videos. In: Proceedings of the 27th ACM International Conference on Multimedia, pp. 2351–2359 (2019)
17. Mazinani, M.R., Ahmadi, K.D.: An adaptive porn video detection based on consecutive frames using deep learning. Rev. d'Intelligence Artif. $35(4)$, 281–290 (2021)
18. Mehmood, A.: Abnormal behavior detection in uncrowded videos with two-stream 3D convolutional neural networks. Appl. Sci. $11(8)$, 3523 (2021)
19. Mehmood, A.: Efficient anomaly detection in crowd videos using pre-trained 2D convolutional neural networks. IEEE Access 9, 138283–138295 (2021)
20. Pu, Y., Wu, X., Wang, S., Huang, Y., Liu, Z., Gu, C.: Semantic multimodal violence detection based on local-to-global embedding. Neurocomputing 514, 148–161 (2022)
21. Reimers, N., Gurevych, I.: Sentence-BERT: sentence embeddings using Siamese BERT-networks. arXiv preprint arXiv:1908.10084 (2019)
22. Shu, Y., Liu, S., Zhou, Y., Xu, H., Jiang, F.: EI 2 SR: Learning an enhanced intra-instance semantic relationship for arbitrary-shaped scene text detection. In: IEEE International Conference on Acoustics, Speech and Signal Processing (ICASSP), pp. 1–5. IEEE (2023)

23. Shu, Y., et al.: Perceiving ambiguity and semantics without recognition: an efficient and effective ambiguous scene text detector. In: Proceedings of the 31th ACM International Conference on Multimedia (2023)

24. Tran, D., Bourdev, L., Fergus, R., Torresani, L., Paluri, M.: Learning spatiotemporal features with 3D convolutional networks. In: Proceedings of the IEEE International Conference on Computer Vision, pp. 4489–4497 (2015)

25. Vaswani, A., et al.: Attention is all you need. In: Advances in Neural Information Processing Systems, vol. 30 (2017)

26. Wang, L., Zhang, J., Wang, M., Tian, J., Zhuo, L.: Multilevel fusion of multimodal deep features for porn streamer recognition in live video. Pattern Recogn. Lett. **140**, 150–157 (2020)

27. Wang, R., et al.: BEVT: BERT pretraining of video transformers. In: Proceedings of the IEEE/CVF Conference on Computer Vision and Pattern Recognition, pp. 14733–14743 (2022)

28. Wu, H., et al.: Fast-VQA: efficient end-to-end video quality assessment with fragment sampling. In: Avidan, S., Brostow, G., Cissé, M., Farinella, G.M., Hassner, T. (eds.) ECCV 2022, Part VI. LNCS, vol. 13666, pp. 538–554. Springer, Cham (2022). https://doi.org/10.1007/978-3-031-20068-7_31

29. Ying, Z., Mandal, M., Ghadiyaram, D., Bovik, A.: Patch-VQ: 'patching up' the video quality problem. In: Proceedings of the IEEE/CVF Conference on Computer Vision and Pattern Recognition, pp. 14019–14029 (2021)

Dynamic Multi-modal Prompting
for Efficient Visual Grounding

Wansen Wu, Ting Liu, Youkai Wang, Kai Xu, Quanjun Yin, and Yue Hu[✉]

College of Systems Engineering, National University of Defense Technology,
Changsha 410003, China
huyue11@nudt.edu.cn

Abstract. Prompt tuning has emerged as a flexible approach for adapting pre-trained models by solely learning additional inputs while keeping the model parameters frozen. However, simplistic prompts are insufficient to effectively address the challenges posed by complex multi-modal tasks such as visual grounding. In this paper, we propose a novel prompting architecture called **D**ynamic **M**ulti-mod**A**l **P**rompting (DMAP) for visual grounding. DMAP incorporates input-dependent prompting to tailor instance-level prompts for more accurate representation and dynamic multi-modal prompting to capture the relationship between the textual and visual inputs. To this end, we design a Dynamic Prompt Network (DPN) to generate multi-modal prompts based on the specific inputs, enhancing both adaptive prompt generation and multi-modal feature fusion. Extensive experimental results demonstrate the superiority of DMAP over competing methods in parameter-efficient settings. Furthermore, DMAP consistently outperforms state-of-the-art VG methods even when fine-tuning all parameters.

Keywords: Visual Grounding · Prompting Tuning · Vision and Language

1 Introduction

Visual grounding (VG) is a fundamental problem in multi-modal learning that aims to align natural language with image regions to identify the region of an image that corresponds to a given description. Most VG methods have exhibited impressive performance across a range of applications, including automated driving systems (ADS) [15], visual question answering (VQA) [1], and vision-language navigation (VLN) [23].

The pretrain-finetune paradigm has achieved great success in addressing vision-language tasks using vast amounts of web data. This involves pretraining a large-scale model with extensive data and fine-tuning it with smaller downstream datasets. However, current high-performance visual grounding models face challenges related to training efficiency. This is primarily due to the computational demands and over-fitting risks associated with fully fine-tuning the entire model.

Q. Liu et al. (Eds.): PRCV 2023, LNCS 14431, pp. 359–371, 2024.
https://doi.org/10.1007/978-981-99-8540-1_29

As a result, to leverage these powerful large models for downstream tasks at a more affordable cost, Parameter-Efficient Fine-Tuning (PEFT) [16] is introduced as an alternative approach. PEFT aims to train a small subset of the existing model parameters or a set of newly added parameters to address the infeasibility and impracticality of fine-tuning large-scale pre-trained models. The typical PEFT method, known as Prompt Tuning [22], involves freezing the parameters of pre-trained models and leveraging learnable continuous prompts to facilitate quick adaptation of these large-scale models. However, all existing PEFT paradigms for visual grounding suffer from two limitations: **1) Input-unrelated prompts.** The text or visual prompts are directly constructed as trainable tensors and optimized through gradient descent as illustrated in Fig. 1a and Fig. 1b. However, in the VG task, different language queries with the same image input should produce significantly different output results. Thus, it is essential to utilize the input information effectively to generate relevant prompts that assist the model in understanding and accomplishing the task. Conversely, unrelated prompts may disrupt predictions and result in a performance decline. **2) Modality-unrelated prompts.** Given that visual grounding involves vision and language modalities, uni-modal pre-trained models lack the ability to effectively integrate and fuse multi-modal features due to their inherent focus on a single modality. As shown in Fig. 1c, the current approach of adding prompt embeddings independently to each modality fails to establish inter-modality connections and utilize the multi-modal information for reasoning effectively.

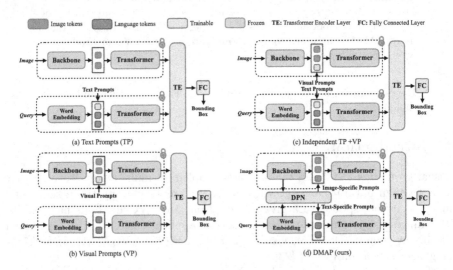

Fig. 1. Comparison of DMAP with standard prompt learning methods. (a) (b) (c) are the architectures of continuous text prompts (TP), visual prompts (VP), and the combination of TP and VP, respectively. (d) is the DMAP (ours).

To tackle these problems, we introduce a novel prompting architecture for visual grounding tasks, *i.e.*, the **D**ynamic **M**ulti-mod**A**l **P**rompting (DMAP).

We propose to build input-dependent prompts for incorporating specific input perturbation patterns combined with a dynamic multi-modal prompting technique to link the prompts learned in vision and language branches. Specifically, we design a Dynamic Prompt Network (DPN) to generate multi-modal prompts by dynamically selecting the prompts from the prompts repository based on the input instance. This allows for efficient yet generalized modeling capabilities and better multi-modal feature fusion (*c.f.* Fig. 1d). With this, we summarize our contributions:

- A novel dynamic multi-modal prompting architecture (DMAP) for visual grounding is proposed, which is a simple yet effective approach that adapts the pre-trained language and object detection models to perform the visual grounding task directly.
- We design the Dynamic Prompt Network (DPN) to generate the dynamic multi-modal instance-level prompts. It leverages image content information to enhance the textual prompt generation and vice versa for generalized modeling capabilities.
- Extensive experimental results show that the performance of DMAP surpasses the other prompt learning methods in parameter-efficient settings by a large margin. Notably, when fine-tuning all parameters, our DMAP still has consistent improvements and gets comparable results with state-of-the-art methods.

2 Related Work

2.1 Visual Grounding

State-of-the-art VG methods can be broadly categorized into two groups: two-stage methods and one-stage methods. Two-stage methods divide the VG task into proposal generation and proposal ranking stages. Initially, a pre-trained detector is used to generate proposals. Then, a multi-modal ranking network evaluates the similarity between the query sentence and the proposals, selecting the most suitable result [5,9,26,28]. On the other hand, one-stage methods directly predict the location of the referred object [24,25,30]. Certain studies treat VG as a conditional object detection problem and employ an end-to-end training approach [6].

2.2 Prompt Learning

Prompt engineering involves creating appropriate instructions for pre-trained models to generate desired outputs. Pretrained language models such as BERT [7] and GPT-3 [3] have shown remarkable generalization abilities when provided with carefully designed prompt templates. In recent work, researchers have proposed methods to automate prompt engineering, such as continuous prompts [2]. Zhou *et al.* [29] argue that using a globally unified prompt may lead to overfitting to the base class and therefore propose conditional prompt learning. In this study, we investigate a dynamic prompting approach to generate input-related multi-modal prompts, aiming to enhance the pre-trained VG models.

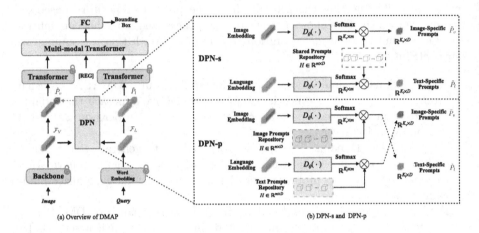

(a) Overview of DMAP   (b) DPN-s and DPN-p

Fig. 2. (a) Overview of our proposed DMAP architecture. (b) Two designs of multi-modal prompts, *i.e.*, DPN-s and DPN-p.

3   Method

The key idea of our DMAP is to generate input-dependent and dynamic multi-modal prompts that benefit the visual grounding process. This section introduces the details of DMAP, and the overall framework is illustrated in Fig. 2.

3.1   Preliminaries

Text Prompts. Prompt learning in the field of NLP provides an efficient method to fine-tune extensive pre-trained models specifically for downstream tasks. The prompt can be designed by humans for a downstream task or learned automatically during the fine-tuning stage, *i.e.*, the hard and the continuous/soft prompts. Formally, we denote the the pre-trained language model as $\delta_L(\cdot)$ and $L = \{w_i \in \mathbb{R}^d \mid n \in \mathbb{N}, 1 \leq i \leq n\}$ as the input language tokens. Conventionally, the model generates a latent language embedding for the provided input, denoted as $h_l = \delta_L([\texttt{CLS}, L])$, where $[\cdot, \cdot]$ denotes the tensor concatenation operation and [CLS] is added global token for classification. Then, latent embedding h_l and [CLS] token can be used to accomplish subsequent tasks. To tailor the model for diverse tasks or datasets, it becomes necessary to refine the extensive pre-trained model $\delta_L(\cdot)$, which may require substantial resources and could be impractical for models with billions of parameters.

To address this problem, prompt learning introduces several learnable tokens $P_l = \{p_i \in \mathbb{R}^d \mid k \in \mathbb{N}, 1 \leq i \leq k\}$ dubbed as language prompts. The language prompts P_l will be concatenated with the inputs and fed to the pre-trained model as follows:

$$\tilde{h}_l = \delta_L([P_l, \text{CLS}, L]), \tag{1}$$

where $\tilde{h}_l$ is the prompted latent embedding and can also be used for solving different downstream tasks. Using the adaptable prompt learning technique, we can tailor the model for novel tasks or datasets by solely training the prompt vectors P_l with a minimal amount of parameters instead of engaging in the laborious process of fine-tuning the burdensome pre-trained model $\delta_L(\cdot)$.

Visual Prompts. Recently, there have been several works introducing prompt learning as an ingenious approach to utilize large-scale pre-trained vision Transformer models efficiently, *e.g.*, object detection and the image classification pre-trained models. The prompts can be inserted in the image pixels or the encoded image feature. In the feature level, let $E(\cdot)$ as the image backbone to extract the visual embedding from inputs. Given the input image $I \in \mathbb{R}^{3 \times W \times H}$, a frozen pre-trained image encoder $\delta_V(\cdot)$ and the visual prompt P_v, the process of prompt learning can be formalized as:

$$\tilde{h}_v = \delta_V([\text{CLS}, (E(I) + P_v)]). \tag{2}$$

3.2 Dynamic Multi-modal Prompting (DMAP)

Based on the application of prompts on transformers [10,13,14], we propose the DMAP to enhance the performance of our models. This approach generates prompts from the DPN by combining textual and visual information at the input instance level. These prompts act as guiding signals that help the model better understand and process the information and generate more accurate and contextually relevant outputs.

Dynamic Prompt Network. Formally, given the language query L and an image I, we exploit the commonly used BERT token embedding layer [7] to get text embedding $\mathcal{F}_L \in \mathbb{R}^{D_l \times N_l}$ and ResNet [8] to get visual embedding $\mathcal{F}_V \in \mathbb{R}^{D_v \times N_v}$, where D_l and D_v is the feature dimensions, and N_l and N_v are the number of text and visual embedding tokens respectively. Note that the original output from ResNet is the 2D feature map. We use the 1×1 convolutional layer to reduce the channel dimension of the feature map and flatten it into the 1D sequence, which is the expected form of the transformer encoder layer.

Concretely, DPN consists of a mapping function and a pre-defined prompts repository with learnable prompt tokens. Based on this prompt repository, it enables the dynamic selection of prompts that are relevant to the input, thereby enhancing flexibility in the process. Let $D_\theta(\cdot)$ denote the mapping function parameterized by θ, and $H = \{h_i \in \mathbb{R}^D \mid m \in \mathbb{N}, 1 \leq i \leq m\}$ denote the prompts repository with m learnable prompt tokens in dimension D. It is important to note that the repository is continuously updated throughout the training process.

Given the visual embedding $\mathcal{F}_V \in \mathbb{R}^{D_v \times N_v}$, we first perform feature mapping, followed by the computation of the correlation coefficient between the features and the prompt tokens in the prompt repository. Finally, we utilize the dot product operation to retrieve the number of K_v prompt tokens from the repository that is relevant to the current input. Formally,

$$\tilde{P}_v = \text{softmax}(D_\theta(\mathcal{F}_V)) \cdot H, \tag{3}$$

where $\tilde{P}_v \in \mathbb{R}^{K_v \times D}$ is the image-specific prompts and will be concatenated with the original inputs to enhance the embeddings:

$$\tilde{h}_v = \delta_V([\text{CLS}, \mathcal{F}_V, \tilde{P}_v]). \tag{4}$$

The prompted visual features $\tilde{h}_v$ will be passed into the multi-modal transformer to predict the referent bounding box.

Similarly, in the language branch, we can calculate the text-specific prompts $\tilde{P}_l \in \mathbb{R}^{K_l \times D}$ conditional on the language query L, and obtain the enhanced text feature embedding $\tilde{h}_l$ as in Eq.(3) and (4), where K_l is the number of text-specific prompt tokens.

Design of Multi-modal Prompts. We argue that commonly used prompt learning methods [18] primarily focused on single modalities, which are comparatively less suitable for visual grounding tasks due to their limited capability to dynamically adapt and align both language and vision representation spaces. Inspired by [12], we develop two novel multi-modal prompting strategies as shown in Fig. 2b, namely DPN-s, and DPN-p. These strategies aim to bridge the gap between the two modalities, facilitating the mutual propagation of gradients for enhanced synergy. We elaborate on the proposed strategies below.

- **DPN-s**, which uses DPN with a shared prompts repository for both visual and language branches. The tokens in the repository are utilized to generate both language-specific and image-specific prompts. Through iterative training and updating, the prompt tokens in the shared repository can acquire knowledge of the cross-modal connections between the two modalities.
- **DPN-p**, which builds private prompt repositories for visual and language branches separately. We construct explicit cross-modal cues by directly concatenating image-specific prompts generated from the image branch onto the text embeddings, and likewise, text-specific prompts generated from the language branch are directly concatenated onto the visual embeddings.

Overall Framework. Inspired by the success of transformers in vision-language tasks, we choose TransVG [6] as the base model for the evaluation of our proposed parameter efficiently fine-tuning architecture. Following [6], we load the pre-trained DETR [4] weights for the visual branch and the BERT [7] weights for the language branch and freeze them during training and testing. We reduce the 6-layer multi-modal transformer encoder to 1-layer, which is followed by an MLP to predict the coordinate of the bounding box directly. As described in Algorithm 1, there are four steps of our proposed DAMP framework: 1) Feature Extraction; 2) Multi-modal Prompts Generation; 3) Cross-modal Fusion; 4) Bounding Box Prediction.

Algorithm 1. Overview of DMAP framework

Input: image backbone E, vision transformer δ_L, language transformer δ_V, position embeddings $\mathbf{Q}_{\mathrm{pos}}$, spatial position embeddings $\mathbf{Q}_{\mathrm{sp}}$, V-L position embeddings $\mathbf{Q}_{\mathrm{vl}}$, mapping function D_θ, prompts repository H, multi-modal transformer δ_M, prediction head MLP, visual prompts $\tilde{P}_v$, text prompts $\tilde{P}_l$, regression token [REG]

Output: Predicted bounding box $\hat{B} = (x_c, y_c, w, h)$

    ▷ Step 1: Feature Extraction

1: $\mathcal{F}_V = E(I)$ ← extract and flatten the visual embedding from image I

2: $\mathcal{F}_L$ ← get the language embedding from L

    ▷ Step 2: Multi-modal Prompts Generation

3: $\tilde{P}_v = \mathrm{softmax}(D_\theta(\mathcal{F}_V)) \cdot H$ ← generating the image-specific prompts

4: $\tilde{P}_l = \mathrm{softmax}(D_\theta(\mathcal{F}_L)) \cdot H$ ← generating the text-specific prompts

    ▷ Step 3: Cross-modal Fusion

5: $\mathcal{F}'_V$ ← apply image-specific prompts $\tilde{P}_v$ to the visual feature $\mathcal{F}_V$ and add sine spatial position embeddings $\mathbf{Q}_{\mathrm{sp}}$

6: $\mathcal{F}'_L$ ← apply text-specific prompts $\tilde{P}_l$ to the language feature $\mathcal{F}_L$ add position embeddings $\mathbf{Q}_{\mathrm{pos}}$

7: $\tilde{h}_v = \delta_V(\mathcal{F}'_V)$ ← utilize the pre-trained vision transformer δ_V to encode the enhanced vision feature $\mathcal{F}'_V$

8: $\tilde{h}_l = \delta_L(\mathcal{F}'_L)$ ← utilize the pre-trained language transformer δ_L to encode the enhanced language feature $\mathcal{F}'_L$

9: $\mathcal{F}_{\mathrm{all}}$ ← integrate the processed language features $\tilde{h}_l$, vision features $\tilde{h}_v$ and regression token [REG]

10: $\mathcal{F}'_{\mathrm{all}} = \delta_M(\mathcal{F}_{\mathrm{all}} + \mathbf{Q}_{\mathrm{vl}})$ ← implement cross-modal fusion through multi-modal transformer δ_M

    ▷ Step 4: Bounding Box Prediction

11: $(x_c, y_c, w, h) = \mathrm{MLP}(\mathcal{F}'_{\mathrm{all}})$ ← prediction head generates the predicted referent bounding box according to cross-modal representation $\mathcal{F}'_{\mathrm{all}}$

4 Experiments

4.1 Experimental Setup

Datasets. We validate the performance of our proposed DAMP on two benchmarked datasets. 1) **RefCOCO.** RefCOCO [27] is collected from MSCOCO [17], which includes 19,994 images with 50,000 referred objects. Each object has more than one referring expression, and there are 142,210 referring expressions in this dataset. 2) **RefCOCOg.** RefCOCOg [20] has 25,799 images with 49,856 referred objects and expressions. We report our performance on RefCOCOg-umd [21] (val-u and test-u).

Evaluation Settings and Metrics. We validated the performance of our proposed DMAP on two different settings: 1) **parameter efficient fine-tuning setting** and 2) **fully fine-tuning setting**. Under the first setting, we use the pre-trained weights from BERT [7] and DERT [4] to initialize the language and visual encoder receptively. The pre-trained parameters will remain frozen throughout the training and testing processes. At the same time, only the prompts, a lightweight multi-modal transformer encoder layer, and an MLP

Table 1. Results of parameter efficient fine-tuning methods on RefCOCO dataset. # Param. denotes the total number of the trainable parameters. We highlight the best and second best performance in the **bold** and underline.

Model	RefCOCO			# Param.
	val	testA	testB	
Baseline	59.46	61.89	54.57	1.8 M (1%)
Baseline+TP	58.71	60.35	54.98	1.8 M (1%)
Baseline+VP	58.88	59.76	54.20	1.8 M (1%)
Baseline+TP+VP	58.92	59.32	53.47	1.8 M (1%)
DMAP-s	<u>62.92</u>	<u>65.35</u>	<u>59.73</u>	2.4 M (1.3%)
DMAP-p	**63.29**	**65.41**	**60.48**	2.5 M (1.4%)
FAOA [25]	72.54	74.35	68.50	182.9 M (100%)

layer will be updated during the training stage. In fully fine-tuning setting, we will load these weights and update all the parameters of the model to achieve optimal performance. We follow the standard metric to report top-1 accuracy (%), where the prediction is correct if the IoU between the predicted box and ground-truth box is above 0.5.

Implementation Details. Following the [6], the training objectives are L1 regression and GIoU loss. We specify the input image size as 640×640 and set the maximum expression length as 20. During image resizing, we maintain the original aspect ratio of each image. The longer edge of an image is resized to 640, while the shorter edge is padded to 640 with the mean value of the RGB channels. We use the AdamW [19] as the optimizer with the learning rate 5e-4. With the exception of the pre-trained weights, the parameters are initialized randomly using Xavier initialization. The batch size is set to 128. For DPN, K_v and K_l are set as 16 and 40, respectively. The number of prompts in the prompts repository is 256 with a dimension of 256. The learning rate of DPN is 0.0025.

4.2 Main Results

Results of Parameter Efficient Fine-Tuning Methods. In the setting of parameter efficient fine-tuning, we compare our proposal with the three classical prompting methods, *i.e.*, TP, VP [2], the combination of TP and VP. The architectures of baselines are depicted in Fig. 1. As shown in Table 1, the intricate nature of the VG task renders the utilization of VP or TP ineffective in enhancing performance. Even the combination of VP and TP fails to offer any discernible benefits to the baseline model. In comparison, both DMAP-s and DMAP-p can bring significant improvements, demonstrating the effectiveness of input-dependent multi-modal prompts. Even compared with the fully fine-tuning model FAOA [25], our method achieves comparable performance by training only 1.4% of its parameters, significantly improving the training efficiency of the model.

Results of Fully Fine-Tuning Methods. In the fully fine-tuning setting, we insert our DMAP into the SOTA method TransVG [6] and fine-tune all parameters. As shown in Table 2, our prompting method benefits remarkably. For example, our DAMP-s leads to 2.0% improvement on RefCOCO testB. The consistent improvement demonstrates the flexibility and extensibility of our proposed DAMP.

Table 2. Comparison of our method with other state-of-the-art methods on Ref-COCO [27] and RefCOCOg [20]. † denotes the results by our re-implementation.

Methods	Backbone	RefCOCO			RefCOCOg	
		val	testA	testB	val-u	test-u
Two-stage Methods						
CMN [11]	VGG16	-	71.03	65.77	-	-
VC [28]	VGG16	-	73.33	67.44	-	-
MAttNet [26]	ResNet-101	76.65	81.14	69.99	66.58	67.27
RvG-Tree [9]	ResNet-101	75.06	78.61	69.85	66.95	66.51
Ref-NMS [5]	ResNet-101	80.70	84.00	76.04	70.55	70.62
One-stage Methods						
FAOA [25]	DarkNet-53	72.54	74.35	68.50	61.33	60.36
ReSC-Large [24]	DarkNet-53	77.63	80.45	72.30	67.30	67.20
TRAR [30]	DarkNet-53	-	81.40	78.60	68.90	68.30
TransVG† [6]	ResNet-50	80.90	82.35	76.35	67.66	67.44
Ours						
TransVG+DMAP-s	ResNet-50	82.27	**84.24**	**78.36**	68.67	67.83
		(+1.4)	(+1.9)	(+2.0)	(+1.0)	(+0.4)
TransVG+DMAP-p	ResNet-50	**82.38**	83.73	77.96	**69.02**	**68.36**
		(+1.5)	(+1.4)	(+1.6)	(+1.4)	(+0.9)

4.3 Ablation Studies

Next, we ablate and analyze the various choices of the DPN architecture. We conduct ablation studies on RefCOCOg under the PEFT setting. As shown in Table 3a, we vary the number of text-specific prompts K_l and K_v image-specific prompts. We find that increasing the number of prompts generally yields better performance, but increasing the length of prompts does not lead to a sustained improvement in performance. In Table 3b, we examine the impact of the size m of the prompts repository on performance. Our findings indicate that a larger prompts repository generally improves performance. However, no

Table 3. (a) Vary the number of prompt length. (b) Vary the size of Prompt Repository. All experiments are conducted on RefCOCO with DMAP-s.

K_v	K_l	val	testA	testB	Size	val	testA	testB
8	20	61.14	64.10	58.45	128	60.32	63.44	57.83
16	40	62.92	65.35	**59.73**	256	**62.92**	65.35	**59.73**
32	80	**63.38**	**65.77**	58.47	512	62.38	**65.46**	58.25
		(a)					(b)	

additional improvement is observed beyond 256 tokens. We hypothesize that the additional prompts tokens beyond 256 solely capture redundant information, thereby offering no further performance benefits.

4.4   Qualitative Results

We analyze the convergence rate of our DMAP and the baseline approaches in RefCOCO val set. To ensure a fair comparison, we maintain consistency in the learning rate, training epochs, and other relevant parameters. The results of these experiments are presented in Fig. 3a, demonstrating that our method effectively facilitates faster and lower loss convergence for the model. Furthermore, we visualize the attention map of the [REG] token at the final layer of the multi-modal transformer layer, both with and without utilizing our DMAP. This visualization provides insights into the positions that the model attends to, as the [REG] token is crucial in generating predictions when passed through the MLP layers. In the first row of Fig. 3b, there are the original input image and the bounding boxes of ground truth (in green), w/o DAMP (in blue) and w/ DMAP (in red), respectively. In the second row, the left side exhibits the attention map w/o DAMP, while the right side demonstrates the attention map w/ DAMP. It is evident that our method effectively provides significant cues to the model, assisting it in achieving accurate localization.

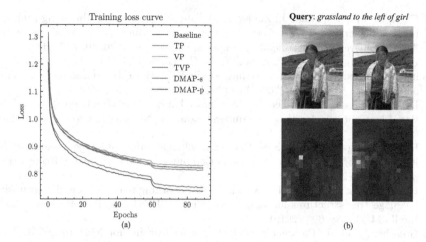

Fig. 3. Loss curve and visualization of bounding boxes and attention map. (Color figure online)

5   Conclusion

In this paper, we proposed a novel dynamic multi-modal prompting architecture (DMAP) for visual grounding. DMAP leverages input-dependent prompts and dynamic multi-modal prompting techniques to effectively capture the intricate relationship between image and language. Extensive experimental results have demonstrated the effectiveness of our DMAP. Our research contributes to advancing the field of visual grounding by offering a novel prompting approach that improves training efficiency and achieves remarkable performance.

Acknowledgement. This research was supported partially by the National Natural Science Fund of China (Grant Nos. 62306329, 62103420, 62103425 and 62103428) and the Natural Science Fund of Hunan Province (Grant Nos. 2021JJ40697, 2021JJ40702, 2022JJ40559 and 2023JJ40676), and Hunan Provincial Innovation Foundation For Postgraduate.

References

1. Antol, S., et al.: VQA: visual question answering. In: Proceedings of the IEEE International Conference on Computer Vision, pp. 2425–2433 (2015)
2. Bahng, H., Jahanian, A., Sankaranarayanan, S., Isola, P.: Exploring visual prompts for adapting large-scale models. arXiv preprint arXiv:2203.17274, vol. 1, no. 3, p. 4 (2022)
3. Brown, T.B., et al.: Language models are few-shot learners. In: NeurIPS (2020)
4. Carion, N., Massa, F., Synnaeve, G., Usunier, N., Kirillov, A., Zagoruyko, S.: End-to-end object detection with transformers. In: Vedaldi, A., Bischof, H., Brox, T., Frahm, J.-M. (eds.) ECCV 2020. LNCS, vol. 12346, pp. 213–229. Springer, Cham (2020). https://doi.org/10.1007/978-3-030-58452-8_13

5. Chen, L., Ma, W., Xiao, J., Zhang, H., Chang, S.F.: Ref-NMS: breaking proposal bottlenecks in two-stage referring expression grounding. In: Proceedings of the AAAI Conference on Artificial Intelligence, vol. 35, pp. 1036–1044 (2021)
6. Deng, J., Yang, Z., Chen, T., Zhou, W., Li, H.: TransVG: end-to-end visual grounding with transformers. In: Proceedings of the IEEE/CVF International Conference on Computer Vision, pp. 1769–1779 (2021)
7. Devlin, J., Chang, M.W., Lee, K., Toutanova, K.: Bert: pre-training of deep bidirectional transformers for language understanding. arXiv preprint arXiv:1810.04805 (2018)
8. He, K., Zhang, X., Ren, S., Sun, J.: Deep residual learning for image recognition. In: Proceedings of the IEEE Conference on Computer Vision and Pattern Recognition, pp. 770–778 (2016)
9. Hong, R., Liu, D., Mo, X., He, X., Zhang, H.: Learning to compose and reason with language tree structures for visual grounding. IEEE Trans. Pattern Anal. Mach. Intell. **44**(2), 684–696 (2019)
10. Houlsby, N., et al.: Parameter-efficient transfer learning for NLP. In: ICML. Proceedings of Machine Learning Research, vol. 97, pp. 2790–2799. PMLR (2019)
11. Hu, R., Rohrbach, M., Andreas, J., Darrell, T., Saenko, K.: Modeling relationships in referential expressions with compositional modular networks. In: Proceedings of the IEEE Conference on Computer Vision and Pattern Recognition, pp. 1115–1124 (2017)
12. Khattak, M.U., Rasheed, H., Maaz, M., Khan, S., Khan, F.S.: Maple: multi-modal prompt learning. In: Proceedings of the IEEE/CVF Conference on Computer Vision and Pattern Recognition, pp. 19113–19122 (2023)
13. Li, M., Chen, L., Duan, Y., Hu, Z., Feng, J., Zhou, J., Lu, J.: Bridge-prompt: towards ordinal action understanding in instructional videos. In: CVPR, pp. 19848–19857. IEEE (2022)
14. Li, X.L., Liang, P.: Prefix-tuning: optimizing continuous prompts for generation. In: ACL/IJCNLP (1), pp. 4582–4597. Association for Computational Linguistics (2021)
15. Li, Y., et al.: A deep learning-based hybrid framework for object detection and recognition in autonomous driving. IEEE Access **8**, 194228–194239 (2020)
16. Lialin, V., Deshpande, V., Rumshisky, A.: Scaling down to scale up: a guide to parameter-efficient fine-tuning. CoRR abs/2303.15647 (2023)
17. Lin, T.-Y., et al.: Microsoft COCO: common objects in context. In: Fleet, D., Pajdla, T., Schiele, B., Tuytelaars, T. (eds.) ECCV 2014. LNCS, vol. 8693, pp. 740–755. Springer, Cham (2014). https://doi.org/10.1007/978-3-319-10602-1_48
18. Loedeman, J., Stol, M.C., Han, T., Asano, Y.M.: Prompt generation networks for efficient adaptation of frozen vision transformers. CoRR abs/2210.06466 (2022)
19. Loshchilov, I., Hutter, F.: Decoupled weight decay regularization. In: ICLR (Poster). OpenReview.net (2019)
20. Mao, J., Huang, J., Toshev, A., Camburu, O., Yuille, A.L., Murphy, K.: Generation and comprehension of unambiguous object descriptions. In: Proceedings of the IEEE Conference on Computer Vision and Pattern Recognition, pp. 11–20 (2016)
21. Nagaraja, V.K., Morariu, V.I., Davis, L.S.: Modeling context between objects for referring expression understanding. In: Leibe, B., Matas, J., Sebe, N., Welling, M. (eds.) ECCV 2016. LNCS, vol. 9908, pp. 792–807. Springer, Cham (2016). https://doi.org/10.1007/978-3-319-46493-0_48
22. Radford, A., et al.: Learning transferable visual models from natural language supervision. In: ICML. Proceedings of Machine Learning Research, vol. 139, pp. 8748–8763. PMLR (2021)

23. Wu, W., Chang, T., Li, X.: Visual-and-language navigation: a survey and taxonomy. arXiv preprint arXiv:2108.11544 (2021)
24. Yang, Z., Chen, T., Wang, L., Luo, J.: Improving one-stage visual grounding by recursive sub-query construction. In: Vedaldi, A., Bischof, H., Brox, T., Frahm, J.-M. (eds.) ECCV 2020. LNCS, vol. 12359, pp. 387–404. Springer, Cham (2020). https://doi.org/10.1007/978-3-030-58568-6_23
25. Yang, Z., Gong, B., Wang, L., Huang, W., Yu, D., Luo, J.: A fast and accurate one-stage approach to visual grounding. In: Proceedings of the IEEE/CVF International Conference on Computer Vision, pp. 4683–4693 (2019)
26. Yu, L., et al.: MAttNet: modular attention network for referring expression comprehension. In: Proceedings of the IEEE Conference on Computer Vision and Pattern Recognition, pp. 1307–1315 (2018)
27. Yu, L., Poirson, P., Yang, S., Berg, A.C., Berg, T.L.: Modeling context in referring expressions. In: Leibe, B., Matas, J., Sebe, N., Welling, M. (eds.) ECCV 2016. LNCS, vol. 9906, pp. 69–85. Springer, Cham (2016). https://doi.org/10.1007/978-3-319-46475-6_5
28. Zhang, H., Niu, Y., Chang, S.F.: Grounding referring expressions in images by variational context. In: Proceedings of the IEEE Conference on Computer Vision and Pattern Recognition, pp. 4158–4166 (2018)
29. Zhou, K., Yang, J., Loy, C.C., Liu, Z.: Conditional prompt learning for vision-language models. In: CVPR, pp. 16795–16804. IEEE (2022)
30. Zhou, Y., et al.: TRAR: routing the attention spans in transformer for visual question answering. In: Proceedings of the IEEE/CVF International Conference on Computer Vision, pp. 2074–2084 (2021)

A Graph-Involved Lightweight Semantic Segmentation Network

Xue Xia, Jiayu You, and Yuming Fang$^{(\boxtimes)}$

Jiangxi University of Finance and Economics, Nanchang 330013, China
leo.fangyuming@foxmail.com

Abstract. To extract cues for pixelwise segmentation in an efficient way, this paper proposes a lightweight model that involves graph structure in the convolutional network. First, a cross-layer module is designed to adaptively aggregate hierarchical features according to the feature relations within multi-scale receptive fields. Second, a graph-involved head is presented, capturing long-range channel and feature dependencies in two sub-domains. Specifically, channel dependency is acquired in a compact spatial domain for context-aware information, while the feature dependency is obtained in the graph feature domain for category-aware representation. Afterwards, by fusing the features with long-range dependencies, the network outputs the segmentation results after a learning-free upsampling layer. Experimental results present that this model remains light while achieving competitive performances in segmentation, proving the effectiveness and efficiency of the proposed sub-modules. (https://github.com/xia-xx-cv/Graph-Lightweight-SemSeg/).

Keywords: Semantic Segmentation · Graph Convolution Network · Light-weight Network

1 Introduction

Being able to support downstream tasks such as scene understanding and autonomous driving, semantic segmentation results are required to be accurate. However, many well-designed semantic segmentation models obtain state-of-the-art performance at the cost of high computation consumption brought by a large number of learnable weights or matrix multiplication. Therefore, these semantic segmentation processes might hinder these tasks from practical application if too much time is spent on computing the masks or storing parameters in memory.

Reducing the parameters without heavily dropping the performance is crucial in designing a lightweight semantic segmentation model. The key points lie in: 1) information reusing to efficiently aggregate information without introducing extra feature extractors, 2) compact feature representations to provide discriminative category-dependency cues, 3) area- or object-level context information for

This work was supported by National Natural Science Foundation 62162029 and 62271237, Jiangxi Natural Science Foundation 20224BAB212010 and Foundation for Distinguished Young Scholars of Jiangxi Province 20224ACB212005.

better scene parsing. The first goal can be implemented by short-connections, while the latter two need specific designing on feature transformation or relationship modeling. Considering these, we intend to make full use of convolution-based spatial and hierarchical features, and we look forward to a graph-based feature representation to model dependencies among pixels for image content understanding.

In this paper, we propose a lightweight graph-involved convolution network for semantic segmentation, providing feature re-using and capturing long-range dependencies. The contributions are three folds,

- We propose a cross-layer feature fusion module to reuse the shallow-to-deep information adaptively according to the attention maps computed within the multi-scale receptive fields across layers.
- We design a graph-involved module to simultaneously extract features in graph domain and a compact spatial domain, separately computing feature and channel dependencies that support category-aware and context-aware cues for semantic segmentation.
- We present a lightweight semantic segmentation network that introduces feature re-using and long-range dependency modeling. The network reaches a trade-off between accuracy and computation.

2 Related Works

2.1 Convolution-Based Semantic Segmentation Models

Full convolutional networks with encoder-decoder structures [2,20] have exhibited comparative capability in segmentation. However, models suffer from accuracy limitation when excluding the global relationship of features. Subsequently, some works focus on capturing object context through global pixel relationships to compensate for convolutional local features. Zhang et al. [39] proposed ENC-Net, in which the encoding semantic loss guided the proposed context encoding modules to understand the global semantics with a very small extra computation cost. Besides, the involved loss enlarged the dependent areas by considering objects rather than isolated pixels during training. Yuan et al. [38] involved non-local [32] to compute pixel relations under category-dependencies, which improved multi-scale modules like ASPP [3]. Fu et al. [9] proposed position and channel attention modules to adaptively combine local features with their global dependencies through the self-attention-based weighted sum. Guo et al. [11] re-emphasized and proved the efficiency and effectiveness of convolution attention for contexts in semantic segmentation. Based on this, we emphasize that convolutions still contributes a lot in high-level vision tasks.

On the other hand, some works endeavour to fuse high-level and low-level features for better scene understanding in semantic segmentation. Ding et al. [7] designed a patch attention module, enabling local attention for improved contextual embedding, and an attention embedding module, fusing high-level and low-level features. Sun et al. [29] proposes a successive pooling attention module to exploit both high-level and low-level features. These works prove that feature re-using benefits the segmentation by combining multiple levels of information in hierarchical networks.

2.2   Graph-Involved Segmentation Models

Graph structures model the non-euclidean relations among data [16] through the aggregation of nodes and edges without having any order, providing ways to hierarchical representations [23] for existing deep networks.

Graph attention mechanism [31] allows nodes to be adaptively assigned with weights according to node relations, providing solutions to both inductive and transductive tasks. Zhang *et al.* [40] presented a dual graph convolution network for simultaneously obtaining channel interdependencies and pixel relations, providing long-range contexts for semantic segmentation. Lu *et al.* [21] adopted a graph model initialized by a fully convolutional network to solve the semantic segmentation problem in a graph node classification way.

Being viewed as an extension of graph attention, the Transformer [30] establishes a fully-connected graph structure [28] among pixels at the cost of a higher computation consumption. To alleviate the memory cost, Huang *et al.* [13] proposed a lightweight module to compute global self-attention within a criss-cross area for semantic segmentation. Nevertheless, this mechanism only captures long-range information in the spatial domain, remaining propagation within a single graph. Wu *et al.* [34] used the bidirectional graph attention mechanism to exploit the interaction between the area sub-graph and the instance sub-graph, enabling more complex graph reasoning for panoptic segmentation. Yet we look forward to a simpler sub-graph structure for semantic segmentation.

2.3   Lightweight Segmentation Models

Lightweight semantic segmentation models [22,24,26,44] shed the light on accurate segmentation on resource-limited applications. Howard *et al.* [12,27] introduced depthwise separable convolutions and inverted residual blocks to reduce model parameters and computational efforts, while Zhang *et al.* [42] proposed adopted pointwise group convolution and channel shuffle as the solution. Gao *et al.* [10] and Wu *et al.* [33] mainly relied on the channel-wise convolutions, thinning the network by the sparse connections and preserving inter-channel information for segmentation accuracy.

Liu *et al.* [19] designed a multi-branch structure with short connections to gather context cues from different feature maps, which indicated the effectiveness of information re-using. To deal with complex objects in remote-sensing images, Li *et al.* [17] proposed lightweight attention modules in both spatial and channel domains using depth-wise separable convolutions and 1D-factorized convolutions. Adopting depth-wise convolutions and dilated convolutions, Li *et al.* [18] captured low-level spatial details and high-level context information using two separate branches, while Zhang *et al.* [43] built a lightweight attention-guided asymmetric network for feature extraction, feature fusion and upsampling. These works prove that capturing local and long-range context information benefits semantic segmentation in scene parsing and relation modeling.

3 Proposed Method

The overall frame work of our method is shown in Fig. 1 consisting of three stages, each of which contains multiple convolutional layers. The first stage works as a stem module, projecting raw pixels into a compact feature space, while the latter two modules exploit hierarchical local features. Based on this backbone, we design a cross-layer aggregation module to adaptively integrate information with the biggest activation values, and present a graph-involved head to capture channel and feature dependencies for pixel classification.

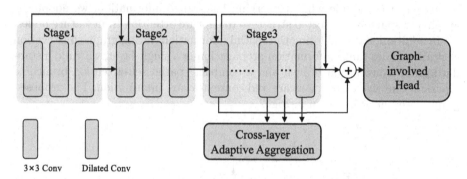

Fig. 1. The overall framework of our method, consisting of a multi-stage backbone with residual connections and a proposed graph-involved head. The gray rectangle with round corners are convolution layers, while the blue ones indicate context layers based on multi-scale convolutional layers. (Color figure online)

3.1 Cross-Layer Attention Supported Adaptive Aggregation

For information reusing, features from different layers are always fused through concatenation or summation before a convolution layer. Although different weights computed by convolution kernels are assigned to different channels, the pixels at different 2D coordinates within a channel share the same weights. Therefore, we argue that these simple fusion ways fail to reflect the importance of variance across areas within one feature channel.

To this end, we present an attention-supported method to aggregating information from existing previous layers while considering spatial relations among pixels. As shown in Eq. (1), $\mathbf{X}_i$ represents features from different layers of the backbone, d stands for the feature dimension and f_i are convolution layers that modify the dimensions.

$$\mathbf{X}_{\mathrm{agg}} = softmax(f_1(\mathbf{X}_1)f_3(\mathbf{X}_2)^T/\sqrt{d})\mathbf{X}_3 + \mathbf{X}_0 \qquad (1)$$

This aggregation strategy computes the relations between pixel pairs at two different depths and applies the relations to feature maps at another depth. In other words, the three sets of feature maps are aggregated in a self-attention way instead of a summation or weighted summation after a concatenation. Especially, the cross-layer attention maps actually aggregate information in a higher order since $\mathbf{X}_i$ already fused the features within the corresponding receptive field from $\mathbf{X}_{i-1}$.

3.2   Graph-Involved Head for Long-Range Dependency

Long-range information contributes to object-level information description. However, Transformers densely model the relations between every pixel, hence matrix multiplication is required, bringing computation and time consumption. In order to simplify the module while preserving long-range information, we adopt the graph as a sparsely-connected structure to capture attention from feature nodes.

In addition, to avoid losing the spatial information during graph construction that allows long-range information propagation in a single layer, we build a standard graph convolution branch and a compact spatial self-attention branch to separately involve category-aware and context-aware information. The proposed graph-based head is shown in Fig. 2.

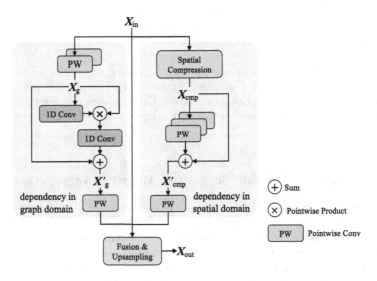

Fig. 2. The proposed graph-involved head contains a graph convolution-based path in the feature domain and a channel dependency path in a compact spatial domain.

Specifically, the input feature $\mathbf{X}_{in}$ is first projected into a graph domain by $\mathbf{Z} = \phi_1(\mathbf{X}_{in})$, where $\phi(\cdot)$ is a convolution layer for feature transform and $\mathbf{Z}$ is the feature that contains information from all nodes. The graph convolution operation is defined as Eq. (2),

$$f(\mathbf{Z}, \mathbf{A}) = \sigma((\mathbf{A} + \mathbf{I})\mathbf{Z}\mathbf{W}) \qquad (2)$$

where $\mathbf{A}$ stands for the adjacent matrix to be learnt, $\sigma(\cdot)$ indicates *Sigmoid* function, $\mathbf{I}$ is an identity matrix for self-loop and $\mathbf{W}$ is the convolution weights to be learnt. Besides, the normalization is omitted here for simplification.

$$\mathbf{X}'_g = \phi_2(\mathbf{Z}f(\mathbf{Z}, \mathbf{A})) \qquad (3)$$

Since we aim to retain the spatial structure, the updated feature is projected back from the graph to the latent space along the feature axis in $\mathbf{Z}$ by Eq. (3). The $\mathbf{X}'_g$ indicates the output feature of this branch and $\phi_2(\cdot)$ refers to a convolution layer for further pulling the feature to the pixel domain.

The above path offers category-aware features by modeling long-range dependency in graph domain through graph convolutions. While the other path presents context-aware features by computing channel dependency and preserving spatial structure in a compact spatial domain, according to Eq. (4).

$$\mathbf{X}'_{cmp} = \sigma(f_q(\mathbf{X}_{cmp})f_k(\mathbf{X}_{cmp}))f_v(\mathbf{X}_{cmp}) + \mathbf{X}_{cmp} \qquad (4)$$

Where the spatially compressed feature $\mathbf{X}_{cmp}$ is obtained by $f_{cmp}(\mathbf{X}_{in})$. $f_q(\cdot), f_k(\cdot)$ and $f_v(\cdot)$ stand for different convolution layers and $\sigma(\cdot)$ stands for the normalization operation. Since the global attention is acquired along channels at every pair of spatial coordinates within a compact spatial space, this module gives cues to area-level contexts for semantic segmentation tasks.

Afterwards, the graph feature and channel dependencies are fused, with a conventional residual path for stability, before being upsampled to become output masks. Noting that learnable modules such as convolutions or deconvolutions are not involved in upsampling, which might bring inaccuracy or miss the potential correction process. In other words, it's a trade-off between accuracy and lightweight.

4 Experimental Analysis

4.1 Implementation Details

Datasets. We evaluate our model on the Cityscapes [5] and Pascal-Person-Parts [4] datasets, both of which are commonly used for semantic segmentation. The Cityscapes dataset was collected from 50 different cities and consisted of 5,000 images with pixel-level annotations at a high resolution of 2048 × 1024. In our experiments, we leverage the training set and validation set containing 2075 images and 500 images, respectively. Pascal-Person-Parts is a human structure segmentation dataset that contains 1,716 training images and 1,817 test images.

The dataset divides the human body into six parts: head, torso, upper arms, lower arms and upper and lower legs.
Implementation Protocol. We implement our experiments on a single Nvidia GeForce RTX 3090 GPU, CUDA11.6, and cuDNN V8 on Ubuntu 18.04.4 LTS. Our net is trained using the ADAM optimizer [15], the batch size is 16 for both datasets, and weight decay is set to 0.0005. We use a varied learning rate with the initial value of 0.001. For data augmentation, we employ the mean subtraction, random scaling with factors {0.5, 0.75, 1, 1.5, 1.75, 2.0} and random flipping. After the augmentation, the data from Cityscapes and Pascal-Person-Parts are cropped into 680 × 680 and 480 × 480, respectively. The proposed network is trained for 450 epochs on the Cityscapes dataset and 600 epochs on the Pascal-Person-Parts dataset from scratch. In the meanwhile, mean intersection-over-union(mIoU) and parameters(Params) are adopted to evaluate segmentation accuracy and model size, respectively.

4.2 Comparisons

The comparison results of semantic segmentation are recorded in Table 1 on Cityscapes and Table 2 on PASCAL-Person-Part. Mean intersection-over-union (mIoU) is computed by averaging the IoU of all categories across testing samples. In Table 1, the mIoUs of ESPNet [22], CGNet [33], LightSeg-ShuffleNet [8], DGCNet(Res101) [40], WaveMix-Lite 128/8 [14], SegFormer(MiT-B0) [35], PPL-LCNet-1x[[6], TopFormer-Tiny [41] and MGD [25] are cited from the paper [1,6,8,14,25,33,36,40]. Noting that we adopt CGNet as our backbone, the performance of CGNet is provided as a baseline instead of comparison. It's clear that our model achieves the second-best result with the 4th-less parameters. Especially, we outperform some models with a lighter model. DGCNet is involved here because it contains a graph-based head after a 101-layer-ResNet backbone, which indicates that the increase of mIoU may rely on a large amount of learnable weights. Figure 3 presents visual outcomes of segmentation performed on the

Table 1. Comparison results in term of mIoU on Cityscapes val set.

Methods	Resolutions	Params(M)	mIoU(%)	Rank
ESPNetv1 [22]	1024 × 512	0.4	61.4	
CGNet [33]	2048 × 1024	0.5	63.5	
LightSeg-ShuffleNet [8]	2048 × 1024	3.5	66.1	
DGCNet(Res101) [40]	2048 × 1024	72.5	80.5	1
WaveMix-Lite 128/8 [14]	512 × 256	2.9	63.3	
SegFormer(MiT-B0) [35]	1024 × 512	7.7	62.6	
PPL-LCNet-1x [6]	1024 × 512	3.0	66.0	
TopFormer-Tiny [41]	1024 × 512	1.4	66.1	3
MGD [25]	800 × 800	1.8	62.6	
Ours	2048 × 1024	1.5	66.8	2

Cityscapes validation set. Notably, the bilateral framework achieves enhanced results in terms of finer details.

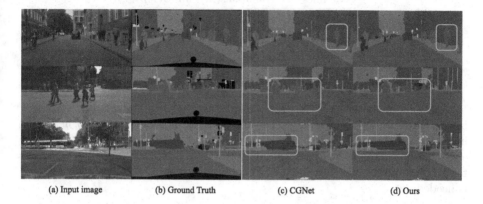

(a) Input image   (b) Ground Truth   (c) CGNet   (d) Ours

Fig. 3. Visual results on the Cityscapes validation set. The differences are marked in the boxes with rounded corners, which indicates that our model achieves clearer objects with better inner consistency.

Table 2. Comparison results in term of mIoU on PASCAL-Person-Part val set in the size of 480 × 480.

Methods	Params(M)	mIoU(%)	Rank
SegNet [2]	29.5	52.9	
ESPNetv1 [22]	0.4	54.8	
BiSeNet-res18 [37]	27.0	62.5	1
CGNet [33]	0.5	56.9	3
LightSeg-ShuffleNet [8]	3.5	58.5	
Ours	1.5	62.1	2

Since image size, image number and categories of dataset PASCAL-Person-Part are smaller than those of Cityscapes, we only involve some latest models and a classical SegNet for comparison in this part. Our model achieves the second-best results among the comparison models. The BiSeNet outperforms our net by 0.4% while it needs more parameters of 25.5M than our model.

4.3 Ablation Study

To gain insights into our design, we present ablation studies on Cityscapes val set and the results are recorded in Table 3. Since the cross-layer aggregation module is very light, the parameter number varies little.

Table 3. Comparison results in the term of mIoU on Cityscapes val set. GiH is short for Graph-involved Head, and CAA is short for Cross-layer Adaptive Aggregation.

Modules	Params(M)	mIoU(%)
Base	0.5	63.5
+ GiH	1.5	64.5
+ GiH + CAA	1.5	66.8

It is obvious that by adding the graph-involved head, the mIoU increases by 1%. Together with cross-layer feature re-using module, the graph head can help increase the mIoU by 3.3%. This does not indicate that cross-layer adaptive aggregation dominant the boosting, since the gradient back propagation across the whole network will be affected. Therefore, further ablation or visualization is needed.

5   Conclusion

We propose a lightweight network that involves graph convolution and a self-attention mechanism for semantic segmentation. To fully reuse information and aggregate features in an efficient way, we present a cross-layer adaptive aggregation module for the backbone. This module integrates features by considering the dependencies among pixels at different layers within multi-scale receptive fields. To exploit long-range information and context features, we also design a graph-involved module that computes channel dependencies in a compact spatial domain and graph dependencies in feature domain. The features from separate sub-domain convey category-aware and localization-aware cues for semantic segmentation. The model achieves relatively high segmenting accuracy with less parameters by relying on depthwise and dilated convolutions and well-designed self-attention-supported sub-modules. Our future work may focus on how to involve interaction between the sub-domains and on how to present classification corrections in upsampling stages.

References

1. Atif, N., Mazhar, S., Sarma, D., Bhuyan, M., Ahamed, S.R.: Efficient context integration through factorized pyramidal learning for ultra-lightweight semantic segmentation. arXiv preprint arXiv:2302.11785 (2023)
2. Badrinarayanan, V., Kendall, A., Cipolla, R.: SegNet: a deep convolutional encoder-decoder architecture for image segmentation. IEEE Trans. Pattern Anal. Mach. Intell. **39**(12), 2481–2495 (2017)
3. Chen, L.C., Papandreou, G., Kokkinos, I., Murphy, K., Yuille, A.L.: DeepLab: semantic image segmentation with deep convolutional nets, atrous convolution, and fully connected CRFs. IEEE Trans. Pattern Anal. Mach. Intell. **40**(4), 834–848 (2017)

4. Chen, X., Mottaghi, R., Liu, X., Fidler, S., Urtasun, R., Yuille, A.: Detect what you can: detecting and representing objects using holistic models and body parts. In: Proceedings of the IEEE Conference on Computer Vision and Pattern Recognition, pp. 1971–1978 (2014)
5. Cordts, M., et al.: The cityscapes dataset for semantic urban scene understanding. In: Proceedings of the IEEE Conference on Computer Vision and Pattern Recognition, pp. 3213–3223 (2016)
6. Cui, C., et al.: PP-LCNet: a lightweight CPU convolutional neural network. arXiv preprint arXiv:2109.15099 (2021)
7. Ding, L., Tang, H., Bruzzone, L.: LANet: local attention embedding to improve the semantic segmentation of remote sensing images. IEEE Trans. Geosci. Remote Sens. **59**(1), 426–435 (2020)
8. Emara, T., Abd El Munim, H.E., Abbas, H.M.: LiteSeg: a novel lightweight convnet for semantic segmentation. In: 2019 Digital Image Computing: Techniques and Applications (DICTA), pp. 1–7. IEEE (2019)
9. Fu, J., et al.: Dual attention network for scene segmentation. In: Proceedings of the IEEE/CVF Conference on Computer Vision and Pattern Recognition, pp. 3146–3154 (2019)
10. Gao, H., Wang, Z., Ji, S.: ChannelNets: compact and efficient convolutional neural networks via channel-wise convolutions. In: Advances in Neural Information Processing Systems, vol. 31 (2018)
11. Guo, M.H., Lu, C.Z., Hou, Q., Liu, Z., Cheng, M.M., Hu, S.M.: SegNeXt: rethinking convolutional attention design for semantic segmentation. In: Proceedings of 36th Conference on Neural Information Processing Systems (NeurIPS), pp. 603–612 (2022)
12. Howard, A.G., et al.: MobileNets: efficient convolutional neural networks for mobile vision applications. arXiv preprint arXiv:1704.04861 (2017)
13. Huang, Z., Wang, X., Huang, L., Huang, C., Wei, Y., Liu, W.: CCNet: criss-cross attention for semantic segmentation. In: Proceedings of the IEEE/CVF International Conference on Computer Vision, pp. 603–612 (2019)
14. Jeevan, P., Viswanathan, K., Sethi, A.: WaveMix-Lite: a resource-efficient neural network for image analysis. arXiv preprint arXiv:2205.14375 (2022)
15. Kingma, D.P., Ba, J.: Adam: a method for stochastic optimization. arXiv preprint arXiv:1412.6980 (2014)
16. Kipf, T.N., Welling, M.: Semi-supervised classification with graph convolutional networks. arXiv preprint arXiv:1609.02907 (2016)
17. Li, H., Qiu, K., Chen, L., Mei, X., Hong, L., Tao, C.: SCAttNet: semantic segmentation network with spatial and channel attention mechanism for high-resolution remote sensing images. IEEE Geosci. Remote Sens. Lett. **18**(5), 905–909 (2020)
18. Li, Y., Li, X., Xiao, C., Li, H., Zhang, W.: EACNet: enhanced asymmetric convolution for real-time semantic segmentation. IEEE Signal Process. Lett. **28**, 234–238 (2021)
19. Liu, J., Zhou, Q., Qiang, Y., Kang, B., Wu, X., Zheng, B.: FDDWNet: a lightweight convolutional neural network for real-time semantic segmentation. In: ICASSP 2020–2020 IEEE International Conference on Acoustics, Speech and Signal Processing (ICASSP), pp. 2373–2377. IEEE (2020)
20. Long, J., Shelhamer, E., Darrell, T.: Fully convolutional networks for semantic segmentation. In: Proceedings of the IEEE Conference on Computer Vision and Pattern Recognition, pp. 3431–3440 (2015)

21. Lu, Y., Chen, Y., Zhao, D., Chen, J.: Graph-FCN for image semantic segmentation. In: Lu, H., Tang, H., Wang, Z. (eds.) ISNN 2019. LNCS, vol. 11554, pp. 97–105. Springer, Cham (2019). https://doi.org/10.1007/978-3-030-22796-8_11

22. Mehta, S., Rastegari, M., Caspi, A., Shapiro, L., Hajishirzi, H.: ESPNet: efficient spatial pyramid of dilated convolutions for semantic segmentation. In: Ferrari, V., Hebert, M., Sminchisescu, C., Weiss, Y. (eds.) ECCV 2018. LNCS, vol. 11214, pp. 561–580. Springer, Cham (2018). https://doi.org/10.1007/978-3-030-01249-6_34

23. Nickel, M., Kiela, D.: Poincaré embeddings for learning hierarchical representations. In: Proceedings of the 31st International Conference on Neural Information Processing Systems, NIPS 2017, pp. 6341–6350. Curran Associates Inc., Red Hook, NY, USA (2017)

24. Paszke, A., Chaurasia, A., Kim, S., Culurciello, E.: ENet: a deep neural network architecture for real-time semantic segmentation. arXiv preprint arXiv:1606.02147 (2016)

25. Qin, J., Wu, J., Li, M., Xiao, X., Zheng, M., Wang, X.: Multi-granularity distillation scheme towards lightweight semi-supervised semantic segmentation. In: Avidan, S., Brostow, G., Cissé, M., Farinella, G.M., Hassner, T. (eds.) European Conference on Computer Vision, vol. 13690. pp. 481–498. Springer, Cham (2022). https://doi.org/10.1007/978-3-031-20056-4_28

26. Romera, E., Alvarez, J.M., Bergasa, L.M., Arroyo, R.: ERFNet: efficient residual factorized convnet for real-time semantic segmentation. IEEE Trans. Intell. Transp. Syst. 19(1), 263–272 (2017)

27. Sandler, M., Howard, A., Zhu, M., Zhmoginov, A., Chen, L.C.: MobileNetV2: inverted residuals and linear bottlenecks. In: Proceedings of the IEEE Conference on Computer Vision and Pattern Recognition, pp. 4510–4520 (2018)

28. Shao, N., Cui, Y., Liu, T., Wang, S., Hu, G.: Is graph structure necessary for multi-hop question answering? In: Proceedings of the 2020 Conference on Empirical Methods in Natural Language Processing (EMNLP), pp. 7187–7192. Association for Computational Linguistics, November 2020. https://doi.org/10.18653/v1/2020.emnlp-main.583. https://aclanthology.org/2020.emnlp-main.583

29. Sun, L., Cheng, S., Zheng, Y., Wu, Z., Zhang, J.: SpaNet: successive pooling attention network for semantic segmentation of remote sensing images. IEEE J. Sel. Top. Appl. Earth Observations Remote Sens. 15, 4045–4057 (2022)

30. Vaswani, A., et al.: Attention is all you need. In: Advances in Neural Information Processing Systems, vol. 30 (2017)

31. Veličković, P., Cucurull, G., Casanova, A., Romero, A., Lio, P., Bengio, Y.: Graph attention networks. arXiv preprint arXiv:1710.10903 (2017)

32. Wang, X., Girshick, R., Gupta, A., He, K.: Non-local neural networks. In: Proceedings of the IEEE Conference on Computer Vision and Pattern Recognition, pp. 7794–7803 (2018)

33. Wu, T., Tang, S., Zhang, R., Cao, J., Zhang, Y.: CGNet: a light-weight context guided network for semantic segmentation. IEEE Trans. Image Process. 30, 1169–1179 (2020)

34. Wu, Y., et al.: Bidirectional graph reasoning network for panoptic segmentation. In: Proceedings of the IEEE/CVF Conference on Computer Vision and Pattern Recognition, pp. 9080–9089 (2020)

35. Xie, E., Wang, W., Yu, Z., Anandkumar, A., Alvarez, J.M., Luo, P.: SegFormer: simple and efficient design for semantic segmentation with transformers. Adv. Neural. Inf. Process. Syst. 34, 12077–12090 (2021)

36. Yang, C., et al.: Pruning parameterization with bi-level optimization for efficient semantic segmentation on the edge. In: Proceedings of the IEEE/CVF Conference on Computer Vision and Pattern Recognition, pp. 15402–15412 (2023)

37. Yu, C., Wang, J., Peng, C., Gao, C., Yu, G., Sang, N.: BiSeNet: bilateral segmentation network for real-time semantic segmentation. In: Ferrari, V., Hebert, M., Sminchisescu, C., Weiss, Y. (eds.) ECCV 2018. LNCS, vol. 11217, pp. 334–349. Springer, Cham (2018). https://doi.org/10.1007/978-3-030-01261-8_20

38. Yuan, Y., Huang, L., Guo, J., Zhang, C., Chen, X., Wang, J.: OCNet: object context network for scene parsing. arXiv preprint arXiv:1809.00916 (2018)

39. Zhang, H., et al.: Context encoding for semantic segmentation. In: Proceedings of the IEEE Conference on Computer Vision and Pattern Recognition, pp. 7151–7160 (2018)

40. Zhang, L., Li, X., Arnab, A., Yang, K., Tong, Y., Torr, P.H.: Dual graph convolutional network for semantic segmentation. In: Proceedings of British Machine Vision Conference (BMVC), pp. 10.1–10.14 (2019)

41. Zhang, W., et al.: TopFormer: token pyramid transformer for mobile semantic segmentation. In: Proceedings of the IEEE/CVF Conference on Computer Vision and Pattern Recognition, pp. 12083–12093 (2022)

42. Zhang, X., Zhou, X., Lin, M., Sun, J.: ShuffleNet: an extremely efficient convolutional neural network for mobile devices. In: Proceedings of the IEEE Conference on Computer Vision and Pattern Recognition, pp. 6848–6856 (2018)

43. Zhang, X., Du, B., Wu, Z., Wan, T.: LAANet: lightweight attention-guided asymmetric network for real-time semantic segmentation. Neural Comput. Appl. **34**(5), 3573–3587 (2022)

44. Zhao, H., Qi, X., Shen, X., Shi, J., Jia, J.: ICNet for real-time semantic segmentation on high-resolution images. In: Ferrari, V., Hebert, M., Sminchisescu, C., Weiss, Y. (eds.) ECCV 2018. LNCS, vol. 11207, pp. 418–434. Springer, Cham (2018). https://doi.org/10.1007/978-3-030-01219-9_25

User-Aware Prefix-Tuning Is a Good Learner for Personalized Image Captioning

Xuan Wang[1], Guanhong Wang[1], Wenhao Chai[2], Jiayu Zhou[1],
and Gaoang Wang[1(✉)]

[1] Zhejiang University, Hangzhou, China
{xuanw,guanhongwang}@zju.edu.cn, {jiayu.21,gaoangwang}@intl.zju.edu.cn
[2] University of Washington, Seattle, USA
wchai@uw.edu

Abstract. Image captioning bridges the gap between vision and language by automatically generating natural language descriptions for images. Traditional image captioning methods often overlook the preferences and characteristics of users. Personalized image captioning solves this problem by incorporating user prior knowledge into the model, such as writing styles and preferred vocabularies. Most existing methods emphasize the user context fusion process by memory networks or transformers. However, these methods ignore the distinct domains of each dataset. Therefore, they need to update the entire caption model parameters when meeting new samples, which is time-consuming and calculation-intensive. To address this challenge, we propose a novel personalized image captioning framework that leverages user context to consider personality factors. Additionally, our framework utilizes the prefix-tuning paradigm to extract knowledge from a frozen large language model, reducing the gap between different language domains. Specifically, we employ CLIP to extract the visual features of an image and align the semantic space using a query-guided mapping network. By incorporating the transformer layer, we merge the visual features with the user's contextual prior knowledge to generate informative prefixes. Moreover, we employ GPT-2 as the frozen large language model. With a small number of parameters to be trained, our model performs efficiently and effectively. Our model outperforms existing baseline models on Instagram and YFCC100M datasets across five evaluation metrics, demonstrating its superiority, including twofold improvements in metrics such as BLEU-4 and CIDEr.

Keywords: Personalized image captioning · Prefix-tuning · Cross-modal

1 Introduction

Image captioning aims to generate descriptive sentences for image content, which has drawn a lot of attention in recent years [5,7,29]. It has a wide range of

---

X. Wang and G. Wang—Equal contribution.

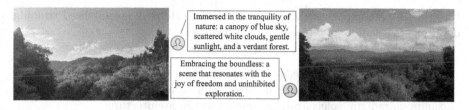

Fig. 1. Owing to the personalities of the users, similar images may exhibit varying descriptions.

applications, such as e-commerce product description, assisting individuals with visual impairments, and automating captioning for social media posts [3,23].

Previous works have explored the generation of captions using an encoder-decoder network with various attention mechanisms [30,31]. However, the captions generated by these methods seem standardized and conservative in practical applications [10]. With the advent of social media, individuals have increasingly opted to express their personalities through posting. Figure 1 illustrates similar images while highlighting the contrasting captions provided by the respective users. Consequently, there is a growing need for research on personalized image captioning. Some works [16,34] have made early attempts to solve this problem by leveraging user context, such as active vocabularies and writing styles. These methods first encode the visual content with user information and then fuse them using memory networks or transformers to generate personalized captions. Although significant improvements have been achieved in personalized image captioning, one major issue remains in most existing methods. Considering the distinct domains of each social media dataset, these approaches require training the entire caption model for each dataset. Additionally, when encountering new samples, they need to update their caption model to generate sentences for the new data. This process requires both time and computing resources.

To address the above issue, we propose a novel personalized image captioning framework based on User-Aware Prefix-Tuning (UAPT). Our approach not only incorporates user-aware prior knowledge to accommodate the distinct characteristics and preferences of individuals, but also leverages the prefix-tuning paradigm [8] to exploit the linguistic capabilities of frozen large language models in order to generate precise captions, thereby bridging the gap between different domains. At first, we use the CLIP [17] image encoder to capture the visual semantics. In addition, a query-guided Mapping Network is used to align the semantic space of vision and language. To enable personalized characteristics of users, we employ transformer layers to fuse visual knowledge with user context prior knowledge and generate a fixed-size embedding sequence denoted as prefixes [8]. These prefixes, concatenated with caption embeddings, are subsequently fed into a GPT-2 model [18] to generate captions. Throughout this process, both the CLIP and GPT-2 models remain frozen. The Mapping Network and Fusion Network are end-to-end training, thus saving time and computing resources. The framework is shown in Fig. 2. This novel approach allows us to effectively capture the user's language style without requiring extensive personalized training data, making it more applicable in real-world scenarios.

The main contributions of this article can be summarized as follows:

- To the best of our knowledge, we are the first to introduce large language models to address personalized image captioning problem with the prefix-tuning paradigm, solely training small network architecture.
- We propose the integration of user-aware prior knowledge with visual information to effectively capture the contextual language style of the user, enhancing personalized image captioning.
- Our model outperforms existing baseline models on both the Instagram and YFCC100M datasets across five evaluation metrics. Remarkably, metrics such as BLEU-4 and CIDEr show twofold improvements, providing substantial evidence of the superiority of our model.

2   Related Work

2.1   Image Captioning

Image captioning serves as a pivotal link between Computer Vision (CV) and Natural Language Processing (NLP), necessitating the generation of contextually and grammatically appropriate textual descriptions for images. Notable progress has been achieved in this field with the advent of deep learning techniques. Earlier approaches utilized convolutional neural network [6] to extract visual features from images and recurrent neural network [13] to generate captions [12,27]. Subsequent studies introduced attention mechanisms, leading to improved caption quality [1,30,31]. In addition, there has been an emergence of large-scale vision and language pre-trained models that can be applied to image captioning [29,33]. However, these models often require significant computational resources and training efforts. To address this issue, Mokady et al. [14] proposed a strategy of keeping the pre-trained weights fixed and focusing on training specific network components to bridge the gap between visual and textual representations. Despite the achievements of these efforts, they are limited to generating neutral captions, which may not fully cater to the diverse needs of users in the era characterized by the rapid emergence of social media.

2.2   Personalized Vision and Language Research

Driven by the rapid proliferation of social media and the growing demand for personalized content, numerous scholars have conducted research with a primary focus on customizing the visual and linguistic domains. Shuster et al. [22] defined 215 potential personality traits to guide caption generation. However, in real-world scenarios, users' styles may not be adequately captured by these traits. Park et al. [16] constructed datasets specifically for personalized image captioning and employed the memory network to store the users' active vocabularies, enabling the generation of personalized captions. Long et al. [10] proposed a model for personalized image captioning that can be applied across different domains, thereby expanding the range of potential application scenarios. Considering users' long-term and short-term stylistic preferences, Zhang et al. [34] proposed an approach that combines these temporal dimensions. However, the implementation of this approach necessitates the collection of more detailed data. To generate captions that align with user preferences, Wang et al. [28] utilized user-provided tags as a foundational element in their methodology. Zeng et al. [32] leveraged user profiles such as age and gender to generate social media comments. Our approach eliminates the need for supplementary information, such as user profiles or posting timestamps, and only requires training a small network.

2.3 Prompt Tuning in Image Captioning

Prompt tuning enables large-scale pre-trained models to adapt to various tasks without requiring parameter adjustments. Jin et al. [4] conducted research demonstrating that even relatively simple hard prompts can enhance the performance of image captioning. Ramos et al. [21] utilized image-based retrieval techniques to extract relevant textual information from a database, thereby generating prompts that subsequently guided the process of caption generation. Taking a distinct approach, Li et al. [7] employed soft prompts by utilizing extracted visual representations as a condition for the large language model. Building upon the insights from these prior works, our research extends the application of prompt tuning to the novel and challenging domain of personalized image captioning.

3   Methods

3.1   Problem Definition

The general personalized image captioning framework can be described as follows: Given a dataset of images, captions, and user context tuples (x^i, c^i, u^i), our objective is to learn a model to generate a caption for a target input image with a specified user language style.

During training, we optimize our model in an autoregressive manner that predicts the next token without considering future tokens by minimizing negative log-likelihood loss $\mathcal{L}$, i.e.,

$$\mathcal{L} = -\max_{\theta} \sum_{i=1}^{N} \sum_{j=1}^{L} \log p_\theta \left(c_j^i \mid x^i, u^i, c_1^i, \ldots, c_{j-1}^i \right), \tag{1}$$

where θ denotes the model trainable parameters; N is the number of samples; L is the maximum length of the generated sentence.

During the process of inference, the generation of the caption is initiated by conditioning it on both the visual prefix and the user context. The subsequent

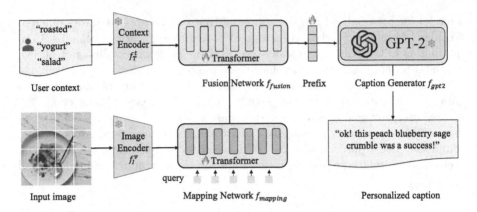

Fig. 2. Overview of our User-Aware Prefix-Tuning Network (UAPT) framework. At first, we utilize a frozen image encoder f_I^v and context encoder f_T^t to extract visual features and user-specific embeddings, respectively. Then, a query-guided mapping network $f_{mapping}$ is exploited to align vision and language semantics. Subsequently, visual knowledge and user prior knowledge are fused by a transformer based fusion network f_{fusion} to output embeddings as prefixes. Finally, the prefixes are input into a frozen large language model f_{gpt2} to generate personalized captions.

tokens are then predicted sequentially, with the assistance of the language model output. At each step, the language model assigns probabilities to all tokens in the vocabulary, and these probabilities are used to determine the next token by employing a beam search algorithm.

In the subsequent section, we introduce our proposed approach, illustrated in Fig. 2. We begin by explaining the utilization of the query-guided visual knowledge mapping module to align visual cues and language semantics. Subsequently, we demonstrate the fusion of user-aware prior knowledge with visual knowledge in order to capture the contextual style of user language for personalized image captioning. Lastly, we employ prefix tuning to prompt a frozen language model for the generation of captions.

3.2   Query-Guided Visual Knowledge Mapping

To acquire visual cues from the images, we utilize the frozen CLIP's image encoder $f_I^v(\cdot)$ to extract visual semantics.

Furthermore, we introduce a set of learnable query vectors q_i to guide the Mapping Network in extracting valuable knowledge from the existing visual feature representations. This approach provides the benefit of directing the model towards extracting feature representations that are more closely aligned with the semantics of the textual space. Simultaneously, it reduces the required training resources and accelerates the training process. Consequently, the resulting mapping representation $\mathcal{V}$ can be reformulated as follows:

$$\mathcal{V} = f_{mapping}(f_I^v(x_i), q_i). \tag{2}$$

Table 1. Statistics of the datasets.

Datasets	Posts	Users	Posts/User	Words/Post	Intra-class	Inter-class
Instagram	721,176	4,820	149.6	8.55	0.0330	0.0088
YFCC100M	462,036	6,197	74.6	6.30	0.1682	0.0044

3.3 User-Aware Prior Knowledge Fusion

Different users often exhibit unique idioms, writing styles, and other linguistic characteristics, which can be considered as personalized traits. To achieve personalized image captioning, it is crucial to incorporate user language style contexts into the network. In this regard, we utilize TF-IDF denoted as $f_{tf-idf}(\cdot)$ to represent the users' distinctive writing habits, which can measure the importance of a word in a document [20]. Specifically, we consider each user's historical postings as a document and apply the TF-IDF to extract personalized keywords for each user. These keywords capture the unique vocabulary and writing style employed by the users in their posts. Subsequently, we input these obtained keywords into the Context Encoder, instantiated as CLIP's text encoder $f_T^t(\cdot)$, to encode them and generate a distinctive embedding vector for each user.

Furthermore, we employ a transformer $f_{fusion}(\cdot)$ to fuse the user's embedded representation with the visual projection representation, enabling a comprehensive interaction between the image information and the user's linguistic style context. This fusion facilitates a stronger correlation between the semantic information of the image and the user's style. The final fusion vector, represented as the output $\mathcal{P}$, can be mathematically expressed as follows:

$$\mathcal{P} = f_{fusion}(f_T^t(f_{tf-idf}(u_i)), \mathcal{V}). \tag{3}$$

3.4 Caption Generation

During the training process, we employ a pre-trained GPT-2 model to generate image captions. Specifically, we input the fusion vectors as prefixes to the GPT-2 model, allowing the model to learn the semantic information of the images and the writing habits of the users to generate the corresponding captions. In the inference process, we utilize the beam search algorithm to generate more accurate and coherent captions. The entire process can be summarized as follows:

$$\mathcal{C} = f_{gpt2}(\mathcal{P}). \tag{4}$$

4 Experiments

4.1 Dataset

We validate the performance of our model using two datasets [16]: Instagram and YFCC100M. The datasets are partitioned following the methodology outlined in [16]. Detailed statistics regarding the datasets are presented in Table 1.

Table 2. Results of evaluation metrics on the Instagram and YFCC100M datasets. The best results for each evaluation metric are marked in bold.

Instagram

Methods	BLEU-1	BLEU-2	BLEU-3	BLEU-4	METEOR	ROUGE-L	CIDEr
ShowTell	0.055	0.019	0.007	0.003	0.038	0.081	0.004
ShowAttTell	0.106	0.015	0.000	0.000	0.026	0.026	0.049
1NN-Im	0.071	0.020	0.007	0.004	0.032	0.069	0.059
Seq2seq	0.050	0.012	0.003	0.000	0.024	0.065	0.034
1NN-Usr	0.063	0.014	0.002	0.000	0.028	0.059	0.025
1NN-UsrIm	0.106	0.032	0.011	0.005	0.045	0.104	0.084
CSMN-P5	**0.171**	0.068	0.029	0.013	0.064	**0.177**	0.214
CSMN-P5-Mul	0.145	0.049	0.022	0.009	0.049	0.145	0.143
Ours	0.161	**0.084**	**0.050**	**0.032**	**0.097**	0.160	**0.343**

YFCC100M

Methods	BLEU-1	BLEU-2	BLEU-3	BLEU-4	METEOR	ROUGE-L	CIDEr
ShowTell	0.027	0.003	0.000	0.000	0.024	0.043	0.003
ShowAttTell	0.088	0.010	0.001	0.000	0.034	0.116	0.076
1NN-Im	0.033	0.006	0.001	0.000	0.020	0.043	0.063
Seq2seq	0.076	0.010	0.000	0.000	0.034	0.066	0.069
1NN-Usr	0.032	0.003	0.001	0.000	0.016	0.051	0.028
1NN-UsrIm	0.039	0.005	0.001	0.000	0.021	0.050	0.050
CSMN-P5	0.106	0.034	0.012	0.004	0.033	0.099	0.064
CSMN-P5-Mul	**0.116**	0.036	0.010	0.003	0.036	**0.111**	0.060
Ours	0.100	**0.042**	**0.018**	**0.008**	**0.064**	0.097	**0.170**

To assess the feasibility and indispensability of personalization, we treat each user as a separate class and evaluate the intra-class and inter-class similarity of the captions. The results of this analysis, presented in the right part of Table 1, unequivocally demonstrate the existence of distinct textual preferences among individual users. This compelling evidence further supports the idea that user characteristics can be effectively captured by considering their posts.

4.2   Implementation Details

In the UAPT, both the Mapping Network and the Fusion Network consist of 5 layers of multi-head self-attention with 8 heads each. The lengths of the learnable query, visual feature representations, and user's embedding vector are $16, 16, 4$ respectively. For the Image Encoder and Context Encoder, we utilize the pre-trained image encoder and text encoder from EVA-CLIP [24]. Additionally, GPT-2 (medium) [19] is employed as the Caption Generator.

We conduct training for 6 epochs using a batch size of 50 on a single NVIDIA GeForce RTX3090. The AdamW [11] optimizer is employed with $\beta_1 = 0.9$, $\beta_2 = $

Table 3. Ablation study on the YFCC100M. The best results for each evaluation metric are marked in bold.

Methods	BLEU-1	BLEU-2	BLEU-3	BLEU-4	METEOR	ROUGE-L	CIDEr
w/o Context	0.074	0.030	0.013	0.006	0.052	0.075	0.121
w/o Mapping	0.083	0.034	0.015	0.007	0.060	0.088	0.154
w/o Fusion	0.085	0.036	0.016	0.007	0.061	0.088	0.161
w/o Query	0.083	0.035	0.015	0.006	0.061	0.089	0.152
Full Model	**0.100**	**0.042**	**0.018**	**0.008**	**0.064**	**0.097**	**0.170**

0.96, and a weight decay rate of 0.005. To schedule the learning rate, we utilize cosine learning rate decay with a peak learning rate of $6e-4$, accompanied by linear warm-up over $6k$ steps. During the inference phase, the temperature of the beam search is set to 0.8, and the beam size is set to 3.

4.3 Baselines

To demonstrate the effectiveness of our model, we select several baseline models for comprehensive comparative analysis, including: 1) **ShowTell** [27]: ShowTell is an encoder-decoder model designed for caption generation, employing an RNN decoder. 2) **ShowAttTell** [30]: ShowAttTell integrates visual attention computation to capture the importance of individual image regions during word decoding. 3) **1NN-Im**, **1NN-Usr**, and **1NN-UsrIm** [16]: These baselines employ a retrieval-based approach by utilizing the captions from the nearest training image, nearest user, or a combination of both as the basis for generating captions. 4) **Seq2seq** [26]: Seq2seq is a recursive neural network with three hidden LSTM layers. This baseline retrieves 60 active words from the querying user in a descending order of TF-IDF weights and predicts the corresponding caption. 5) **CSMN-P5** and **CSMN-P5-Mul** [16]: These baselines integrated the user's active vocabulary as part of their memory network. CSMN-P5-Mul extended CSMN-P5 by incorporating multi-layer CNNs into its architecture.

4.4 Quantitative Analysis

We utilize BLEU [15], METEOR [2], ROUGE-L [9], and CIDEr [25] as evaluation metrics. Table 2 presents the performance of UAPT and the compared baselines on the two datasets. The first part of the table corresponds to architectures without personalization, while the remaining results incorporate personalization. Across both datasets, our model outperforms the existing baseline models across five evaluation metrics. In the Instagram dataset, our model achieves significant improvements, surpassing other baseline models by more than double in metrics such as BLEU-3, BLEU-4, and CIDEr. Similarly, in the YFCC100M dataset, our model demonstrates substantial enhancements, more than doubling the performance of the baseline models in terms of BLEU-4 and CIDEr scores. These findings conclusively demonstrate the superior performance of our model.

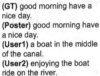

(GT) good morning have a nice day.
(Poster) good morning have a nice day.
(User1) a boat in the middle of the canal.
(User2) enjoying the boat ride on the river.
(User3) i love the view from the boat.

(GT) day 222 365 custom designed numbers everyday.
(Poster) day 355 365 custom designed numbers everyday.
(User1) just a few more days to see more of this beautiful design.
(User2) enjoyed the simplicity of this design.
(User3) i love this design.

(GT) picture 3 give the best caption photo by @username.
(Poster) picture 24 give the best caption photo credits to @username.
(User1) i'm a little bit of a squirrel.
(User2) this is my favorite picture of a squirrel.
(User3) when you see a squirrel in the park.

(GT) happy father's day.
(Poster) happy father's day.
(User1) happy father's day to you.
(User2) happy father's day to the best dad in the world.
(User3) happy father's day to my dad.

(GT) i love this fish.
(Poster) i love this fish.
(User1) a pair of fish in the aquarium at the miamisburg aquarium.
(User2) some of the fish that were in the aquarium at the time of this photo.
(User3) some fish in the water.

(GT) starry night on top of the world.
(Poster) starry night at the rim of the world.
(User1) the night sky at the dawn of the day.
(User2) moonlight over the mountains.
(User3) the moon at the top of the mountain.

(GT) tern in flight.
(Poster) snowy tern in flight.
(User1) a tern in flight.
(User2) caspian tern at dusk.
(User3) the king of all terns.

(GT) golden temple at the center of angkor wat.
(Poster) a view of the main temple at angkor wat.
(User1) another view of the upper tiers of the palace.
(User2) sunset at the top of a temple at angkor wat.
(User3) sunrise at angkor wat.

Fig. 3. Examples from Instagram (top) and YFCC100M (bottom).

4.5   Ablation Study

We perform ablation experiments to evaluate the effectiveness of each component in the model, as shown in Table 3. These experiments included the following variations: 1) **"w/o Context"** which removes the Context Encoder; 2) **"w/o Mapping"** which eliminates the Mapping Network; 3) **"w/o Fusion"** which excludes the Fusion Network; 4) **"w/o Query"** where the value of learnable query is set to 0 and the weights are frozen.

The exclusion of any component from the model results in a decline in the evaluation metrics, highlighting the indispensability of each constituent. Among the variations, the impact of "w/o Context" is the most pronounced, emphasizing the importance of user-aware prior knowledge in generating higher-quality captions. This finding further supports the notion that our model successfully constructs unique embedding sequences for each user, enabling accurate guidance in generating personalized captions.

4.6   Qualitative Analysis

Figure 3 presents authentic examples of generated captions from the Instagram and YFCC100M datasets. Each instance includes the ground truth (GT) caption, captions generated by the original posting user (Poster) and other users (User1, User2, and User3).

By employing our model, we can generate captions for Posters that are either identical or remarkably similar to the GT. It is noteworthy that even in situations where the correlation between images and text is relatively weak, as demonstrated in the second and third examples, our model can still generate captions that align with the users' preferences by acquiring knowledge of their writing patterns. Moreover, these examples reveal user-specific vocabulary tendencies. For example, User2 often uses "enjoy" in captions, and User3 frequently expresses emotions through "i love". This demonstrates our model's proficiency in capturing users' characteristics and generating captions that align with their preferences and styles. Moreover, different users produce captions in their distinctive styles for the same image, further substantiating our model's personalized generation capability.

In the YFCC100M dataset, it is noteworthy that our model demonstrates the capability to generate captions that exhibit a higher level of precision and specificity compared to the GT, as exemplified in the third example. This observation underscores the remarkable performance and potential of our model.

5   Conclusions

In this paper, we have proposed a novel personalized image captioning framework based on user-aware prefix-tuning. Our method utilizes query-guided visual knowledge mapping for aligning vision and language semantics, while also fusing user-aware prior knowledge to consider the characteristics and preferences of users. Finally, we leverage prefix-tuning to extract knowledge from frozen large language model, narrowing the gap between different semantic domains. With a small number of parameters to be trained, our model boosts the performance efficiently and effectively. The experiments conducted on two publicly available datasets demonstrate the superiority of our model and validate the contributions of our method.

In the future, we will explore how to integrate more fine-grained user context to generate more captivating and tailored descriptions for personalized image captioning.

Acknowledgments. This work is supported by the Fundamental Research Funds for the Central Universities (No. 226-2023-00045) and National Natural Science Foundation of China (No. 62106219).

References

1. Anderson, P., et al.: Bottom-up and top-down attention for image captioning and visual question answering. In: Proceedings of the IEEE Conference on Computer Vision and Pattern Recognition, pp. 6077–6086 (2018)
2. Banerjee, S., Lavie, A.: METEOR: an automatic metric for MT evaluation with improved correlation with human judgments. In: Proceedings of the ACL Workshop on Intrinsic and Extrinsic Evaluation Measures for Machine Translation and/or Summarization, pp. 65–72 (2005)
3. Hossain, M.Z., Sohel, F., Shiratuddin, M.F., Laga, H.: A comprehensive survey of deep learning for image captioning. ACM Comput. Surv. (CsUR) 51(6), 1–36 (2019)
4. Jin, W., Cheng, Y., Shen, Y., Chen, W., Ren, X.: A good prompt is worth millions of parameters? Low-resource prompt-based learning for vision-language models. arXiv preprint arXiv:2110.08484 (2021)
5. Karpathy, A., Fei-Fei, L.: Deep visual-semantic alignments for generating image descriptions. In: Proceedings of the IEEE Conference on Computer Vision and Pattern Recognition, pp. 3128–3137 (2015)
6. Krizhevsky, A., Sutskever, I., Hinton, G.E.: ImageNet classification with deep convolutional neural networks. Commun. ACM 60(6), 84–90 (2017)
7. Li, J., Li, D., Savarese, S., Hoi, S.: BLIP-2: bootstrapping language-image pre-training with frozen image encoders and large language models. arXiv preprint arXiv:2301.12597 (2023)
8. Li, X.L., Liang, P.: Prefix-tuning: optimizing continuous prompts for generation. arXiv preprint arXiv:2101.00190 (2021)
9. Lin, C.Y.: ROUGE: a package for automatic evaluation of summaries. In: Text Summarization Branches Out, pp. 74–81 (2004)
10. Long, C., Yang, X., Xu, C.: Cross-domain personalized image captioning. Multimedia Tools Appl. 79, 33333–33348 (2020)
11. Loshchilov, I., Hutter, F.: Decoupled weight decay regularization. arXiv preprint arXiv:1711.05101 (2017)
12. Mao, J., Xu, W., Yang, Y., Wang, J., Yuille, A.L.: Explain images with multimodal recurrent neural networks. arXiv preprint arXiv:1410.1090 (2014)
13. Mikolov, T., Karafiát, M., Burget, L., Cernocký, J., Khudanpur, S.: Recurrent neural network based language model. In: Interspeech, vol. 2, pp. 1045–1048, Makuhari (2010)
14. Mokady, R., Hertz, A., Bermano, A.H.: ClipCap: clip prefix for image captioning. arXiv preprint arXiv:2111.09734 (2021)
15. Papineni, K., Roukos, S., Ward, T., Zhu, W.J.: BLEU: a method for automatic evaluation of machine translation. In: Proceedings of the 40th Annual Meeting of the Association for Computational Linguistics, pp. 311–318 (2002)
16. Park, C.C., Kim, B., Kim, G.: Towards personalized image captioning via multimodal memory networks. IEEE Trans. Pattern Anal. Mach. Intell. 41(4), 999–1012 (2018)
17. Radford, A., et al.: Learning transferable visual models from natural language supervision. In: International Conference on Machine Learning, pp. 8748–8763. PMLR (2021)
18. Radford, A., Narasimhan, K., Salimans, T., Sutskever, I., et al.: Improving language understanding by generative pre-training (2018)

19. Radford, A., Wu, J., Child, R., Luan, D., Amodei, D., Sutskever, I., et al.: Language models are unsupervised multitask learners. OpenAI Blog **1**(8), 9 (2019)
20. Ramos, J., et al.: Using TF-IDF to determine word relevance in document queries. In: Proceedings of the First Instructional Conference on Machine Learning, vol. 242, pp. 29–48. Citeseer (2003)
21. Ramos, R., Martins, B., Elliott, D., Kementchedjhieva, Y.: SmallCap: lightweight image captioning prompted with retrieval augmentation. In: Proceedings of the IEEE/CVF Conference on Computer Vision and Pattern Recognition, pp. 2840–2849 (2023)
22. Shuster, K., Humeau, S., Hu, H., Bordes, A., Weston, J.: Engaging image captioning via personality. In: Proceedings of the IEEE/CVF Conference on Computer Vision and Pattern Recognition, pp. 12516–12526 (2019)
23. Stefanini, M., Cornia, M., Baraldi, L., Cascianelli, S., Fiameni, G., Cucchiara, R.: From show to tell: a survey on deep learning-based image captioning. IEEE Trans. Pattern Anal. Mach. Intell. **45**(1), 539–559 (2022)
24. Sun, Q., Fang, Y., Wu, L., Wang, X., Cao, Y.: EVA-CLIP: improved training techniques for clip at scale. arXiv preprint arXiv:2303.15389 (2023)
25. Vedantam, R., Lawrence Zitnick, C., Parikh, D.: CIDEr: consensus-based image description evaluation. In: Proceedings of the IEEE Conference on Computer Vision and Pattern Recognition, pp. 4566–4575 (2015)
26. Vinyals, O., Kaiser, L., Koo, T., Petrov, S., Sutskever, I., Hinton, G.: Grammar as a foreign language. In: Advances in Neural Information Processing Systems, vol. 28 (2015)
27. Vinyals, O., Toshev, A., Bengio, S., Erhan, D.: Show and tell: a neural image caption generator. In: Proceedings of the IEEE Conference on Computer Vision and Pattern Recognition, pp. 3156–3164 (2015)
28. Wang, L., Chu, X., Zhang, W., Wei, Y., Sun, W., Wu, C.: Social image captioning: exploring visual attention and user attention. Sensors **18**(2), 646 (2018)
29. Wang, W., et al.: Image as a foreign language: BEiT pretraining for all vision and vision-language tasks. arXiv preprint arXiv:2208.10442 (2022)
30. Xu, K., et al.: Show, attend and tell: neural image caption generation with visual attention. In: International Conference on Machine Learning, pp. 2048–2057. PMLR (2015)
31. You, Q., Jin, H., Wang, Z., Fang, C., Luo, J.: Image captioning with semantic attention. In: Proceedings of the IEEE Conference on Computer Vision and Pattern Recognition, pp. 4651–4659 (2016)
32. Zeng, W., Abuduweili, A., Li, L., Yang, P.: Automatic generation of personalized comment based on user profile. arXiv preprint arXiv:1907.10371 (2019)
33. Zhang, P., et al.: VinVL: revisiting visual representations in vision-language models. In: Proceedings of the IEEE/CVF Conference on Computer Vision and Pattern Recognition, pp. 5579–5588 (2021)
34. Zhang, W., Ying, Y., Lu, P., Zha, H.: Learning long-and short-term user literal-preference with multimodal hierarchical transformer network for personalized image caption. In: Proceedings of the AAAI Conference on Artificial Intelligence, vol. 34, pp. 9571–9578 (2020)

An End-to-End Transformer with Progressive Tri-Modal Attention for Multi-modal Emotion Recognition

Yang Wu, Pai Peng, Zhenyu Zhang, Yanyan Zhao[✉], and Bing Qin

Harbin Institute of Technology, Harbin, China
{ywu,ppeng,zyzhang,yyzhao,qinb}@ir.hit.edu.cn

Abstract. Recent works on multi-modal emotion recognition move towards end-to-end models, which can extract the task-specific features supervised by the target task compared with the two-phase pipeline. In this paper, we propose a novel multi-modal end-to-end transformer for emotion recognition, which can effectively model the tri-modal features interaction among the textual, acoustic, and visual modalities at the low-level and high-level. At the low-level, we propose the progressive tri-modal attention, which can model the tri-modal feature interactions by adopting a two-pass strategy and can further leverage such interactions to significantly reduce the computation and memory complexity through reducing the input token length. At the high-level, we introduce the tri-modal feature fusion layer to explicitly aggregate the semantic representations of three modalities. The experimental results on the CMU-MOSEI and IEMOCAP datasets show that ME2ET achieves the state-of-the-art performance. The further in-depth analysis demonstrates the effectiveness, efficiency, and interpretability of the proposed tri-modal attention, which can help our model to achieve better performance while significantly reducing the computation and memory cost (Our code is available at https://github.com/SCIR-MSA-Team/UFMAC.).

Keywords: Multi-modal emotion recognition · Multi-modal transformer · Feature fusion

1 Introduction

Multi-modal emotion recognition aims at detecting emotion in the utterance, which consists of three modal inputs, including textual, visual, and acoustic. It has gained increasing attention from people since it is vital for many downstream applications such as natural human-machine interaction and mental healthcare.

Most of the existing studies [11,12] generally use a two-phase pipeline, first extracting uni-modal features from input data and then performing the proposed multi-modal feature fusion method on the features. Different from this line of works, some works train the feature extraction and multi-modal feature fusion modules together in an end-to-end manner since they consider that this approach enables the model to extract the task-specific features supervised by the

© The Author(s), under exclusive license to Springer Nature Singapore Pte Ltd. 2024
Q. Liu et al. (Eds.): PRCV 2023, LNCS 14431, pp. 396–408, 2024.
https://doi.org/10.1007/978-981-99-8540-1_32

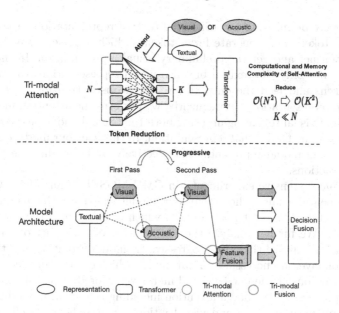

Fig. 1. Illustration of ME2ET and the proposed tri-modal attention. Our proposed tri-modal attention can reduce the complexity of self-attention from $\mathcal{O}(N^2)$ to $\mathcal{O}(K^2)$, which makes ME2ET memory and computation-efficient.

emotion prediction loss. Dai *et al.* [3] propose the MESM model, which takes the CNN blocks as the feature extraction module and uses the cross-modal attention to capture bi-modal feature interactions. However, MESM only captures the features interactions between the textual and either acoustic or visual modalities, ignoring leveraging the feature interactions between the acoustic and visual modalities, which is very useful for understanding the sentiment semantics [11].

To address this problem, we propose the multi-modal end-to-end transformer (ME2ET) shown in Fig. 1, which can capture the tri-modal feature interactions among the textual, acoustic, and visual modalities effectively at the low-level and high-level. At the high-level, we propose the tri-modal feature fusion layer to sufficiently fuse the three uni-modal semantic representations. At the low-level, we propose the progressive tri-modal attention. The main idea of it is *to generate fewer but more informative visual/acoustic tokens based on the input visual/acoustic tokens by leveraging the tri-modal feature interactions*. Through reducing the length of the input tokens, we can significantly reduce the computation and memory complexity. If the length of the input tokens is N and the length of the generated tokens is $K(K \ll N)$, which is a hyper-parameter, we can reduce the computation and memory complexity of the self-attention block in the visual/acoustic transformer from $\mathcal{O}(N^2)$ to $\mathcal{O}(K^2)$.

To be more specific, we adopt a simple two-pass strategy to capture the tri-modal feature interactions in the proposed progressively tri-modal attention. In the first pass, considering the visual data contains more noise as some parts of the face may be missing, we utilize the textual representation to attend the input visual tokens and generate fewer tokens, which are then passed into the visual transformer and get the preliminary visual representation. In the second pass, we use not only the textual but also visual representations to attend the acoustic tokens and feed the outputs into the acoustic transformer producing the acoustic representation. Subsequently, we perform the attention mechanism on the original visual tokens again using both the textual and acoustic representations and get the final visual representation. In this way, our model can obtain the semantic uni-modal representations effectively by leveraging the tri-modal feature interactions.

We conduct extensive experiments on CMU-MOSEI [1] and IEMOCAP [2]. The experimental results show that our model surpasses the baselines and achieves the state-of-the-art performance, which demonstrates the effectiveness of our model. We further analyze the contribution of each part of our model, and the results indicate that the progressive tri-modal attention and tri-modal feature fusion layer are necessary for our model, which can capture the tri-modal feature interactions at the low-level and high-level. Moreover, the in-depth analysis of the progressive tri-modal attention including the ablation study, computation and memory analysis, and visualization analysis shows its effectiveness, efficiency, and interpretability.

The main contributions of this work are as follows:

- We propose the progressive tri-modal attention, which can help our model to achieve better performance while significantly reducing the computation and memory cost by fully exploring the tri-modal feature interactions.
- We introduce the multi-modal end-to-end transformer (ME2ET), which is a simple and efficient *purely* transformer-based multi-modal emotion recognition framework.
- We evaluate ME2ET on two public datasets and ME2ET obtains the state-of-the-art results. We also conduct an in-depth analysis to show the effectiveness, efficiency and interpretability of ME2ET.

2   Related Work

There are two lines of works conducted on utterance-level multi-modal emotion recognition. One line of works adopts a two-phase pipeline, first extracting features and then fusing them. MulT [11] uses the cross-modal attention to capture the bi-modal interactions. Self-MM [12] proposes to leverage the estimated uni-modal labels to help the model learn better uni-modal representations. DMD [9] presents a decoupled multimodal distillation to enhance the discriminative features of each modality. The other line of works trains the whole model in an end-to-end manner since they consider the extracted features may not be suitable for the target task and can not be fine-tuned, which may lead to sub-optimal performance. MESM [3] applies the VGG blocks to extract the visual and acoustic features and proposes the cross-modal attention to make the model focus on the important features. TSL-Net [14] explores finding which parts of the video

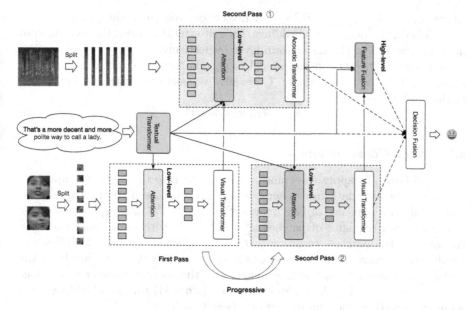

Fig. 2. An illustration of the proposed ME2ET model. ME2ET uses the progressive tri-modal attention and the tri-modal feature fusion layer to fully explore the tri-modal feature interactions at the low-level and high-level.

convey sentiment. Zhang et al. [13] propose a cross-modal temporal erasing network to encode the whole video context instead of the keyframes. In this paper, we focus on the end-to-end multi-modal emotion recognition and propose an efficient multi-modal end-to-end transformer (ME2ET).

3 Approach

In this section, we introduce the proposed multi-modal end-to-end transformer (ME2ET) in detail, which is shown in Fig. 2. ME2ET adopts the progressive tri-modal attention to leverage the tri-modal feature interactions at the low-level to generate fewer but more useful tokens based on the input tokens, which significantly reduces the memory and computation cost. In addition, ME2ET also uses the tri-modal feature fusion layer to model the tri-modal feature interactions at the high-level.

Transformer. We first introduce the transformer architecture. The transformer architecture can be described as follows.

$$\mathbf{z}_0 = [\mathbf{x}_{cls}; \mathbf{t}] + \mathbf{E}_{pos}$$
$$\mathbf{z}'_l = \mathrm{MSA}(\mathrm{LN}(\mathbf{z}_{l-1})) + \mathbf{z}_{l-1} \tag{1}$$
$$\mathbf{z}_l = \mathrm{FFN}(\mathrm{LN}(\mathbf{z}'_l)) + \mathbf{z}'_l$$

where $\mathbf{x}_{cls}$, $\mathbf{t}$, $\mathbf{E}_{pos}$, MSA, FFN, and LN denote the embedding of [CLS], the input tokens, the position embeddings, the self-attention layer, the feed-forward neural network, and the LayerNorm layer respectively.

We can use the transformer model to encode the input tokens $\mathbf{t}$ producing $\mathbf{z}_L$, where L is the layer number. Generally, we take the representation of $\mathbf{x}_{cls}$ as the output.

$$\mathbf{z}_L = \text{Transformer}(\mathbf{t}) \tag{2}$$

3.1   Token Construction

Visual Tokens. To apply the transformer architecture to the visual input, we transform the raw data into tokens, which can be passed to the transformer model. Given a sequence of face images $\mathbf{x}^{\mathbf{v}} \in \mathbb{R}^{J \times H \times W \times C}$, we split each image in them into a sequence of patches $\mathbf{x}_{\mathbf{p}}^{\mathbf{v}} \in \mathbb{R}^{P \times P \times C}$ without overlap, and then concatenate the patches, resulting in Q patches, where $Q = JHW/P^2$ is the number of patches. J, H, W, and C are the image number, the height of the face image, the width of the face image, and the channel number of the face image respectively. We then map each patch into a 1D token embedding using a linear projection layer, producing Q tokens $\mathbf{t}^{\mathbf{v}} \in \mathbb{R}^{Q \times d}$.

Acoustic Tokens. For the acoustic input, we first convert the input audio waveform into the spectrogram $\mathbf{x}^{\mathbf{a}} \in \mathbb{R}^{F \times T \times 1}$ and then split it into rectangular patches $\mathbf{x}_{\mathbf{p}}^{\mathbf{a}} \in \mathbb{R}^{F \times 2 \times 1}$ in the temporal order to keep the timing information, where F is the dimension of features and T is the frame number. Finally, we apply a linear layer to transform the patches into tokens, producing M tokens $\mathbf{t}^{\mathbf{a}} \in \mathbb{R}^{M \times d}$.

Textual Tokens. We construct the textual tokens following BERT [6]. We then pass the textual tokens to BERT and obtain the representation of [CLS], denoted as v_l.

3.2   Progressive Tri-Modal Attention

To capture tri-modal feature interactions at the low-level, we propose the progressive tri-modal attention, which consists of two passes. In the first pass, we obtain the preliminary visual representation by leveraging the textual information. Note that, we do not directly take this representation as the final visual representation, as we consider that incorporating the acoustic information can further improve the visual representation. Specifically, the preliminary visual representation is obtained as follows. Given the visual tokens $\mathbf{t}^{\mathbf{v}} \in \mathbb{R}^{Q \times d}$ and acoustic tokens $\mathbf{t}^{\mathbf{a}} \in \mathbb{R}^{M \times d}$, instead of passing the transformed tokens directly into the transformer block, we first use the proposed attention mechanism to generate new visual tokens $\mathbf{z}^{\mathbf{v}} \in \mathbb{R}^{K \times d}$ leveraging the textual information and the token number K is a hyper-parameter and is smaller than Q and M. This approach can significantly reduce the computation and memory complexity. Then we pass the $\mathbf{z^v}$ to the visual transformer and obtain the preliminary visual representation v_{one}.

$$\mathbf{z^v} = \text{softmax}([\mathbf{t^v};\mathbf{v^l}]W_{lv} + b_{lv})^\top \mathbf{t^v}$$
$$v_{one} = \text{Transformer}_v(\mathbf{z^v}) \tag{3}$$

where $\mathbf{v^l}$ is obtained by repeating v_l, $W_{lv} \in \mathbb{R}^{2 \cdot d \times K}$ and $b_{lv} \in \mathbb{R}^K$.

In the second pass, we utilize the textual representation v_l and the preliminary visual representation v_{one} to guide the model to address the important acoustic tokens and obtain the acoustic representation a.

$$\mathbf{z^a} = \text{softmax}([\mathbf{t^a};\mathbf{v^v};\mathbf{v^l}]W_{la} + b_{la})^\top \mathbf{t^a}$$
$$a = \text{Transformer}_a(\mathbf{z^a}) \tag{4}$$

where $\mathbf{v^l}$, $\mathbf{v^v}$ are obtained by repeating v_l and v_{one} respectively, $W_{la} \in \mathbb{R}^{3 \cdot d \times K}$, and $b_{la} \in \mathbb{R}^K$.

Subsequently, we perform the attention over the original visual tokens again since in the first pass the acoustic information is not utilized, which is useful for selecting informative visual tokens. The generated tokens are then passed into **the same visual transformer** used in the first pass, producing the refined visual representation v. Finally, we get the visual representation v and acoustic representation a by leveraging the tri-modal feature interactions.

$$\mathbf{z^{\hat{v}}} = \text{softmax}([\mathbf{t^v};\mathbf{v^a};\mathbf{v^l}]W_{l\hat{v}} + b_{l\hat{v}})^\top \mathbf{t^v}$$
$$v = \text{Transformer}_v(\mathbf{z^{\hat{v}}}) \tag{5}$$

where $\mathbf{v^l}$, $\mathbf{v^a}$ are obtained by repeating v_l and a respectively, $W_{l\hat{v}} \in \mathbb{R}^{3 \cdot d \times K}$, and $b_{l\hat{v}} \in \mathbb{R}^K$.

3.3   Multi-modal Feature Fusion

To capture the tri-modal feature interactions at the high-level, we propose the tri-modal feature fusion layer to fuse the semantic representations obtained by the transformer models and predict the results. Finally, we apply the decision fusion layer to aggregate the predicted results and generate the final prediction label.

$$p_{fusion} = [v;a;v_l]W_{fusion} + b_{fusion}$$
$$p_v = vW_{fv} + b_{fv}$$
$$p_a = aW_{fa} + b_{fa} \tag{6}$$
$$p_l = v_lW_{fl} + b_{fl}$$
$$p = ([p_v^\top;p_a^\top;p_l^\top;p_{fusion}^\top]W_{decision})^\top$$

where $W_{fv}, W_{fa}, W_{fl} \in \mathbb{R}^{d \times C}$, $W_{fusion} \in \mathbb{R}^{3 \cdot d \times C}$, $W_{decision} \in \mathbb{R}^{4 \times 1}$, $b_{fv}, b_{fa}, b_{fl} \in \mathbb{R}^C$, $b_{fusion} \in \mathbb{R}^C$, and C is the class number.

Table 1. Results on the IEMOCAP dataset. †indicates the results are copied from [3]. The best results are in **bold**.

Models	Angry		Excited		Frustrated		Happy		Neutral		Sad		Average	
	ACC.	F1	ACC.	F1	ACC.	F1	ACC.	F1	ACC.	F1	ACC.	F1	ACC.	F1
LF-LSTM†	71.2	49.4	79.3	57.2	68.2	51.5	67.2	37.6	66.5	47.0	78.2	54.0	71.8	49.5
LF-TRANS†	81.9	50.7	85.3	57.3	60.5	49.3	85.2	37.6	72.4	49.7	87.4	57.4	78.8	50.3
EmoEmbs†	65.9	48.9	73.5	58.3	68.5	52.0	69.6	38.3	73.6	48.7	80.8	53.0	72.0	49.8
MulT†	77.9	60.7	76.9	58.0	72.4	57.0	80.0	46.8	74.9	53.7	83.5	65.4	77.6	56.9
FE2E†	88.7	63.9	89.1	61.9	71.2	57.8	90.0	44.8	**79.1**	58.4	89.1	65.7	84.5	58.8
MESM†	88.2	62.8	88.3	61.2	74.9	58.4	89.5	47.3	77.0	52.0	88.6	62.2	84.4	57.4
FE2E-BERT	**90.6**	64.4	87.3	59.9	72.4	59.8	90.6	43.4	77.3	54.0	90.0	65.2	84.7	57.8
MESM-BERT	86.6	63.0	89.2	57.6	75.6	58.5	90.7	43.1	77.5	54.6	89.2	65.3	84.8	57.0
FE2E-TRANS	88.5	66.4	86.8	62.1	77.7	60.7	**92.3**	46.4	77.1	55.5	91.5	69.7	85.7	60.1
MESM-TRANS	87.3	64.6	87.8	63.2	78.4	62.2	90.0	46.3	77.1	55.2	92.0	71.2	85.4	60.5
ELRL	90.1	**66.8**	88.5	**66.8**	77.7	57.0	90.5	**48.5**	78.1	56.6	90.7	69.6	85.9	60.9
Ours	89.8	65.9	**89.2**	63.9	**79.2**	**60.7**	90.0	44.7	78.8	**58.7**	**92.4**	**73.8**	**86.5**	**61.3**

Table 2. Results on the MOSEI dataset. †indicates the results are copied from [3]. The best results are in **bold**.

Models	Angry		Disgusted		Fear		Happy		Sad		Surprised		Average	
	WACC.	F1	WACC.	F1	WACC.	F1	WACC.	F1	WACC.	F1	WACC.	F1	WACC.	F1
LF-LSTM†	64.5	47.1	70.5	49.8	61.7	22.2	61.3	73.2	63.4	47.2	57.1	20.6	63.1	43.3
LF-TRANS†	65.3	47.7	74.4	51.9	62.1	24.0	60.6	72.9	60.1	45.5	62.1	24.2	64.1	44.4
EmoEmbs†	66.8	49.4	69.6	48.7	63.8	23.4	61.2	71.9	60.5	47.5	63.3	24.0	64.2	44.2
MulT†	64.9	47.5	71.6	49.3	62.9	25.3	**67.2**	**75.4**	64.0	48.3	61.4	25.6	65.4	45.2
FE2E†	67.0	49.6	**77.7**	**57.1**	63.8	26.8	65.4	72.6	65.2	49.0	**66.7**	**29.1**	67.6	47.4
MESM†	66.8	49.3	75.6	56.4	65.8	28.9	64.1	72.3	63.0	46.6	65.7	27.2	66.8	46.8
FE2E-BERT	66.8	49.4	72.8	55.0	66.2	29.1	66.5	71.4	64.4	48.4	62.5	28.1	66.5	46.9
MESM-BERT	66.5	49.2	77.6	54.0	69.2	28.4	62.7	72.2	63.6	47.7	60.5	26.3	66.7	46.3
FE2E-TRANS	65.7	48.1	74.4	54.5	62.3	27.1	64.3	72.2	64.8	48.3	64.2	27.7	66.0	46.3
MESM-TRANS	65.4	47.8	73.1	55.4	62.2	27.4	61.8	72.6	64.2	48.0	63.4	28.1	65.0	46.6
ELRL	67.5	50.2	76.3	57.0	69.0	29.0	63.0	72.6	65.5	49.2	65.7	27.6	67.8	47.6
Ours	**67.9**	**51.1**	76.4	56.4	**69.3**	**29.3**	66.4	73.2	**66.2**	**50.0**	63.3	27.7	**68.3**	**48.0**

4   Experiments

4.1   Datasets

We evaluate our model on CMU-MOSEI [1] and IEMOCAP [2]. We use the reorganized datasets released by the previous work [3] since the original datasets do not provide the aligned raw data. IEMOCAP consists of 7,380 annotated utterances and each utterance is labeled with an emotion label from the set {*angry, happy, excited, sad, frustrated,* and *neutral*}. The CMU-MOSEI dataset consists of 20,477 annotated utterances and each utterance is annotated with multiple emotion labels from the set {*happy, sad, angry, fearful, disgusted,* and *surprised*}. Each utterance in the datasets consists of three modalities: an audio file, a text transcript, and a sequence of sampled image frames. Following prior works [3], we use the Accuracy (Acc.) and F1-score to evaluate the models on IEMOCAP. For the CMU-MOSEI dataset, we take the weighted Accuracy (WAcc.) and F1-score as the metrics.

4.2 Training Details

We use Adam as the optimizer. We use the binary cross-entropy loss to train our model on both datasets. The learning rate is set to 1e−4. The epoch number and batch size are set to 40 and 8 respectively. The token number K is set to 256. The max lengths of the visual and textual tokens are set to 576 and 300 respectively. The max lengths of the acoustic tokens are set to 512 and 1024 for IEMOCAP and MOSEI respectively, considering the average duration of the utterances in CMU-MOSEI is much longer than IEMOCAP. For the hyper-parameters of the transformer models, d, k, d_h, and L are 768, 12, 64, and 12 respectively. For the visual modality, we adopt MTCNN to detect faces from the input images. The image resolution of the obtained face image is 128 by 128. We split the face images into 16 × 16 patches following DeiT [10]. For the acoustic modality, we convert the input audio waveform into a sequence of 128-dimensional log-Mel filterbank (fbank) features computed with a 25ms Hamming window every 10ms and split it into 128 × 2 patches following AST [5].

4.3 Baselines

We compare our proposed model with the following two lines of baselines. One line of models adopts a two-phase pipeline. **LF-LSTM** first uses LSTMs to encode the input features and then fuses them. **LF-TRANS** uses the transformer models to encode the input features and fuses them for prediction. **EmoEmbs** [4] leverages the cross-modal emotion embeddings for multi-modal emotion recognition. **MulT** [11] utilizes the cross-modal attention to fuse multi-modal features. This line of models take the extracted features as the inputs. The other baselines are trained in an end-to-end manner. **FE2E** [3] first uses two VGG models to encode visual and acoustic inputs and then utilizes ALBERT [7] to encode the textual input. Finally, it uses a decision fusion layer to fuse the predicted results. **MESM** [3] utilizes the cross-modal sparse CNN block to capture bi-modal interactions. **FE2E-BERT** and **MESM-BERT** take the BERT models as the textual encoders. **FE2E-TRANS** and **MESM-TRANS** utilize BERT, AST, and DeiT as the backbone models. **ELRL** [8] proposes a combination of emotion-level embedding over the fused multi-modal features to learn the emotion-related features from multi-modal representation. These baselines take the raw data as the input.

4.4 Experimental Results

We compare our model with baselines, and the experimental results are shown in Table 1 and Table 2. We can see from the results that our model surpasses all baseline models on two datasets. We attribute the success to strong model

Table 3. Results of the ablation study on the IEMOCAP and MOSEI datasets. The best results are in **bold**.

Models	IEMOCAP		MOSEI	
	Avg.Acc	Avg.F1	Avg.Acc	Avg.F1
ME2ET	**86.5**	**61.3**	**68.3**	**48.0**
w/o Two-pass	86.2	60.7	68.1	47.5
w/o Feature Fusion	85.9	60.1	66.9	47.3
w/o Attention	86.2	58.1	66.8	47.1

capacity and effective feature fusion. Specifically, our model uses three transformer models to model the raw input data, which can capture the global intramodal dependencies. Moreover, we propose the progressive tri-modal attention and tri-modal feature fusion layer to model the tri-modal feature interactions, which enables the model to predict labels by leveraging the tri-modal information effectively. Comparing with the two-phase pipeline models, we find that the end-to-end models can achieve better performance, and we attribute it to that the end-to-end models can extract more task-discriminative features supervised by the target task loss. The comparison between MESM and FE2E shows that although MESM utilizes the bi-modal attention to capture the interactions between textual and acoustic/visual based on FE2E, MESM obtains worse results than FE2E on two datasets. This observation indicates that MESM fails to balance the computation cost with the model performance. Besides, we replace the backbone networks of baselines with the powerful pre-trained transformers for a fair comparison and they obtain worse results than our proposed method.

5   Analysis

5.1   Ablation Study

We conduct the ablation study to distinguish the contribution of each part. There are several variants of our model. **ME2ET** is our proposed full model. **ME2ET w/o Attention** does not use the progressive tri-modal attention to capture tri-modal interactions, which uses three transformer models separately to encode the inputs and applies the feature fusion and decision fusion layers to predict the result. **ME2ET w/o Two-pass** does not perform the attention over the visual tokens in the second pass to obtain the refined visual representation. **ME2ET w/o Feature Fusion** does not use the tri-modal feature fusion layer. The result of the ablation study is shown in Table 3. We observe that ablating the progressive tri-modal attention hurts the model performance, which demonstrates the importance of our proposed attention. Comparing ME2ET w/o Two-pass with ME2ET, we find that performing the attention over the visual tokens again by leveraging the tri-modal information is useful for model prediction. We also observe that utilizing the proposed tri-modal feature fusion layer can boost the model performance, which can enable ME2ET to model the tri-modal feature interactions at high-level.

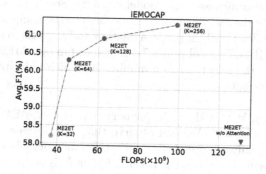

Fig. 3. Comparison of ME2ET w/o Attention and ME2ET with different token number K.

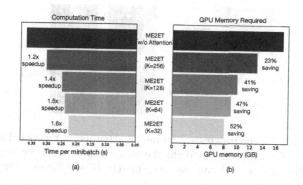

Fig. 4. Computation time **(a)** and GPU memory required **(b)** for a combined forward and backward pass on IEMOCAP.

5.2 Computation and Memory Analysis

Computation Efficiency. Firstly, we adopt the number of float-point operations (FLOPs) as our metric to measure the computation complexity. The performance of ME2ET w/o Attention and ME2ET with different token number K on IEMOCAP is shown in Fig. 3. Comparing ME2ET with ME2ET w/o Attention, we can find that our models not only significantly reduce the computation complexity but also obtain better results than ME2ET w/o Attention. Specifically, when the token number K is set to 32, our proposed model achieves better performance than ME2ET w/o Attention while only requiring about 3 times less computation. We also observe that our model obtains better results as we increase the token number. This is in line with our expectations because a larger K enables the model to address the important information from more perspectives and also increases the model capacity. Secondly, in order to more accurately evaluate the speeds of models, we analyze the computation time performance of ME2ET and ME2ET w/o Attention and show the result in Fig. 4(a). We can see that ME2ET can obtain 1.2x–1.6x speedup over ME2ET w/o Attention depending on the selected token number K, which indicates that ME2ET is computation-efficient.

Memory Efficiency. We list the GPU memory required for ME2ET and ME2ET w/o Attention in Fig. 4(b). We can observe that with our proposed tri-modal attention, ME2ET only requires 48%−−77% GPU memory while achieves better performance. Specifically, when the token number K is set to 32, ME2ET only uses 48% of the memory required by ME2ET w/o Attention. This analysis shows that ME2ET is memory-efficient.

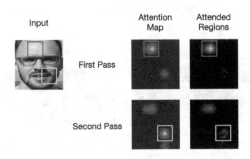

Fig. 5. Visualization of the progressive tri-modal attention of the visual modality. We use the red and yellow boxes to highlight the image regions with the larger attention weights in the first pass and second pass respectively. (Color figure online)

5.3   Visualization Analysis

To have an intuitive understanding of our proposed model, we visualize the attention weights of our progressive tri-modal attention. Specifically, we average the attention weights on the input tokens produced by different attention heads and use the obtained attention map to highlight the important regions. Based on the visualization result for the visual modality in Fig. 5, we observe that the proposed progressive tri-modal attention successfully addresses the important image regions. In the first pass, our model only uses the textual and visual information and fails to address the important patches. In the second pass, our model pays more attention to the raised lip corners by capturing the tri-modal feature interactions, which are useful for predicting emotion.

6   Conclusion

In this paper, we propose the multi-modal end-to-end transformer (ME2ET), which utilizes the progressive tri-modal attention and tri-modal feature fusion layer to capture the tri-modal feature interactions at the low-level and high-level. We conduct extensive experiments on two datasets to evaluate our proposed model and the results demonstrate its effectiveness. The computation and memory analysis shows that ME2ET is computation and memory-efficient. With the proposed progressive tri-modal attention, ME2ET can achieve better performance by fully leveraging the tri-modal feature interactions, while obtaining 1.2x–1.6x speedup and saving 23%–52% GPU memory during training. The visualization analysis shows the interpretability of our model, which can successfully address the informative tokens.

References

1. Bagher Zadeh, A., Liang, P.P., Poria, S., Cambria, E., Morency, L.P.: Multimodal language analysis in the wild: CMU-MOSEI dataset and interpretable dynamic fusion graph. In: Proceedings of the 56th Annual Meeting of the Association for Computational Linguistics (Volume 1: Long Papers), July 2018
2. Busso, C., et al.: IEMOCAP: interactive emotional dyadic motion capture database. Lang. Resour. Eval. **42**(4), 335–359 (2008)
3. Dai, W., Cahyawijaya, S., Liu, Z., Fung, P.: Multimodal end-to-end sparse model for emotion recognition. In: Proceedings of the 2021 Conference of the North American Chapter of the Association for Computational Linguistics: Human Language Technologies, pp. 5305–5316, June 2021
4. Dai, W., Liu, Z., Yu, T., Fung, P.: Modality-transferable emotion embeddings for low-resource multimodal emotion recognition. In: Proceedings of the 1st Conference of the Asia-Pacific Chapter of the Association for Computational Linguistics and the 10th International Joint Conference on Natural Language Processing, pp. 269–280, December 2020
5. Gong, Y., Chung, Y.A., Glass, J.: AST: audio spectrogram transformer. In: Proceedings Interspeech 2021, pp. 571–575 (2021)
6. Kenton, J.D.M.W.C., Toutanova, L.K.: BERT: pre-training of deep bidirectional transformers for language understanding. In: Proceedings of NAACL-HLT, pp. 4171–4186 (2019)
7. Lan, Z., Chen, M., Goodman, S., Gimpel, K., Sharma, P., Soricut, R.: ALBERT: a lite BERT for self-supervised learning of language representations. In: International Conference on Learning Representations (2020)
8. Le, H.D., Lee, G.S., Kim, S.H., Kim, S., Yang, H.J.: Multi-label multimodal emotion recognition with transformer-based fusion and emotion-level representation learning. IEEE Access **11**, 14742–14751 (2023)
9. Li, Y., Wang, Y., Cui, Z.: Decoupled multimodal distilling for emotion recognition. In: Proceedings of the IEEE/CVF Conference on Computer Vision and Pattern Recognition, pp. 6631–6640 (2023)
10. Touvron, H., Cord, M., Douze, M., Massa, F., Sablayrolles, A., Jégou, H.: Training data-efficient image transformers & distillation through attention. In: International Conference on Machine Learning, pp. 10347–10357. PMLR (2021)

11. Tsai, Y.H.H., Bai, S., Liang, P.P., Kolter, J.Z., Morency, L.P., Salakhutdinov, R.: Multimodal transformer for unaligned multimodal language sequences. In: Proceedings of the 57th Annual Meeting of the Association for Computational Linguistics, pp. 6558–6569, July 2019

12. Yu, W., Xu, H., Yuan, Z., Wu, J.: Learning modality-specific representations with self-supervised multi-task learning for multimodal sentiment analysis. In: Proceedings of the AAAI Conference on Artificial Intelligence, vol. 35, pp. 10790–10797 (2021)

13. Zhang, Z., Wang, L., Yang, J.: Weakly supervised video emotion detection and prediction via cross-modal temporal erasing network. In: Proceedings of the IEEE/CVF Conference on Computer Vision and Pattern Recognition, pp. 18888–18897 (2023)

14. Zhang, Z., Yang, J.: Temporal sentiment localization: listen and look in untrimmed videos. In: Proceedings of the 30th ACM International Conference on Multimedia, pp. 199–208 (2022)

Target-Oriented Multi-criteria Band Selection for Hyperspectral Image

Huijuan Pang[1], Xudong Sun[1], Xianping Fu[1,2], and Huibing Wang[1(✉)]

[1] School of Information Science and Technology, Dalian Maritime University, Dalian 116026, China
huibing.wang@dlmu.edu.cn
[2] Peng Cheng Laboratory, Shenzhen 518000, China

Abstract. Band selection is an effective method to reduce the dimensionality of hyperspectral data and has become a research hotspot in the field of hyperspectral image analysis. However, existing BS methods mostly rely on a single band prioritization, resulting in incomplete band evaluation. Furthermore, many BS methods use generic criteria without considering the spectral characteristics of specific targets, leading to poor application capabilities of the selected bands in target detection tasks. Therefore, this paper proposes an innovative target-oriented MCBS method for selecting the most suitable detection bands for specific objectives. Firstly, based on several baseline criteria proposed in this study, a standard decision matrix is constructed to evaluate the bands from multiple perspectives, forming a target-oriented band priority sequence. Then the method divides the original data into subspaces with low correlation and select the highest-scoring band from each subspace to form the band subset. Thus, a low-correlation subset of bands specifically designed for object detection is obtained. Experimental results on three datasets demonstrate that the proposed method outperforms several widely used cutting-edge BS methods in terms of target detection.

Keywords: Band Selection · Target Detection · Multi-criteria Decision · Band Prioritization

1 Introduction

In recent years, hyperspectral image (HSI) processing has attracted a notable surge of interest. Compared to RGB images, HSIs can capture richer discriminative physical clues through narrow continuous spectral bands. Consequently, HSIs excel at capturing and describing the spectral properties of objects, making them highly applicable in tasks such as detection and recognition of targets [1]. Nevertheless, the large dimensionality also causes severe redundancy, inefficient information processing, and Hughes phenomenon [2]. Thus, the dimensionality reduction (DR) [3,4] of HSI becomes a necessary step in addressing these issues.

This work is supported by National Natural Science Foundation of China (No. 62002041 and 62176037), Dalian Science and Technology Bureau (No. 2021JJ12GX028 and 2022JJ12GX019).

Band selection (BS) is a commonly used and effective approach in hyperspectral DR. The current BS strategy can be roughly divided into three main methods based on whether the samples are labeled or not: supervised [5], semisupervised [6], and unsupervised [7]. Due to the exorbitant cost of sample labeling, unsupervised BS methods are more favored in practical applications [8,9].

Unsupervised methods typically select a subset from the hyperspectral bands based on a specific band prioritization (BP) criterion, such as information divergence (ID) [10], variance [11], or information entropy (IE) [12]. For instance, Wang et al. introduced an adaptive subspace partition strategy (ASPS) that divides continuous bands into low-correlation subspaces and then selects bands according to the noise level and entropy [12].

Despite the promising performance achieved by the above methods, there are still several limitations that require attention. First, the majority of existing methods suffer from a limitation where bands are evaluated based on a single BP criterion, resulting in incomplete evaluation, which further leads to poor robustness and weak generalization ability when applied across different datasets. Therefore, Sun et al. proposed multi-criteria-based BS (MCBS), which utilizes entropy, variance, and noise level to construct a decision matrix for evaluating bands from different perspectives [13]. The comprehensive evaluation of bands enables improved classification performance. However, for target detection tasks, the performance of MCBS is not stable enough because it relies on the common BP criteria ignoring the specific spectral characteristics of the target.

To address the aforementioned problems, this paper innovatively puts forward a target-oriented MCBS method (TOMCBS). First, a multi-criteria decision matrix is constructed to comprehensively evaluate the bands in terms of information content, noise level and target discrimination ability. Thus, a target-oriented band priority sequence is obtained. Then, the initial hyperspectral image bands is divided into multiple contiguous subspaces based on the similarity between bands to reduce redundancy among the selected subset. Finally, the optimal band subset is composed of the bands with the highest target-oriented priority selected from each subspace. Experimental results illustrate that the bands selected by TOMCBS exhibits superior target discrimination ability compared to several cutting-edge methods.

2   Related Works

Currently, typical unsupervised BS methods can generally be divided into ranking-based methods, clustering-based methods, and other pattern methods, depending on the selection strategy employed for the application.

Ranking-based methods primarily measure the significance of each band through BP criteria, and then quantify the results of bands built upon the chosen count of bands to obtain a subset of bands in a sorted order. Representative methods include information divergence [10], constrained band selection (CBS) [10], information entropy (IE) [12], maximum variance principal component analysis (MVPCA) [11], etc. In the aforementioned methods, information divergence is frequently employed to evaluate image quality based on a

Gaussian probability distribution. It commonly opts for the maximum count of non-Gaussian bands to constitute a subset, thereby effectively representing the information content within the bands. MVPCA constructs a covariance matrix using the hyperspectral data, from which eigenvalues and eigenvectors are computed to form a loading factor matrix. Ultimately, the variances of all bands are sorted in a descending order. These methods have achieved promising results, but all of these methods share a common drawback, which is the relatively high correlation among the chosen bands.

Clustering-based methods group bands into clusters based on similarity and the band subset consists of bands that are close to the cluster center. The typical methods are hierarchical clustering [14], enhanced fast density-peak-based clustering (E-FDPC) [15], and optimal clustering framework (OCF) [16], etc. Among these methods, the hierarchical clustering structure maximizes the ratio of between-class variance to within-class variance as well as the ratio between bands. This method uniquely utilizes mutual information or Kullback-Leibler divergence to quantify the similarity between bands. However, while considering both metrics, it's not ideal for noisy datasets. The OCF utilizes the scores obtained from the fast density-peak clustering method to individually select one band within each segment determined by dynamic programming partitioning. Compared to other algorithms, its limitations become apparent when a greater quantity of bands is chosen. It introduces significant errors by solely relying on the variance-based band power ratio for band recommendation. Furthermore, clustering-based BS methods may also lead to the omission of important information bands.

To tackle the previously mentioned concerns, some spatial partition-based band selection methods that merge the benefits of ranking and clustering approaches have emerged. Firstly, the raw data is partitioned into different subspaces. Then, the most indicative bands are chosen based on the provided criteria. Typical approaches comprise the adaptive subspace partition strategy (ASPS) [12] and constrained target BS with subspace partition (CTSPBS) [17]. However, these methods are still constrained by the BP criteria selection, leading to inadequate evaluation of bands and limited generalization capability across diverse datasets.

As a result, the incorporation of multiple BP criteria into the band selection process has garnered widespread attention. For instance, hypergraph spectral clustering BS (HSCBS) [18] and sparsity regularized deep subspace clustering (SRDSC) [19] BS thoroughly illustrate that the amalgamation of diverse BP criteria can lead to a more all-encompassing and exhaustive band evaluation, ultimately enhancing classification accuracy. However, these methods are unable to freely select the chosen criteria and cannot quantify the importance of each criterion. The MCBS [13] method innovatively introduces a weighting measure between different criteria. However, it relies on common band partitioning criteria and does not consider the spectral characteristics specific to particular objectives.

In conclusion, by combining the advantages of the MCBS and ASPS approaches and incorporating the consideration of target-specific spectral information, we introduce a target-oriented MCBS method for target detection.

3   Methodology

3.1   Target-Oriented Band Priority by MCBS

The basic principle of TOMCBS is to regard hyperspectral BS as a multiple criteria decision-making (MCDM) problem, thus obtaining a target-oriented band prioritization sequence. Specifically, a MCDM matrix integrating multiple BP criteria is first constructed to integrate multiple BP criteria so as to comprehensively evaluate the bands from different perspectives. In TOMCBS, IE, noise estimation (NE) and maximum variance (MaxV) are selected as the three benchmark BP of the decision matrix to evaluate the information content, noise level and target recognition ability of the band respectively.

IE is commonly employed to quantify the amount of information in information theory, which is computed as

$$H(\mathbf{x}_l) = -\sum_{\zeta \in \Phi} p(\zeta) \log p(\zeta), \tag{1}$$

where $\mathbf{x}_l$ represents the l-th band, ζ is the grayscale value, Φ denotes the collection of all grayscale values associated with $\mathbf{x}_l$ and $p(\zeta)$ represents the probability density function. It is worth mentioning that $p(\zeta)$ can be calculated more efficiently by utilizing the grayscale histogram.

NE estimates the global noise level based on the local variance. To begin with, every band image is segmented into miniature segments consisting of E*E pixels. Let S_i denotes the value of the i-th pixel. The mean density (MD) of each block is determined by $MD = \frac{1}{E^2} \sum_{i=1}^{E^2} S_i$. Consequently, the global density (LD) of each block is calculated as

$$LD = \frac{1}{(E^2 - 1)} \sum_{i=1}^{E^2} (S_i - MD)^2. \tag{2}$$

Next, for each band image, the range between the maximum variance ($maxLD$) and minimum variance ($minLD$) across all blocks is divided into C equal-width intervals, $C = (maxLV - minLV)/\alpha$. Here, α represents the partition granularity, which is set to 3 in this article. Then, by allocating the blocks based on the values of LV, the NE value for each band can be obtained, $N = (N_1, N_2, \ldots, N_L)$.

MaxV is based on the concept of constrained energy minimization (CEM) and measures the priority of bands based on the variance induced by specific targets, which is defined by

$$V(\mathbf{x}_l) = \left(\mathbf{d}_{\mathbf{x}_l}^T \mathbf{Y}_{\mathbf{x}_l}^{-1} \mathbf{d}_{\mathbf{x}_l}\right)^{-1}, \tag{3}$$

where V represents the variance generated by the specific target, $\mathbf{x}_l$ symbolizes the column vector of the l-th band image, prior information $\mathbf{d}$ denotes the spectral characteristics of the specific target intended for detection, and $\mathbf{Y}_{\mathbf{x}_l}$ is the sample autocorrelation matrix of the l-th band image. Furthermore, MaxV is used to measure the priority relationship between two bands. Specifically, let $\mathbf{X}$ be the subset of all bands, MaxV is defined as

$$\mathbf{x}_i \prec \mathbf{x}_j \Leftrightarrow V\left(\mathbf{X} - \mathbf{x}_i\right) < V\left(\mathbf{X} - \mathbf{x}_j\right), \tag{4}$$

where the symbol "$\prec$" is denotes "superior to".

Secondly, the conceptual model of MCBS is defined as a multi-criteria decision problem with L bands and K BP criteria ($K = 3$ in this paper). The decision matrix $\mathbf{D}$ is obtained as

$$\mathbf{D} = \begin{array}{c} \\ \mathrm{BP}_1 \\ \mathrm{BP}_2 \\ \mathrm{BP}_3 \end{array} \begin{array}{c} \mathbf{x}_1 \ \mathbf{x}_2 \ \dots \ \mathbf{x}_L \\ \left[\begin{array}{cccc} a_{11} & a_{12} & \dots & a_{1L} \\ a_{21} & a_{22} & \dots & a_{2L} \\ a_{31} & a_{32} & \dots & a_{3L} \end{array} \right] \end{array}, \tag{5}$$

where BP_1, BP_2 and BP_3 respectively represent the three baseline criteria mentioned earlier in this section. a_{kl} corresponds to the evaluation score of the l-th band based on the k-th BP criterion. Taking into account the negative correlation between NE and the desired objective, each score of the evaluation criterion in the MCDM matrix requires to be positivized. Additionally, in order to facilitate comparisons between the different criteria, matrix $\mathbf{D}$ is further normalized as

$$\hat{a}_{kl} = \frac{a_{kl}}{\sum_{l=1}^{L} a_{kl}} \quad s.t. \ 1 \le l \le L, 1 \le k \le K. \tag{6}$$

The normalized decision matrix is created by

$$\hat{\mathbf{D}} = \begin{array}{c} \\ \mathrm{BP}_1 \\ \mathrm{BP}_2 \\ \mathrm{BP}_3 \end{array} \begin{array}{c} \mathbf{x}_1 \ \mathbf{x}_2 \ \dots \ \mathbf{x}_L \\ \left[\begin{array}{cccc} \hat{a}_{11} & \hat{a}_{12} & \dots & \hat{a}_{1L} \\ \hat{a}_{21} & \hat{a}_{22} & \dots & \hat{a}_{2L} \\ \hat{a}_{31} & \hat{a}_{32} & \dots & \hat{a}_{3L} \end{array} \right] \end{array}. \tag{7}$$

In order to rank the bands, positive ideal solution (**PIS**) and negative ideal solution (**NIS**) are introduced, which are defined as

$$\mathbf{PIS} = \left(\hat{a}_1^+, \hat{a}_2^+, \cdots, \hat{a}_K^+\right)^T \ s.t. \ \hat{a}_k^+ = \max_{l \in [1,L]} \left\{\hat{a}_{kl}^+\right\}, 1 \le k \le K, \tag{8}$$

$$\mathbf{NIS} = \left(\hat{a}_1^-, \hat{a}_2^-, \cdots, \hat{a}_K^-\right)^T \ s.t. \ \hat{a}_k^- = \min_{l \in [1,L]} \left\{\hat{a}_{kl}^-\right\}, 1 \le k \le K. \tag{9}$$

Then, distances between each band and **PIS** as well as **NIS** are calculated. Specifically, the distances d_l^+ and d_l^- between the l-th band and **PIS** and **NIS**, respectively, can be calculated by

$$d_l^+ = \left(\sum_{i=1}^{K} |\mathbf{w} \odot (\mathbf{PIS} - \hat{a}_{il})|^2\right)^{\frac{1}{2}}, 1 \le l \le L, \tag{10}$$

$$d_l^- = \left(\sum_{i=1}^{K} |\mathbf{w} \odot (\hat{a}_{il} - \mathbf{NIS})|^2 \right)^{\frac{1}{2}}, 1 \le l \le L, \tag{11}$$

where "$\odot$" denotes the Hadamard product, indicating elementwise multiplication. $\mathbf{w}$ is the weight vector. A detailed description of the weighting calculation method will be provided in the following. Finally, all the bands are sorted in ascending order based on their proximity to the ideal solutions. The proximity measure r_l can be calculated as

$$r_l = \frac{d_l^+}{d_l^+ + d_l^-}, 1 \le l \le L. \tag{12}$$

A smaller value of r_l corresponds to a higher priority for the band. Based on the value of r_l, MCBS can derive a final sequence of bands that balances multiple BP criterions.

It should be noted that each criterion possesses varying capabilities to assess the bands. In order to address this issue, a weight estimation method is employed to quantify the contributions of different BP criteria, which is calculated as

$$w_k = \frac{div_k}{\sum_{k=1}^{K} div_k} \; s.t. \; 1 \le k \le K, \tag{13}$$

where div_k is the degree of diversity, it can be represented as

$$div_k = 1 - e_k \; s.t. \; 1 \le k \le K, \tag{14}$$

where e_k expresses the entropy of all bands under the k-th BP criterion and it is specifically denoted by

$$e_k = -\frac{1}{\ln L} \sum_{l=1}^{L} \hat{a}_{kl} \ln(\hat{a}_{kl}) \; s.t. \; 1 \le k \le K. \tag{15}$$

Then, a weight vector $\mathbf{w} = (w_1, w_2, \cdots, w_K), 1 \le k \le K$ is obtained to describe the relative significance of each BP criterion, where w_k is the weight for the k-th evaluation criterion. In general, as the dispersion of information entropy increases, the corresponding criterion exhibits a stronger ability to differentiate between bands, leading to an increase in its associated weight.

3.2   The Implementation of TOMCBS

Due to the high correlation between bands, there is a significant redundancy that does not meet the requirements of fast object detection. Inspired by ASPS, the TOMCBS method utilizes the SP method to partition the initial bands into M subspaces and selects the highest-scoring bands to form an optimal band subset.

Firstly, the original hyperspectral data $\mathbf{x}_1, \mathbf{x}_2, ..., \mathbf{x}_L \in \mathbb{R}^{WH \times L}$ is divided into several subcubes $\mathbf{P}_i$ using the coarse partitioning strategy, where W and H

represent the width and height of each band, respectively. The count of bands is still represented by the variable L. Let M be the number of final selected bands, and then determine the number of bands in each subspace $\mathbf{P}_i$ by $Z = L/M$.

Then, a fine partitioning strategy is employed to further divide the subcubes $\mathbf{P}_i$, which can accurately assigns each spectral band to a specific subcube. The key to the fine partitioning strategy lies in identifying the division points of adjacent subspaces. Prior to this, it is necessary to obtain the similarity between bands. Therefore, the similarity matrix $\mathbf{Dis}_{ij}$ can be constructed as

$$\mathbf{Dis}_{ij} = \sqrt{\sum_{n=1}^{WH} (\mathbf{b}_{ni} - \mathbf{b}_{nj})^2}. \tag{16}$$

The precise partitioning points u between two adjacent subcubes ($\mathbf{P}_i$ and $\mathbf{P}_{i+1}$) can be obtained by

$$u = \arg\left\{ \max_u \frac{Dis^{inter}}{Dis^{intra}} \right\}, \tag{17}$$

where Dis^{inter} represents the distance between adjacent subspaces $\mathbf{P}_i$ and $\mathbf{P}_{i+1}$. Dis^{intra} denotes the distance within a subspace, which consists of two components. Specifically, Dis^{inter} and Dis^{intra} are defined as follows:

$$Dis^{inter} = \max \left| \mathbf{Dis}_{ij} \right| \ \ s.t. \ 1 \le i < u < j \le 2Z, \tag{18}$$

$$Dis^{intra} = \frac{1}{u(u-1)} \sum_{i=1}^{u} \sum_{j=1}^{u} \mathbf{Dis}_{ij} + \frac{1}{(2Z-u-1)(2Z-u-2)} \sum_{i=u+1}^{2Z} \sum_{j=u+1}^{2Z} \mathbf{Dis}_{ij}. \tag{19}$$

Next, the Eq. (17) is used to continuously update u in to obtain refined subcube partitioning points. Thus, all bands are divided into subspaces $\mathbf{P}_i \in \{\mathbf{P}_1, \mathbf{P}_2, ..., \mathbf{P}_M\}$. Within each subspace, the bands are highly correlated, while the correlation between subspaces is low. Finally, the highest-scoring bands are selected from each subspace $\mathbf{P}_i$ to obtain an ideal band subset $\mathbf{X}_{BS}$ with reduced internal redundancy and robust representational capability. Specifically, the implementation of TOMCBS is described as follows

---

Algorithm. TOMCBS

Input: HSI data $\mathbf{X} = \{\mathbf{x}_1, \mathbf{x}_2, ..., \mathbf{x}_L\}$, the sample spectral of target $\mathbf{d}$.
Output: The selected band subset $\mathbf{X}_{BS}$.
1. The standardized matrix $\widehat{\mathbf{D}}$ is constructed based on three BP evaluation criterion.
2. Calculate and acquire the weight vector $\mathbf{w}$.
3. A target-oriented band priority sequence is obtained based on the degree of proxim -ity to the **PIS** and **NIS**.
4. The initial HSI data $\mathbf{X}$ is adaptively divided into band subset $\mathbf{P}_i$ by ASPS.
5. Select the band with the smallest r_l value in $\mathbf{P}_i$ to make up the optimal band subs -et $\mathbf{X}_{BS}$.
Output: $\mathbf{X}_{BS}$

---

4   Experiments

To validate the efficacy of the TOMCBS method, this study performed comparative experiments on three publicly available hyperspectral datasets using six existing methods. Next, we will introduce the experimental configuration, covering various perspectives such as comparison methods employed, band number selection, and evaluation criteria. Furthermore, based on the experimental data, the performance of the seven BS methods used will be qualitatively and quantitatively analyzed to demonstrate the strengths of the TOMCBS method.

4.1   Datasets

The San Diego 1 scene, captured from the Airborne Visible/Infrared Imaging Spectrometer (AVIRIS) images, depicts the landscape surrounding the San Diego airport area in California, USA. The scene encompasses a size of 100 * 100 pixels. After excluding bands with lower signal-to-noise ratio (SNR) and those affected by water absorption, the remaining spectral bands in this scene amount to 189. Similarly, the ABU-airport 2 scene and ABU-airport 3 scene were also manually extracted from AVIRIS images, both possessing a resolution of 100 * 100 pixels. Both the ABU-airport 2 and ABU-airport 3 scenes consist of 205 spectral bands.

4.2   Parameter Settings and Classifier

Based on the calculation of VD using the Harsanyi-Farrand-Chang (HFC) method [20]. In this article, M is chosen by using HFC with $PF = 10^{-3}$, where $M = 10$ for San Diego 1, $M = 15$ for ABU-airport 2, and $M = 12$ for ABU-airport 3.

This experiment utilizes a target-oriented backpropagation criterion to select the most suitable bands for object detection. Therefore, the CEM detector is implemented. Additionally, the efficiency of different BS methods is compared through visual inspection and the area under the curve (AUC) values. To facilitate a more intuitive comparison of experimental results, three AUC metrics are set in this experiment to quantitatively analyze the detection results in Fig. 1. Among them, the AUC value of (P_D, P_F) is an important metric for assessing the overall performance of target detection, where a value closer to 1 indicates better performance of the method in target detection. The AUC value of (P_D, τ) represents the ability to detect target pixels, and a higher value is desirable. Conversely, the AUC value of (P_F, τ) denotes the background suppression capability, where a smaller value is preferred.

4.3   Comparison Experiments

To evaluate the effectiveness of TOMOBS on each dataset, a comparative analysis and research were conducted using six cutting-edge BS methods. First-ly, IE, NE, and MVPCA were compared as simple and effective BS methods. Then,

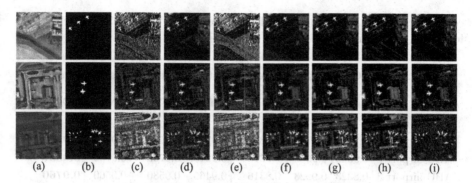

(a)   (b)   (c)   (d)   (e)   (f)   (g)   (h)   (i)

Fig. 1. Detection results using three different band subsets selected by (c) IE, (d) NE, (e) MVPCA, (f) MaxV, (g) E-FDPC, (h) ONR, (i) TOMCBS. The detection result images in the first row, second row, and third row correspond to the San Diego1, AUB-airport2, and AUB-airport3 datasets, respectively. (a) and (b) are the corresponding False-color image and ground truth.

a constraint-based BS method, MaxV was chosen as a comparative method. Additionally, the proposed method was compared with the latest BS methods, E-FDPC and ONR.

Figure 1 illustrates the target detection results of seven different BS methods, allowing for a visual comparison of their performance in terms of target response and background pixel suppression. As shown in Fig. 1, the first row displays the detection result images of the band subsets picked by the seven methods for the San Diego 1 dataset. The middle row showcases the detection outcomes of the chosen band subsets on the AUB-airport 2 dataset, and the third row corresponds to the AUB-airport 3 dataset. When $M = 10$, the band subsets chosen by IE and MVPCA are unable to detect the target objects effectively. On the other hand, the band subsets chosen by the NE and ONR show improved responsiveness towards the targets compared to IE and MVPCA, but they exhibit poorer ability to suppress the background. On the contrary, the bands selected by MaxV and E-FDPC exhibit better target detection performance and stronger background suppression capabilities. However, they still fall short compared to TOMCBS. When $M = 15$, the target detection performance of the bands selected by IE and MVPCA has significantly improved compared to when the number of bands was 10. However, they still remain at a disadvantage compared to other methods. The bands selected by ONR show almost no improvement compared to previous results. The bands selected by NE exhibit a decreasing trend in target response with an increase in the number of bands. Similarly, the band subset selected by E-FDPC has the drawback of insufficient background suppression. Only MaxV and TOMCBS exhibit good target response and background pixel suppression in their selected bands. When M is set to 12, the advantage of the MaxV method is also lost. Overall, the band subset selected by TOMCBS not only exhibits a strong response to the targets but also effectively suppresses background pixels compared to the other six methods. This suggests that TOMCBS is more suitable for target detection tasks.

Table 1. The AUC values of the seven distinct BS methods across three datasets.

Dataset	IE	NE	MVPCA	MaxV	E-FDPC	ONR	TOMCBS
(P_D, P_F) ↑							
San Diego 1	0.8419	0.9758	0.8342	0.9774	0.9773	0.9736	**0.9830**
ABU-airport 2	0.9029	0.9688	0.9313	0.9810	0.9739	0.9752	**0.9837**
ABU-airport 3	0.8323	0.9523	0.8316	0.9465	0.9580	0.9560	**0.9760**
(P_D, τ) ↑							
San Diego 1	0.3579	0.4890	0.3735	0.4986	0.4851	0.4837	**0.5004**
ABU-airport 2	0.2448	0.2537	0.2669	0.3417	0.3750	**0.3942**	0.3799
ABU-airport 3	0.3365	0.3551	0.3585	0.2949	0.3174	0.2543	**0.3783**
(P_F, τ) ↓							
San Diego 1	0.1086	0.0390	0.1299	0.0434	**0.0385**	0.0410	0.0440
ABU-airport 2	0.0368	**0.0257**	0.0425	0.0330	0.0417	0.0535	0.0390
ABU-airport 3	0.1325	0.0498	0.1386	**0.0326**	0.0591	0.0357	0.0501

Table 1 presents the target detection results of different BS methods on three datasets: San Diego 1, AUB-airport 2, and AUB-airport 3. It can be observed that on the San Diego 1 dataset, the MaxV and E-FDPC methods achieved relatively high AUC values of (P_D, P_F). However, the TOMCBS method outperformed them with a (P_D, P_F) AUC value of 0.9830, which is 0.0056 higher than the best-performing MaxV among the six comparative methods. Similarly, on the AUB-airport 2 dataset, the MaxV method performed well with a (P_D, P_F) AUC value of 0.9810, which is 0.0071 and 0.0058 higher than the (P_D, P_F) AUC values of the E-FDPC method (0.9739) and the ONR method (0.9752), respectively. However, it is still lower than the (P_D, P_F) AUC value of the TOMCBS method by 0.0027. The same trend is observed on the AUB-airport 3 dataset, where the TOMCBS method achieved a (P_D, P_F) AUC value 0.01802 higher than the best-performing E-FDPC method. Similarly, the TOMCBS method achieves the highest AUC values of (P_D, τ) in both the San Diego 1 and AUB-airport 3 datasets. Although it slightly lags behind ONR in the AUB-airport 2 dataset, its AUC value of (P_D, τ) is still higher than other methods. It is worth mentioning that the TOMCBS method does not exhibit a prominent performance in background suppression. This could be attributed to its integration of multiple criteria; the diverse nature of these criteria in TOMCBS leads to a more comprehensive assessment of bands, which, in turn, might not be as precise as individual criteria. While the E-FDPC method exhibits superior background suppression capability with smaller AUC value of (P_F, τ), its advantage in AUC value of (P_D, τ) is not significant. The efficiency of target pixel detection is low, resulting in its overall performance being inferior to TOMCBS. In conclusion, the results across multiple datasets indicate that regardless of the number of bands being 10, 12, or 15, the TOMCBS method consistently obtains higher AUC values of (P_D, P_F) compared to other methods. This implies that the TOMCBS method has better proficiency in target detection compared to existing methods.

Table 2. The AUC values of the BS Methods with Different Modules on Two Datasets.

IE	NE	MaxV	ASPS	San Diego 1	ABU-airport 3
✗	✓	✓	✓	0.9815	0.9767
✓	✗	✓	✓	0.9821	0.9770
✓	✓	✗	✓	0.9802	0.9411
✓	✓	✓	✗	0.9718	0.9795
✓	✓	✓	✓	**0.9831**	**0.9796**

4.4 Ablation Experiments

In order to enhance the performance of the baseline, this study primarily introduced two modules: MaxV and ASPS. To validate the effectiveness of these two modules, we conducted five ablation experiments by removing the IE, NE, MaxV, and ASPA modules individually. Table 2 presents the AUC values of the five ablation experiments conducted on San Diego 1 and AUB-airport 3, respectively. The results of experiments in Groups 1, 2 and 3 indicate a significant improvement in target detection accuracy after the addition of MaxV. The enhanced target detection performance suggests that the inclusion of MaxV renders the band selection method more suitable for the task of target detection. Similarly, the experiments in Groups 3 and 5 also demonstrate the superiority of the band selection method with the addition of MaxV from another perspective. The experiments in Group 4 exhibited worse results compared to Group 5, providing strong evidence for the effectiveness of the ASPS module in reducing spectral redundancy.

5 Conclusion

This article proposes a target-oriented MCBS method (TOMCBS) for detecting specific targets. This method innovatively combines the ASPS framework and the MCBS method while incorporating the BP of specific target constraints, which effectively reduces the correlation among the selected bands, making them more suitable for target detection task. The results of the experiments conducted on multiple real-world datasets provide compelling evidence that TOMCBS exhibits significant promise as a specific target detection method within the field of BS.

References

1. Zhang, L., Zhang, L., Tao, D., Huang, X., Du, B.: Hyperspectral remote sensing image subpixel target detection based on supervised metric learning. IEEE Trans. Geosci. Remote Sens. **52**(8), 4955–4965 (2013)

2. Sun, X., Zhu, Y., Fu, X.: RGB and optimal waveband image fusion for real-time underwater clear image acquisition. IEEE Trans. Instrum. Meas., 1 (2023)

3. Feng, L., Meng, X., Wang, H.: Multi-view locality low-rank embedding for dimension reduction. Knowl.-Based Syst. **191**, 105172 (2020)

4. Jiang, G., Wang, H., Peng, J., Chen, D., Fu, X.: Graph-based multi-view binary learning for image clustering. Neurocomputing **427**, 225–237 (2021)

5. Yang, H., Du, Q., Su, H., Sheng, Y.: An efficient method for supervised hyperspectral band selection. IEEE Geosci. Remote Sens. Lett. **8**(1), 138–142 (2010)

6. Feng, J., Jiao, L., Liu, F., Sun, T., Zhang, X.: Mutual-information-based semi-supervised hyperspectral band selection with high discrimination, high information, and low redundancy. IEEE Trans. Geosci. Remote Sens. **53**(5), 2956–2969 (2014)

7. Zhang, M., Ma, J., Gong, M.: Unsupervised hyperspectral band selection by fuzzy clustering with particle swarm optimization. IEEE Geosci. Remote Sens. Lett. **14**(5), 773–777 (2017)

8. Wang, H., Yao, M., Jiang, G., Mi, Z., Fu, X.: Graph-collaborated auto-encoder hashing for multiview binary clustering. IEEE Trans. Neural Netw. Learn. Syst. (2023)

9. Wang, H., Feng, L., Meng, X., Chen, Z., Yu, L., Zhang, H.: Multi-view metric learning based on KL-divergence for similarity measurement. Neurocomputing **238**, 269–276 (2017)

10. Chang, C.I., Wang, S.: Constrained band selection for hyperspectral imagery. IEEE Trans. Geosci. Remote Sens. **44**(6), 1575–1585 (2006)

11. Chang, C.I., Du, Q., Sun, T.L., Althouse, M.L.: A joint band prioritization and band-decorrelation approach to band selection for hyperspectral image classification. IEEE Trans. Geosci. Remote Sens. **37**(6), 2631–2641 (1999)

12. Wang, Q., Li, Q., Li, X.: Hyperspectral band selection via adaptive subspace partition strategy. IEEE J. Sel. Top. Appl. Earth Observations Remote Sens. **12**(12), 4940–4950 (2019)

13. Sun, X., Shen, X., Pang, H., Fu, X.: Multiple band prioritization criteria-based band selection for hyperspectral imagery. Remote Sens. **14**(22), 5679 (2022)

14. Ji, H., Zuo, Z., Han, Q.L.: A divisive hierarchical clustering approach to hyperspectral band selection. IEEE Trans. Instrum. Meas. **71**, 1–12 (2022)

15. Jia, S., Tang, G., Zhu, J., Li, Q.: A novel ranking-based clustering approach for hyperspectral band selection. IEEE Trans. Geosci. Remote Sens. **54**(1), 88–102 (2015)

16. Wang, Q., Zhang, F., Li, X.: Optimal clustering framework for hyperspectral band selection. IEEE Trans. Geosci. Remote Sens. **56**(10), 5910–5922 (2018)

17. Sun, X., Zhang, H., Xu, F., Zhu, Y., Fu, X.: Constrained-target band selection with subspace partition for hyperspectral target detection. IEEE J. Sel. Top. Appl. Earth Observations Remote Sens. **14**, 9147–9161 (2021)

18. Wang, J., Wang, H., Ma, Z., Wang, L., Wang, Q., Li, X.: Unsupervised hyperspectral band selection based on hypergraph spectral clustering. IEEE Geosci. Remote Sens. Lett. **19**, 1–5 (2021)

19. Das, S., Pratiher, S., Kyal, C., Ghamisi, P.: Sparsity regularized deep subspace clustering for multicriterion-based hyperspectral band selection. IEEE J. Sel. Top. Appl. Earth Observations Remote Sens. **15**, 4264–4278 (2022)
20. Yu, C., Lee, L.C., Chang, C.I., Xue, B., Song, M., Chen, J.: Band-specified virtual dimensionality for band selection: an orthogonal subspace projection approach. IEEE Trans. Geosci. Remote Sens. **56**(5), 2822–2832 (2018)

Optimization and Learning Methods

Pairwise Negative Sample Mining for Human-Object Interaction Detection

Weizhe Jia(iD) and Shiwei Ma(✉)(iD)

School of Mechatronic Engineering and Automation, Shanghai University, Shanghai 200444, China
{jia_wei_zhe,masw}@shu.edu.cn

Abstract. In recent years, Human-Object Interaction (HOI) detection has been riding the wave of the development of object detectors. Typically, one-stage methods exploit them by instantiating HOI and detecting them in a coarse end-to-end learning scheme. With the invention detection transformer (DETR), more studies followed and addressed HOI detection in this novel set-prediction manner and achieved decent performance. However, the scarcity of positive samples in the dataset, especially among object-dense images, hinders learning and leads to lower detection quality. To alleviate this issue, we propose a sample mining technique that utilizes non-interactive human-object pairs to generate negative samples containing HOI features, enriching the sample sets to help the learning of queries. We also introduce interactivity priors, namely filtering out insignificant background objects to inhibit their disturbance. Our technique can be seamlessly integrated into an end-to-end training scheme. Additionally, we propose an unorthodox two-stage transformer-based method that separates pairwise detection and interaction inference to be handled by two cascade decoders, further exploiting this technique. Experimental results on mainstream datasets demonstrate that our approach achieves new state-of-the-art performance, surpassing both one-stage and traditional two-stage methods. Our study reveals the potential to convert between the two method types by adjusting data utilization.

Keywords: Human-object interaction · Transformer · Sample mining

1 Introduction

Human-object interaction (HOI) detection has long attracted research interest, which aims to identify interactions between people and objects in the image. HOI instances are typically represented as triplets, such as <Human, Hit, Ball>. Traditional HOI detection methods take a two-stage approach [2–4,12,18], sequentially detecting objects and recognizing interaction. With the evolution of object detection, one-stage methods [7,19] have emerged, which instantiate HOI instances and jointly detect them by utilizing a single detector. This allows end-to-end training while achieving superior performance. Later, DETR [1] provided a novel insight which formulated instance detection into a set prediction problem. This inspired the spawning of many transformer-based methods [8,16,20],

Q. Liu et al. (Eds.): PRCV 2023, LNCS 14431, pp. 425–437, 2024.
https://doi.org/10.1007/978-981-99-8540-1_34

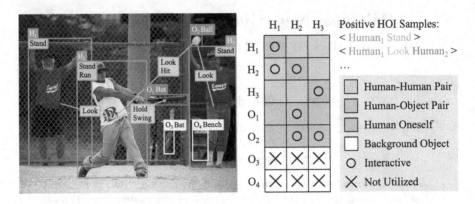

Fig. 1. Illustration of our negative sample mining technique. Typical one-stage methods use only positive samples, while we use additional auto-formed negative HO pairs to provide data from which the model can learn interaction-related features. For instance, the triplet <Human$_1$, Swing, Bat$_1$> is a negative sample that contains postural and spatial information about the verb swing. Small background objects containing no useful visual features are being filtered.

which achieved the new state-of-the-art thanks to the powerful DETR. These methods usually mimic DETR's training while following the one-stage ideology, which instantiates all positive HOI triplets and fine-tunes the model for detection. However, a key difference between object detection and HOI detection went unnoticed. Specifically, a typical object detection training image provides dozens of positive samples (e.g. MS-COCO [11]), whereas images from HOI datasets usually have very few. Hence, during the training, most detection slots are matched onto padded empty samples that contain no meaningful data that aid in the queries' learning of interaction features. This slower the learning and lowers the detection quality. The small magnitude of mainstream datasets [2,5] further worsened the issue, leading to the underutilization of DETR's potential.

While painstakingly enlarging the dataset provides more positive samples overall, the lack of samples per image still hinders the learning, which requires a solution. Hence we propose a cheap, easy-to-perform sample mining technique (Fig. 1). The core idea is inspired by contrastive pre-training [13], which reminds us that HOI data can be used similarly. Besides all the positive samples labelled, the rest of the human-object pairs can automatically form negative samples, which contain some remaining value to exploit. Further, we inject interactivity priors by filtering small background objects with little value. This technique can provide the model with more rich data to learn from and discover features that aid in interaction detection. Yet, it can still be easily integrated into an end-to-end training scheme, without any extra procedure. Additionally, we design an unorthodox two-stage transformer based-method that derives from a one-stage method and performs human-object pair (HO pair) detection and interaction recognition in series. When combined with our training scheme, the model is encouraged to pick up as many HO pairs as possible before distinguishing the interactive pairs. In short, **look harder and think deeper.** Moreover, the attention mechanism in the transformer encoder can guarantee an aggregation of complex contextual information of the image background, which the traditional two-stage method cannot utilize. Finally, experiments on two main benchmarks show that our training scheme introduces healthy two-stage-like inner behavior. Meanwhile, our method shares all the advantages of one-stage methods.

To sum up: (1) We propose a simple data mining technique enriching samples containing useful HOI features. (2) We designed an unorthodox two-stage transformer-based method that jointly detects HO pair and then recognizes interaction. (3) Our method reaches new state-of-the-art on popular HOI datasets.

2 Related Work

Earlier two-stage methods are a type of logical approach to HOI detection. The first two-stage method [2] was proposed to solve object detection and interaction classification in a cascade manner. The first-stage object detector provides visual features and locations of individual objects (including humans); The second-stage multi-branch network infers interaction class for all human-object pairs by aggregating related features. Many works improved this framework. Typical ones include using human visual features to help locate objects [4], learning with interaction relation graphs [3,12], and mining human pose [18].

In contrast, one-stage methods take a holistic approach. By exploiting state-of-the-art object detectors, one-stage methods [7,19] solve HOI detection by instantiating interaction and training object detectors to pick them up, then match them to human and object instances with a lightweight matching scheme. Interaction can be represented by midpoints between the human centers and object center [19] or the bounding box of the human and object union [7], depending on the object detector implementation. Recently, the success of DETR [1] in object detection spawned many one-stage HOI detection methods based on set prediction [8,20]. DETR uses a transformer encoder to aggregate image-wide features; Then a decoder uses learnable queries to turn a set of initial embeddings (zero vectors) into object embeddings for extracting a fixed number of predictions using feed-forward networks; Finally, the results are matched with the sample set to train the model. Typically, Tamura *et al.* proposed QPIC [16], a model heavily similar to DETR with only a few detection heads added to generate HOI predictions. In their practice, co-occur verbs are merged so that each prediction slots capture one interactive HO pair. Despite the simplicity, their method performs surprisingly well and reaches new state-of-the-art.

Compared to previous transformer-based methods, our method is two-staged, as the instance detection and interaction recognition are handled by two decoders in series. Yet it still allows end-to-end training like any one-stage method. It's neither a conventional two-stage method, as the instances are detected pairwise. Overall, our method holds many advantages from both method types.

3   Methods

3.1   Model Architecture

We define one HOI instance as a human-object pair with one or more interaction(s) between them. It consists of a human-box vector $b^{(h)} \in [0,1]^4$, an object-box vector $b^{(o)} \in [0,1]^4$, an object-class one-hot vector $c \in \{0,1\}^{N_{obj}}$ and an action-class many-hot vector $a \in \{0,1\}^{N_{act}}$, denoted as $T = (b^{(h)}, b^{(o)}, c, a)$.

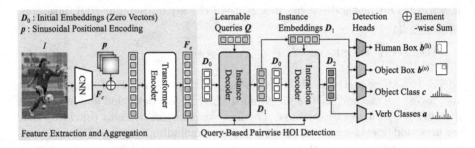

Fig. 2. Our HOI detection model. Given an image, a convolutional neural network (CNN) extracts features, and a transformer encoder aggregates image-wide features. Two cascaded decoders generate instance and interaction embeddings, and a group of feed-forward networks (FFN) extract predictions from them.

Given an input image $I \in \mathbb{R}^{H \times W \times 3}$, the model produces N_q HOI predictions, one per query (Fig. 2). First, the backbone network extracts a downscaled visual feature map $F_c \in \mathbb{R}^{H' \times W' \times D_c}$ with the channel dimension match the encoder input shape. Then, F_c is flattened into an embedding sequence and passes through a standard transformer encoder [17], obtaining the feature map $F_e \in \mathbb{R}^{H'W' \times D_c}$:

$$F_e = f_{enc}(F_c; p), \tag{1}$$

where $f_{enc}(\cdot; \cdot)$ is a stack of transformer encoder layers, p is a fixed positional encoding [1]. Next, two cascade decoders perform two-stage HOI detection. With N_q learnable query vectors $Q = \{q_j | q_j \in \mathbb{R}^{D_c}\}_{j=1}^{N_q}$, the first decoder transforms the initial zero embeddings D_0 into instance embeddings D_1 that each contains information about a specific HO pair, which are used as the queries for the second decoder to generate interaction embeddings D_2. They are obtained as:

$$D_1 = f_{dec1}(D_0; Q, F_e, p), \tag{2}$$

$$D_2 = f_{dec2}(D_0; D_1, F_e, p). \tag{3}$$

where $f_{dec1}(\cdot; \cdot, \cdot, \cdot)$ and $f_{dec2}(\cdot; \cdot, \cdot, \cdot)$ are stacks of transformer decoder layers [17]. The positional encoding p is also used here to complement spatial information. All embedding sets have the same shape as Q. Afterward, D_1 and D_2 are fed into the detection heads to provide HOI predictions, with D_1 for human and object detection and D_2 for interaction(s) recognition. Two embeddings at the same position generate one complete prediction. Specifically, four FFNs with ReLU activation f_h, f_o, f_c, f_a are employed for predicting the four components of an HOI instance $(\hat{b}^{(h)}, \hat{b}^{(o)}, \hat{c}, \hat{a}) \in [0,1]^{(4,4,N_{obj}+1,N_{act})}$, according to $(\hat{b}^{(h)}, \hat{b}^{(o)}, \hat{c}) = (\sigma \circ f_h, \sigma \circ f_o, \varsigma \circ f_c)(d_{1,j})$, $\hat{a}_j = \sigma \circ f_a (d_{2,j})$, where $j \in \{1, \cdots, N_q\}$ is the slot index. σ, ς are the Sigmoid and Softmax activation. The $(N_{obj}+1)$-th element of $\hat{c}_j$ means empty, while $\hat{a}_j$ is many-hot and require no empty class. Finally, all the N_q HOI instances are obtained as: $\hat{T} = \{\hat{T}_j | \hat{T}_j = (\hat{b}_j^{(h)}, \hat{b}_j^{(o)}, \hat{c}_j, \hat{a}_j)\}_{j=1}^{N_q}$.

During inferencing, filter heads are leveraged to reserves high-confidence results. For the j-th slot, the object index $\hat{k}_j$ and HOI triplet scores vector $\hat{s}_j$ are calculated in turn as $\hat{k}_j = \arg\max_k(\hat{c}_j[k])$, $\hat{s}_j = \hat{c}_j[\hat{k}_j] \cdot \hat{a}_j$. All triplets with a score greater than the preset threshold τ are preserved. The same procedure is performed on all slots, obtaining the final inference results of the image.

3.2 Training Scheme

We adopt end-to-end training using richened sample sets formed by our mining technique. Given an image I, its object labels B and HOI labels T. First, we generate a padded sample set $\tilde{T}$ using Algorithm 1, where all HO pairs are utilized as samples, besides small background objects filtered by area threshold α. Additionally, empty samples $\varnothing$ whose object being **empty** and action vector being $\mathbf{0}$ are padded into $\tilde{T}$ to match the number of queries N_q. Then, we find the optimal bipartite matching between predictions $\hat{T}$ and padded samples $\tilde{T}$ and sum up losses for all matched pairs. Specifically, the matching cost between the j-th result $\hat{T}_j$ and the i-th sample T_i is calculated as:

$$\mathcal{H}_{i,j} = \mathbf{1}_{T_i \neq \varnothing} \left(\eta_b \mathcal{H}_{i,j}^{(b)} + \eta_u \mathcal{H}_{i,j}^{(u)} + \eta_c \mathcal{H}_{i,j}^{(c)} + \eta_a \mathcal{H}_{i,j}^{(a)} \right), \tag{4}$$

where $\mathbf{1}_p(x)$ denotes the indicator function, η_b, η_u, η_c and η_a are the hyperparameters for weight adjusting. The cost of box regression, IoU, object class and action class, denoted as $\mathcal{H}_{i,j}^{(b)}$, $\mathcal{H}_{i,j}^{(u)}$, $\mathcal{H}_{i,j}^{(c)}$, and $\mathcal{H}_{i,j}^{(a)}$, are calculated as follows:

$$\mathcal{H}_{i,j}^{(b)} = \max \left\{ \left\| b_i^{(h)} - \hat{b}_j^{(h)} \right\|_1, \left\| b_i^{(o)} - \hat{b}_j^{(o)} \right\|_1 \right\}, \tag{5}$$

$$\mathcal{H}_{i,j}^{(u)} = -\min_{x \in \{h,o\}} \left\{ \text{GIoU} \left(b_i^{(x)}, \hat{b}_j^{(x)} \right) \right\}, \tag{6}$$

$$\mathcal{H}_{i,j}^{(c)} = -c_i^\top \hat{c}_j = -\hat{c}_j[k] \quad \text{s.t.} \quad c_i[k] = 1, \tag{7}$$

$$\mathcal{H}_{i,j}^{(a)} = -\frac{1}{2} \left(\frac{a_i^\top \hat{a}_j}{\|a_i\|_1 + \epsilon_a} + \frac{(1-a_i)^\top (1-\hat{a}_j)}{\|1-a_i\|_1 + \epsilon_a} \right), \tag{8}$$

where $\text{GIoU}(\cdot, \cdot)$ is the generalized IoU [15], $v[k]$ denotes the element of vector v with index k (starting at 0), and ϵ_a is a small positive value for avoiding zero division. With pair-by-pair costs obtained, the Hungarian Algorithm [9] is applied to find the optimal bipartite matching permutation $\hat{\omega} =$

Algorithm 1: Pairwise Negative Samples Padding

Parameters: Number of queries N_q, small object area threshold α
Input: Objects samples $\mathcal{B} = \{(b,c)\}_{i=1}^{|\mathcal{B}|}$, HOI samples $\mathcal{T} = \{(H,O,a)\}_{i=1}^{|\mathcal{T}|}$
Output: Padded ground truth set $\tilde{\mathcal{T}}$

$\mathcal{P} \leftarrow \{(T.H, T.O) \mid T \in \mathcal{T}\};$ /* Get interactive HO-pairs */
$\mathcal{A} \leftarrow \{T.H \mid T \in \mathcal{T}\} \cup \{T.O \mid T \in \mathcal{T}\};$ /* Get active instances */
$\mathcal{H} \leftarrow \mathcal{B}.\text{Filter}(c = \text{Human} \wedge \text{area} > \alpha);$ /* Get non-background humans */
$\tilde{\mathcal{T}} \leftarrow \mathcal{T};$ /* Initialize by positive HOI samples */
for $H \in \mathcal{H}$ **do**
    for $O \in \mathcal{B}$ **do**
        if $(O.\text{area} < \alpha \wedge O \notin \mathcal{A}) \vee (H,O) \in \mathcal{P}$ **then**
            | **continue**;   /* Skip through unwanted pairs */
        end
        $\tilde{\mathcal{T}}.\text{Add}((H,O,\mathbf{0}));$ /* Pad the negative sample */
        if $|\tilde{\mathcal{T}}| = N_q$ **then**
            | **return** $\tilde{\mathcal{T}};$ /* Stop when reaching number of queries */
        end
    end
end
return $\tilde{\mathcal{T}} \cup \{\varnothing\}_{i=1}^{N_q - |\tilde{\mathcal{T}}|};$ /* Pad empty instances to match #queries */

$\arg\min_{\omega \in \Omega_{N_q}} \sum_{i=1}^{N_q} \mathcal{H}_{i,\omega(i)}$, where Ω_{N_q} is the set of all possible permutations of N_q elements. Thus, the matching results are obtained as $j = \hat{\omega}(i)$. Finally, the overall loss of the model is the evaluation of the similarity of two sets of matched instances, which is a combination of four individual terms similar to (4), formulated as follows:

$$\mathcal{L} = \lambda_b \mathcal{L}_b + \lambda_u \mathcal{L}_u + \lambda_c \mathcal{L}_c + \lambda_a \mathcal{L}_a. \tag{9}$$

λ_b, λ_u, λ_c and λ_a are the hyperparameters. The loss of box regression, IoU, object class and action class, denoted as $\mathcal{L}_b$, $\mathcal{L}_u$, $\mathcal{L}_c$, and $\mathcal{L}_a$, is calculated as:

$$\mathcal{L}_b = \frac{1}{N_T} \sum_{T_i \neq \varnothing} \left(\left\| b_i^{(h)} - \hat{b}_{\hat{\omega}(i)}^{(h)} \right\|_1 + \left\| b_i^{(o)} - \hat{b}_{\hat{\omega}(i)}^{(o)} \right\|_1 \right), \tag{10}$$

$$\mathcal{L}_u = 2 - \frac{1}{N_T} \sum_{T_i \neq \varnothing} \sum_{x \in \{h,o\}} \text{GIoU} \left(b_i^{(x)}, \hat{b}_{\hat{\omega}(i)}^{(x)} \right), \tag{11}$$

$$\mathcal{L}_c = -\frac{1}{N_q} \sum_{i=1}^{N_q} \ln c_i^\top \hat{c}_{\hat{\omega}(i)}, \tag{12}$$

$$\mathcal{L}_a = \frac{1}{\sum_{T_i \neq \varnothing} \|a_i\|_1} \sum_{i=1}^{N_q} l_f \left(a_i, \hat{a}_{\hat{\omega}(i)} \right), \tag{13}$$

where $l_f(\cdot,\cdot)$ is the focal loss [10]. Note that slots matched onto $\varnothing$ are excluded from the calculation of $\mathcal{L}_b$ and $\mathcal{L}_u$ since $\varnothing$ has undefined bounding boxes.

4   Experiments and Results

4.1   Datasets and Evaluation Metrics

We evaluate our method on two mainstream HOI detection datasets V-COCO [5] and HICO-DET [2]. V-COCO is a subset of MS-COCO with 5400/4946 images in the train/test sets; HICO-DET has 37633/9546 images in the train/test sets. Both datasets inherit the 80 object classes from MS-COCO. V-COCO covers 29 verb classes with four no-object verbs (stand, walk, run, smile) and one rare class (point); HICO-DET covers 117 verb classes, forming 600 interaction (verb-object) types, among which 138 are rare (with <10 samples).

Following the standard setup, mean average precision (mAP) is adopted as the metric. True positive is defined by criteria: IoU>0.5 with corresponding ground truth for human and object boxes, and correct classification of objects and verbs. For V-COCO, mAP is calculated for all verb classes and 24 classes that exclude no-object and rare classes (Setup 2). For the completeness of evaluation, we also evaluate the mAP of difficult classes excluded from Setup 2. For literature lacking their $mAP_{\overline{S2}}$, an equivalent value can be acquired by:

$$mAP_{\overline{S2}} = \frac{N}{N - N_2}mAP - \frac{N_2}{N - N_2}mAP_{S2}, \tag{14}$$

where $N = 29, N_2 = 24$; For HICO-DET, mAP is calculated for all interaction classes, rare classes and non-rare classes respectively. Additionally, we evaluate the mAP of HO pair detection, denoted as mAP_P.

4.2   Implementation Details

Due to V-COCO having no-object verbs, we add a virtual object class indicating the person oneself and set the ground truth bounding box as the human bounding box. HICO-DET does not come with background object labels, hence we obtain them by pseudo-labelling with a stand-alone object detector and update the ground truths using Non-Maximum Suppression [14]:

$$\mathcal{B}_{det} = \text{Detect}(\boldsymbol{I}).\text{Filter}(s > 0.9), \quad \mathcal{B} = \text{NMS}(\mathcal{B}_{org} \cup \mathcal{B}_{det}). \tag{15}$$

We choose ResNet-50 [6] as our backbone CNN. We use sinusoidal positional embedding, a 6-layer encoder and two 3-layer decoders for the transformer. The number of queries N_q is 100. Embedding dimension D_c is 256. The number of heads for multi-head attention, FFN hidden dimension and dropout probability is 8, 2048, and 0.1, following [1]. All FFNs have a hidden dimension of 256. Bounding-box FFNs f_h and f_o have three linear layers with ReLU activation; Classification FFNs f_c and f_a have one linear layer. To test scalability, a variant with ResNet-101 and two 6-layer decoders is also implemented.

Data augmentation during training includes random horizontal flipping, scaling, cropping and color jittering, following [16]. The small object area threshold α is 0.04. The matching-cost and training-loss hyperparameters (η_b, η_u, η_c, η_a)

and $(\lambda_b, \lambda_u, \lambda_c, \lambda_a)$ are both $(2.5, 1, 1, 1)$. The small values ϵ_a is 10^{-4}. γ for focal loss is 2. We apply AdamW optimizer with learning rate, backbone learning rate, weight decay and gradient clipping value of 10^{-4}, 10^{-5}, 10^{-4} and 0.1.

The model is trained on two GeForce RTX 2080 Ti with a batch size of 2 and batch interval of 4 for gradient accumulation. We use pre-trained DETR weights to initialize the backbone, encoder, instance decoder and object-detection FFNs. We train the model in two stages. In stage 1, all interaction labels are set to **0**, so the model purely focuses on detecting HO pairs. Stage 2 applies regular training. The model is trained in stage 1 for E_1 epochs, then decreases all learning rates tenfold and trains another E_1' epochs, followed by the same pattern in stage 2, denoted as (E_1, E_1', E_2, E_2'). The model trained on V-COCO has $(30, 10, 30, 20)$ epochs, and that on HICO-DET has $(20, 10, 40, 20)$ epochs. Note that $(0, 0, E_2, E_2')$ implies end-to-end training. See comparisons in Sect. 4.4.

4.3 Overall Comparison

Our method outperforms various state-of-the-art HOI detection methods, including all transformer-based approaches (column T with ✓ in Table 1) and the powerful one-stage method QPIC, with 3.7%/6.5% relative increment on V-COCO/HICO-DET compared to the previous best. This performance boost demonstrates how our method leverages negative samples to enhance HOI learning at a macro level. Note that we are only slightly behind HOTR on mAP$_{S2}$ of V-COCO. However, this method struggles with difficult classes excluded from S2 evaluation. In contrast, our method exhibits a more balanced performance across S2 and $\overline{S2}$. Overall, our model exhibits leading and consistent performance across all evaluation metrics. Some qualitative results are shown in Fig. 4.

4.4 Ablation Studies

To optimize the utilization of training resources, we conduct an ablation study on training setup (Table 2a). We divide 80 epochs among 4 training segments and adjust batch interval b_i for gradient accumulation to optimize mAP. Results show that $b_i = 4$ to simulate a batch size 16 across two GPUs is optimal (row 2 vs 3). The results also show that pre-training with zero vectors as interaction labels boosts performance (row 1 vs 2). Without pre-training, the model initially performs reasonably well but is eventually outperformed. We interpret this as a sign that the instance decoders' better-learned embeddings aid in subsequent interaction detection, optimizing performance for both stages of HOI detection. This proves the advantage of dividing decoders and splitting training stages.

We further examine the impact of the area threshold in our sample padding algorithm (Table 2b). Results show that including moderately sized background objects improves the final performance (row 3, $\alpha = 0.04$). This confirms our intuition of avoiding excessive use of small objects, which lack clear visual features that aid interaction inference. Without this filter, numerous unhelpful samples spam into the ground truth set, disrupt the learning of HOI features and lead to poor model performance (row 1, $\alpha = 0$).

Table 1. mAP(%) Comparison against state-of-the-art methods.

Methods	Backbone	T	HICO-DET [2]			V-COCO [5]		
			Full	Rare	N.R	Full	S2	S̄2
Two-stage Methods								
HO-RCNN [2]	VGG16		7.81	5.37	8.54	-	-	-
InteractNet [4]	R50-FPN		9.94	7.16	10.77	40.0	-	-
GPNN [12]	R-DCN152		13.11	9.34	14.23	44.0	-	-
PMFNet [18]	R50-FPN		17.46	15.65	18.00	52.0	-	-
DRG [3]	R50-FPN		24.53	19.47	26.04	51.0	-	-
One-stage Methods								
UnionDet [7]	R50-FPN		17.58	11.72	19.33	47.5	56.2	5.7
IP-Net [19]	H104		19.56	12.79	21.58	52.3	-	-
HOI-Trans [20]	R50	✓	23.46	16.91	25.41	52.9	-	-
HOTR [8]	R50	✓	25.10	17.34	27.42	55.2	**64.4**	11.0
QPIC [16]	R50	✓	29.07	21.85	31.23	58.8	61.0	48.2
QPIC	R101	✓	29.90	23.92	31.69	58.3	60.7	46.8
Ours	R50	✓	29.82	22.28	32.38	59.4	60.3	**55.2**
Ours	R101	✓	**31.84**	**25.93**	**33.86**	**61.0**	62.2	**55.2**

Table 2. Ablation studies conducted on V-COCO.

(a) On training setup

E_1, E_1', E_2, E_2'	b_i	@epoch30 mAP$_P$	mAP	@epoch80 mAP$_P$	mAP
(0, 0, 60, 20)	4	15.8	**46.6**	18.5	53.1
(30, 0, 30, 20)	4	**21.4**	3.7	**22.0**	**56.3**
(30, 0, 30, 20)	1	16.3	4.1	18.6	54.0

(b) On the value of α

α	HO-pair mAP	mAP$_{N.R.}$	Verb mAP	mAP$_{S2}$
0	**22.1**	**20.1**	56.3	56.8
0.01	19.2	17.6	56.5	56.9
0.04	16.7	14.7	**58.2**	**58.9**

Table 3. Detailed comparison against QPIC on (V-COCO, HICO-DET).

Model	Backbone	#Enc.	#Dec.	Gflops	#params	#epochs	mAP$_P$	mAP
QPIC	ResNet50	6	6	**95**	**41M**	150	(9.1, 14.3)	(58.2, 29.1)
	ResNet101	6	6	175	60M	150	(10.0, 15.4)	(57.8, 29.9)
Ours	ResNet50	6	3+3	**95**	**41M**	90	(16.7, 17.3)	(59.4, 29.8)
	ResNet101	6	6+6	175	70M	90	(17.5, 18.1)	(**61.0, 31.8**)

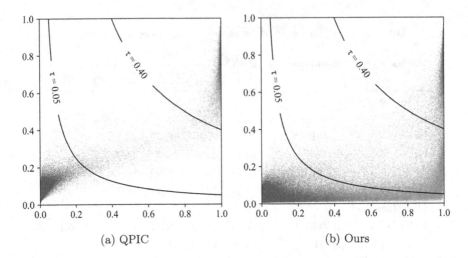

(a) QPIC   (b) Ours

Fig. 3. Score state space of models on V-COCO test set. **Horizontal axis**: object score; **Vertical axis**: maximum verb score. Each dot marks a predicted HO pair, with red for invalid, pink for non-interactive, and blue for interactive. (Color figure online)

4.5 Model Behavior Analysis (vs QPIC)

We compare our model vs QPIC, the one-stage state-of-the-art, under our experiment conditions (Table 3). Our model outperforms QPIC in fewer epochs and without extra computational costs. Moreover, our model has better scalability, especially on V-COCO where QPIC suffers a performance downgrade. Both models perform query-based HOI detection. The fixed number of queries caps time complexity at $\mathcal{O}(1)$ instead of increasing with instances like traditional two-stage methods.

Further, we display all model predictions on the V-COCO test set in the score state space (Fig. 3). Our results are well-separated and evenly spread across both dimensions, unlike QPIC's highly coupled results. This aligns with real-world scenarios where numerous contextually vague instances occupy mid-score regions. This behavior implies that our model effectively decouples detection and inference, with only a weak remaining prior that interactive pairs usually have high object confidence. These results once again highlight the effectiveness of our sample mining. By providing more samples per image, more queries can get paired up and improved. Even at worst, the detector still captures high-object-confidence HO pairs, which have the potential of spawning interactivity.

Fig. 4. High confidence results ($s > 0.4$) among all 100 queries from images in HICO-DET (rows 1, 2) and V-COCO test set - zoom in for details.

5   Conclusions

In this paper, we propose a query-based pairwise detector for HOI detection. Our negative sample mining technique enables a clean two-stage behavior, outperforming our one-stage counterpart. Our study reveals the common ground between the two method types and shows how a simple adjustment in the data utilization allows for easy conversion between them. In future work, we aim to leverage text-to-image methods to cost-effectively acquire positive training samples to expand HOI datasets.

References

1. Carion, N., Massa, F., Synnaeve, G., Usunier, N., Kirillov, A., Zagoruyko, S.: End-to-end object detection with transformers. In: Vedaldi, A., Bischof, H., Brox, T., Frahm, J.-M. (eds.) ECCV 2020. LNCS, vol. 12346, pp. 213–229. Springer, Cham (2020). https://doi.org/10.1007/978-3-030-58452-8_13
2. Chao, Y.W., Liu, Y., Liu, X., Zeng, H., Deng, J.: Learning to detect human-object interactions. In: IEEE Winter Conference on Applications of Computer Vision (WACV), pp. 381–389 (2018)
3. Gao, C., Xu, J., Zou, Y., Huang, J.-B.: DRG: dual relation graph for human-object interaction detection. In: Vedaldi, A., Bischof, H., Brox, T., Frahm, J.-M. (eds.) ECCV 2020. LNCS, vol. 12357, pp. 696–712. Springer, Cham (2020). https://doi.org/10.1007/978-3-030-58610-2_41
4. Gkioxari, G., Girshick, R., Dollár, P., He, K.: Detecting and recognizing human-object interactions. In: Proceedings of IEEE/CVF Conference on Computer Vision and Pattern Recognition (CVPR), pp. 8359–8367 (2018)
5. Gupta, S., Malik, J.: Visual semantic role labeling. arXiv preprint arXiv:1505.04474 (2015)
6. He, K., Zhang, X., Ren, S., Sun, J.: Deep residual learning for image recognition. In: Proceedings of IEEE/CVF Conference on Computer Vision and Pattern Recognition (CVPR), pp. 770–778 (2016)
7. Kim, B., Choi, T., Kang, J., Kim, H.J.: UnionDet: union-level detector towards real-time human-object interaction detection. In: Vedaldi, A., Bischof, H., Brox, T., Frahm, J.-M. (eds.) ECCV 2020. LNCS, vol. 12360, pp. 498–514. Springer, Cham (2020). https://doi.org/10.1007/978-3-030-58555-6_30
8. Kim, B., Lee, J., Kang, J., Kim, E.S., Kim, H.J.: HOTR: end-to-end human-object interaction detection with transformers. In: Proceedings of IEEE/CVF Conference on Computer Vision and Pattern Recognition (CVPR), pp. 74–83 (2021)
9. Kuhn, H.W.: The Hungarian method for the assignment problem. Nav. Res. Logistics Quart. **2**(1–2), 83–97 (1955)
10. Lin, T.Y., Goyal, P., Girshick, R., He, K., Dollár, P.: Focal loss for dense object detection. In: Proceedings of IEEE/CVF International Conference on Computer Vision, (ICCV), pp. 2980–2988 (2017)
11. Lin, T.-Y., et al.: Microsoft COCO: common objects in context. In: Fleet, D., Pajdla, T., Schiele, B., Tuytelaars, T. (eds.) ECCV 2014. LNCS, vol. 8693, pp. 740–755. Springer, Cham (2014). https://doi.org/10.1007/978-3-319-10602-1_48
12. Qi, S., Wang, W., Jia, B., Shen, J., Zhu, S.C.: Learning human-object interactions by graph parsing neural networks. In: Proceedings of the European Conference on Computer Vision (ECCV), pp. 401–417 (2018)
13. Radford, A., et al.: Learning transferable visual models from natural language supervision. In: International Conference on Machine Learning (ICML), pp. 8748–8763 (2021)
14. Redmon, J., Divvala, S., Girshick, R., Farhadi, A.: You only look once: unified, real-time object detection. In: Proceedings of IEEE/CVF Conference on Computer Vision and Pattern Recognition, (CVPR), pp. 779–788 (2016)
15. Rezatofighi, H., Tsoi, N., Gwak, J., Sadeghian, A., Reid, I., Savarese, S.: Generalized intersection over union: a metric and a loss for bounding box regression. In: Proceedings of the IEEE/CVF Conference on Computer Vision and Pattern Recognition (CVPR), pp. 658–666 (2019)

16. Tamura, M., Ohashi, H., Yoshinaga, T.: QPIC: query-based pairwise human-object interaction detection with image-wide contextual information. In: Proceedings of the IEEE/CVF Conference on Computer Vision and Pattern Recognition (CVPR), pp. 10410–10419 (2021)

17. Vaswani, A., et al.: Attention is all you need. In: Advances in Neural Information Processing Systems (NeurIPS), vol. 30 (2017)

18. Wan, B., Zhou, D., Liu, Y., Li, R., He, X.: Pose-aware multi-level feature network for human object interaction detection. In: Proceedings of the IEEE/CVF International Conference on Computer Vision (ICCV), pp. 9469–9478 (2019)

19. Wang, T., Yang, T., Danelljan, M., Khan, F.S., Zhang, X., Sun, J.: Learning human-object interaction detection using interaction points. In: Proceedings of the IEEE/CVF Conference on Computer Vision and Pattern Recognition (CVPR), pp. 4116–4125 (2020)

20. Zou, C., et al.: End-to-end human object interaction detection with hoi transformer. In: Proceedings of the IEEE/CVF Conference on Computer Vision and Pattern Recognition (CVPR), pp. 11825–11834 (2021)

An Evolutionary Multiobjective Optimization Algorithm Based on Manifold Learning

Jiaqi Jiang, Fangqing Gu$^{(\boxtimes)}$, and Chikai Shang

School of Mathematics and Statistics, Guangdong University of Technology,
Guangzhou 510006, China
fqgu@gudt.edu.com

Abstract. Multi-objective optimization problem is widespread in the real world. However, plenty of typical evolutionary multi-objective optimization (EMO) algorithms are extremely tough to deal with large-scale optimization problems (LSMOPs) due to the curse of dimensionality. In reality, the dimension of the manifold representing the Pareto solution set is much lower than that of the decision space. This work proposes a decision space reduction technique based on manifold learning using locality-preserving projects. The critical insight is to improve search efficiency through decision space reduction. The high-dimensional decision space is first mapped to a low-dimensional subspace for a more effective search. Subsequently, a transformation matrix which is Pseudo-inverse of the projection matrix, maps the resultant offspring solutions back to the primal decision space. The proposed decision space reduction technique can be integrated with most multi-objective evolutionary algorithms. This paper integrates it with NSGA-II, namely LPP-NSGA-II. We compare the proposed LPP-NSGA-II with four state-of-the-art EMO algorithms on thirteen test problems. The experimental results reveal the effectiveness of the proposed algorithm.

Keywords: Manifold learning · Evolutionary algorithm · Large-scale multi-objective optimization · Decision space reduction

1 Introduction

Multi-objective optimization problems (MOPs) are common in people's daily lives. Under mild conditions, a multi-objective optimization problem can be formulated as follows:

$$
\min_{\mathbf{x} \in [u_i, v_i]^D} F(\mathbf{x}) = (f_1(\mathbf{x}), f_2(\mathbf{x}), \cdots, f_M(\mathbf{x}))
$$
$$
s.t. \quad g_i(\mathbf{x}) \le 0, i = 1, 2, \cdots, q;
$$
$$
h_j(\mathbf{x}) = 0, j = 1, 2, \cdots, p. \tag{1}
$$

This research was funded by the Natural Science Foundation of Guangdong Province (2021A1515011839).

where u_i and v_i are the upper and lower bounds of x_i for $i = 1, 2, \cdots, D$, $M \geq 2$ is the number of objectives. $g_i(x) \leq 0$ and $h_j(x) = 0$ determine the feasible region of the solution. Usually, when $D \geq 100$, problem (1) is termed a large-scale multi-objective problem (LSMOP).

Over the past few decades, the use of evolutionary algorithms (EAs) to deal with MOPs has been extensively studied [4,19], and plenty of evolutionary multi-objective optimization (EMO) algorithms have been proposed, such as well-known dominance-based NSGA-II [3], NSGA-III [2], decomposition-based MOEA/D [23], M2M [11], and indicator-based SRA2 [8], etc. Although a great number of EAs have been successfully applied to solve MOPs, they still face many challenges in dealing with practical problems with large-scale decision variables. The scalability of decision space is a problem that has not been researched deeply in the multi-objective research community. Generally, the main difficulty of EMO on LSMOPs consists in that the research space of a problem grows exponentially with the dimension, that is, "the curse of dimensions". Accordingly, the landscape of the search space may become more complex, and conflicts among multiple objectives might be more serious, resulting in rapid degradation of existing EMO.

Historically, a variety of EMO algorithms have been designed to handle LSMOPs. These typical algorithms can be categorized as decomposition approaches based on cooperative co-evolution (CC) methods and non-decomposition approaches that solve the LSMOP as a whole. In the decomposition methods, the intuitive idea is through divide-and-conquer strategies. The whole LSMOP is split into a set of low-dimensional sub-problems that are easier to be processed by existing solvers. The CC method is first used for GA, termed CCGA [12]. In addition, numerous improved DE and PSO are integrated with the CC method, such as CCPSO2 [10], DECC-DG [14], and so on. Due to CC methods being sensitive to decomposition strategies and poor performance on non-separable functions, many researchers consider non-decomposition methods to tackle LSMOPs. These methods mainly enhance their performance in exploring complex search space by developing new effective operators, represented by the following algorithms: CPSO [1,13], etc.

We notice that most of the EA-based methods for LSMOPs mainly focus on designing new algorithms tailored to the problem, aiming to improve their performance in tackling intricate optimization issues. However, there is a lack of research from the perspective of changing the problem. To fill this gap, we consider tackling the complex problem into a familiar and easy-to-solve problem. The Pareto Set (PS) is usually a straight line for a bi-objective optimization problem and a surface for a three-objective problem. It turns out that the PS is almost low-dimensional. Hence, it is feasible to consider transforming the large-scale problem into a problem with lower dimensions through dimensionality reduction to some extent.

With the development of dimensionality reduction technology, many mature methods have been proposed and widely used in various fields. These methods can be divided into linear methods and non-linear learning methods.

Linear methods include classical PCA, LDA [21] and ICA [6], etc. For non-linear methods, two subtypes can also be categorized according to whether there is an explicit non-linear projection function between the original data and their representations: kernel methods [22] and manifold learning [9].

Since 2000, manifold learning has been a branch of non-linear dimensionality reduction. In 2000 two articles in Science magazine: Isomap [16] and LLE [15], led to the rapid development of this field. Manifold learning is a class of dimensionality reduction methods that borrow from the concept of topological popularity. In recent years, manifold learning as one of the core research directions has attracted much attention from experts. Among them, one of the most representative algorithms is Locality Preserving Projects (LPP) [5]. LPP is a linear dimensionality reduction method proposed by He et al. in 2003 based on the Laplacian Eigenmaps (LE) algorithm [5]. Compared to PCA, LPP retains global information and can then be considered an alternative to PCA. Apart from having the characteristics of preserving the geometric relationship and local structure of data in space transformation, LPP also has the advantages of simple, intuitive, and fast computation as a linear method.

Because PS is usually a low-dimensional manifold, we utilize the manifold learning method to find a hyperplane that aims to accelerate the search efficiency of the algorithm and improve its convergence performance. Consequently, we propose a manifold learning dimensionality reduction-based EMO for LSMOPs. To begin with, we apply the dimensionality reduction method of manifold learning to obtain the low-dimensional representation of the high-dimensional decision variable and determine the dimension of the decision space after dimensionality reduction through a reconstruction error. Afterward, crossover and mutation operators are conducted in the low-dimensional subspace. Ultimately, the offspring in subspace is inversely mapped back to the original space by a transformation matrix. The proposed decision space reduction technique is integrated with the most popular multi-objective evolutionary algorithm NSGA-II, namely LPP-NSGA-II. We compare the proposed LPP-NSGA-II with four state-of-the-art EMO algorithms on test suites (ZDT and SMOP). The experimental results show that the proposed algorithm can effectively improve the search efficiency. The main contributions can be summarized as follows:

(1) Propose a dimensionality reduction strategy for decision variables based on manifold learning to reduce the dimension of decision space. It can effectively avoid the problem of the "curse of dimensionality" and improve the computational efficiency of the algorithm.

(2) Integrate the dimensionality reduction technique with NSGA-II and propose an EMO for solving LSMOPs, named LPP-NSGA-II. We compare the proposed LPP-NSGA-II with its counterparts. The experimental results show that the proposed dimensionality reduction technique improves the search efficiency of the algorithm.

The remainder of this article is organized as follows. Section 2 provides the main procedures of the proposed algorithm. Section 3 presents the design of experiments and results to evaluate the performance empirically. Section 4 draws conclusions and points to our future work.

2   Proposed Algorithm

In this section, we proposed a novel large-scale multi-objective evolutionary algorithm based on decision variable reduction. And the decision space reduction is achieved by manifold learning. We combine the proposed dimensional reduction technique with NSGA-II, named LPP-NSGA-II. Additionally, we describe LPP-NSGA-II in detail.

2.1   Dimensionality Reduction of Decision Variables via LPP

The generic problem of linear dimensionality reduction can be formalized as follows. Let $P = \{\mathbf{x}_1, \mathbf{x}_2, \cdots, \mathbf{x}_N\}$ be a solution set obtained by an EMO in the original decision space $\mathbb{R}^s$ and $X = (\mathbf{x}_1, \mathbf{x}_2, \cdots, \mathbf{x}_N)$ is a matrix composed of these solutions. We aim to find a transformation matrix U that projects these N points to a low-dimensional data points $\mathbf{y}_1, \mathbf{y}_2, \cdots, \mathbf{y}_N$ in $\mathbb{R}^l (l \ll s)$, such that $\mathbf{x}_i$ can be represented by $\mathbf{y}_i$ as much as possible, where $\mathbf{y}_i = U^T \mathbf{x}_i$ and U denotes a transformation matrix.

Based on [5], the detailed algorithmic procedure of LPP is below:

Constructing the Adjacency Graph G: Construct a graph G with N nodes. An edge will be put between nodes i and j if $\mathbf{x}_i$ and $\mathbf{x}_j$ are "close", and the criterion of "close" can be defined by ϵ neighborhoods, (i.e., $\|\mathbf{x}_i - \mathbf{x}_j\|^2 < \epsilon$) or k-nearest neighbors.

Choosing the Weights: $W = (w_{ij})_{N \times N}$ is a sparse, symmetric matrix. The value of weight w_{ij} is calculated by

$$w_{ij} = \begin{cases} e^{-\frac{\|\mathbf{x}_i - \mathbf{x}_j\|^2}{t}} & \text{if } \mathbf{x}_i \text{ and } \mathbf{x}_j \text{ are close}, \\ 0 & \text{otherwise}. \end{cases} \tag{2}$$

Eigenmaps: The mapping matrix can be obtained by solving the following minimization problem based on the standard spectral graph theory:

$$J = \min \frac{1}{2} \sum_{ij} \|\mathbf{y}_i - \mathbf{y}_j\|^2 w_{ij} \tag{3}$$

The objective function can further be derived and eventually be expressed as:

$$J = \min U^T X L X^T U$$
$$s.t. \quad U^T X D X^T U = 1 \tag{4}$$

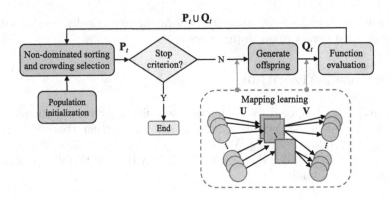

Fig. 1. The basic framework of LPP-NSGA-II.

where $L = D - W$ is called a Laplacian matrix and D is a diagonal matrix with $D_{ii} = \sum_j W_{ij}$. The larger the value of D_{ii}, the more "important" $\mathbf{y}_i$ is. where $Y = (\mathbf{y}_1, \mathbf{y}_2, \cdots, \mathbf{y}_N)$ is a matrix composed of the projection of the solutions. Accordingly, the above minimization problem can be simply transformed into an eigenvalue decomposition to solve:

$$XLX^T U = \lambda X D X^T U \tag{5}$$

Let $\{\mathbf{u}_1, \mathbf{u}_2, \cdots, \mathbf{u}_D\}$ be the generalized eigenvector of (5) sorted according to their eigenvalues $\lambda_1 < \lambda_2 < \cdots < \lambda_D$. In general, the less the value of l, the greater the information loss of the original data will be. Consequently, we determine the dimension of the reduced decision space according to the reconstruction error. There may be no inverse matrix for $U = (\mathbf{u}_1, \mathbf{u}_2, \cdots, \mathbf{u}_l)$ since it is singular or non-square matrices in most cases. We can find its pseudo-inverse to reconstruct the solutions. Accordingly, the reconstruction error is defined as follows:

$$loss(l) = \frac{\left\| V U^T X - X \right\|_2}{\|X\|_2} \tag{6}$$

where V is the Pseudo-inverse of the projection matrix U^T.

$$l^* = \min l$$
$$s.t. \ loss(l) < \delta \tag{7}$$

where $\delta = 0.05$ is the maximum allowable reconstruction error. Then we can obtain the transformation matrix $U = (\mathbf{u}_1, \mathbf{u}_2, \cdots, \mathbf{u}_{l^*})$. Furthermore, we obtain the projection embedded for each data point as follows:

$$\mathbf{x}_i \to \mathbf{y}_i = U^T \mathbf{x}_i, \ U = (\mathbf{u}_1, \mathbf{u}_1, \cdots, \mathbf{u}_{l^*}). \tag{8}$$

Algorithm 1: LPP-NSGA-II

Input:
- N: the population size;
- FE_{max}: Maximum number of iterations;
- Genetic operators and their associated parameters.

Output: Nondominated solutions of P_t.

1 Randomly generate N solutions in the primal space to initialize the population $P_t = \{\mathbf{x}_1, \mathbf{x}_2, \cdots, \mathbf{x}_N\}$, set $t = 0$;

2 **while** $t < FE_{max}$ **do**

3      $U, V \leftarrow$ Mapping learning.

4      $P'_t = \{\mathbf{y}_1, \mathbf{y}_2, \cdots, \mathbf{y}_N\} \leftarrow$ Map to subspace via U.

5      $\mathbf{c}_i \leftarrow$ Crossover operation on $\mathbf{y}_i$ and $\mathbf{y}_r$ which is randomly selected from P'_t, $i, r = 1, 2, \cdots, N$ ($i \neq r$).

6      $\mathbf{c}_i \leftarrow$ Mutation operator on $\mathbf{c}_i$, $i = 1, 2, \cdots, N$.

7      $Q'_t = \{\mathbf{c}_1, \mathbf{c}_2, \cdots, \mathbf{c}_N\} \leftarrow$ Resultant offspring in subspace.

8      $Q_t = \{\mathbf{x}_1, \mathbf{x}_2, \cdots, \mathbf{x}_N\} \leftarrow$ Inversely mapping back to the primal space via V.

9   Evaluation(Q_t) $\leftarrow$ Assess the fitness values.

10     $R_t = P_t \bigcup Q_t \leftarrow$ Merge populations of parent and offspring.

11   Non-dominated sorting and crowding distance sorting.

12     $P_t \leftarrow$ Environment selection.

13     $t = t + 1$.

14 **end**

15 **return** P_t

2.2 Integrating the Proposed Dimensional Reduction Technique into NSGA-II: LPP-NSGA-II

The proposed dimensional reduction technique can be integrated into most existing multi-objective optimization algorithms. We integrate it into NSGA-II, one of the most popular EMO algorithms, named LPP-NSGA-II. The basic framework of LPP-NSGA-II is shown in Fig. 1.

Algorithm 1 presents the steps of the proposed LPP-NSGA-II in detail. First, we randomly generate N initial solutions in the D-dimensional primal space as initial populations P_t and calculate their objectives. Then, a mapping learning process (as shown in Fig. 1) is carried out by LPP, and we can obtain the transformation matrix U and its Pseudo-inverse V. Then, P_t can be mapped into P'_t in the subspace via matrix U. Since the range of data will change after dimensionality reduction, the upper and lower bounds of each decision variable are supposed to be modified, and here we use the maximum and minimum values of each dimension as their adjusted bounds. Afterward, crossover and mutation operators are performed in low-dimensional subspace to obtain offspring Q'_t in subspace. Finally, the offspring solutions are inversely mapped back to the primal space for solution evaluation, yielding Q'_t. These are iterated until the maximum number of evaluations (FE_{max}) is reached, which is the criterion for the program termination in Fig. 1.

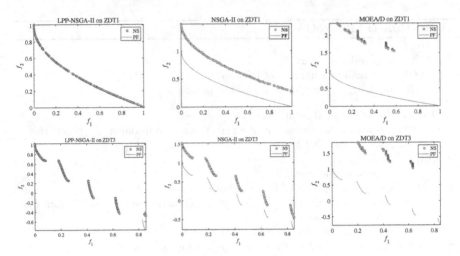

Fig. 2. Final NSs with the median of the IGD value found by LPP-NSGA-II, NSGA-II, and MOEA/D for ZDT1 and ZDT3.

2.3 Inversely Map Back to Primal Space for Environment Selection

According to Algorithm 1, the offspring Q'_t in subspace is mapped back to the primal D-dimensional space, which can be achieved by transformation matrix V, that is, $Q_t = VQ'_t$. After solution evaluation, the final offspring populations Q_t is obtained.

Subsequently, a combined population $R_t = P_t \bigcup Q_t$ is formed. The population R_t is of size $2N$. Then, perform fast nondominated sorting and crowding distance on R_t. Based on each solution's rank and crowding distance, we perform environmental selection on R_t to obtain the new population P_{t+1}.

3   Experiments and Results

In this section, we will focus on experimental studies to investigate the performance of the proposed algorithm. Initially, we briefly introduce the experimental settings, including test problems, compared algorithms, and performance metrics adopted in this article. Following this, we present the parameter settings of the compared algorithms. Eventually, we provide the experimental results and relevant analysis.

3.1   Experimental Settings

1) **Test Problems**: For the sake of demonstrating the performance of the proposed algorithm, several suits of benchmark problems, i.e., the ZDT test suits [7], and the SMOP test suits [18] are considered in this paper. Due to the fact that ZDT5 is a Boolean function that needs binary encoding, it is

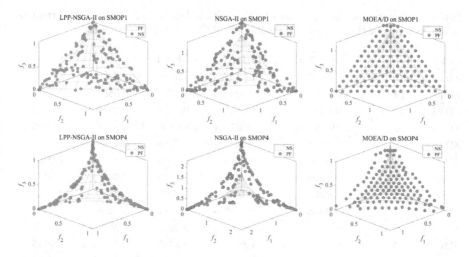

Fig. 3. Final NSs with the median of the IGD value found by LPP-NSGA-II, NSGA-II, and MOEA/D for SMOP1 and SMOP4.

omitted in our study. We compared LPP-NSGA-II with four state-of-the-art EMO algorithms on these benchmark problems.

2) **Compared Algorithms**: During the experiment, we compared the proposed LPP-NSGA-II with the following four representative EMO algorithms: Pareto-based NSGA-II [3], decomposition-based MOEA/D [23], indicator-based evolutionary SRA2 [8], and hybrid EMO algorithm Two_Arch2 [20]. Here, we introduce the main idea of these algorithms.

3) **Performance Metrics**: In order to evaluate the performance of the compared algorithms, two widely used metrics: 1) the inverted generational distance (IGD) and 2) the hypervolume indicator (HV), are adopted in the following experimental studies. They are both capable of assessing convergence and uniformity performance simultaneously.

3.2   Parameter Settings

The experiments are conducted on PlatEMO [17], a completely MATLAB-based framework for multi-objective and many-objective optimization. Since the test algorithms are stochastic, the algorithms are performed on each test problem with 30 independent runs.

1) The population size N for each algorithm (including the proposed algorithm) is set to 100 for two-objective test problems ZDT1-ZDT4, ZDT6, and 150 for three-objective test problems SMOP1-SMOP8. Besides, the number of decision variables D for all test problems is 100.

2) The maximum number of iterations is 250 for ZDT3, 500 for ZDT6, and 200 for the rest. All algorithms will terminate when the number of iterations reaches the maximum number.

Table 1. Performance comparison of LPP-NSGA-II with four state-of-the-art algorithms in terms of the average IGD values on ZDT and SMOP test suits.

Benchmark	Objectives	LPP-NSGA-II	NSGA-II	MOEA/D	SRA2	Two_Arch2
ZDT1	2	**0.0163**	0.2592 +	1.3136 +	0.2938 +	0.4406 +
ZDT2	2	**0.1301**	0.5222 +	2.6272 +	0.6884 +	1.3959 +
ZDT3	2	**0.0242**	0.1641 +	0.9020 +	0.1942 +	0.2565 +
ZDT4	2	**16.8277**	173.6379 +	428.0146 +	254.0287 +	522.5893 +
ZDT6	2	**0.7535**	2.6814 +	5.7086 +	3.4776 +	5.1792 +
SMOP1	3	**0.0745**	0.2125 +	0.2200 +	0.2108 +	0.2467 +
SMOP2	3	**0.1295**	0.7898 +	0.6365 +	0.6511 +	0.8172 +
SMOP3	3	1.8639	**0.9723** −	1.0405 −	1.0009 −	1.0016 −
SMOP4	3	**0.0436**	0.2319 +	0.1848 +	0.1268 +	0.1728 +
SMOP5	3	**0.1943**	0.2757 +	0.2809 +	0.2531 +	0.2470 +
SMOP6	3	0.1393	0.0981 −	0.2385 +	**0.0688** −	0.0715 −
SMOP7	3	**0.3507**	0.4274 +	0.6627 +	0.4208 +	0.6416 +
SMOP8	3	**0.5550**	2.5722 +	2.1164 +	2.3197 +	2.8798 +
+/ ≈ /−	-	-	11/0/2	12/0/1	11/0/2	11/0/2

Notes: "+", "−", and "≈" in two tables denote the performance of the proposed algorithms is superior to, inferior to, or similar to that of the compared algorithms under Wilcoxon's rank-sum test with a 0.05 significance level

3) During dimension reduction using LPP, the neighborhood size k is set to 10 for ZDT test suits and 15 for three-objective SMOP test suits. The parameter t is set to 1 for all test suits adopted in this section.
4) The polynomial mutation rate used in algorithms is formulated to $0.1/D$ where D denotes the number of decision variables. Additionally, the other parameters of the algorithms are determined by the default settings given on PlatEMO.

3.3 Experimental Results and Analysis

The algorithms are run independently 30 times in each test problem, and the average values of the obtained IGD and HV metrics are shown in Table 1 and Table 2, respectively. In these tables, the best results of the five algorithms are highlighted in boldface. The IGD and HV of the compared algorithms are also analyzed in accordance with Wilcoxon's rank-sum test, a non-parametric statistical hypothesis testing, which is employed here to compare the mean IGD (HV) of LPP-NSGA-II with that of the other algorithms for highlighting the significance of the findings. Where "+", "−" and "≈" in two tables denote the performance of the proposed algorithm is superior to, inferior to, or similar to that of the compared algorithms, respectively. Moreover, due to space limitations, we only plotted the nondominated solution sets (NSs) with the median IGD value obtained by the proposed LPP-NSGA-II, Pareto-based NSGA-II, and decomposition-based MOEA/D over 30 independent runs.

Table 2. Performance comparison of LPP-NSGA-II with 7 state-of-the-art algorithms in terms of the average HV values on ZDT and SMOP test suits.

Benchmark	Objectives	LPP-NSGA-II	NSGA-II	MOEA/D	SRA2	Two_Arch2
ZDT1	2	**0.7027**	0.4137 +	0.0000 +	0.3759 +	0.2407 +
ZDT2	2	**0.1301**	0.0425 +	0.0000 +	0.0011 +	0.0000 +
ZDT3	2	**0.5973**	0.4856 +	0.0420 +	0.4610 +	0.4177 +
ZDT4	2	0.0000	0.0000 ≈	0.0000 ≈	0.0000 ≈	0.0000 ≈
ZDT6	2	**0.0364**	0.0000 ≈	0.0000 ≈	0.0000 ≈	0.0000 ≈
SMOP1	3	**0.7860**	0.6142 +	0.6234 +	0.63098 +	0.5814 +
SMOP2	3	**0.7170**	0.0492 +	0.1498 +	0.1293 +	0.0433 +
SMOP3	3	0.0000	0.0094 ≈	0.0079 ≈	0.0073 ≈	0.0071 ≈
SMOP4	3	**0.9718**	0.8227 +	0.8613 +	0.8994 +	0.8778 +
SMOP5	3	**0.8600**	0.7686 +	0.6943 +	0.8021 +	0.8192 ≈
SMOP6	3	0.9075	0.9382 ≈	0.8140 +	0.9377 ≈	**0.9501** −
SMOP7	3	**0.1220**	0.0623 +	0.0034 +	0.0775 ≈	0.0076 +
SMOP8	3	**0.0281**	0.0000 ≈	0.0000 ≈	0.0000 ≈	0.0000 ≈
+/ ≈ /−	-	-	8/5/0	9/4/0	7/6/0	7/5/1

Notes: "+", "−", and "≈" in two tables denote the performance of the proposed algorithms is superior to, inferior to, or similar to that of the compared algorithms under Wilcoxon's rank-sum test with a 0.05 significance level

The results in Table 1 and Table 2 report that the comprehensive performance of our LPP-NSGA-II is better than the comparison algorithm in terms of IGD and HV metrics. As shown in Fig. 2, for ZDT1 and ZDT3, the approximate PFs obtained by the proposed LPP-NSGA-II almost converge to the true PF, with a limit of 200 iterations. These indicate that the proposed algorithm has better performance than the compared algorithms in terms of convergence speed. In the meantime, it can be noted that our proposed algorithm still has favorable distribution in ZDT1 and ZDT3 test problems.

The results in Table 1 and Table 2 also show that LPP-NSGA-II attains the best performance on ZDT and SMOP test suits except for ZDT3, ZDT6, SMOP3 and SMOP6. For further observation, Fig. 3 plots the NSs with average IGD values among 30 independent runs obtained by LPP-NSGA-II, NSGA-II, and MOEA/D on SMOP1 and SMOP4 with 100 decision variables. It can be seen from Fig. 3 that LPP-NSGA-II can almost converge to the real PF under the condition of a few iterations in SMOP1 and SMOP4, and still has promising distribution.

3.4 Sensitivity to the Parameter k

To study the influence of parameter k on the performance of the proposed algorithm, we conduct the parameter sensitivity study with $k \in \{10, 15, 20\}$. In this experiment, all parameters except k remained consistent with the aforementioned experiments. On account of page space limitation, only the average of the IGD metric found in 30 independent runs for each k on ZDT1, SMOP1, and SMOP4 are presented. It can be seen from Fig. 4 that the best parameter value of k is pertinent to the number of objectives.

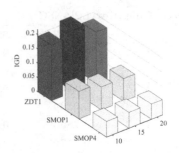

Fig. 4. Parameter sensitivity studies of k on ZDT1, SMOP1, and SMOP4.

Nevertheless, the performance of LPP-NSGA-II is relatively stable on parameter k.

4   Conclusion

In this article, we have proposed an EMO based on decision space reduction. The decision space reduction technique is based on manifold learning using LPP. Through mapping learning, the high-dimensional decision space is mapped to a low-dimensional subspace for search, and then the pseudo-inverse of the projection matrix maps the resultant offspring back to the primal decision space, which improves the computational efficiency and search efficiency of the algorithm. The proposed decision space reduction technique can be combined with most EMO algorithms. We have integrated the proposed dimension reduction technique with NSGA-II, named LPP-NSGA-II. We have tested the proposed LPP-NSGA-II on 13 different test problems, also have compared it with four state-of-the-art EMO algorithms. The experimental results indicated the efficacy of the proposed approach.

References

1. Cheng, R., Jin, Y.: A competitive swarm optimizer for large scale optimization. IEEE Trans. Cybern. **45**(2), 191–204 (2014)
2. Deb, K., Jain, H.: An evolutionary many-objective optimization algorithm using reference-point-based nondominated sorting approach, part i: solving problems with box constraints. IEEE Trans. Evol. Comput. **18**(4), 577–601 (2014)
3. Deb, K., Pratap, A., Agarwal, S., Meyarivan, T.: A fast and elitist multiobjective genetic algorithm: NSGA-II. IEEE Trans. Evol. Comput. **6**(2), 182–197 (2002)
4. Gu, F., Liu, H.L., Cheung, Y.M., Zheng, M.: A rough-to-fine evolutionary multiobjective optimization algorithm. IEEE Trans. Cybern. **52**(12), 13472–13485 (2021)
5. He, X., Niyogi, P.: Locality preserving projections. In: Advances in Neural Information Processing Systems, vol. 16, pp. 153–160 (2003)

6. Huang, D.S., Mi, J.X.: A new constrained independent component analysis method. IEEE Trans. Neural Networks **18**(5), 1532–1535 (2007)
7. Huband, S., Hingston, P., Barone, L., While, L.: A review of multiobjective test problems and a scalable test problem toolkit. IEEE Trans. Evol. Comput. **10**(5), 477–506 (2006)
8. Li, B., Tang, K., Li, J., Yao, X.: Stochastic ranking algorithm for many-objective optimization based on multiple indicators. IEEE Trans. Evol. Comput. **20**(6), 924–938 (2016)
9. Li, B., Li, Y.R., Zhang, X.L.: A survey on laplacian eigenmaps based manifold learning methods. Neurocomputing **335**, 336–351 (2019)
10. Li, X., Yao, X.: Cooperatively coevolving particle swarms for large scale optimization. IEEE Trans. Evol. Comput. **16**(2), 210–224 (2011)
11. Liu, H.L., Gu, F., Zhang, Q.: Decomposition of a multiobjective optimization problem into a number of simple multiobjective subproblems. IEEE Trans. Evol. Comput. **18**(3), 450–455 (2013)
12. Ma, X., et al.: A survey on cooperative co-evolutionary algorithms. IEEE Trans. Evol. Comput. **23**(3), 421–441 (2019)
13. Mohamed, A.W., Almazyad, A.S.: Differential evolution with novel mutation and adaptive crossover strategies for solving large scale global optimization problems. Appl. Comput. Intell. Soft Comput. **2017** (2017)
14. Omidvar, M.N., Li, X., Mei, Y., Yao, X.: Cooperative co-evolution with differential grouping for large scale optimization. IEEE Trans. Evol. Comput. **18**(3), 378–393 (2013)
15. Roweis, S.T., Saul, L.K.: Nonlinear dimensionality reduction by locally linear embedding. Science **290**(5500), 2323–2326 (2000)
16. Tenenbaum, J.B., Silva, V.D., Langford, J.C.: A global geometric framework for nonlinear dimensionality reduction. Science **290**(5500), 2319–2323 (2000)
17. Tian, Y., Cheng, R., Zhang, X., Jin, Y.: Platemo: a matlab platform for evolutionary multi-objective optimization [educational forum]. IEEE Comput. Intell. Mag. **12**(4), 73–87 (2017)
18. Tian, Y., Lu, C., Zhang, X., Tan, K.C., Jin, Y.: Solving large-scale multiobjective optimization problems with sparse optimal solutions via unsupervised neural networks. IEEE Trans. Cybern. **51**(6), 3115–3128 (2020)
19. Trivedi, A., Srinivasan, D., Sanyal, K., Ghosh, A.: A survey of multiobjective evolutionary algorithms based on decomposition. IEEE Trans. Evol. Comput. **21**(3), 440–462 (2016)
20. Wang, H., Jiao, L., Yao, X.: Two_arch2: an improved two-archive algorithm for many-objective optimization. IEEE Trans. Evol. Comput. **19**(4), 524–541 (2014)
21. Xanthopoulos, P., Pardalos, P.M., Trafalis, T.B., Xanthopoulos, P., Pardalos, P.M., Trafalis, T.B.: Linear discriminant analysis. In: Robust Data Mining, pp. 27–33 (2013)
22. Yekkehkhany, B., Safari, A., Homayouni, S., Hasanlou, M.: A comparison study of different kernel functions for SVM-based classification of multi-temporal polarimetry SAR data. Int. Arch. Photogram. Remote Sens. Spatial Inf. Sci. **40**(2), 281 (2014)
23. Zhang, Q., Li, H.: MOEA/D: a multiobjective evolutionary algorithm based on decomposition. IEEE Trans. Evol. Comput. **11**(6), 712–731 (2007)

Path Planning of Automatic Parking System by a Point-Based Genetic Algorithm

Zijia Li and Fangqing Gu$^{(\boxtimes)}$

Guangdong University of Technology, Guangzhou, China
fqgu@gdut.edu.cn

Abstract. The path planning module in automatic parking systems plays a crucial role in enhancing performance and user experience. To tackle this vital aspect, we propose a path planning algorithm that utilizes a point-based genetic algorithm (POGA) in automatic parking systems. Initially, we generate some control points randomly within the parking range and then establish trajectories between each pair of control points using a cubic spline curve. In addition, we incorporate a penalty term to increase the distance between the vehicle and obstacles along the collision path. By implementing this technique, we successfully obtain the initial shortest path that ensures safety, along with the corresponding control points. Finally, POGA is proposed to conduct an optimal search for control points, in which each control points is an individual. This algorithm combines the well-established Dijkstra algorithm with a genetic algorithm, thereby improving stability and convergence. Through comprehensive comparisons with a path-based genetic algorithm, our experimental results demonstrate the superior stability and faster convergence rate exhibited by POGA across various parking scenarios.

Keywords: Automatic parking · Path planning · Separation axis theorem · Point-based genetic algorithm

1 Introduction

In recent years, automatic parking technology has become a crucial intelligent technology applied to automobiles [9]. Path planning plays a significant role in the performance, length, and safety of vehicle parking [18]. Therefore, continuous improvement of path planning algorithms is a current research focus to enable vehicles to plan safe shortest paths in various parking scenarios [14].

Automatic parking path planning algorithms can be roughly divided into two categories [2]. One is based on artificial intelligence, using heuristic methods to induce the drivers' parking behavior rules and convert the drivers' behavior rules into the behavior of the central controller. For example, Zhao Liang et

This research was funded by the Natural Science Foundation of Guangdong Province (2021A1515011839).

al. [12] proposed a fuzzy control algorithm that offers improved performance compared to the PID control method, resulting in a smaller error and faster response speed. Furthermore, a vehicle parking control methodology using fuzzy logic is proposed in [4].

The other is to design a safe and effective parking path in advance based on parking environment constraints and vehicle movement status. They can be further divided into the following three categories. Traditional algorithms are generally simple in structure and fast in the calculation but difficult to adapt to a complex environment, including RRT [6,16], A* [3], APF [5] and Dijkstra algorithms and so on. Based on the hybrid A* algorithm, the heuristic term of hybrid A* is added a penalty term of obstacle distance to the Reeds-shepp curve as the heuristic term, which makes the trajectory as far away from obstacles as possible in a vast space in [18].

The intelligent algorithm can adapt to the complex environment well, but the computation time is long, and the convergence speed is slow, including different neural networks [8], RL [19], PL [10,13], ACO [11], PSO and GA and so on. Vieira et al. [17] applied genetic algorithms to the path planning of automatic parking. However, there is no effective genetic operator designed for specific parking problems, which requires considerable time to run the algorithm. The traditional crossover and mutation operators of GA have been re-designed to increase the searching speed, and the effectiveness of the evolution test by the improved GA has been validated in [7].

In order to give full play to the advantages of different algorithms, many scholars try to combine traditional algorithms with intelligent algorithms and have made some achievements [1]. Roberge at el. [15] used a genetic algorithm to solve the path planning problem of unmanned aerial vehicles and proposed a technology that parallelizes GA and PSO while minimizing communication between processes to improve computing speed.

Based on the abovementioned literature, this paper proposes a path-planning algorithm combining Dijkstra and a genetic algorithm. Firstly, the state variables of the initial control points are randomly generated. The vehicle trajectories between each pair of control points are obtained according to cubic spline curve fitting. Then, the initial distance matrix is obtained by calculating the length of the curve. Secondly, based on the separation axis theorem, the vehicle is detected whether there is a collision with the obstacle. A penalty term is added to the initial distance matrix if there is a collision. Finally, Dijkstra is used to obtain the initial approximate solution, and a point-based genetic algorithm is used to search for a global optimal solution.

The remainder of this article is organized as follows: In Sect. 2, we use cubic spline interpolation to curve fit the control points and build a vehicle model in the process of automatic parking. Sect. 3 analyzes the collision detection of the parking track. We propose specific selection, crossover, and mutation operators for parking path planning and combine the genetic operations into the Dijkstra algorithm in Sect. 4. In Sect. 5, the path-based genetic algorithm and the point-based genetic algorithm are compared under different parking scenarios and different initial heading angles. Finally, we conclude this paper in Sect. 6.

2   Path Planning Model of Vehicle

Due to the low speed of the vehicle during parking, the vehicle can be considered a rigid body without considering the lateral slip of the wheels. We assume that the wheels only perform pure rolling and steering motions. In order to simplify the collision detection process in path planning, the vehicle is simplified into a rectangle whose size is the maximum overall dimension of the vehicle body, as shown in Fig. 1, where m is the length of rear overhang, n is the length of front overhang, l is the wheelbase of the vehicle, L is the length of the vehicle, $2b$ is the width of the vehicle, θ is the vehicle heading angle, and (x, y) is the coordinates of the center of the rear wheel of the vehicle.

We designate the center of the rear wheel as the control point, with state variables represented by $\mathbf{x} = (x, y, \theta)^T$. Additionally, we define its coordinates as $\mathbf{y} = (x, y)^T$. Assuming that the vehicle's steering wheel remains stationary, cubic spline curve fitting is used to obtain the vehicle motion trajectory between two control points. For each parking trajectory, given the state variables of the two control points $\mathbf{x}_i = (x_i, y_i, \theta_i)^T$ and $\mathbf{x}_j = (x_j, y_j, \theta_j)^T$, the curve equation from the ith control point to the jth control

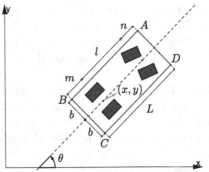

Fig. 1. Vehicle motion diagram.

point can be expressed as $S_{ij}(x) = s_1(x - x_i)^3 + s_2(x - x_i)^2 + s_3(x - x_i) + s_4$, where, s_1, s_2, s_3 and s_4 are undetermined parameters, and the derivative of S_{ij} is expressed as $S'_{ij}(x) = 3s_1(x - x_i)^2 + 2s_2(x - x_i) + s_3$.

Based on the definition of cubic spline interpolation $S_{ij}(x_i) = y_i$, $S_{ij}(x_j) = y_j$, and natural boundary conditions $S'_{ij}(x_i) = \tan\theta_i$, $S'_{ij}(x_j) = \tan\theta_j$, let $h = x_j - x_i$, the coefficients of the cubic spline interpolation function can be obtained as shown in Eq. (1). Finally, the trajectory length can be solved by curve integral.

$$\begin{cases} s_1 = 3\frac{y_j - h\tan\theta_i - y_i}{h^3} - 2\frac{\tan\theta_j - \tan\theta_i}{h} \\ s_2 = \tan\theta_j - \tan\theta_i \\ s_3 = \tan\theta_i \\ s_4 = y_i \end{cases} \tag{1}$$

Then $l_{ij} = \int_{x_i}^{x_j} \sqrt{1 + [S'_{ij}(x)]^2}\,dx$, where, l_{ij} is the trajectory length between the ith control point and the jth control point, $x \in [x_i, x_j]$. However, since the original function of the integral function $\sqrt{1 + [S'_{ij}(x)]^2}$ does not exist, the numerical integral method is used to solve it, which will cause large overhead in the iteration of genetic algorithm. Therefore, we select s points uniformly between every two control points and use the length of the interpolation line segment to approximate the trajectory length by $l_{ij} \approx \sum_{k=1}^{s+1} \sqrt{(x_k - x_{k-1})^2 + (y_k - y_{k-1})^2}$, where, $(x_0, y_0) = (x_i, y_i)$, $(x_{s+1}, y_{s+1}) = (x_j, y_j)$.

3   Collision Detection Based on Separation Axis Theorem

To determine the shortest parking path, we compute the track distance between control points, resulting in the distance matrix $\mathbf{D} = (d_{ij})_{(N+2)*(N+2)}$, where $d_{ij} = l_{ij}$. Here, N denotes the population size, representing the number of control points, while $N+2$ accounts for the total number of control points, including the start and end points. On this basis, collision detection ensures that the shortest path obtained through Dijkstra algorithm is safe.

This study focuses on static collision detection for safe and efficient parking paths in known scenarios. Existing methods suffer from drawbacks like high time complexity and limited adaptability. To address these issues, we propose a collision detection algorithm based on the separation axis theorem, striking a balance between accuracy and efficiency.

The separation axis theorem is a collision detection algorithm based on geometric shapes. It utilizes pro-

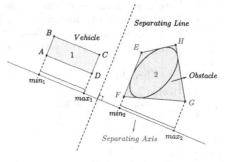

Fig. 2. Diagram of the separation axis theorem.

jection axes between polygons to judge whether there is an intersection. If a projection axis exists where the projections of two convex polygons do not overlap, they are considered non-intersecting. Conversely, if there is overlap on all projection axes, the polygons are determined to intersect. Figure 2 shows the collision detection diagram between vehicles and obstacles, where $\mathbf{AB} = (x_B - x_A, y_B - y_A)$ is one side of the vehicle, and the vector perpendicular $(y_A - y_B, x_B - x_A)$ to it is the separation axis. For obstacles of any shape, we can always surround it with a convex quadrilateral.

Therefore, the length of the parking path from the ith control point to the jth control point after adding the collision penalty item can be obtained by Eq. (2).

$$d_{ij} = l_{ij} + M e_{ij} \tag{2}$$

where

$$e_{ij} = \min(|max_2 - min_1|, |max_1 - min_2|) \tag{3}$$

represents the extent to which the vehicle trajectory collides with the obstacle from the ith control point to the jth control point. It is the minimum overlap distance on the separation axis. If there is no collision, then $e_{ij} = 0$. M is the penalty constant, usually a large positive integer. The collision detection algorithm is shown in Algorithm 1.

Algorithm 1: Collision Detection

Input:
- $A^{(k)}$, $B^{(k)}$, $C^{(k)}$, $D^{(k)}$: The vehicle vertex coordinates between the ith and jth control point correspond to the k interpolation point;
- E, F, G, H: The obstacle vertex coordinates;

Output: d_{ij}: The length of parking path from the ith to the jth control point.

1 $e_{ij} \leftarrow \infty$;
2 **for** $k = 1 : s$ **do**
3   calculate the *Separating Axis* for each side of the vehicle and the obstacle;
4   **for** *every Separating Axis* **do**
5   calculate the minimum (min_1) and maximum (max_1) values of the vehicle projected onto the *Separating Axis*, as well as the minimum (min_2) and maximum (max_2) values of the obstacle projected onto the same *Separating Axis*;
6   **if** $(max_1 < min_2)$ $||$ $(min_1 > max_2)$ **then**
7       $e_{ij} \leftarrow 0$;
8   **break**;
9   **else**
10   calculate e'_{ij} by Eq. (3);
11       $e_{ij} \leftarrow \min(e_{ij}, e'_{ij})$;
12   **end**
13   **end**
14   $e_{ij}^{(k)} \leftarrow e_{ij}$;
15 **end**
16 $e_{ij} \leftarrow \max(e_{ij}^{(1)}, \ldots, e_{ij}^{(k)})$;
17 d_{ij} is obtained by Eq. (2);
Result: d_{ij}

---

4 The Proposed Point-Based Genetic Algorithm

4.1 Population Initialization

Considering the continuity of parking space, we use real number coding for the state variables of control points and randomly generate control points outside the obstacle area of the parking space. The initial population is denoted as $\mathbf{X}_0 = \{\mathbf{x}_1, \ldots, \mathbf{x}_N\}$.

4.2 Genetic Operator

Selection Operator. If traditional selection operators are used to select paths, problems such as redundancy, regression, and volatility arise, making it challenging to find the optimal path. To address these issues, we propose a new selection operator designed specifically to overcome the limitations of traditional selection operators. We select $\frac{N}{2}$ control points according to the fitness of the individuals based on the parking track distance between the control points. Specifically, we select n_0 initial control points to the next generation using the Dijkstra algorithm with **D**. When $n_0 < \frac{N}{2}$, parts of control points with a size of $N' = \eta N$, $\eta \in (0, 1)$ are randomly selected from the initial control points and the distance matrix **D'** which is a sub-matrix. Then, we use the Dijkstra algorithm on **D'** to identify n_1 control points traversing the shortest path. This process is repeated p times to obtain p shortest paths. The control points with the highest number of occurrences, and in case of a tie, the control points with the shortest path distance, are selected to form a set of state variables for the child control points, denoted as $child^s$. This new selection operator allows us to effectively choose promising control points for the next generation.

Crossover Operator. We propose a crossover operator based on the vehicle heading angle obtained by linear interpolating the front and rear control points. In continuous space, for the crossover operator of control point coordinates, the search space of the traditional uniform crossover operator is the rectangle, with the diagonal being the line segment formed by these two control points. Considering the actual parking situation, the shortest path should meet two conditions: smooth enough and no collision with obstacles. Therefore, we select three control points from the population of control points participating in the crossover operation and use the triangle area $\Delta \mathbf{y}_1 \mathbf{y}_2 \mathbf{y}_3$ formed by these three control points as the search space, as shown in Fig. 3 and Fig. 4. Equation (4) shows the specific crossover operation.

$$\mathbf{x}^c = \begin{pmatrix} x^c \\ y^c \\ \theta^c \end{pmatrix} = \begin{pmatrix} x_2 \\ y_2 \\ \theta_2 \end{pmatrix} + \begin{pmatrix} \lambda_1(x_1 - x_2) + \lambda_2(x_3 - x_2) \\ \lambda_1(y_1 - y_2) + \lambda_2(y_3 - y_2) \\ \lambda_3(\theta_1 - \theta_2) \end{pmatrix} \quad (4)$$

where, $\mathbf{x}^c$ represents the control points obtained through crossover operation, λ_1, λ_2, λ_3 indicate random numbers in the range $(0, 1)$.

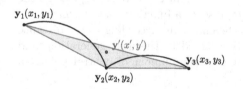

Fig. 3. Before crossing.

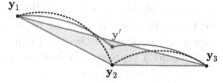

Fig. 4. After crossing.

Mutation Operator. We conduct the non-uniform mutation operator on the new control points obtained by the crossover operator. The mutation results are shown in Eq. (5) and Eq. (6), where $rand$ is a random number in $[0, 1]$, $\mathbf{x}^m$ represents the control points obtained through mutation operation, $\Delta(t, q) = q \cdot (1 - r^{(1 - \frac{t}{T}) \cdot d})$, t is the current generation number, T is the maximum number of generation, r is a random number with uniform distribution in $[0, 1]$, d is the system parameter, $\mathbf{x}_{\max} = (x_{max}, y_{max}, \theta_{max})^T$, and $\mathbf{x}_{\min} = (x_{min}, y_{min}, \theta_{min})^T$. If $rand > 0.5$, $\mathbf{x}^m$ is given by Eq. (5).

$$\mathbf{x}^m = \mathbf{x}^c + \Delta(t, \mathbf{x}_{\max} - \mathbf{x}^c) \tag{5}$$

Otherwise, $\mathbf{x}^m$ is given by Eq. (6).

$$\mathbf{x}^m = \mathbf{x}^c + \Delta(t, \mathbf{x}^c - \mathbf{x}_{\min}) \tag{6}$$

Algorithm 2: Genetic Algorithm

Input:
- N: The size of the population;
- T: The maximum number of generation;

Output:
- $dist_{min}$: The shortest safe parking path length;
- $\mathbf{X}_T$: The final control point set obtained by genetic algorithm.

1 randomly initialize $\mathbf{X}_0 = \{\mathbf{x}_1, \ldots, \mathbf{x}_N\}$, $t = 0$;
2 **while** $t < T$ **do**
3 $\quad$ initialize $child\_size = 0$, $\mathbf{child}^s = \emptyset$;
4 $\quad$ calculate D with Algorithm 1 based on $\mathbf{X}_t$;
5 $\quad$ **while** $child\_size < \frac{N}{2}$ **do**
6 $\quad\quad$ $\mathbf{D}' \leftarrow$ randomly selected N' rows and N' columns in $\mathbf{D}$;
7 $\quad\quad$ $[dist_{min}, \mathbf{X}^s] \leftarrow$ **Dijkstra** $(\mathbf{D}')$;
8 $\quad\quad$ $\mathbf{child}^s \leftarrow \mathbf{child}^s \cup \mathbf{X}^s$;
9 $\quad\quad$ $child\_size \leftarrow$ **Size** $(\mathbf{child}^s)$;
10 $\quad$ **end**
11 $\quad$ $\mathbf{X}^c$ is obtained by crossing $\mathbf{child}^s$ using Eq. (4);
12 $\quad$ $\mathbf{X}^m$ is obtained by mutating $\mathbf{X}^c$ using Eq. (5) and Eq. (6);
13 $\quad$ $t \leftarrow t + 1$;
14 $\quad$ $\mathbf{X}_t \leftarrow \mathbf{child}^s \cup \mathbf{X}^m$;
15 **end**

Result: $dist_{min}$, $\mathbf{X}_T$

The pseudocode of POGA is shown in Algorithm 2, where $\mathbf{X}_t$ represents the population of the t iteration, **Dijkstra** represents a function that calculates the shortest path from the starting point to the end point according to the distance matrix, $\mathbf{X}^s$ represents the control points selected to the next generation, and **Size** $(\mathbf{child}^s)$ represents a function that gets the number of control points in $\mathbf{child}^s$.

5   Experimentation and Results

The obstacles around the target parking space are different in different parking scenarios. Even some vehicles, limited by the driver's parking technology and complex parking conditions, may have non-standard or even beyond the parking space, significantly increasing the difficulty of automatic parking. Therefore, this paper selected three common parking scenarios for experiments, as shown in Fig. 5, and the shortest safe parking trajectories of vehicles with different initial heading angles are studied in different scenarios.

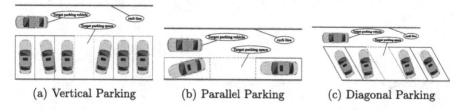

(a) Vertical Parking   (b) Parallel Parking   (c) Diagonal Parking

Fig. 5. Diagram of three common different parking scenarios.

For the convenience of expression, we use GA to represent the path-based genetic algorithm and POGA to represent the point-based genetic algorithm. In this paper, the POGA and GA proposed above are used for comparative experiments to compare the smoothness, rationality, and shortest length of the parking trajectory of the two algorithms under the same parking scene and the same initial heading angle. The experiments were conducted on MATLAB R2020b, and the parameters of the vehicle and parking scenes are provided in Tables 1, 2, 3, and 4.

Table 1. The vehicle parameters selected in the experiment.

Parameters	Values
n	0.960
m	0.929
l	2.800
b	0.971

Table 2. Vertical parking scene parameters.

Parameters	Values
Initial control point	$(-9.969, 4, 0), (-9.969, 4, \frac{\pi}{12}), (-9.969, 4, -\frac{\pi}{12})$
Target control point	$(-1.155, -1.021, -\frac{\pi}{2})$
Obstructing	$(-3.14, -2.35), (-3.14, 2.19)$
vehicle 1	$(-4.85, 2.19), (-4.85, -2.35)$
Obstructing	$(2.17, -2.58), (3.62, 2.04)$
vehicle 2	$(1.82, 2.61), (0.37, -2.02)$

5.1 Vertical Parking

The parameter settings of POGA are as follows: the number s of interpolation points between any two control points is 50, population size N is 50, $\eta = 0.8$, mutation probability P_m is 0.1, crossover probability P_c is 1, system parameter d in mutation operator is 4, and the maximum number of iterations T is 50. As shown in Fig. 6(a)(d), when iteration times are both 50, both POGA and GA can obtain a smooth and reasonable parking path. It shows that genetic algorithms are feasible for automatic parking path planning. However, the distance between the vehicle in Fig. 6(a) and the obstacle on the left side is very close. In contrast,

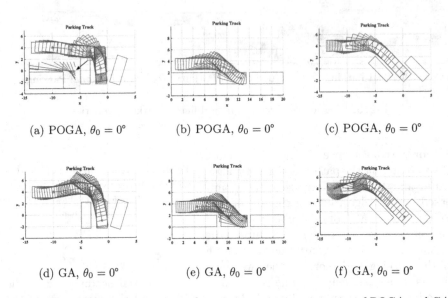

(a) POGA, $\theta_0 = 0°$ (b) POGA, $\theta_0 = 0°$ (c) POGA, $\theta_0 = 0°$

(d) GA, $\theta_0 = 0°$ (e) GA, $\theta_0 = 0°$ (f) GA, $\theta_0 = 0°$

Fig. 6. The initial heading angle is $0°$, and the parking trajectories of POGA and GA in different parking scenarios.

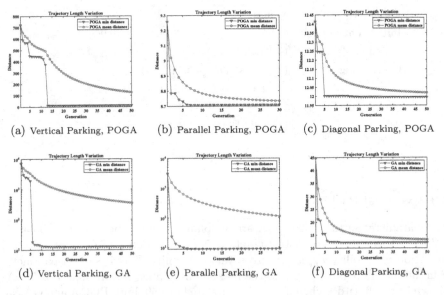

(a) Vertical Parking, POGA   (b) Parallel Parking, POGA   (c) Diagonal Parking, POGA

(d) Vertical Parking, GA   (e) Parallel Parking, GA   (f) Diagonal Parking, GA

Fig. 7. When the initial heading angle is $0°$, the trajectory length variations of POGA and GA in different scenarios.

that between the vehicle in Fig. 6(d) and the obstacle on the left side is still a certain distance, so the parking path length obtained by POGA is shorter. It shows that POGA has stronger convergence. Figure 7 visually illustrates the iterative process in different cases.

5.2 Parallel Parking

The POGA parameter settings are the same as in vertical parking, except for $T = 30$. As shown in Fig. 6(b)(e), both POGA and GA can find the shortest safe parking path under different initial heading angles. However, because the parallel parking scene is relatively simple, there is no obvious difference between the POGA and GA in terms of trajectory length.

Table 3. Parallel parking scene parameters.

Parameters	Values
Initial control point	$(4.5, 3.5, 0), (4.5, 4, \frac{\pi}{12}), (4.5, 4, -\frac{\pi}{12})$
Target control point	$(12.5, 1, 0)$
Obstructing vehicle 1	$(0, 0), (8, 0), (8, 2), (0, 2)$
Obstructing vehicle 2	$(14, 0), (20, 0), (20, 2), (14, 2)$

5.3 Diagonal Parking

The POGA parameter settings are the same as in vertical parking. As shown in Fig. 6(c)(f), when iteration times are both 50, both POGA and GA can obtain relatively smooth and reasonable parking paths. Both algorithms have good effects in this case, and it is impossible to see directly whether the POGA is meaningful. Therefore, we conducted 30 repeated experiments and counted the experimental results to discuss the effect of the POGA further, as shown in Table 5, where Average Time(s) represents the average running time of the algorithm for 30 runs, Minimum Distance(m) represents the average length of the shortest parking trajectory for 30 runs. Distance Var represents the variance of the length of the parking trajectory during 30 runs.

From Table 5, we can see that both POGA and GA can search the parking path with better results under different parking scenarios and the initial heading angles of vehicles. The average operation time of POGA is 10–40 s longer than that of GA, and the average shortest path length obtained by POGA is always 0.2–1 m shorter than that of GA, with a smaller variance. It shows that compared with GA, POGA has stronger convergence and stability when the operation time is not increased much and can find the approximate solution of the optimal solution when the number of iterations is less.

Table 4. Diagonal parking scene parameters.

Parameters	Values
Initial control point	$(-9.969, 4, 0)$, $(-9.969, 4, \frac{\pi}{12})$, $(-9.969, 4, -\frac{\pi}{12})$
Target control point	$(0.15, -0.97, -\frac{\pi}{4})$
Obstructing vehicle 1	$(-5, 2.13)$, $(-6.2, 0.92)$ $(-2.99, -2.29)$, $(-1.79, -1.08)$
Obstructing vehicle 2	$(1.5, 2.13)$, $(0.3, 0.92)$ $(3.51, -2.29)$, $(4.71, -1.08)$

Table 5. The POGA and GA were running thirty times to compare the results under different parking environments and different initial course angles.

Algorithm	Initial Vehicle Heading Angle($^\circ$)	Vertical Parking			Parallel Parking			Diagonal Parking		
		Average Time(s)	Minimum Distance(m)	Distance Var	Average Time(s)	Minimum Distance(m)	Distance Var	Average Time(s)	Minimum Distance(m)	Distance Var
GA	-15	229.69	12.94	0.3262	83.45	9.38	0.1447	106.08	12.26	0.2615
POGA	-15	270.88	12.00	0.0102	92.66	8.88	0.0528	131.09	11.82	0.0138
GA	0	213.24	12.82	0.0830	66.26	9.40	0.1507	105.35	12.15	0.1298
POGA	0	241.74	12.02	0.0271	82.81	8.72	0.0287	125.10	11.89	0.0447
GA	15	227.86	12.97	0.2487	83.06	9.33	0.1109	109.19	12.08	0.1182
POGA	15	271.81	12.01	0.0198	94.76	8.87	0.0492	128.92	11.84	0.0155

6   Conclusion

We have proposed a point-based genetic algorithm (POGA) to address the path-planning problem in automated parking. We have employed the traditional Dijkstra algorithm to obtain an initial solution for the safe parking trajectory. Subsequently, we have introduced improved selection and crossover operators tailored explicitly for the automated parking problem, aiming to generate high-quality solutions. We have applied POGA to vertical, parallel, and diagonal parking scenarios and compared its performance with traditional genetic algorithms (GA). Simulation results have demonstrated that POGA exhibits faster convergence rates and stronger stability. POGA has found the shortest and safest parking path while demonstrating remarkable migration capability across different scenarios. Notably, POGA is highly valuable for automated parking in complex environments.

References

1. Chai, R., Tsourdos, A., Savvaris, A., Chai, S., Xia, Y.: Two-stage trajectory optimization for autonomous ground vehicles parking maneuver. IEEE Trans. Industr. Inf. **15**(7), 3899–3909 (2019)
2. Chen, X., Mai, H., Zhang, Z., Gu, F.: A novel adaptive pseudospectral method for the optimal control problem of automatic car parking. Asian J. Control **24**(3), 1363–1377 (2021)
3. Choi, J.W., Huhtala, K.: Constrained global path optimization for articulated steering vehicles. IEEE Trans. Veh. Technol. **65**(4), 1868–1879 (2016)

4. Cigánek, J.: Automatic parking control using fuzzy logic. In: 2019 6th International Conference on Advanced Control Circuits and Systems (ACCS) & 2019 5th International Conference on New Paradigms in Electronics & Information Technology (PEIT), pp. 203–208 (2019)
5. Dong, Y., Zhang, Y., Ai, J.: Experimental test of artificial potential field-based automobiles automated perpendicular parking. Int. J. Veh. Technol. **2016**, 1–10 (2016)
6. Feng, Z., Chen, S., Chen, Y., Zheng, N.: Model-based decision making with imagination for autonomous parking. In: 2018 IEEE Intelligent Vehicles Symposium (IV), pp. 2216–2223 (2018)
7. Gao, F., Zhang, Q., Han, Z., Yang, Y.: Evolution test by improved genetic algorithm with application to performance limit evaluation of automatic parallel parking system. IET Intel. Transport Syst. **15**(6), 754–764 (2021)
8. Grigorescu, S., Trasnea, B., Cocias, T., Macesanu, G.: A survey of deep learning techniques for autonomous driving. J. Field Robot. **37**(3), 362–386 (2020)
9. Hanafy, M., Gomaa, M.M., Taher, M., Wahba, A.M.: Path generation and tracking for car automatic parking employing swarm algorithm. In: The 2011 International Conference on Computer Engineering & Systems (2012)
10. Li, L., Lin, Y., Zheng, N., Wang, F.Y.: Parallel learning: a perspective and a framework. IEEE/CAA J. Automatica Sinica **4**(3), 389–395 (2017)
11. Li, Y., Ming, Y., Zhang, Z., Yan, W., Wang, K.: An adaptive ant colony algorithm for autonomous vehicles global path planning. In: 2021 IEEE 24th International Conference on Computer Supported Cooperative Work in Design (CSCWD), pp. 1117–1122 (2021)
12. Liang, Z., Zheng, G., Li, J.: Application of fuzzy control strategy in automatic parking path planning. In: 2012 4th International Conference on Intelligent Human-Machine Systems and Cybernetics, vol. 2, pp. 132–135 (2012)
13. Lin, Y.L., Li, L., Dai, X.Y., Zheng, N.N., Wang, F.Y.: Master general parking skill via deep learning. In: 2017 IEEE Intelligent Vehicles Symposium (IV), pp. 941–946 (2017)
14. Ma, S., Jiang, H., Han, M., Xie, J., Li, C.: Research on automatic parking systems based on parking scene recognition. IEEE Access **5**, 21901–21917 (2017)
15. Roberge, V., Tarbouchi, M., Labonte, G.: Comparison of parallel genetic algorithm and particle swarm optimization for real-time UAV path planning. IEEE Trans. Industr. Inf. **9**(1), 132–141 (2013)
16. Tang, W., Yang, M., Le, F., Yuan, W., Wang, B., Wang, C.: Micro-vehicle-based automatic parking path planning. In: 2018 IEEE International Conference on Real-time Computing and Robotics (RCAR), pp. 160–165 (2018)
17. Vieira, R.P., Revoredo, T.C.: Path planning for automobile urban parking through curve parametrization and genetic algorithm optimization. In: 2022 26th International Conference on Methods and Models in Automation and Robotics (MMAR), pp. 152–157 (2022)
18. Xiong, L., Gao, J., Fu, Z., Xiao, K.: Path planning for automatic parking based on improved hybrid a* algorithm. In: 2021 5th CAA International Conference on Vehicular Control and Intelligence (CVCI), pp. 1–5 (2021)
19. Zhang, P., Xiong, L., Yu, Z., Fang, P., Zhou, Y.: Reinforcement learning-based end-to-end parking for automatic parking system. Sensors **19**(18), 3996 (2019)

Penalty-Aware Memory Loss for Deep Metric Learning

Qian Chen[1], Run Li[2], and Xianming Lin[3(✉)]

[1] Institute of Artificial Intelligence, Xiamen University, Xiamen 361005,
People's Republic of China
[2] Youtu Lab, Tencent Technology (Shanghai) Co. Ltd., Shanghai, China
[3] Key Laboratory of Multimedia Trusted Perception and Efficient Computing,
Ministry of Education of China, Xiamen University, Xiamen 361005,
People's Republic of China
linxm@xmu.edu.cn

Abstract. This paper focus on the deep metric learning, which has been widely used in various multimedia tasks. One popular solution is to learn a suitable distance metric by using triplet loss terms with sophisticated sampling strategies, However, existing schemes usually do not well consider global data distributions, which makes the learned metric suboptimal. In this paper, we address this problem by proposing a *Penalty-Aware Memory Loss* (PAML) that fully utilizes the expressive power of the combination of both global data distribution and local data distribution to learn a high-quality metric. In particular, we first introduce a memory bank to build the category prototype, that can capture the global geometric structure of data from the training data. The memory bank allows imposing a penalty regularizer during the training procedure without significantly increasing computational complexity. Subsequently, a new triplet loss with softmax is defined to learn a new metric space via the classical SGD optimizer. Experiments on four widely used benchmarks demonstrate that the proposed PAML outperforms state-of-the-art methods, which is effective and efficient to improve image-level deep metric learning.

Keywords: Metric learning · Fine-grained · Convolutional neural network

1 Introduction

Deep metric learning [21,23,25,28] aims to learn a distance metric for comparing deep features, which has been used in many multimedia applications, such as multimedia retrieval [10], person Re-ID [6,17], and face recognition [14,15].

Its purpose is to train a convolutional neural network (CNN) that maps input images to a learned metric space. This learned metric space could reflect well the actual semantic distance among categories. Therefore, to improve the semantic accuracy of the similarity measurement [7], several approaches based on training tuples (*e.g.*, pairs [22] , triplets [7,10], or quadruples [30,31]) have been proposed.

Q. Liu et al. (Eds.): PRCV 2023, LNCS 14431, pp. 462–474, 2024.
https://doi.org/10.1007/978-981-99-8540-1_37

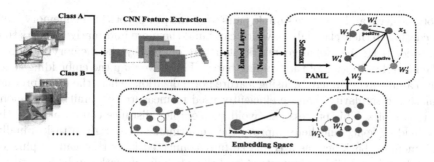

Fig. 1. The pipeline of the proposed framework. The deep features are extracted by the CNN, and are then constrained in the unit hypersphere by the embedding layer and the normalized operations. Then, we build a memory bank to capture the category prototype W_1 (the red circle points), that is further constrained with its corresponding deep features (the blue circle points) extracted from each training batch. We then derive the new prototype W_1' in the memory using our updating scheme explained in Sect. 2.3. Finally, a new triplet loss is imposed on the deep features and the new prototypes from memory bank, which can be directly optimized with the classical SGD operator. (Color figure online)

In this paper, we mainly focus on triplet-based tuples that pursuits the small intra-class distances as well as large inter-class distances among instances.

However, collecting such triplet tuples among all training data is expensive and time-consuming. To deal with this inefficiency, the prior works [7,10,13] generally established triplets in each individual training batch, rather than the entire training set. The representative works include, (semi-) hard triplets [7], smart triplet mining [2], and others [8,10,13]. Moreover, another line of works [2,7,34] proposed optimizing the triplet loss construction procedure, or alternatively adopting the proxy-based loss [29], which optimizes the feature embedding space by comparing each sample with proxy samples.

Proxy-based triplets learning has made great progress, but it has some problems. It only considers the local data distribution, not the global one. It also re-samples some data points too much or too little. This can cause model overfitting or underfitting. Adding proxy-based learning to deep model training makes it worse, because only a small batch is used to update the model.

Aims to overcome the aforementioned problems, we introduce a new deep metric learning method, named Penalty-Aware Memory Loss (PAML), which is based on supervised learning that is designed to realize the fine-gained retrieval. First, we introduce a new memory bank module to characterize the global structure of the whole training data, which stores the prototypes well positioned to capture the corresponding semantic information of each class. Second, in order to make the memory bank well, we propose a penalty-aware prototype updating scheme that is based on the information accumulation of local data in each batch. By using this scheme, the prototypes in the memory bank have both global data distribution information and local data distribution information. Then, we select both positive and negative sample pairs from the prototypes in the memory bank according to their class labels, which can build the more informative triplets that consider both the global geometric structure offered by the memory bank and the local distribution of the training batch. Third, we map the embedding space onto a unit hypersphere by using feature normalization, where the inner product (or cosine similarity) can be efficiently used to measure the similarities among data samples. This process can avoid the complex distance computation in the embedding space and help improve the performance of the given task. Finally, the memory back acts as a trainable fully-connected layer that can be plugged into the original CNN architecture, and the whole framework can be learned via classical stochastic gradient descent (SGD).

The Fig. 1 shows the framework of the proposed method, and the contributions of our paper are summarized as follows:

- Our method aims to combine global information and local information, which allows us to build triplets using prototypes with more information.
- Different from previous methods that consider global cues, we introduce a memory bank as a new global measure to store the prototypes for each category and updating prototypes based on the information accumulation of the local data in each batch.
- We efficiently construct triplets with the prototypes in the memory bank, and propose a penalty-aware loss to enhance the expressive power of the learned model.
- We extensively evaluate our method on fine-grained classification task, and our method can outperform the state-of-the-arts on four datasets.

2   The Proposed Method

2.1   Notations and Problem Formulations

Let $\mathcal{X} = \{(x_1, l_1), (x_2, l_2), ..., (x_N, l_N)\}$ denote the original image set, containing M images attributed to c categories, where x_i is the image and l_i is the corresponding objective label. Then, we can randomly sample a subset $\hat{\mathcal{X}} = \{(x_i, l_i)|i = 1, ..., n\}$ that contains $n \ll N$ training data, and this subset can serve as a training batch.

The key of metric learning is to learn a new distance metric $D(y_i, y_j)$: $\mathbb{R}^k \times \mathbb{R}^k \to \mathbb{R}$, which maps two k-dimensional representations (e.g., y_i and y_j) to a distance score between two corresponding images. In general, the calculation of such distance is always by the function $D(y_i, y_j) = \sqrt{(y_i - y_j)^T M(y_i - y_j)}$, where M is a positive semi-definite matrix that can reflect the optimal metric among features. M can be further factorized into two matrix multiplication, which makes the new distance $D(y_i, y_j) = \|y_i^T L - y_j^T L\|$, where L embeds the representation into the a new vector space. Therefore, we define a new function $g(\cdot; \theta_\psi)$ that is to embed feature y_i into the learned metric space $z_i = g(y_i; \theta_\psi) \in \mathbb{R}^d$, where θ_ψ is the parameter of the mapping function and z_i denotes the feature embedded into the new metric space.

The deep model aims to learn a non-linear transformation of an input image of the form $f(\cdot; \theta_\phi)$ with parameter θ_ϕ to extract its visual representation $y_i = f(x_i; \theta_\phi) \in \mathbb{R}^k$, where the transformation is usually fulfilled by a convolutional module with x_i input. With the above definition, we can directly combine the deep model with the classical metric learning. In particular, the function g can be seen as a fully-connected (FC) layer, which can be further appended after the convolutional module as a feature embedding layer.

This FC layer transforms high-dimensional deep features into a learned metric space to measure the semantic similarity/dissimilarity between images. As a result, we define the distance metric $D(\cdot, \cdot)$ between images x_i and x_j in the learned metric space as follows:

$$D(x_i, x_j) = ||g(f(x_i; \theta_\phi)); \theta_\psi) - g(f(x_j; \theta_\phi)); \theta_\psi)|| = ||z_i - z_j||, \qquad (1)$$

where $|| \cdot ||$ denotes the l_2-norm, and $z_i = g(f(x_i; \theta_\phi); \theta_\psi)$.

In order to achieve high intra-class compactness and inter-class sparseness, recent deep metric learning methods [5,10,13,27,32] resort to optimizing the above-mentioned convolutional module $f(\cdot)$ and FC layer $g(\cdot)$ with a triplet loss, as defined below:

$$\mathcal{L}_T = \sum_{(z_i, z_j, z_k) \in \mathrm{T}} (||z_i - z_j||^2 - ||z_i - z_k||^2 + \gamma)_+, \qquad (2)$$

where (z_i, z_j, z_k) denotes a triplet tuple, in which (z_i, z_j) is the positive pair with samples from the same category, and (z_i, z_k) is the negative pair with samples from two different categories, γ is a predefined margin, and $(\cdot)_+$ denotes $\max(0, \cdot)$. Moreover, to further improve the performance, recent methods [3,35] focus on the feature normalization in the metric space, which transform the original features into a hypersphere. The features in the metric space are all normalized in a hypersphere:

$$\hat{z}_i = \alpha \times \frac{z_i}{||z_i||} \qquad (3)$$

where α is a normalization constant. With this process, the Euclidean distance $||\hat{z}_i - \hat{z}_j||^2$ can be naturally transformed to the cosine similarity $2\alpha^2 - 2\hat{z}_j^T \hat{z}_i$.

Moreover, since the function $g(\cdot)$ serve as an FC layer, we can incorporate the classification loss function $\mathcal{L}_C$ (Softmax Loss) to keep different categories as far away as possible. As a result, given a triplet tuple $(\hat{z}_i, \hat{z}_j, \hat{z}_k)$ with normalized features, we combine the classic triplet loss with the classification loss. So the total loss function $\mathcal{L}$ becomes:

$$\mathcal{L} = \lambda_1 \mathcal{L}_T + \lambda_2 \mathcal{L}_C$$
$$= \lambda_1 \sum_i^N \sum_{j \neq l_i}^C \left(2\hat{z}_j^T \hat{z}_i - 2\hat{z}_k^T \hat{z}_i + \gamma \right)_+ - \lambda_2 \sum_{i=1}^N \log \frac{e^{\hat{z}_{l_i}^T \hat{z}_i}}{\sum_j e^{\hat{z}_j^T \hat{z}_i}}. \qquad (4)$$

where λ_1 and λ_2 are the weight parameters for balancing the two terms, $\hat{z}_i$ and $\hat{z}_j$ are same-class samples, while $\hat{z}_k$ is from a different class, and $\hat{z}_{l_i}$ is the random selected feature attributed to l_i-th class.

2.2 Penalty-Aware Memory Loss

Enumerating the full set of triplet combinations from training samples for triplet loss optimization in Eq. (2) and Eq. (4) would usually consume a huge amount of computation. To reduce the complexity of optimization, existing triplet loss optimization schemes usually rely on triplet sampling techniques to largely reduce the number of triplets for training. However, such mining strategies often lead to the deficiency that the sampled triplets are not sufficient to accurately characterize the global geometric structure of data.

To address the this problem, we propose using a memory bank to better capture the global geometric structure of training data. To this end, following the feature embedding layer g, we add a memory bank that stores a set of category prototypes $W \in \mathbb{R}^{d \times c}$, at the same time, it can also be viewed as a set of c vectors of dimension $d \times 1$. The i-th column of the memory bank contains the prototype $W_i \in \mathbb{R}^{d \times 1}$ attributed to the i-th category of training samples, and W_{l_i} represents the prototype of the corresponding category of z_i. We define $\{W - W_{l_i}\}$ the prototypes of all the remaining categories. Like Eq. (3), the prototypes in the memory bank are all normalized and then mapped into the same hypersphere to convert subsequent calculations to cosine distance:

$$\hat{W}_i = \alpha \times \frac{W_i}{||W_i||} \tag{5}$$

where α is the normalization constant according to Eq. (3).

Since each prototype in memory bank contains semantic features and labels, we use W_{l_i} and $W_j \in \{W - W_{l_i}\}$ to replace z_j and z_k in the original triplet in Eq. (4), respectively, to construct new triplet loss $\mathcal{L}_T$ and new classification loss $\mathcal{L}_C$. So, we replace $\hat{z}_j$ and $\hat{z}_k$ in Eq. (4) with $\hat{W}_{l_i}$ and $\hat{W}_j$, respectively, and we have

$$\begin{aligned}
\mathcal{L} &= \lambda_1 \mathcal{L}_T + \lambda_2 \mathcal{L}_C \\
&= \lambda_1 \sum_i^N \sum_{j \neq l_i}^C \left(\hat{W}_j^T \hat{z}_i - \hat{W}_{l_i}^T \hat{z}_i + \gamma \right)_+ - \lambda_2 \sum_{i=1}^N \log \frac{e^{\hat{W}_{l_i}^T \hat{z}_i}}{\sum_j e^{\hat{W}_j^T \hat{z}_i}}.
\end{aligned} \tag{6}$$

Note that, this memory bank W can be built as a memory layer, which can be directly integrated into a CNN module using a FC layer with zero bias. We observe that, using the memory bank can capture the global structure of data to alleviate the problem of complexity caused by the mining strategies to certain extent. However, directly using SGD optimizer to update such memory bank will lead to the loss of local characteristics in training batches. We hope the global information in the memory bank and local information in the batch should be well balanced. Therefore, the updating scheme is a very important step.

Penalty-Aware Scheme. To better update memory bank, we propose a penalty-aware scheme, which not only comprehensively considers local data distributions but also represents global data geometric structure based on the accumulated information of local data in individual batches. Inspired by the center loss, for samples belonging to category $m \in \{1, ..., n | n < c\}$ in a training batch, its center feature C_m is applied to penalize its corresponding prototype in the memory bank to obtain a more accurate representation of the prototype of the category, where C_m is defined as the mean of the embedded features of samples attributed to the m-th category:

$$C_m = \frac{\sum_i^n (1\{l_i = m\}\hat{z}_i)}{\sum_i^n 1\{l_i = m\}}, \quad (7)$$

where $1\{l_i = m\}$ is an indicator function whose value is one, if the i-th sample of the batch belongs to the m-th category; otherwise, the value is zero. n is the batch size. After calculating C_m, the next step is to update the corresponding prototype in the memory bank using weighted calculation in a penalty-aware manner:

$$\tilde{W}_m = \alpha \times \frac{(1 - \beta)\hat{W}_m + \beta C_m}{||(1 - \beta)\hat{W}_m + \beta C_m||}, \quad (8)$$

where hyperparameter β is set to weight local information and global information.

To select an appropriate β, we design three schemes as follows:

- The simplest one is to set β as a fixed value (e.g., $\beta = 0.5$).
- The second method is to linearly attenuate β with the training time:

$$\beta = 1 - \frac{e_i}{E}, \quad (9)$$

where time variables e_i and E, respectively denoting the current epoch number and the total epoch number required for the training, are used to control the decay of β.
- Thirdly, according to the change of training time, β is nonlinearly attenuated based on the Gaussian function with zero mean and unit variance as follows:

$$\beta = G\left(\omega(1 - \frac{e_i}{E})\right), \quad (10)$$

where $G(\cdot)$ is the cumulative distribution function of the Gaussian distribution with zero mean and unit variance, and set hyperparameter ω to 3 based on the normal distribution table and experience.

At last, after updating $\tilde{W}$ by Eq. (8), we can rewrite our final objective function as follows:

$$\mathcal{L} = \lambda_1 \mathcal{L}_T + \lambda_2 \mathcal{L}_C$$
$$= \lambda_1 \sum_i^N \sum_{j \neq l_i}^C \left(\tilde{W}_j^T \hat{z}_i - \tilde{W}_{l_i}^T \hat{z}_i + \gamma\right)_+ - \lambda_2 \sum_{i=1}^N \log \frac{e^{\tilde{W}_{l_i}^T \hat{z}_i}}{\sum_j e^{\tilde{W}_j^T \hat{z}_i}}. \quad (11)$$

We can use an alternating optimization scheme to optimize the whole framework. The first step is to use Eq. (8) update the memory bank, and the next step is to use classical SGD optimizer to update the parameters $theta_\phi$ and θ_ψ.

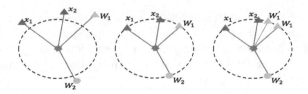

Fig. 2. Illustration of the penalty-aware memory bank updating process. The red triangles indicate the same-category pair x_1 and x_2 in the embedded space, the yellow triangle indicates the prototype W_1 of the same corresponding category, and the yellow circle indicates the prototype of the other category. From left to right: the initial features obtained by the model, the locations of the normalized features and prototypes on the unit hypersphere surface, and W_1' of x_1 and x_2 obtained by correcting the yellow triangle. (Color figure online)

2.3 Analysis for Penalty-Aware Memory Loss

In this subsection, we will explain why the memory bank can be used as a penalty during the metric learning. To this end, we first simplify the formula in Eq. (11) as

$$\mathcal{L}_i = \underbrace{\sum_{j \neq l_i}^{C} \left(\tilde{W}_j^T \hat{z}_i - \tilde{W}_{l_i}^T \hat{z}_i \right)}_{\mathcal{L}_T} - \underbrace{\log \frac{e^{\tilde{W}_{l_i}^T \hat{z}_i}}{\sum_j e^{\tilde{W}_j^T \hat{z}_i}}}_{\mathcal{L}_C}. \tag{12}$$

Here we explain again that $\tilde{W}_i^T \hat{z}_i$ measures the similarity between the positive samples $\hat{z}_i$ and $\tilde{W}_i^T$, and $\tilde{W}_j^T \hat{z}_i$ measures the similarity between the negative samples $\hat{z}_i$ and $\tilde{W}_j^T$. In this way, we use $\frac{\partial \mathcal{L}_C}{\partial \tilde{W}_i}$ and $\frac{\partial \mathcal{L}_T}{\partial \tilde{W}_i}$ to illustrate the role of $\frac{\partial \mathcal{L}}{\partial \tilde{W}_i}$ in backward propagation, which will greatly facilitate our analysis of backward propagation, in which the gradient of the loss w.r.t. W_{l_i} can be obtained by the chain-rule:

$$\begin{aligned} \frac{\partial \mathcal{L}_i}{\partial W_{l_i}} &= \left(\frac{\partial \mathcal{L}_T}{\partial \tilde{W}_{l_i}} + \frac{\partial \mathcal{L}_C}{\partial \tilde{W}_{l_i}} \right) \frac{\partial \tilde{W}_{l_i}}{\partial W_{l_i}} \\ &= \left(-\hat{z}_i - (1 - \frac{e^{\tilde{W}_{l_i}^T \hat{z}_i}}{\sum_j e^{\tilde{W}_j^T \hat{z}_i}}) e^{\hat{z}_i \tilde{W}_i^T} \hat{z}_i \right) \frac{\partial \tilde{W}_{l_i}}{\partial W_{l_i}}. \end{aligned} \tag{13}$$

Similarly, the gradient w.r.t. W_j can be obtained by

$$\begin{aligned} \frac{\partial \mathcal{L}_i}{\partial W_j} &= \left(\frac{\partial \mathcal{L}_T}{\partial \tilde{W}_j} + \frac{\partial \mathcal{L}_C}{\partial \tilde{W}_j} \right) \frac{\partial \tilde{W}_j}{\partial W_j} \\ &= \left(\hat{z}_i + \frac{e^{\tilde{W}_{l_i}^T \hat{z}_i}}{\sum_j e^{\tilde{W}_j^T \hat{z}_i}} e^{\hat{z}_i \tilde{W}_j^T} \hat{z}_i \right) \frac{\partial \tilde{W}_j}{\partial W_j}. \end{aligned} \tag{14}$$

Both $\frac{\partial \tilde{W}_{l_i}}{\partial W_{l_i}}$ in $\frac{\partial \mathcal{L}_i}{\partial W_{l_i}}$ and $\frac{\partial \tilde{W}_j}{\partial W_j}$ in $\frac{\partial \mathcal{L}_i}{\partial W_j}$ contain hyperparameter β and their corresponding prototypes, which shows that β will promote the W update in the back propagation.

Moreover, the gradient w.r.t. z_i can also be obtained by the chain-rule:

$$
\begin{aligned}
\frac{\partial \mathcal{L}_i}{\partial \hat{z}_i} &= \frac{\partial \mathcal{L}_T}{\partial \hat{z}_i} + \frac{\partial \mathcal{L}_C}{\partial \hat{z}_i} \\
&= \sum_{j \neq i}^{C} \left(1 + \frac{e^{\tilde{W}_j^T \hat{z}_i}}{\sum_j e^{\tilde{W}_j^T \hat{z}_i}} \right) - C\tilde{W}_i + \frac{e^{\tilde{W}_{l_i}^T \hat{z}_i}}{\sum_j e^{\tilde{W}_j^T \hat{z}_i}} \tilde{W}_{l_i}.
\end{aligned}
\tag{15}
$$

As shown in Eq. (13) and Eq. (14), in the training step, $\frac{\partial \mathcal{L}_T}{\partial W_{l_i}}$ and $\frac{\partial \mathcal{L}_T}{\partial W_j}$ are used to directly optimize the model in the embedding layer, whereas $\frac{\partial \mathcal{L}_C}{\partial W_{l_i}}$ and $\frac{\partial \mathcal{L}_C}{\partial W_j}$ are used to indirectly optimize the model in the final FC layer. From the perspective of updating the prototype of $\tilde{W}_{l_i}$, $\frac{\partial \mathcal{L}_C}{\partial W_{l_i}}$ is a penalty term for $\frac{\partial \mathcal{L}_T}{\partial W_{l_i}}$. Specifically, $\frac{\partial \mathcal{L}_C}{\partial W_{l_i}}$ and $\frac{\partial \mathcal{L}_T}{\partial W_{l_i}}$ have the same sign, that is, the gradient directions of $\frac{\partial \mathcal{L}_C}{\partial W_{l_i}}$ and $\frac{\partial \mathcal{L}_T}{\partial W_{l_i}}$ are the same, which accelerates $\frac{\partial \mathcal{L}_T}{\partial W_{l_i}}$ to reach the real prototype of the category. By contrast, the negative sample pair $\hat{z}_i$ and $\tilde{W}_j^T$ makes the opposite effect of what made by the positive sample pair. Equation (15) shows that $\hat{z}_i$ will be pulled closer by $\tilde{W}_{l_i}$, $\tilde{W}_j$ retreat. This step plays the role of the moving process of W_1 toward W_1' as shown in Fig. 2 in the back propagation.

3 Experiments

3.1 Datasets and Evaluation Protocols

We evaluated our method and other representative methods on the following four popular datasets:**CUB-200-2011** [24], **CARS196** [11], **In-shop Clothes Retrieval** [18] and **Stanford Online Products** [10]. We use the same strategy in [9] to partition the training and test sets for all datasets.

For all datasets, we adopt the standard evaluation protocol Recall@K to evaluate the performances of the compared methods. Recall@K is the average recall score over all query images in the test set. Specifically, given the K most similar pictures returned for each query image, the recall score is 1 if there exists at least one positive image in the top K returned images, and 0 otherwise.

3.2 Settings

We adopt the widely-used Resnet-50 pre-trained on ImageNet as the backbone network. The input images are first resized to 281×281 pixels and then cropped to 256×256 pixels. To augment training data, all 256×256 images are further randomly cropped to 224×224 and also flipped horizontally for data augmentation. Moreover, we normalize the image features using the channel means and

Table 1. Retrieval performance on CUB-200-2011 and CARS196. Note that, the baseline results on CUB-200-2011 and CARS196 are cited from (HDML [34], TML [32], DCES [1], MIC [12], DGCRL [35]), which we have also reproduced. All accuracy retes are averaged number.

Method	CUB-200-2011				CARS196			
	R@1	R@2	R@4	R@8	R@1	R@2	R@4	R@8
Contrastive [22]	26.4	37.3	49.8	62.3	21.7	32.3	46.1	59.9
Triplet [7]	36.1	48.6	59.3	70.0	39.1	50.4	63.3	74.5
LiftedStruct [10]	47.2	58.9	59.3	70.0	49.0	60.3	72.1	81.5
N-pairs [13]	45.3	58.4	69.5	79.4	53.9	66.7	77.7	86.3
CTL [26]	53.5	66.4	78.0	86.9	77.5	85.2	90.6	94.3
BIER [19]	55.3	67.2	76.9	85.1	78.0	85.8	91.1	95.1
A-BIER [20]	57.5	68.7	78.3	86.2	82.0	89.0	93.2	96.1
ABE-8 [27]	60.6	71.5	79.8	87.4	85.2	90.5	94.0	96.1
HDML [34]	64.4	75.9	84.0	90.6	80.1	86.3	91.7	94.9
TML [32]	64.5	73.9	83.0	89.4	86.3	92.3	95.4	97.3
DCES [1]	65.9	76.6	84.4	90.6	72.6	80.3	85.1	89.8
MIC [12]	66.1	78.8	86.6	–	82.6	89.1	93.2	–
DGCRL [35]	67.9	79.1	86.2	91.8	75.9	83.9	89.7	94.0
Ours, BS = 64	71.1	80.6	87.0	92.1	85.0	90.3	94.4	96.8
Ours, BS = 256	**72.8**	**83.7**	**88.5**	**92.9**	**86.9**	**93.1**	**95.8**	**97.6**

standard deviations. We use the same hyper-parameters in all experiments without fine-tuning. The batch size (BS) is set to 64 and 256 for CUB-200-2011 and CARS196. In addition to datase CUB-200-2011 and CARS196, it is worth noting that we did not use the bounding boxes provided by the data set. For dataset Stanford Online Products and In-shop Clothes Retrieval, due to the limited memory size of GPU, we set the batch size to 64 and 128. Stochastic gradient descent (SGD) with a momentum of 0.9 is used for training the CNN models in all experiments. Besides, the initial learning rate and the weight decay are set to be 0.001 and 0.0005, respectively. We set the hyper-parameters $\lambda_1 = 0.3$ and $\lambda_2 = 0.1$ empirically.

3.3   Comparison with Baselines and State-of-the-Arts

We compare our model with several representative baseline methods, including Contrastive [22], Triplet [7], LiftedStruct [10] and N-pairs [13], and the latest state-of-the-arts: CTL [26], BIER [19], A-BIER [20], ABE-8 [27], DAML [4], HDML [34], TML [32], DCES [1], MIC [12], DGCRL [35], and FastAP [5]. As mentioned earlier, we apply the proposed framework to triplet loss. We use the same pre-trained CNN model to evaluate all of the above methods for fair comparisons. Table 1 and Table 2 compare the performances of our method with the

Table 2. Retrieval performances of various schemes on In-shop Clothes and Stanford Online Products Retrieval. Note that, the baseline results on In-shop Clothes and Stanford Online Products Retrieval are cited from (DCES [1], FastAP [5], MIC [12], TML [32]), which we have also reproduced. All accuracy retes are averaged number.

Method	Stanford Online Products				In-shop Clothes			
	R@1	R@10	R@100	R@1000	R@1	R@10	R@20	R@30
LiftedStruct [10]	62.1	79.8	91.3	97.4	-	-	-	-
N-pairs [13]	67.7	83.8	93.0	97.8	-	-	-	-
Hard-aware Cascade [33]	70.1	84.9	93.2	97.8	62.1	84.9	89.0	91.2
A-BIER [20]	74.2	88.3	94.8	98.2	83.1	95.1	96.9	97.5
Hierarchical Triplets [8]	74.8	88.3	94.8	98.4	80.9	94.3	95.8	97.2
ABE-8 [27]	76.3	88.4	94.8	98.2	87.3	96.7	97.9	98.2
DCES [1]	75.9	88.4	94.9	98.1	85.7	95.5	96.9	97.5
FastAP [5]	76.4	89.0	95.1	98.2	**90.9**	**97.7**	**98.5**	**98.8**
MIC [12]	77.2	89.4	95.6	-	88.2	97.0	-	98.0
TML [32]	**78.0**	**91.2**	**96.7**	**99.0**	-	-	-	-
Ours, BS = 64	76.3	88.7	95.1	98.3	88.1	96.8	97.5	98.1
Ours, BS = 128	77.6	90.3	**96.7**	**99.0**	89.3	97.1	98.2	**98.8**

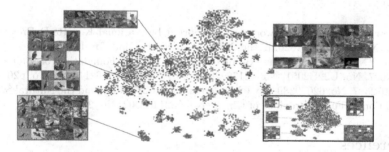

Fig. 3. Visualization of our method using t-SNE [16] on CUB-200-2011. The enlarged local parts clearly demonstrate that our method successfully groups similar images together, while separating dissimilar images far apart. Specifically, compared to the results with HDML [34] indicated by the black boxes, our method obviously achieves better grouping in the metric space.

baselines and state-of-the-arts on the four datasets. Our model consistently outperforms the state-of-the-arts in terms of Recall@K, achieving 71.1% Recall@1 on CUB-200-2011 and 87.0% Recall@1 on CARS196. Specifically, on the dataset CUB-200-2011, our proposed method is relatively improved by 4.9% compared with the method DGCRL [35], and the method MIC [12] is relatively improved by 6.7%, which has achieved a significant improvement. On the dataset CARS196, our newly proposed Penalty-Aware Memory Loss also has a significant improvement, which is a relative increase of 0.6% compared with method TML [32] and a relative increase of 1.7% compared with method ABE-8 [27]. Besides, we observe our proposed framework can achieve very competitive performance on the other two datasets, e.g., achieving 77.6% Recall@1 with only 0.4% difference from the best-performance method TML [32] on Stanford Online Products and achieving 89.3% Recall@1 with only 1.6% difference from the best-performance method TML FastAP [5] on In-shop Clothes Retrieval. Such difference mainly comes from the characteristics of data distribution of the training set. Overall, our proposed Penalty-Aware Memory loss can better balance the local and global distributions of data while learning from a limited amount of training set compared to the other methods. This advantage is even more pronounced in small data sets such as CUB-200-2011 and Cars196 (Fig. 3).

4   Conclusion

In this paper, we proposed a penalty-aware memory loss that fully utilizes the expressive power of the combination of both global data distribution and local data distribution to learn a high-quality metric. We have demonstrated the effectiveness of this method in the image retrieval task on four widely used datasets. In the future, we will further study the more general model of this method, so that it can be used to improve various machine learning methods beyond metric learning.

Acknowledgement. This work was supported by National Key R&D Program of China (No. 2022ZD0118202), the National Science Fund for Distinguished Young Scholars (No. 62025603), the National Natural Science Foundation of China (No. U21B2037, No. U22B2051, No. 62176222, No. 62176223, No. 62176226, No. 62072386, No. 62072387, No. 62072389, No. 62002305 and No. 62272401), and the Natural Science Foundation of Fujian Province of China (No. 2021J01002, No. 2022J06001).

References

1. Artsiom, S., Vadim, T., Uta, B., Bjorn, O.: Divide and conquer the embedding space for metric learning. In: CVPR, pp. 471–480 (2019)
2. Ben, H., BG, K., Gustavo, C., Ian, R., Tom, D., et al.: Smart mining for deep metric learning. In: ICCV, pp. 2821–2829 (2017)
3. Deng, J., Guo, J., Xue, N., Stefanos, Z.: Arcface: additive angular margin loss for deep face recognition. In: CVPR, pp. 4690–4699 (2019)
4. Duan, Y., Zheng, W., Lin, X., Lu, J., Zhou, J.: Deep adversarial metric learning. In: CVPR, pp. 2780–2789 (2018)
5. Fatih, C., He, K., Xia, X., Brian, K., Stan, S.: Deep metric learning to rank. In: CVPR, pp. 1861–1870 (2019)
6. Feng, J., Wu, A., Zheng, W.S.: Shape-erased feature learning for visible-infrared person re-identification. In: CVPR, pp. 22752–22761 (2023)
7. Florian, S., Dmitry, K., James, P.: Facenet: a unified embedding for face recognition and clustering. In: CVPR, pp. 815–823 (2015)
8. Ge, W.: Deep metric learning with hierarchical triplet loss. In: ECCV, pp. 269–285 (2018)

9. Hong, X., Richard, S., Robert, P.: Deep randomized ensembles for metric learning. In: ECCV, pp. 723–734 (2018)
10. Hyun, O., Xiang, Y., Stefanie, J., Silvio, S.: Deep metric learning via lifted structured feature embedding. In: CVPR, pp. 4004–4012 (2016)
11. Jonathan, K., Michael, S., Deng, J., Li, F.: 3D object representations for fine-grained categorization. In: ICCV Workshops, pp. 554–561 (2013)
12. Karsten, R., Biagio, B., Bjorn, O.: MIC: mining interclass characteristics for improved metric learning. In: ICCV, pp. 8000–8009 (2019)
13. Kihyuk, S.: Improved deep metric learning with multi-class n-pair loss objective. In: NeurIPS, pp. 1857–1865 (2016)
14. Kim, M., Jain, A.K., Liu, X.: Adaface: quality adaptive margin for face recognition. In: CVPR, pp. 18750–18759 (2022)
15. Kim, M., Liu, F., Jain, A., Liu, X.: DCFace: synthetic face generation with dual condition diffusion model. In: CVPR, pp. 12715–12725 (2023)
16. Laurens, V.: Accelerating t-SNE using tree-based algorithms. JMLR **15**(93), 3221–3245 (2014)
17. Lin, X., et al.: Learning modal-invariant and temporal-memory for video-based visible-infrared person re-identification. In: CVPR, pp. 20973–20982 (2022)
18. Liu, Z., Luo, P., Qiu, S., Wang, X., Tang, X.: Deepfashion: powering robust clothes recognition and retrieval with rich annotations. In: CVPR, pp. 1096–1104 (2016)
19. Michael, O., Georg, W., Horst, P., Horst, B.: Bier-boosting independent embeddings robustly. In: ICCV, pp. 5189–5198 (2017)
20. Michael, O., Georg, W., Horst, P., Horst, B.: Deep metric learning with bier: Boosting independent embeddings robustly. IEEE TPAMI **42**(2), 276–290 (2018)
21. Shen, Y., Sun, X., Wei, X.S.: Equiangular basis vectors. In: CVPR, pp. 11755–11765 (2023)
22. Sumit, C., Raia, H., Yann, L., et al.: Learning a similarity metric discriminatively, with application to face verification. In: CVPR, vol. 1, pp. 539–546 (2005)
23. Sun, Y., et al.: Circle loss: a unified perspective of pair similarity optimization. In: CVPR, pp. 6398–6407 (2020)
24. Wah, C., Branson, S., Welinder, P., Perona, P., Belongie, S.: The Caltech-UCSD birds-200-2011 dataset (2011)
25. Wang, C., Zheng, W., Li, J., Zhou, J., Lu, J.: Deep factorized metric learning. In: CVPR, pp. 7672–7682 (2023)
26. Wang, F., Xiang, X., Cheng, J., Yuille, A.L.: Normface: L2 hypersphere embedding for face verification. In: ACM MM, pp. 1041–1049 (2017)
27. Wonsik, K., Bhavya, G., Kunal, C., Jungmin, L., Keunjoo, K.: Attention-based ensemble for deep metric learning. In: ECCV, pp. 736–751 (2018)
28. Xiong, J., Lai, J.: Similarity metric learning for RGB-infrared group re-identification. In: CVPR, pp. 13662–13671 (2023)
29. Yair, M., Alexander, T., K, L.T., Sergey, I., Saurabh, S.: No fuss distance metric learning using proxies. In: ICCV, pp. 360–368 (2017)
30. Yang, H., Chu, X., Zhang, L., Sun, Y., Li, D., Maybank, S.J.: Quadnet: quadruplet loss for multi-view learning in baggage re-identification. Pattern Recogn. **126**, 108546 (2022)
31. Yang, M., Huang, Z., Hu, P., Li, T., Lv, J., Peng, X.: Learning with twin noisy labels for visible-infrared person re-identification. In: CVPR, pp. 14308–14317 (2022)
32. Yu, B., Tao, D.: Deep metric learning with tuplet margin loss. In: ICCV, pp. 6490–6499 (2019)
33. Yuan, Y., Yang, K., Zhang, C.: Hard-aware deeply cascaded embedding. In: ICCV, pp. 814–823 (2017)

34. Zheng, W., Chen, Z., Lu, J., Zhou, J.: Hardness-aware deep metric learning. In: CVPR, pp. 72–81 (2019)
35. Zheng, X., Ji, R., Sun, X., Y. Wu, Y.Y., Huang, F.: Towards optimal fine grained retrieval via decorrelated centralized loss with normalize-scale layer. In: AAAI, vol. 33, pp. 9291–9298 (2019)

Central and Directional Multi-neck Knowledge Distillation

Jichuan Chen[1,2], Ziruo Liu[1,2], Chao Wang[1], Bin Yang[1], Renjie Huang[1], Shunlai Xu[2], and Guoqiang Xiao[1(✉)]

[1] School of Computer and Information Science, Southwest University, Chongqing 400715, China
gqxiao@swu.edu.cn
[2] Chongqing Academy of Animal Sciences, Chongqing 402460, China
http://www.swu.edu.cn/

Abstract. There are already many mature methods for single-teacher knowledge distillation in object detection models, but the development of multi-teacher knowledge distillation methods has been slow due to the complexity of knowledge fusion among multiple teachers. In this paper, we point out that different teacher models have different detection capabilities for a certain category, and through experiments, we find that for individual target instances, there are differences in the main feature regions of the target, and the detection results also have randomness. Networks that are weak in overall detection performance for a certain category sometimes perform well, so it is not appropriate to simply select the best learning based on detection results. Therefore, we propose a novel multi-teacher distillation method that divides the central and boundary features of the instance target region through the detection of teacher models, and uses clustering to find the center of the response features of each category of teacher models as prior knowledge to guide the learning direction of the student model for the category as a whole. Since our method only needs to calculate the loss on the feature map, FGD can be applied to multiple teachers with the same components. We conducted experiments on various detectors with different backbones, and the results show that our student detector achieved excellent mAP improvement. Our distillation method achieved an mAP of 39.4% on COCO2017 based on the ResNet-50 backbone, which is higher than the single-teacher distillation learning method of the baseline model. Our code and training logs can be obtained at https://github.com/CCCCPRCV/Multi_Neck.

Keywords: Knowledge distillation · Multi-teacher · Object detection

1 Introduction

As a fundamental task in computer vision, object detection has received significant attention and made great progress in recent years [1–4]. It also has wide

This work was supported by the Chongqing Academy of Animal Sciences and the Special Project for Technological Innovation and Application Development of Chongqing Municipality (cstc2021jscx-dxwtBX0008).

© The Author(s), under exclusive license to Springer Nature Singapore Pte Ltd. 2024
Q. Liu et al. (Eds.): PRCV 2023, LNCS 14431, pp. 475–486, 2024.
https://doi.org/10.1007/978-981-99-8540-1_38

applications in the real world, such as autonomous driving [5,6] and intelligent surveillance [7,8].

Due to the exponential development of convolutional neural networks, deep neural networks (DNNs) have been applied in various fields of computer vision [7, 9,10]. To achieve success on mobile devices, visual tasks need to overcome limited computing resources [11,12]. Model compression has been a key research task to address this issue, and knowledge distillation is a prominent technique.

  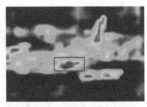

(a) image   (b) Cascade Res101   (c) Faster Rcnn-Res101

Fig. 1. Visualization of class activation maps for two different complex teacher models, the red boxed area shows the difference of the two teachers on the same region of the image. (Color figure online)

Knowledge distillation is a learning method for extracting and learning compact knowledge, With the continuous application of object detection in real life, more and more researchers have been applying it to the simplification of object detection models. They have improved the learning effectiveness of student models by distilling knowledge from multiple module features of teacher models [13], balancing target and background features [14,15], and other methods. However, there is little analysis of features within the target region, and the knowledge is limited to individual teacher models.

In recent years, researchers have also started exploring the training of a student model with diversified knowledge by learning from multiple teacher models. Li et al. [16] shares a common backbone feature and maintains diverse knowledge by utilizing multiple heads to learn corresponding teacher models; Liu et al. [17] proposed a novel adaptive multi-teacher multi-level knowledge distillation learning framework, which adaptively learns the instance-level teacher importance weight for optimal learning. These methods provide valuable references for multi-teacher knowledge distillation. However, occasionally, a certain teacher model may achieve better detection results on classes with lower mAP (mean Average Precision) scores, neglecting the overall direction of the student model learning.

In order to further explore the differences in focus points and response to edge features within the object region on the multi-teacher knowledge distillation method, we analyzed the features of the same image in different teacher models, as shown in Fig. 1, it is easy to find that they have different focus points on the same target feature, and have a smaller response to the edge features within

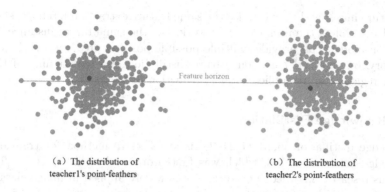

(a) The distribution of
teacher1's point-feathers

(b) The distribution of
teacher2's point-feathers

Fig. 2. The cluster of the same category features using different teacher models.

the target region. Therefore, it is necessary to pay more attention to the central features of the target. We propose a novel Multi-Neck knowledge distillation learning framework, which learns an intermediate student model for the student model through multi-teacher knowledge distillation. The intermediate student model learns diverse knowledge from different teacher models through multiple neck branches, and the student model directly learns the knowledge fused from this intermediate student model. At the same time, we further optimize the learning focus feature regions of the student model. To ensure the consistency of student models learning from a specific teacher's direction, we employed statistical analysis and clustering on different teacher models' features, as shown in Fig. 2. For each category, we identified a feature center and evaluated the learning weights, enabling each teacher to effectively transmit their excellent features related to different target categories to the student model.

2   Related Work

2.1   Object Detection

Modern object detection methods can be roughly divided into two categories: two-stage detectors and one-stage detectors. Two-stage detectors mainly include Faster R-CNN [18] and its variants [9,19], which use region proposal networks (RPN) to generate proposal regions and then make predictions for each region. One-stage detectors can be further divided into anchor-based [20–22] and anchor-free one-stage detectors [1,2]. One-stage detectors directly obtain the classification and bounding box of the target on the feature map. In contrast, two-stage detectors usually achieve better results by using RPN and RCNN heads, but the number of anchor boxes is much greater than the number of targets, which brings additional computational cost and time. Anchor-free detectors present a method of directly predicting the key points and position of the target, predicting the class and location of the target for each point on the feature map. Although these two types of networks have different detection heads, their inputs are feature maps, and they can use the same feature extraction modules, such as using FPN [23] as the neck of the network. For the same target instance, both detection models can provide multiple possible position predictions. Therefore, in theory, our method based on feature distillation and partitioning of target feature regions can be applied to most detection networks.

2.2 Knowledge Distillation

Knowledge distillation [6,13–17,24–26] is an effective method for transferring knowledge between models, which was first proposed by Hinton et al. [24] In previous research, knowledge extraction was widely applied in image classification tasks and is now frequently used in object detection tasks [13,14,25].

Some works have successfully applied knowledge distillation to detectors. Chen et al. [13] applied knowledge distillation to detection by extracting knowledge of neck features, classification head, and regression head. However, directly learning too much knowledge from the teacher model may also learn noise or other redundant knowledge in the teacher model. GID [6] learns regions where the student and teacher perform differently. Guo et al. [27] experimentally found that both foreground and background of images play important roles in distillation, and separating them is more beneficial for the student. Liu et al. [15] separated images and used the teacher's attention mask to select key parts for distillation, while also considering the global relationship between different pixels.

These knowledge distillation methods provide various solutions for the student model to learn from the teacher model's knowledge, further promoting researchers' exploration in multi-teacher knowledge distillation. Li et al. [28] applied multi-teacher learning to multi-task learning, where each teacher corresponds to a task. Fukuda et al. [26] manually adjusted the weight of each teacher, but this manual setting of weights is very expensive. Liu et al. [17] adaptively learned instance-level teacher importance weights for optimal learning, but may overlook the consistency of the student model's learning of a single class target and the learning of important target regions.

In this paper, we use the teacher model to divide the center and boundary attention masks of multiple possible proposals for a single target, and select key parts for distillation. At the same time, considering the consistency of the student model's learning of a single class target from a global perspective, we bring improvements to multi-teacher knowledge distillation.

3   Method

3.1   MultiNeck-Teacher Knowledge Distillation

The overall framework of the proposed method is shown in Fig. 3. For each example in the given training set, we first utilize several well-trained teacher networks (for ease of description, only 2 teacher models are shown in the figure

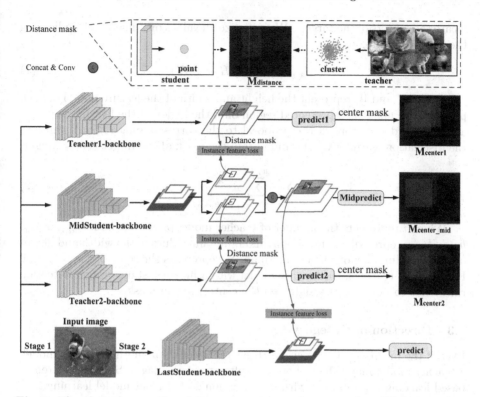

Fig. 3. The clustering results of the same category features using different teacher models.

as an example) to generate corresponding features as high-level knowledge. We employ the neck part of the model (FPN [23] is used as an example in the figure) for knowledge distillation learning, while modifying the neck structure of the intermediate student model to have multiple branches, with each branch learning the knowledge of a complex teacher model. Finally, a diversified FPN with multiple knowledge is obtained by concatenating and downsampling features of the same dimension, which is used for the learning of the final student model. During the distillation learning, we adopt the method proposed in FGD [15] as the baseline, and also incorporate the supervision methods of center region learning and category direction learning, which will be detailed in the following two subsections.

3.2 Center of Instance Object

On the feature map, the model exhibits smaller high-response regions on the instance targets compared to the target regions, and also tends to make more predictions within these high-response regions. Therefore, we employ a lower IoU threshold to obtain a greater number of target prediction proposals in order to locate the centers of the targets. We initialize all values on a mask to 1, indicating the background region.

$$M_{x,y} = 1, (x < H, y < W) \tag{1}$$

Where H and W represent the height and width of the feature map. For each proposal, we map the represented region onto the region of the feature map size, and then add the score of the proposal to the corresponding region's value as the distillation learning weight and obtain the mask of the target center weight.

$$M_{center} = \sum_{i=0}^{n} \sum_{x=w_{min}}^{w_{max}} \sum_{y=h_{min}}^{h_{max}} (S_i + M_{x,y}) \tag{2}$$

Where i represents the number of teacher model proposals, and w_{min}, w_{max}, h_{min}, h_{max} represent the maximum and minimum values of the width and height of the bounding box on the feature map, S_i represents the score of the proposal. For multiple teachers, the teacher model with a higher instance target detection rate tends to have larger weights on the central region mask.

3.3   Direction of Category

Even if the mean average precision (mAP) is lower than another teacher model, a teacher model may achieve better results on a few images. Simple preference-based learning can interfere with the directional of student model learning.

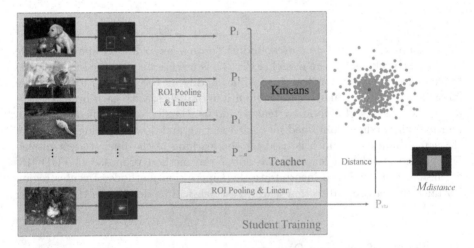

Fig. 4. The clustering results of the same category features using different teacher models.

Before training, we extract the neck features of all category instances from the teacher model and use ROIPooling to standardize these features to the same size. As shown in Fig. 4, these features are normally distributed around the center of the category in high-dimensional space. Therefore, we use clustering algorithms to find the center of each category as prior knowledge to guide the student model's learning. Then, on the feature map of an image, the distance mask between all category centers can be represented as:

$$\vec{P}_i^{center} = Kmeans(\vec{P}_{1\_i}, \vec{P}_{2\_i}, \vec{P}_{3\_i}, \cdots) \tag{3}$$

Where $\vec{P}_i^{center}$ represents the clustering center, i represents the number of categories, The calculation of feature point vectors is as follows.

$$\vec{P}_i = Linear(ROIPooling(B_{gt,i})) \tag{4}$$

Where B_{gt} represents the region of the ground truth bounding box on the feature map, The distance mask between all category centers on a single image can be represented as:

$$M_{distance} = \sum_{j=0}^{k} \sum_{x=w_{min}}^{w_{max}} \sum_{y=h_{min}}^{h_{max}} \begin{cases} \frac{1}{D_j}, & \text{if } M_{distance,x,y} < \frac{1}{D_j} \\ M_{distance,x,y}, & \text{if } M_{distance,x,y} \geq \frac{1}{D_j} \end{cases} \tag{5}$$

$$D_j = Euclidean(\vec{P}_{i,j}, \vec{P}_i^{center}) \tag{6}$$

Where k represents the total number of targets in an image, Dj represents the Euclidean distance between the feature point and category feature center.

3.4 Overall Loss

Our experimental method is based on the FGD method, and the distilled loss function is also compatible with other distillation supervision methods. Overall, the total loss used to train the student model is as follows:

$$L = L_{original} + \alpha L_{center} + \beta L_{distance} \tag{7}$$

$L_{original}$ represents the original loss for baseline detectors, α and β are hyperparameters used to balance two losses.

$$L_{center} = \sum_{k=1}^{C} \sum_{i=1}^{H} \sum_{j=1}^{W} M_{center} A_{i,j}^S A_k^C (F_{k,i,j}^T - F_{k,i,j}^S)^2 \tag{8}$$

$$L_{distance} = \sum_{k=1}^{C} \sum_{i=1}^{H} \sum_{j=1}^{W} M_{distance} A_{i,j}^S A_k^C (F_{k,i,j}^T - F_{k,i,j}^S)^2 \tag{9}$$

where F^T and F^S is the feature of teachers and students, A^S, A^C denote the spatial and channel attention of teacher.

$$A^S = C \cdot softmax((\frac{1}{HW} \cdot \sum_{i=1}^{H} \sum_{j=1}^{W} |F_{i,j}^T|)/t) \tag{10}$$

$$A^C = H \cdot W \cdot softmax((\frac{1}{C} \cdot \sum_{k=1}^{C} |F_k^T|)/t) \tag{11}$$

Where t is the temperature hyper-parameter, the same loss method is employed when the student model learns from the intermediate student model.

4   Experiments

4.1   Datasets and Evaluation Metrics

We conducted experiments on the COCO2017 dataset [29], which includes 118,287 training images and 5,000 validation images, covering 80 categories. Additionally, we also conducted experiments on the PASCAL VOC dataset [30] for object detection and achieved promising results. We used mAP as the evaluation metric for the detection task.

4.2   Implementation Details

In order to minimize the impact of experimental settings, most of our implementation details are kept consistent with the original FGD [15]. We use ResNet-50 as the backbone for both the student model and the intermediate teacher model. All experiments are conducted using mmdetection [31] and Pytorch. Two hyper-parameters, $\alpha = 5 \times 10^{-5}$ and $\beta = 2 \times 10^{-5}$, are also set to balance the weight of the center distance and the loss of the center region. A total of 16 epochs are trained in the experiment, and the parameters are updated once per epoch.

4.3   Experiment on COCO Datasets

Comparison with Other Methods. As shown in Table 1 and Table 2, our method achieves higher detection accuracy for each student model compared to the baseline model FGD [15]. Compared to other multi-teacher models, our method also achieves better detection results.

Table 1. Comparisons with baseline on the datasets of COCO.

Method	Teacher	Student	AP	AP_S	AP_M	AP_L
Ours multi-teacher	Faster RCNN-Res101	Faster RCNN-Res50	**39.4**	22.1	43.3	52.4
	Cascade RCNN-res101	Cascade RCNN-res50	**39.6**	23.4	43.5	52.3
	RetinaNet-Res101	RetinaNet-Res50	**38.5**	21.4	42.5	51.8
	GFL-Res101	GFL-Res50	**41.2**	23.8	45.1	52.9
FGD single-teacher	Faster RCNN-Res101	Faster RCNN-Res50	38.2	21.4	42.1	50.3
	Cascade RCNN-res101	Cascade RCNN-res50	39.0	22.9	42.6	51.1
	RetinaNet-Res101	RetinaNet-Res50	37.8	20.9	41.3	50.1
	GFL-Res101	GFL-Res50	40.5	23.3	44.5	52.4

Table 2. Comparisons with other multi-teacher distillation on the datasets of COCO.

Method	Teacher	Student	AP	AP_S	AP_M	AP_L
AvgMKD(2T)	Faster RCNN-Res101 Cascade RCNN-res101	Stu-Resnet50	38.7	21.3	42.5	51.2
AMTML-KD(2T)	Faster RCNN-Res101 Cascade RCNN-res101	Stu-Resnet50	39.2	22.1	43.0	51.8
Multi-Neck-KD	Faster RCNN-Res101 Cascade RCNN-res101	Faster RCNN-Res50	**39.4**	**22.1**	**43.3**	**52.4**

Ablation Study. In this section, we conducted ablation study to validate the individual contributions of our proposed method. The results are shown in Table 3. The first row displays the performance of the original model trained with basic classification loss, achieving an AP of 39.4. After sequentially incorporating our proposed methods, the detection results of the model have all been improved. These results validate the effectiveness of our approach.

Table 3. Ablation study of the components in our method on the COCO.

Method	M_{center}	$M_{distance}$	AP	AP_S	AP_M	AP_L
Multi-Neck-KDFaster-RcnnRes50			37.4	20.1	41.3	50.1
	✓		38.6	21.5	42.3	51.4
		✓	38.1	21.0	41.7	50.8
	✓	✓	**39.4**	**22.1**	**43.3**	**52.4**

Visualization of Results. We selected a subset of images to visualize the results. As show in Fig. 5 The second row in the figure shows the visualization results obtained by distilling Faster R-CNN using the FGD method, while the third row shows the visualization results obtained by distilling Cascade using the FGD method. The fourth row shows the results obtained by using both of these teacher distillation methods, indicating that our approach leads to more accurate object predictions by the model.

4.4 Experiment on PASCAL VOC Datasets

In order to further validate the effectiveness of our method, as show in Table 4, we conducted experiments on the PASCAL VOC datasets. By comparing with various methods and conducting ablation experiments to analyze the effects of each proposed method separately, we still achieved good experimental results.

Image

AvgMKD(2T)

Ours

Fig. 5. Visualization of class activation map, our method exhibits a larger response on the target object.

Table 4. Comprehensive comparisons on the datasets of PASCAL VOC.

Method	Teacher	Student	AP	AP_s	AP_m	AP_l
Ours multi-teacher	Faster RCNN-Res101	Faster RCNN-Res50	**77.2**	68.3	74.5	84.3
	Cascade RCNN-res101	Cascade RCNN-res50	**78.2**	69.2	75.7	85.4
	RetinaNet-Res101	RetinaNet-Res50	**76.2**	66.8	73.6	82.7
	GFL-Res101	GFL-Res50	**78.9**	69.2	76.2	86.0
FGD single-teacher	Faster RCNN-Res101	Faster RCNN-Res50	75.6	65.4	72.2	82.1
	Cascadek RCNN-res101	Cascade RCNN-res50	76.8	67.2	73.9	83.5
	RetinaNet-Res101	RetinaNet-Res50	73.4	64.5	70.2	81.1
	GFL-Res101	GFL-Res50	77.1	68.6	74.2	84.6

5   Conclusion

Object detection is one of the most important branches of computer vision. The localization of objects is a key task in object detection. We have identified the shortcomings of existing single-teacher knowledge distillation and multi-teacher knowledge distillation methods. Accordingly, we propose our method based on the division of difficult points. We learn from the central regions of each instance object, which are determined by the detection results of the teacher models. We also calculate the feature centers of each category for different teacher models, which collectively guide the learning direction of the student model for each instance object. We conduct extensive experiments on the COCO dataset and the PASCAL VOC dataset to evaluate our method. Our model achieves the highest accuracy on most evaluation metrics.

For future work, we plan to utilize more diverse forms of data to achieve better detection, such as improving the strategy for label assignment, enhancing distillation in dense object detection, and using higher-resolution feature maps. Additionally, we will explore more methods to bridge the gap between knowledge distillation methods and practical applications.

References

1. Zhou, X., Wang, D., Krähenbühl, P.: Objects as points, arXiv preprint arXiv:1904.07850 (2019)
2. Tian, Z., Shen, C., Chen, H., He, T.: FCOS: fully convolutional one-stage object detection. In: Proceedings of the IEEE/CVF International Conference on Computer Vision, pp. 9627–9636 (2019)
3. Carion, N., Massa, F., Synnaeve, G., Usunier, N., Kirillov, A., Zagoruyko, S.: End-to-end object detection with transformers. In: Vedaldi, A., Bischof, H., Brox, T., Frahm, J.-M. (eds.) ECCV 2020, Part I. LNCS, vol. 12346, pp. 213–229. Springer, Cham (2020). https://doi.org/10.1007/978-3-030-58452-8_13
4. Liu, Z., et al.: Swin transformer: hierarchical vision transformer using shifted windows. In: Proceedings of the IEEE/CVF International Conference on Computer Vision, pp. 10 012–10 022 (2021)
5. Cai, Y., et al.: YOLOv4-5D: an effective and efficient object detector for autonomous driving. IEEE Trans. Instrum. Meas. **70**, 1–13 (2021)
6. Dai, X., et al.: General instance distillation for object detection. In: Proceedings of the IEEE/CVF Conference on Computer Vision and Pattern Recognition, pp. 7842–7851 (2021)
7. Tao, X., Zhang, D., Wang, Z., Liu, X., Zhang, H., Xu, D.: Detection of power line insulator defects using aerial images analyzed with convolutional neural networks. IEEE Trans. Syst. Man Cybernet. Syst. **50**(4), 1486–1498 (2018)
8. Chen, S.-H., Tsai, C.-C.: SMD LED chips defect detection using a YOLOv3-dense model. Adv. Eng. Inform. **47**, 101255 (2021)
9. Cai, Z., Vasconcelos, N.: Cascade R-CNN: delving into high quality object detection. In: Proceedings of the IEEE Conference on Computer Vision and Pattern Recognition, pp. 6154–6162 (2018)
10. Zhang, W., Li, X., Ding, Q.: Deep residual learning-based fault diagnosis method for rotating machinery. ISA Trans. **95**, 295–305 (2019)
11. Howard, A.G., et al.: MobileNets: efficient convolutional neural networks for mobile vision applications, arXiv preprint arXiv:1704.04861 (2017)
12. Zhang, X., Zhou, X., Lin, M., Sun, J.: ShuffleNet: an extremely efficient convolutional neural network for mobile devices. In: Proceedings of the IEEE Conference on Computer Vision and Pattern Recognition, pp. 6848–6856 (2018)
13. Chen, G., Choi, W., Yu, X., Han, T., Chandraker, M.: Learning efficient object detection models with knowledge distillation. In: Advances in Neural Information Processing Systems, vol. 30 (2017)
14. Wang, T., Yuan, L., Zhang, X., Feng, J.: Distilling object detectors with fine-grained feature imitation. In: Proceedings of the IEEE/CVF Conference on Computer Vision and Pattern Recognition, pp. 4933–4942 (2019)
15. Yang, Z., et al.: Focal and global knowledge distillation for detectors. In: Proceedings of the IEEE/CVF Conference on Computer Vision and Pattern Recognition, pp. 4643–4652 (2022)

16. Tran, L., et al.: Hydra: preserving ensemble diversity for model distillation, arXiv preprint arXiv:2001.04694 (2020)
17. Liu, Y., Zhang, W., Wang, J.: Adaptive multi-teacher multi-level knowledge distillation. Neurocomputing **415**, 106–113 (2020)
18. Ren, S., He, K., Girshick, R., Sun, J.: Faster R-CNN: towards real-time object detection with region proposal networks. In: Advances in Neural Information Processing Systems, vol. 28 (2015)
19. Zhang, H., Chang, H., Ma, B., Wang, N., Chen, X.: Dynamic R-CNN: towards high quality object detection via dynamic training. In: Vedaldi, A., Bischof, H., Brox, T., Frahm, J.-M. (eds.) ECCV 2020, Part XV. LNCS, vol. 12360, pp. 260–275. Springer, Cham (2020). https://doi.org/10.1007/978-3-030-58555-6_16
20. Liu, W., et al.: SSD: single shot MultiBox detector. In: Leibe, B., Matas, J., Sebe, N., Welling, M. (eds.) ECCV 2016, Part I. LNCS, vol. 9905, pp. 21–37. Springer, Cham (2016). https://doi.org/10.1007/978-3-319-46448-0_2
21. Lin, T.-Y., Goyal, P., Girshick, R., He, K., Dollár, P.: Focal loss for dense object detection. In: Proceedings of the IEEE International Conference on Computer Vision, pp. 2980–2988 (2017)
22. Redmon, J., Farhadi, A.: YOLOv3: an incremental improvement, arXiv preprint arXiv:1804.02767 (2018)
23. Lin, T.-Y., Dollár, P., Girshick, R., He, K., Hariharan, B., Belongie, S.: Feature pyramid networks for object detection. In: Proceedings of the IEEE Conference on Computer Vision and Pattern Recognition, pp. 2117–2125 (2017)
24. Hinton, G., Vinyals, O., Dean, J.: Distilling the knowledge in a neural network, arXiv preprint arXiv:1503.02531 (2015)
25. Mishra, A., Marr, D.: Apprentice: using knowledge distillation techniques to improve low-precision network accuracy, arXiv preprint arXiv:1711.05852 (2017)
26. Fukuda, T., Suzuki, M., Kurata, G., Thomas, S., Cui, J., Ramabhadran, B.: Efficient knowledge distillation from an ensemble of teachers. In: Interspeech, pp. 3697–3701 (2017)
27. Guo, J., et al.: Distilling object detectors via decoupled features. In: Proceedings of the IEEE/CVF Conference on Computer Vision and Pattern Recognition, pp. 2154–2164 (2021)
28. Tan, X., Ren, Y., He, D., Qin, T., Zhao, Z., Liu, T.-Y.: Multilingual neural machine translation with knowledge distillation, arXiv preprint arXiv:1902.10461 (2019)
29. Lin, T.-Y., et al.: Microsoft COCO: common objects in context. In: Fleet, D., Pajdla, T., Schiele, B., Tuytelaars, T. (eds.) ECCV 2014, Part V. LNCS, vol. 8693, pp. 740–755. Springer, Cham (2014). https://doi.org/10.1007/978-3-319-10602-1_48
30. Everingham, M., Eslami, S.A., Van Gool, L., Williams, C.K., Winn, J., Zisserman, A.: The pascal visual object classes challenge: a retrospective. Int. J. Comput. Vision **111**, 98–136 (2015)
31. Chen, K., et al.: MMdetection: open MMLab detection toolbox and benchmark, arXiv preprint arXiv:1906.07155 (2019)

Online Class-Incremental Learning in Image Classification Based on Attention

Baoyu Du[1,2], Zhonghe Wei[1,2], Jinyong Cheng[1,2(✉)], Guohua Lv[1,2], and Xiaoyu Dai[1,2]

[1] Key Laboratory of Computing Power Network and Information Security, Ministry of Education, Shandong Computer Science Center (National Supercomputer Center in Jinan), Qilu University of Technology (Shandong Academy of Sciences), Jinan, China
cjy@qlu.edu.cn
[2] Shandong Provincial Key Laboratory of Computer Networks, Shandong Fundamental Research Center for Computer Science, Jinan, China

Abstract. Online class-incremental learning aims to learn from continuous and single-pass data streams. However, during the learning process, catastrophic forgetting often occurs, leading to the loss of knowledge about previous classes. Most of the current methods to solve the problem of catastrophic forgetting do not fully exploit the semantic information in single-pass data streams. To effectively address this issue, we propose an Attention-based Dual-View Consistency (ADVC) strategy. Specifically, we generate an attention score for each region through the attention mechanism to identify those important features that are useful for classification. This enables the model to thoroughly explore the semantic information in the single-pass data stream and improve prediction accuracy. At the same time, we consider the effectiveness of sample retrieval from another perspective (the samples themselves) by using data augmentation methods in memory to generate variants of old classes of samples to alleviate the problem of catastrophic forgetting. Through extensive experimental results, we demonstrate that our approach outperforms state-of-the-art methods on a range of benchmark datasets.

Keywords: Class-incremental learning · attention · ADVC · data augmentation

1 Introduction

In recent years, although deep learning has been rapidly developing in areas such as image classification and face recognition, most deep learning algorithms are based on the assumption of independent identical distribution and can only be oriented to static data. However, in practical applications, many industries

This work was supported by the Natural Science Foundation of Shandong Province, China, under Grant Nos. ZR2020MF041 and ZR2022MF237, and the National Natural Science Foundation of China under Grant No. 11901325.

require dynamic streaming data. For example, the streams of images and videos uploaded by social media, the streams of information from banking systems, the streams of management data from operators, etc. The characteristics of these streams of data are in a constant state of change. Therefore, the model needs to continuously learn new information and be able to handle new tasks.

Since the model has a catastrophic forgetting problem in learning new tasks, i.e., the model always forgets the previously learned knowledge when learning new knowledge. To address this problem, researchers have proposed incremental learning [9,22,30] methods. These methods allow trained models to continuously learn new knowledge while retaining previously acquired knowledge. At present, there are three types of class-incremental learning methods: replay-based methods [11,21,23,29] that store prior samples, parameter isolation-based methods [10,16,17], and regularization-based methods [4,5,19]. Parameter isolation-based methods introduce a parameter isolation mechanism in the model, storing parameters of different tasks or categories separately, thereby allowing retention of previous task parameters while learning new tasks. A novel incremental learning framework based on stage isolation is proposed in [26], which utilizes a series of stage-isolated classifiers to perform the learning task in each stage without interference from other stages. Regularization-based methods adjust the regularization of the model to balance the influence between new and old tasks, thus protecting the knowledge learned previously. In [9], a method based on Bayesian optimization is proposed, which can automatically determine the optimal regularization strength for each learning task. Since the replay-based method provides samples of data from the previous task, which helps to maintain the model's understanding of the old knowledge, it has been shown to be a more effective method in class-incremental learning. For example, [12] proposed a novel Bayesian continual learning method called posterior meta-replay to mitigate forgetting. Meanwhile, [18] utilized off-policy learning and behavioral cloning to greatly reduce catastrophic forgetting in multi-task reinforcement learning. Despite the commendable performance of these methods, they often overlook the exploration of semantic information in single-pass data streams, which leads to lower classification accuracy.

To solve this problem, we propose ADVC, which exploits the attention mechanism [8,28] to comprehensively explore the semantic information between two different views, thus improving the classification accuracy. Specifically, ADVC transforms incoming images into different pairs of views and sends them into the network model. During this process, the attention mechanism generates an attention score for each region, enabling the model to recognize important features that are useful for classification. Subsequently, the model trains using the original view and the augmented view, both weighted with attention scores. The goal of training is to maximize the mutual information between these two views to help the model better understand the semantic information of the images. Meanwhile, some current sample retrieval strategies tend to focus more on the design of the selection method and do not start from the samples themselves. For this reason, we thought of keeping enough information about the old classes by subjecting the stored samples to the CutMix [31] data augmentation operation. The contributions of this paper can be summarized as follows:

• Unlike most existing methods for class-incremental learning, we focus not on the storage and replay of samples but on the full exploration of semantic information in single-pass data streams through attention-based dual view consistency (ADVC) to improve classification accuracy.

• We study the effectiveness of the sample retrieval strategy from another perspective. Starting from the samples themselves, we use the data enhancement operation of CutMix to enhance the stored samples to improve the effectiveness of the retrieved samples.

• Extensive empirical results show that our method performs significantly better than existing methods on several benchmark datasets.

2   Proposed Method

In this section, we provide a comprehensive description of the overall framework of attention-based dual view consistency. First, we introduce the problem definition of online class-incremental learning and then detail a strategy of data augmentation combined with maximum gradient interference (MGI) [11] to preserve old class performance. Finally, we focus on how the Attention-based Dual View Consistency effectively mines information from the single-pass data stream.

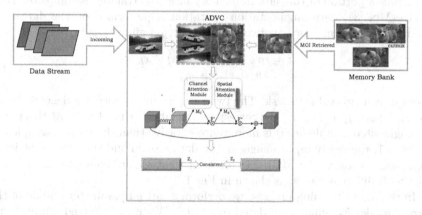

Fig. 1. This is the flow of our method. At the time $t + 1$, the incoming images in the data stream and the augmented images retrieved from memory are converted into image pairs with different views. The image pairs are then sent to ADVC to fully explore the semantic information between them using attention mechanisms to maximize their representation consistency.

2.1 Problem Definition

In incremental learning, we consider a more general and realistic setting where samples in a data stream are seen only once and the model needs to keep learning new classes in a single-pass through the data stream, a setting called online class incremental learning. We assume that the data stream is indicated by $D = \{D_1, D_2, \ldots, D_N\}$ and the data set $D_i \in (X, Y)$ does not satisfy the independent identical distribution condition, where X denotes the sample, Y denotes the label of X, and N denotes the total number of tasks. In online class-incremental learning, the model completes training for each task sequentially, and the data stream is only seen once during the training phase. $\{D_t\}$ indicates the set of data for task t. The goal of task t is to train a model that can correctly classify the classes in D_t.

2.2 Sample Augmentation and Retrieval

We think about the effectiveness of the retrieval strategy from a new perspective (the samples themselves). For this reason, we thought of using data augmentation operations to enhance the samples in memory so that the samples retain better information about the old classes. We chose the simplest and most effective way of data augmentation by mixing labels. In this way, the old classes of samples in memory are mixed to make the samples more representative.

At time t, the model receives a batch of input images from the data stream and stores a portion of the data in memory after the training is complete. Then the CutMix [31] data augmentation operation is performed on the data temporarily stored in memory for two samples (x_a, y_a) and (x_b, y_b):

$$\begin{aligned}
\tilde{x} &= \mathbf{m} \odot x_a + (\mathbf{1} - \mathbf{m}) \odot x_b \\
\tilde{y} &= \lambda y_a + (1 - \lambda) y_b,
\end{aligned} \tag{1}$$

where m is a two-valued mask. The two samples in memory are denoted by x_a and x_b, where y_a and y_b are the corresponding labels. λ is set to 1. At this point, the augmented sample $(\tilde{x}, \tilde{y})$ is more representative than the normal sample. At time t + 1, the newly input images in the data stream and the images retrieved from memory using the MGI [11] retrieval strategy are converted into image pairs with different views, as shown in Fig. 1.

In the model training process, we perform a virtual parameter update of the current model F_θ, which is updated to $F_v (\theta_v)$. When the received sample is x_i, the $\theta_v = \theta - \alpha \nabla \mathcal{L} (F (x_i), y_i)$, where α is the learning rate. We randomly select L candidate samples $\tilde{x}$ from the memory bank and then calculate the gradient change of that sample in the current model and the virtual update model, respectively, according to Eqs. (2). Finally, we ranked the gradient changes in Sort $(G (\tilde{x}; \theta_v) - G (\tilde{x}; \theta))$ descending order and selected the top k samples as the retrieved samples.

$$\begin{aligned}
G (\tilde{x}; \theta) &= \|\nabla_\theta \mathcal{L} (y_i, F (\tilde{x}, \theta))\|_1 \\
G (\tilde{x}; \theta_v) &= \|\nabla_{\theta_v} \mathcal{L} (y_i, F_v (\tilde{x}, \theta_v))\|_1
\end{aligned} \tag{2}$$

2.3   Attention-Based Dual View Consistency

In online class-incremental learning, each new incoming image from the tasks can only be seen once. In order to more fully utilize the semantic information of single-channel data streams, our proposed Attention-based Dual View Consistency (ADVC) strategy not only compels the network to learn to classify input images but also emphasizes guiding the model on which regions to seek consistency through the attention mechanism. The introduction of the attention mechanism not only reduces unnecessary computations but also enables better utilization of valuable semantic information.

Generating Augmented Views. Firstly, data augmentation techniques (such as vertical flipping) are applied to each input image, and the images are retrieved from memory to generate augmented views. These views visually differ from the original images, but retain the semantic content of the original images.

Applying Attention Mechanism. During the training of the model, we introduce an attention layer at an appropriate location in the model (after the convolutional layers). The attention layer aims to generate a weight for each input data point, determining where the model should "focus". During the forward propagation process of the model, the attention layer produces an attention map, indicating which parts of the input the model should focus on.

$$\mathbf{F}'_{x^1} = \mathbf{M_c}(\mathbf{F}_{x^1}) \otimes \mathbf{F}_{x^1},$$
$$\mathbf{F}''_{x^1} = \mathbf{M_s}(\mathbf{F}'_{x^1}) \otimes \mathbf{F}'_{x^1}. \tag{3}$$

$\mathbf{F}_{x^1} \in \mathbb{R}^{C \times H \times W}$ is an intermediate feature map of one of the views of F_x. The channel attention map $\mathbf{M_c} \in \mathbb{R}^{C \times 1 \times 1}$ is used to enhance channel features. $\mathbf{M_s} \in \mathbb{R}^{1 \times H \times W}$ is the spatial attention map, which can explore important areas more accurately. F''_{x^1} is the final output. The intermediate feature map of the other view of F_x is represented by $\mathbf{F}_{x^2} \in \mathbb{R}^{C \times H \times W}$. The output F''_{x^2} of the other view image can be obtained according to Eqs. (3). After getting the channel attention and spatial attention, use the broadcast mechanism to refine the information of the original feature map, and finally get the refined feature map.

Training with Dual View Consistency. Once the attention map is obtained, it can be used to guide the maximization of mutual information between two different views. The attention mechanism plays a key role here as it can guide the model where to seek consistency. According to DVC [11], by forcing the joint distribution to be the same as the marginal distributions, $I(X_1; X_2)$ can be approximated as follows:

$$I(X_1; X_2) \approx \frac{1}{3}(H(X_1) + H(X_2) + H(X_1, X_2)). \tag{4}$$

Specifically, we only calculate and maximize the mutual information in the areas indicated by the attention map to better utilize valuable information. For each input image x, we now use the feature F''_{x1} after two rounds of attention transformation, the joint probability matrix $\boldsymbol{P} \in \mathbb{R}^{C \times C}$ can be calculated as:

$$P = \frac{1}{n} \sum_{i=1}^{n} F'' \left(\boldsymbol{x}_i^1\right) \cdot F'' \left(\boldsymbol{x}_i^2\right)^{\top} \tag{5}$$

where x_i^1 and x_i^2 are two transformed versions of the same image x_i, and n is the batch size at time t.

Optimization Objective Function. Our aim is to maximize $I(z_1; z_2)$, thus the MI loss can be formulated as follows:

$$\mathcal{L}_{MI} = -I\left(\boldsymbol{z}_1; \boldsymbol{z}_2\right) \tag{6}$$

where z_1 and z_2 are the representations of the dual-view image pairs. In addition, to constrain the difference between the joint distribution and marginal distributions, as well as the distance between the features after attention transformation and the original features, we use the following two loss terms, respectively.

$$\mathcal{L}_{DL} = L_1 \left(p\left(z_1, z_2\right), p\left(z_1\right)\right) + L_1 \left(p\left(z_1, z_2\right), p\left(z_2\right)\right), \tag{7}$$

$$\mathcal{L}_{AT} = ||F''(\boldsymbol{x}) - F(\boldsymbol{x})||_2^2, \tag{8}$$

where L_1 denotes Mean Absolute Error (MAE) loss, $p(z_1)$ and $p(z_2)$ represent the marginal distributions of z_1 and z_2, respectively, and $p(z_1, z_2)$ denotes the joint distribution of z_1 and z_2. As shown in Eqs. (8), we use the L_2 norm (square Euclidean distance) to calculate the distance between features, where F''_x is the feature after two rounds of attention transformation, and F_x is the original feature. The total loss function $\mathcal{L}$ can be formulated as follows:

$$\mathcal{L} = \lambda_1 \mathcal{L}_{CE} + \lambda_2 \mathcal{L}_{MI} + \lambda_3 \mathcal{L}_{DL} + \lambda_4 \mathcal{L}_{AT}, \tag{9}$$

where $\mathcal{L}_{CE}$ represents the Cross-Entropy loss, and $\lambda 1$, $\lambda 2$, $\lambda 3$, and $\lambda 4$ are the balance coefficients of the four losses.

Class-Incremental Learning Update. At each update stage, ADVC will use the new categories and data to update the model. Both the attention mechanism and dual view consistency can help the model not forget old knowledge while introducing new data.

3   Experiments

In this section, we present the data set used for the experiments, the evaluation metrics, and the associated benchmark model, as well as the experimental setup. We compare our method with existing techniques through various experiments and further investigate the effectiveness of each component of the method.

3.1 Datasets

We use three datasets that are more commonly found in class incremental learning to facilitate comparison with other class incremental learning models: Split CIFAR-10, split CIFAR-100, and split Mini-ImageNet. CIFAR-10 [15] consists of 10 classes, with 5000 training samples and 1000 test samples for each class. Split CIFAR-10 is a splitting of the CIFAR-10 dataset into 5 smaller datasets for 5 tasks, where each task has 2 classes. CIFAR-100 [15] has 100 classes containing 600 images each. Split CIFAR-100 is a splitting of the CIFAR-100 dataset into 10 disjoint sub-datasets of 10 tasks, where each task contains 10 classes. Mini-ImageNet [24] contains a total of 60,000 color images in 100 classes, of which there are 600 samples in each class. Split Mini-ImageNet is a task that splits the Mini-ImageNet dataset into 10 sub-datasets for 10 non-overlapping classes, each with 10 classes.

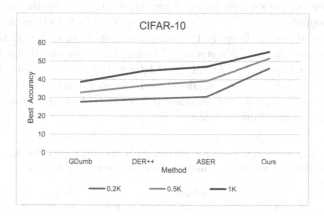

Fig. 2. Best Accuracy (higher is better). On the CIFAR-10 dataset, our method achieves better results than the baseline method.

3.2 Metrics

Online class-incremental learning refers to continual learning in a continuous stream of data. The average accuracy rate and the forgetting rate of old tasks are the two most common evaluation metrics for online class-incremental learning. The average accuracy can reflect the performance of the incremental model, and [6] proposed an estimation method for the average accuracy. Where $a_{ij} \in [0, 1]$ means the classification performance of the model on the test set of task j after the model has been trained on task i. The average accuracy A_t for task t:

$$\text{Average Accuracy } (A_t) = \frac{1}{t} \sum_{j=1}^{t} a_{t,j}. \tag{10}$$

The average forgetting rate indicates how well the model performs on older tasks. [20] introduced forgetting rates to measure the extent to which the model forgets old tasks. For a task t the average forgetting rate F_t:

$$\text{Average Forgetting } (F_t) = \frac{1}{t-1} \sum_{j=1}^{t-1} f_{t,j},$$

$$\text{where } f_{i,j} = \max_{l \in \{1,...,i-1\}} a_{l,j} - a_{i,j}.$$

(11)

3.3 Baselines

We use these popular class-incremental learning methods listed below as our baseline. ER [7] (Experience Replay) is a method of randomly sampling and updating memory. MIR [1] (Maximally Interfered Retrieval) selects the old sample that interferes the most with the new data (the most increase in loss), i.e., the sample whose prediction will be most negatively affected by the update of the foreseen parameters. GSS [2] (Gradient-based Sample Selection) focuses more on memory updates to diversify the sample gradients in memory. DER++ [3] (Dark Experience Replay) is a stronger baseline that uses knowledge distillation to preserve past experience. ASER [21] (Adversarial Shapley Value Experience Replay) Use SV for Memory Retrieval and Memory Update. DVC [11] (Dual View Consistency) explores the semantic information of a single-pass data stream by maximizing mutual information.

Table 1. Average Accuracy (higher is better), M is the memory buffer size. The bolded font indicates the best results. All numbers are the average of 15 runs.

Method	Mini-ImageNet			CIFRA-100			CIFRA-10		
	M = 1K	M = 2K	M = 5K	M = 1K	M = 2K	M = 5K	M = 0.2K	M = 0.5K	M = 1K
ER	10.2 ± 0.5	12.9 ± 0.8	16.4 ± 0.9	11.6 ± 0.5	15.0 ± 0.5	20.5 ± 0.8	23.2 ± 1.0	31.2 ± 1.4	39.7 ± 1.3
MIR	10.1 ± 0.6	14.2 ± 0.9	18.5 ± 1.0	11.3 ± 0.3	15.1 ± 0.3	22.2 ± 0.7	24.6 ± 0.6	32.5 ± 1.5	42.8 ± 1.4
GSS	9.3 ± 0.8	14.1 ± 1.1	15.0 ± 1.1	9.7 ± 0.2	12.4 ± 0.6	16.8 ± 0.8	23.0 ± 0.9	28.3 ± 1.7	37.1 ± 1.6
GDumb	7.3 ± 0.3	11.4 ± 0.2	19.5 ± 0.5	10.0 ± 0.2	13.3 ± 0.4	19.2 ± 0.4	26.6 ± 1.0	31.9 ± 0.9	37.5 ± 1.1
DER++	10.9 ± 0.6	15.0 ± 0.7	17.4 ± 1.5	11.8 ± 0.4	15.7 ± 0.5	20.8 ± 0.8	28.1 ± 1.2	35.4 ± 1.3	42.8 ± 1.9
ASER	11.5 ± 0.6	13.5 ± 0.8	17.8 ± 1.0	14.3 ± 0.5	17.8 ± 0.5	22.8 ± 1.0	29.6 ± 1.0	38.2 ± 1.0	45.1 ± 2.0
DVC	15.4 ± 0.7	17.2 ± 0.8	19.1 ± 0.9	19.7 ± 0.7	22.1 ± 0.9	24.1 ± 0.8	**45.4 ± 1.4**	50.6 ± 2.9	52.1 ± 2.5
Ours	**17.3 ± 0.8**	**19.4 ± 1.2**	**23.6 ± 0.9**	19.4 ± 1.2	21.8 ± 0.8	**24.5 ± 0.3**	45.1 ± 1.0	**50.8 ± 0.7**	**54.7 ± 0.5**

3.4 Implementation Detail

We use a single-head setup where the model has a shared output layer, which is needed to classify all labels without a task ID. Furthermore, based on existing class incremental learning methods [11,21], we use ResNet18 as the network architecture for CIFAR 10, CIFAR 100, and Mini-ImageNet. We train the network using a random gradient descent (SGD) optimizer with the retrieval batch size k set to 10. For CIFAR-100 and Mini-ImageNet, $\lambda 1 = \lambda 2 = 1$, $\lambda 3 = \lambda 4 = 4$. For CIFAR-10, $\lambda 1 = \lambda 2 = 1$, $\lambda 3 = \lambda 4 = 2$.

3.5 Comparative Performance Evaluation

As depicted in Fig. 2, our method exhibits outstanding performance on the CIFAR-10 dataset. Regardless of the memory size, ADVC consistently demonstrates the highest accuracy rate, particularly when M = 0.2K, where the improvement effect of ADVC is most pronounced. This suggests that our method can produce good results even under conditions of memory scarcity.

Table 1 shows the average accuracy at the end of the CIFAR-10, CIFAR-100, and Mini-ImageNet data streams. As can be seen from the results in Table 1, the improvement of our method on Mini-ImageNet is significantly higher than that on CIFAR-10. More specifically, the best performance of our method on Mini-ImageNet is 4.5% better than the strongest baseline DVC, and the average performance is 2.9% improved. The best score on the CIFAR-100 increased by 0.4%, and on the CIFAR-10, by 2.6%. Due to the higher complexity and resolution of images in Mini-ImageNet compared to CIFAR-100, images in Mini-ImageNet often contain more detailed information. The attention mechanism is better at identifying important features and areas in clearer images, hence, ADVC achieves better results on Mini-ImageNet.

Table 2. Average Forgetting (lower is better). The bolded font indicates the best results. M is the memory buffer size. All numbers are an average of 15.

Method	Mini-ImageNet			CIFRA-100			CIFRA-10		
	M = 1K	M = 2K	M = 5K	M = 1K	M = 2K	M = 5K	M = 0.2K	M = 0.5K	M = 1K
ER	32.7 ± 0.9	29.1 ± 0.7	26.0 ± 1.0	39.1 ± 0.9	34.6 ± 0.9	30.6 ± 0.9	60.9 ± 1.0	50.2 ± 2.5	39.5 ± 1.6
MIR	31.5 ± 1.2	25.6 ± 1.1	20.4 ± 1.0	39.5 ± 0.6	33.3 ± 0.8	28.3 ± 0.7	61.8 ± 1.0	51.5 ± 1.4	38.0 ± 1.5
GSS	33.5 ± 0.8	28.0 ± 0.7	27.5 ± 1.2	38.2 ± 0.7	34.3 ± 0.6	30.2 ± 0.8	62.2 ± 1.3	55.3 ± 1.3	44.9 ± 1.4
DER++	33.8 ± 0.8	28.6 ± 0.8	27.1 ± 1.3	41.9 ± 0.6	36.7 ± 0.5	33.5 ± 0.8	55.9 ± 1.8	45.0 ± 1.0	34.6 ± 2.8
ASER	33.8 ± 1.3	30.5 ± 1.3	25.1 ± 0.8	43.0 ± 0.5	37.9 ± 0.6	29.6 ± 0.9	56.4 ± 1.6	47.5 ± 1.3	39.6 ± 2.0
DVC	25.1 ± 0.7	23.1 ± 0.7	21.9 ± 0.8	30.6 ± 0.7	27.8 ± 1.0	**26.1 ± 0.5**	27.2 ± 2.5	**21.3 ± 3.1**	**19.7 ± 2.9**
Ours	**24.6 ± 0.9**	**22.6 ± 0.6**	**17.8 ± 1.2**	**29.2 ± 0.5**	**26.7 ± 0.8**	28.7 ± 1.0	**25.1 ± 1.0**	30.5 ± 1.7	24.6 ± 0.5

The average forgetting rate of the baseline method with our proposed method for CIFAR-10, CIFAR-100, and Mini-ImageNet is shown in Table 2. Our method also demonstrates excellent performance in terms of forgetting rate. As shown in Table 2, ADVC achieves the largest improvement on the Mini-ImageNet dataset, especially at M = 5K, where our method outperforms MIR [1] by a reduction of 2.6%. This indicates that our method effectively mitigates the problem of catastrophic forgetting [14,20] through sample augmentation and retrieval (Sect. 2.2).

Table 3. Ablation studies on CIFAR-10. "Baseline" represents the model with the DVC method. All numbers are the average of 15 runs.

Method	M = 0.2K	M = 0.5K	M = 1K
Baseline	45.4 ± 1.4	50.6 ± 2.9	52.1 ± 2.5
Baseline + SE	47.6 ± 1.0	54.1 ± 1.2	56.1 ± 1.8
Baseline + ECA	46.8 ± 0.9	54.0 ± 1.2	57.4 ± 0.8
Baseline + CBAM	**49.1 ± 1.3**	**56.0 ± 0.8**	**58.9 ± 1.9**

3.6 Ablation Study

Attention. To verify that the method we adopted can better explore the semantic information between the dual views, we conducted a series of experiments. As shown in Table 3, we used the more classical channel attention modules SENet [13] and ECANet [25] for comparison with the CBAM [27] module we used, with DVC [11] as the baseline. From the experimental results, it can be seen that the method we used is remarkably better than the improvements of SE and ECA, and the results are more satisfactory at M = 0.5K and M = 1K than at M = 0.2K. It can be seen that the larger the memory, the richer the semantic information in the single-pass data stream explored by the attention mechanism, and this leads to higher classification accuracy.

Each Component. In addition to the attention mechanism, we also verified the effectiveness of the various components of our proposed method. As shown in Table 4, both CutMix and CBAM have demonstrated certain improvement effects. Notably, when M = 5K, the model's accuracy improved by nearly 3% points, reaching 21.9%. This indicates that the CutMix data augmentation can assist the model in better understanding and remembering the training data, thereby enhancing the model's performance. After incorporating the CBAM, the accuracy rose to 23.6%. This suggests that the introduction of the CBAM attention module enables the model to focus on and extract important features in the images more effectively, thus further enhancing the model's performance.

Table 4. Ablation studies on Mini-ImageNet. "Baseline" represents the model with the DVC method. All numbers are the average of 15 runs.

Method	M = 1K	M = 2K	M = 5K
Baseline	15.4 ± 0.7	17.2 ± 0.8	19.1 ± 0.9
Baseline + CutMix	15.9 ± 1.2	19.1 ± 1.1	21.9 ± 1.0
Baseline + CutMix + CBAM	**17.3 ± 0.8**	**19.4 ± 1.2**	**23.6 ± 0.9**

4   Conclusion

In this paper, we propose an Attention-based Dual View Consistency (ADVC) strategy to explore more semantic information between views. On one hand, our method can reduce or even filter out some unimportant information through the attention mechanism, in order to improve classification accuracy. On the other hand, we use an effective data augmentation technique to maintain sufficient information of the old classes, making the retrieved samples more representative, thereby mitigating the catastrophic forgetting problem. We conducted extensive experiments on three commonly used benchmark datasets to demonstrate that our method can not only enhance classification accuracy but also effectively mitigate the catastrophic forgetting problem, enabling the model to retain memory of old tasks while learning new ones.

References

1. Aljundi, R., et al.: Online continual learning with maximally interfered retrieval. arXiv preprint arXiv:1908.04742 **2** (2019)
2. Aljundi, R., Lin, M., Goujaud, B., Bengio, Y.: Gradient based sample selection for online continual learning. In: Advances in Neural Information Processing Systems, vol. 32 (2019)
3. Buzzega, P., Boschini, M., Porrello, A., Abati, D., Calderara, S.: Dark experience for general continual learning: a strong, simple baseline. Adv. Neural. Inf. Process. Syst. **33**, 15920–15930 (2020)
4. Cha, S., Hsu, H., Hwang, T., Calmon, F.P., Moon, T.: CPR: classifier-projection regularization for continual learning. arXiv preprint arXiv:2006.07326 (2020)
5. Chaudhry, A., Dokania, P.K., Ajanthan, T., Torr, P.H.S.: Riemannian walk for incremental learning: understanding forgetting and intransigence. In: Ferrari, V., Hebert, M., Sminchisescu, C., Weiss, Y. (eds.) ECCV 2018. LNCS, vol. 11215, pp. 556–572. Springer, Cham (2018). https://doi.org/10.1007/978-3-030-01252-6_33
6. Chaudhry, A., et al.: Continual learning with tiny episodic memories (2019)
7. Chaudhry, A., et al.: On tiny episodic memories in continual learning. arXiv preprint arXiv:1902.10486 (2019)
8. Dan, Z., Fang, Y.: Deliberate multi-attention network for image captioning. In: Yu, S., et al. (eds.) PRCV 2022, Part I. LNCS, vol. 13534, pp. 475–487. Springer, Cham (2022). https://doi.org/10.1007/978-3-031-18907-4_37
9. Gok, E.C., Yildirim, M.O., Kilickaya, M., Vanschoren, J.: Adaptive regularization for class-incremental learning. arXiv preprint arXiv:2303.13113 (2023)
10. Golkar, S., Kagan, M., Cho, K.: Continual learning via neural pruning. arXiv preprint arXiv:1903.04476 (2019)
11. Gu, Y., Yang, X., Wei, K., Deng, C.: Not just selection, but exploration: online class-incremental continual learning via dual view consistency. In: Proceedings of the IEEE/CVF Conference on Computer Vision and Pattern Recognition, pp. 7442–7451 (2022)
12. Henning, C., et al.: Posterior meta-replay for continual learning. Adv. Neural. Inf. Process. Syst. **34**, 14135–14149 (2021)
13. Hu, J., Shen, L., Sun, G.: Squeeze-and-excitation networks. In: Proceedings of the IEEE Conference on Computer Vision and Pattern Recognition, pp. 7132–7141 (2018)

14. Kirkpatrick, J., et al.: Overcoming catastrophic forgetting in neural networks. Proc. Nat. Acad. Sci. **114**(13), 3521–3526 (2017)
15. Krizhevsky, A., Hinton, G., et al.: Learning multiple layers of features from tiny images (2009)
16. Lee, S., Ha, J., Zhang, D., Kim, G.: A neural dirichlet process mixture model for task-free continual learning. arXiv preprint arXiv:2001.00689 (2020)
17. Mallya, A., Lazebnik, S.: PackNet: adding multiple tasks to a single network by iterative pruning. In: Proceedings of the IEEE Conference on Computer Vision and Pattern Recognition, pp. 7765–7773 (2018)
18. Rolnick, D., Ahuja, A., Schwarz, J., Lillicrap, T., Wayne, G.: Experience replay for continual learning. In: Advances in Neural Information Processing Systems, vol. 32 (2019)
19. Rudner, T.G., Smith, F.B., Feng, Q., Teh, Y.W., Gal, Y.: Continual learning via sequential function-space variational inference. In: International Conference on Machine Learning, pp. 18871–18887. PMLR (2022)
20. Serra, J., Suris, D., Miron, M., Karatzoglou, A.: Overcoming catastrophic forgetting with hard attention to the task. In: International Conference on Machine Learning, pp. 4548–4557. PMLR (2018)
21. Shim, D., Mai, Z., Jeong, J., Sanner, S., Kim, H., Jang, J.: Online class-incremental continual learning with adversarial shapley value. In: Proceedings of the AAAI Conference on Artificial Intelligence, vol. 35, pp. 9630–9638 (2021)
22. Tao, X., Chang, X., Hong, X., Wei, X., Gong, Y.: Topology-preserving class-incremental learning. In: Vedaldi, A., Bischof, H., Brox, T., Frahm, J.-M. (eds.) ECCV 2020. LNCS, vol. 12364, pp. 254–270. Springer, Cham (2020). https://doi.org/10.1007/978-3-030-58529-7_16
23. Tiwari, R., Killamsetty, K., Iyer, R., Shenoy, P.: GCR: gradient coreset based replay buffer selection for continual learning. In: Proceedings of the IEEE/CVF Conference on Computer Vision and Pattern Recognition, pp. 99–108 (2022)
24. Vinyals, O., Blundell, C., Lillicrap, T., Wierstra, D., et al.: Matching networks for one shot learning. In: Advances in Neural Information Processing Systems, vol. 29 (2016)
25. Wang, Q., Wu, B., Zhu, P., Li, P., Zuo, W., ECA-Net, Q.H.: Efficient channel attention for deep convolutional neural networks. arXiv preprint arXiv:1910.03151 (2019)
26. Wang, Y., Ma, Z., Huang, Z., Wang, Y., Su, Z., Hong, X.: Isolation and impartial aggregation: a paradigm of incremental learning without interference. arXiv preprint arXiv:2211.15969 (2022)
27. Woo, S., Park, J., Lee, J.-Y., Kweon, I.S.: CBAM: convolutional block attention module. In: Ferrari, V., Hebert, M., Sminchisescu, C., Weiss, Y. (eds.) ECCV 2018. LNCS, vol. 11211, pp. 3–19. Springer, Cham (2018). https://doi.org/10.1007/978-3-030-01234-2_1
28. Xie, N., Yu, W., Yang, L., Guo, M., Li, J.: Attention-based fusion of directed rotation graphs for skeleton-based dynamic hand gesture recognition. In: Yu, S., et al. (eds.) PRCV 2022, Part I. LNCS, vol. 13534, pp. 293–304. Springer, Cham (2022). https://doi.org/10.1007/978-3-031-18907-4_23
29. Yan, Q., Gong, D., Liu, Y., van den Hengel, A., Shi, J.Q.: Learning bayesian sparse networks with full experience replay for continual learning. In: Proceedings of the IEEE/CVF Conference on Computer Vision and Pattern Recognition, pp. 109–118 (2022)

30. Yin, S.Y., Huang, Y., Chang, T.Y., Chang, S.F., Tseng, V.S.: Continual learning with attentive recurrent neural networks for temporal data classification. Neural Netw. **158**, 171–187 (2023)
31. Yun, S., Han, D., Oh, S.J., Chun, S., Choe, J., Yoo, Y.: CutMix: regularization strategy to train strong classifiers with localizable features. In: Proceedings of the IEEE/CVF International Conference on Computer Vision, pp. 6023–6032 (2019)

Online Airline Baggage Packing Based on Hierarchical Tree A2C-Reinforcement Learning Framework

Pan Zhang[1,2], Ming Cui[1], Wei Zhang[1,2,3（✉）], Jintao Tian[1], and Jiulin Cheng[1]

[1] College of Aeronautical Engineering, Civil Aviation University of China, Tianjin, China
`weizha_2001@163.com`
[2] Key Laboratory of Smart Airport Theory and System, CAAC, Tianjin, China
[3] Aviation Ground Special Equipment Research Base, Tianjin, China

Abstract. In this study, we propose a hierarchical tree A2C-reinforcement learning framework for online packing of airline baggage. Our approach aims to solve the critical loading algorithm issue and pave the way for automation in the loading process. The framework integrates a hierarchical tree search for processing external environment data with a deep reinforcement learning module designed around the A2C framework. This module incorporates airline baggage constraints, an enhanced reward function, and defined action and state spaces. To refine the packing strategy, the tree search adjusts leaf node selection probabilities based on complete baggage information, while baggage classification utilizes the Family Unity concept. We validate our algorithm's efficacy by contrasting it with BR, LSAH, discrete deep reinforcement learning and Online BPH algorithm, utilizing the established airline baggage dataset. Our experimental findings reveal that our proposed method's average loading filling rate is 71.2%, superior by 8.5% compared to similar selected algorithms, with faster calculation time. We set up an experimental platform and the results affirm the proposed method's efficacy.

Keywords: air transport · airline baggage · online packing · hierarchical tree · deep reinforcement learning · A2C

1 Introduction

The loading and transportation of checked baggage play a crucial role in ensuring efficient airport operations. Current procedures primarily involve manual labor, leading to issues such as low efficiency, high labor intensity, and problems with baggage damage and suboptimal loading space utilization. Given the rapid increase in air passenger traffic volume, automating baggage handling has become a critical need for airports [1]. The online baggage packing algorithm represents a core challenge in achieving automatic loading.

The loading problem of airline baggage can be categorized as a 3D Bin Packing Problem, which is an NP-complete problem [2]. Depending on whether the information about all items is known in advance and the loading order, the 3D packing problem

© The Author(s), under exclusive license to Springer Nature Singapore Pte Ltd. 2024
Q. Liu et al. (Eds.): PRCV 2023, LNCS 14431, pp. 500–513, 2024.
https://doi.org/10.1007/978-981-99-8540-1_40

can be further divided into offline and online packing. Currently, the research on 3D packing algorithm mainly focuses on the offline algorithm [3], where the information about items is known, and the loading order can be rearranged arbitrarily. Some scholars have studied offline algorithms for baggage loading planning. Xing et al. [4] utilized a dynamic quadtree search algorithm and constructed a quadtree by grouping airline baggage into composite strips, enabling dynamic selection of code placement. Hong et al. [5] employed a reinforcement learning algorithm with "trial and error" learning to design multiple constraints for baggage loading planning. Although offline algorithms can achieve high filling rates, their applicability is limited due to insufficient consideration of real application scenarios.

The loading problem of airline baggage in the transport line pertains to online packing. Research on online algorithms primarily focuses on heuristic and deep reinforcement learning algorithms. Heuristic algorithms in this field commonly employ genetic algorithms, particle swarm algorithms, and hybrid algorithms. Ha et al. [6] developed an online 3D packing algorithm by combining two heuristic algorithms. Zhang et al. [7] proposed a multi-layer heuristic algorithm. Zhang et al. [8] proposed an improved particle swarm optimization algorithm to enhance convergence speed and space utilization. Zhang et al. [9] proposed a heuristic algorithm based on the "fill point" strategy, allowing for simultaneous planning of multiple baggage locations to improve algorithm performance. Heuristic algorithms demonstrate good results, their convergence is often limited, and improvements require designing complex loading rules, which hampers breakthroughs in algorithm performance.

The deep reinforcement learning algorithm combines deep learning and reinforcement learning. By utilizing the powerful representation capabilities of deep learning and the action-state function of the reinforcement learning agent, the packing problem can be abstracted as a "stacking" process of irregular boxes, enabling trial and error learning for obtaining palletizing strategies. The field of deep reinforcement learning is mainly used for combinatorial optimization problems [10, 11]. Yang et al. [12] combined the deep reinforcement learning algorithm with the greedy search strategy to greatly improve the performance of the algorithm. Hu et al. [13] optimized the loading sequence of items based on the pointer network to improve the filling rate; Jiang et al. [14] used the multi-modal encoder and the search framework in constraint program to improve the solution quality. Verma et al. [15], based on DQN deep reinforcement learning algorithm, solved the direction and position of loaded items and proposed a packing algorithm based on the robotic arm packing scene, which was superior to the most advanced online heuristic algorithm in terms of volumetric efficiency index. But it did not take into account the interference problems in the actual palletizing process. Although deep reinforcement learning can be used to deal with packing problems, the discretization representation of action-state space used in most current algorithms is limited by the degree of spatial discretization, and can't realize the high-precision solution of loading position. Zhao et al. [16] employed a heuristic algorithm based on tree search, called PCT, to represent the action-state space. This approach overcomes the training difficulties of deep reinforcement learning models caused by limited degrees of action-state space discretization. Combining the PCT algorithm with the ACKTR algorithm, they achieved online loading of goods. Additionally, most algorithms focus on the loading of various containers or logistics vehicles, with limited research on airline baggage transportation considering aviation space constraints. Under the unique context of airline baggage loading, current deep reinforcement learning algorithms struggle to integrate multiple constraints and achieve convergence simultaneously.

Aiming at the limitations of the research, this paper proposes a hierarchical tree A2C-reinforcement learning framework for online airline baggage packing. The framework aims to optimize the filling rate of the baggage cart. The paper has the following contributions:

1) We develop a hierarchical tree reinforcement learning framework specifically for online airline baggage, leveraging the hierarchical tree for state-action representation and the A2C for model training.
2) We enhance the loading strategy of the hierarchical tree search model, adjusting leaf node selection based on comprehensive baggage information.
3) The reward function has been refined, considering several factors like volume, positioning, and interference during robotic arm stacking, using the Family Unity concept for baggage categorization.
4) We introduce bespoke constraints for airline baggage, ensuring baggage security during packing and accounting for their value and minimal transportation loss tolerance.

The remaining contents of the paper are arranged as follows: In Sect. 2, the hierarchical tree deep reinforcement learning framework is proposed for online airline baggage packing. In Sect. 3, experiments are arranged to verify the effectiveness of the model, followed by the conclusions in Sect. 4.

2   Hierarchical Tree Deep Reinforcement Learning Framework

The hierarchical tree deep reinforcement learning framework proposed in this paper comprises two main components: the hierarchical tree search model and the deep reinforcement learning framework.

The hierarchical tree search module utilizes a hierarchical tree search strategy to represent the external environmental information required for deep reinforcement learning. Its output serves as the input for the deep reinforcement learning module. The deep reinforcement learning model is responsible for determining the current location of the target baggage.

2.1   Algorithm Framework

This paper introduces the hierarchical tree deep reinforcement learning algorithm for online airline baggage packing. The proposed algorithm framework is depicted in Fig. 1.

In Fig. 1, the environment is defined, including the stacking information of the current baggage cart and the information of the baggage waiting to be stacked.

A hierarchical tree search model is defined, utilizing a hierarchical tree search strategy to represent the external environmental information required for deep reinforcement learning. The strategy characterizes the action-state space through internal nodes and leaf nodes. The internal nodes represent the completed baggage placement, including the three-dimensional dimensions of the current baggage and its position in the baggage cart coordinate system.

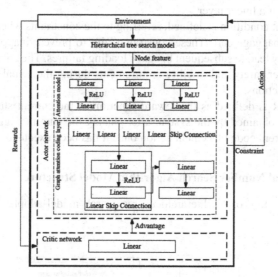

Fig. 1. Overall framework of the algorithm

We define action A, representing the set of actions performed by the agent, which includes the combination of the current baggage's stacking position and stacking mode. The stacking position refers to the three-dimensional coordinates of the baggage in the baggage cart, and the stacking mode encompasses vertical and horizontal placement. Upon executing the current action, the algorithm's environment information is updated.

The state S is defined to represent the set of all external environment states, including the current stacking state and the baggage state. The stacking status reflects the overall state of the baggage that has been loaded onto the baggage cart, while the baggage status represents the three-dimensional information of the baggage waiting to be loaded. The state serves as the representation of the algorithm's external environment, and it undergoes changes after action execution.

The agent is defined, which refers to the action entity of the algorithm. It determines the optimal action based on the current state and executes the action to transition to the next state. In this paper, the agent adopts the A2C algorithm framework, consisting of two modules: the Actor network and the Critic network.

Actor network module: The Actor network generates an action distribution based on the environment, node characteristic information, and Advantage value, providing guidance for action decision-making. The module comprises two parts: the Attention model and the graph attention coding layer. The Attention model includes three attention modules, each containing two Linear layers connected by ReLU activation functions. The graph attention coding layer consists of a Skip Connection layer, a Linear Skip

Connection layer, and two Linear layers. The linear skip connection layer comprises two linear layers connected by a ReLU activation function.

Critic network module: The Critic network evaluates the value of an action based on the reward and state characteristic information, namely the Advantage value. The module consists of a linear layer.

The constraint condition is defined, referring to the constraints when placing airline baggage on the baggage cart. These constraints aim to prevent inappropriate loading positions that may lead to subsequent baggage loading failures. The constraints comprehensively consider safety and stability requirements during the actual loading process to avoid unsafe actions.

The reward R is defined as the reward function in the current state, representing the reward value obtained by the system after placing the current baggage. The Agent evaluates the current code placement action once it has been executed.

2.2 Hierarchical Number Search Algorithm Model Structure

The network structure of the hierarchical tree search model adopted in this paper is shown in Fig. 2.

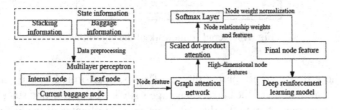

Fig. 2. Structure diagram of hierarchical tree search model

The hierarchical tree search model comprises three independent components: Multi-Layer Perceptron (MLP), Graph Attention Networks (GAT), scaled dot-product attention networks, and SoftMax layers. By processing each network structure, the final node characteristics of each leaf node can be outputted to the deep reinforcement learning model based on the input information of pile type and baggage.

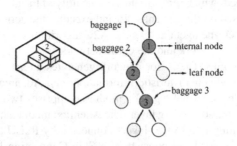

Fig. 3. Schematic diagram of hierarchical tree search model

MLP describes the information of internal nodes I_k, leaf nodes L_k, and the current baggage to be loaded m_k information. It is represented by an expression $t = \{MLP(I_k), MLP(L_k), MLP(m_k)\}$, where feature quantity $F = I_k + L_k + m_k + 1$ and varies with the number of nodes. The hierarchical tree search model for baggage loading process is shown in Fig. 3.

When the baggage cart is empty, there are no internal nodes in the baggage cart, only a few leaf nodes (white). When baggage 1 is placed in the planned loading position, the target leaf node is updated to internal node 1 (gray), and a leaf node is generated from this node. As the loading process progresses, these nodes are iteratively updated to form a dynamic hierarchical tree T_k.

The traditional three-dimensional packing strategy $\pi(L_k | T_k, m_k)$ solely considers the information of the leaf node and the current baggage to be palletized. However, in practical applications, the position of the already loaded baggage may change due to collisions during the stacking process of the newly loaded baggage, potentially affecting the stacking pattern of the existing baggage. Therefore, this paper fully considers the given state T_k of the layered tree and the new baggage m_k to be stacked. The loading strategy is designed as follows:

$$\pi(L_k | T_k, m_k) \tag{1}$$

During the baggage loading process, when the leaf node is covered by the position of the airline baggage, this leaf node will be removed from L_k to ensure the algorithm's feasibility.

As the hierarchical tree grows with the time step k, this paper adopts GAT to describe the hierarchical tree structure. GAT transforms the node features into high-dimensional node features, while the scaled dot product attention calculates the relationship weight between each node and the node features. The Softmax layer is used to normalize the weights between nodes.

2.3 Agent: A2C Algorithm

This paper utilizes the Advantage Actor Critic (A2C) network framework for training. The A2C algorithm employs a parallel architecture, employing multiple parallel independent workers to gather independent sampling experience, thus effectively enhancing training efficiency. In the A2C algorithm, the advantage function replaces the traditional feedback value in the Critic network, serving as an indicator to assess the quality of the current selected action. The advantage function $A_\pi(s, a)$ is defined as follows:

$$A_\pi(s, a) = Q_\pi(s, a) - V_\pi(s) \tag{2}$$

where $Q_\pi(s, a)$ represents the value function corresponding to action a performed under policy π, while $V_\pi(s)$ represents the sum of the product of the function of all possible action values and the action probability taken under the policy π. The goal of the algorithm is to discover the optimal strategy $\pi(a_k | s_k)$ that maximize the cumulative reward.

The A2C algorithm takes the state features $s_k = (T_k, m_k)$ extracted from the GAT layer as input. It weights the leaf nodes through the Actor network, generating the current policy distribution π_θ. Additionally, it maps the global features $\bar{i}\prime$ to the state value function through the Critic network to calculate the cumulative reward obtained during the current baggage loading process. The Actor network is trained to obtain the optimal parameter values.

2.4 Action Constraints

Actions generated by the actor network need to be designed to fulfill the requirements of airline baggage palletizing. Considering the high value and low tolerance for transportation loss of airline baggage, real-world scenarios involve baggage with handles, moving wheels, and other structural components. To ensure that the baggage remains stable, without falling, tilting, or moving during the stacking process, several constraints must be applied to the airline baggage.

Symbol Description and Objective Function. To accurately represent the loading position of airline baggage, the lower right corner point of the baggage cart's head is taken as the coordinate origin. The long side of the baggage cart represents the x axis direction, the short side represents the y axis direction, and the height represents the z axis direction. The dimensions of the baggage cart are denoted as L, W and H; The dimensions of the i-th airline baggage are denoted as l_i, w_i and h_i. The coding point P, which represents the left front lower corner of the baggage planned by the algorithm, is expressed as $P(x_i, y_i, z_i)$ in the baggage cart coordinate system.

Given a set of baggage to be loaded $M = (m_1, m_2, \ldots, m_n)$, the loading process aims to maximize the loading rate of the baggage cart while adhering to the loading constraints. The objective function F is defined as:

$$\left[F = \frac{\sum_{i=1}^{i=n} l_i \cdot w_i \cdot h_i}{L \cdot W \cdot H} \rightarrow max \right] \tag{3}$$

Constraints. Taking into account the requirements of loading safety and stability, the following constraint conditions are defined:

1) Volume constraint: The total volume of loaded baggage must not exceed the volume of the baggage cart, that is:

$$\sum_{i=1}^{i=n} l_i \cdot w_i \cdot h_i \leq L \cdot W \cdot H \tag{4}$$

2) Direction constraint: During the palletizing process, the baggage can only be placed parallel to the bottom surface of the baggage cart with the largest surface area. The sides of the bottom surface must be parallel to the length and width of the baggage cart's bottom surface, that is, h_i should be parallel to the z axis, and l_i should be parallel to either the x axis or the y axis.

3) No overlap constraint: No compression or overlap is allowed between bag-gage items, that is:

$$\begin{cases} \left| x_i + \frac{1}{2}l_i - \left(x_j + \frac{1}{2}l_j\right) \right| \geq \frac{1}{2}(l_i + l_j) \\ \left| y_i + \frac{1}{2}w_i - \left(y_j + \frac{1}{2}w_j\right) \right| \geq \frac{1}{2}(w_i + w_j) \\ \left| z_i + \frac{1}{2}h_i - \left(z_j + \frac{1}{2}h_j\right) \right| \geq \frac{1}{2}(h_i + h_j) \end{cases} \tag{5}$$

4) Stability constraints: The contact area between the bottom of each palletized baggage item and other baggage or the baggage cart surface must be at least 70% of the bottom area, that is:

$$s_i \geq 0.7 \cdot l_i \cdot w_i \tag{6}$$

where s_i denotes the bottom contact area of the i-th baggage.

5) Sequence constraints: Baggage must be placed in the order of conveyor transport.

2.5 Rewards

Rewards play a crucial role in evaluating the actions performed by the agent. In order to comprehensively address the objectives of the airline baggage loading problem, the reward function in this paper takes into account three components:

Volume Reward: The primary goal of the packing algorithm is to achieve a higher filling rate while maintaining loading safety. Hence, after each time step k, the reward function encourages the algorithm to stack the baggage effectively. The volume reward is defined as follows:

$$r_v = a \cdot \frac{V_k}{V_c} \tag{7}$$

where r_v represents the volume reward of the current baggage, a is a constant set to 10, V_k represents the volume of the current baggage $k(V_k = l_k * w_k * h_k)$, and V_c is the volume of the baggage cart ($V_c = L * W * H$).

Position Reward: In the actual loading process, the robotic arm is typically positioned on the side of the baggage cart rather than the top. Placing baggage near the robotic arm initially may lead to collisions between the arm and the baggage, as depicted in Fig. 4.

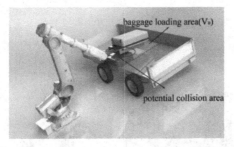

Fig. 4. Impact phenomenon during baggage loading

Weighted Reward for Categorized Layout: To ensure the stability of the loaded baggage, a reward function is designed to mitigate the likelihood of uneven height within the same plane, which can lead to a deterioration in flatness or even falls. The algorithm adopts the Family Unity strategy [17], which groups items belonging to the same family into the same bins to handle the loading process. Based on this idea, airline lines are categorized into five size ranges. The algorithm assigns weight w_{kt} to each leaf node based on the baggage category and stacking subspace information, and the node is selected using deep reinforcement learning. The weight w_{kt} of leaf nodes is calculated as:

$$w_{kt} = V_k - c \cdot dist(l_k, l_n) \tag{8}$$

where w_{kt} represents the weight of baggage k at leaf node t, V_k is the volume of baggage k, and $dist(l_k, l_n)$ represents the average distance between the current baggage k and the same type of baggage already loaded.

Incorporating the above three components, the reward function obtained by the algorithm when planning the current baggage k is designed as follows:

$$R_k = w_{kt} \left(a \cdot \frac{V_k}{V_c} - \bar{r} + r_p \right) \tag{9}$$

where w_{kt} represents the weight the weight of baggage k at leaf node t, a is a constant set to 10, V_k is the volume of baggage k currently being loaded ($V_k = l_k \cdot w_k \cdot h_k$). V_c refers to the total volume of the baggage cart ($V_c = L \cdot W \cdot H$), and $\bar{r}$ is the average value of the reward function for all iterations since training.

3 Experiment and Verification

3.1 Experimental Environment

Simulation Environment. The simulation environment consists of both hardware and software components. The hardware setup includes an NVIDIA GeForce RTX3090 GPU and an Intel(R) Xeon(R) CPU E5-2460 v4@2.40 GHz. The operating system employed is Windows 10 and the algorithm implementation is carried out using the PyTorch framework.

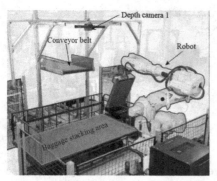
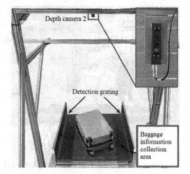

(a) Experiment platform   (b) Baggage information collection

Fig. 5. Experiment platform for automated baggage stacking

Hardware Environment. The research team has established an experimental platform for automatic baggage loading, as shown in Fig. 5(a). This platform comprises a robot, a baggage cart, conveyor belt, depth camera 1, control unit, etc. The robot autonomously picks up baggage from the conveyor belt and stacks it onto the baggage cart. Depth camera 1 positioned above the baggage cart captures information regarding the stacking process. The equipment for baggage information collection is shown in Fig. 5(b). Upon reaching the designated position on the conveyor belt and triggering the detection grating, the conveyor belt halts, initiating the baggage information collection process. Depth camera 2 is utilized as the device for baggage information collection.

3.2 Data Set and Parameter Selection

Data Set. The algorithm selects a baggage cart with dimensions of 288 cm * 147 cm * 84 cm. The training and test data sets consist of size information from 200 sets of real baggage entering and exiting an airport.

Parameter Selection. The hyperparameters utilized in the training phase are determined through a comparison of training results. The selected hyperparameters are as follows: a learning rate of $1e-6$, a discount factor of 0.99, a threshold of 100 for the number of internal nodes, a threshold of 60 for the number of leaf nodes, and a total of 120,000 training iterations.

3.3 Simulation Experiment

Algorithm performance comparison and analysis. To evaluate the optimization effectiveness of the proposed algorithm, a comparison is conducted with the BR algorithm [6], LSAH algorithm [13], deep reinforcement learning algorithm with spatial discretization [18], and Online BPH heuristic algorithm [6]. Baggage data and baggage cart size are the same. Each algorithm is tested 20 times, and the average results are considered. The loading performance of the algorithms is presented in Table 1.

Table 1. Algorithm loading effect comparison

Algorithm type	Average filling rate (%)	Average quantity	Average load time (s)
BR	57.0	22.10	0.12
LSAH	54.3	21.05	0.03
Spatial discretization	61.3	23.16	0.05
Online BPH	62.7	23.50	0.05
Method of this paper	71.2	28.70	0.04

Based on the experimental data in Table 1, the proposed method achieves an average filling rate of 71.2% and loads an average of 28.7 pieces of baggage. It outperforms other intelligent algorithms in terms of filling rate. Its performance is superior due to the unique characteristics of the airline baggage loading scenario. Compared to the BR algorithm, LSAH algorithm, discrete space deep reinforcement learning algorithm, and Online BPH algorithm, the proposed method increases the filling rate by 14.2%, 16.9%, 9.9%, and 8.5%. In terms of loading efficiency, the proposed method has an average computing time of 0.04 s, higher than the Online BPH algorithm, BR algorithm, and spatial discretization method, but similar to the LSAH algorithm.

Family Unity Strategy Control Experiment. To assess the impact of the Family Unity strategy on algorithm performance while keeping other parameters unchanged, a controlled experiment is conducted, and the results are shown in Fig. 6.

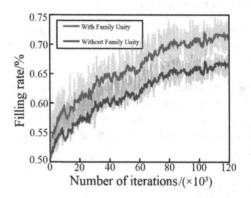

Fig. 6. Training effect curve

As observed in Fig. 6, when combined with the Family Unity idea, the proposed method exhibits a higher filling rate and better training performance. This outcome is attributed to the Family Unity concept, which emphasizes placing similar baggage close together to avoid wasted space caused by loose placement.

The proposed algorithm is applied to the loading simulation experiment, and the visualization results are shown in Fig. 7. The simulation demonstrates sufficient bottom contact area for the loaded baggage positions due to the constraints imposed. The loading effect is stable. The Family Unity concept ensures that baggage of the same type is grouped closely together, resulting in a compact and stable loading process.

3.4   Field Experiment Verification

To further validate the feasibility of the proposed algorithm, the research group has established an experimental platform for automatic baggage loading, as shown in Fig. 5. The experiment employs an airline baggage cart with dimensions of 288 cm * 147 cm * 84 cm. The baggage sizes range from 13 in to 28 inches, with specific dimensions varying from 36 cm * 33 cm * 18 cm to 70 cm * 47 cm * 28 cm. In order to observe the real-time changes in the palletizing pattern during baggage loading, the palletizing process of a specific experiment is shown in Fig. 8.

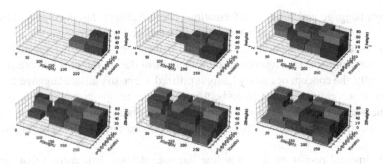

Fig. 7. Visualization of simulation effects

Fig. 8. Experimental results

From Fig. 8, it can be observed that the proposed hierarchical tree deep reinforcement learning algorithm prioritizes the corners, bottom layer, back outer layer, and top layer during the loading process. By incorporating the Family Unity algorithm, which groups similar-sized baggage together, excessive gaps between baggage are avoided, thereby minimizing space wastage. The multiple constraints introduced in this study ensure that the upper baggage remains in a stable position throughout the palletizing process, preventing slipping or tumbling and ensuring the safety of the baggage palletizing operation.

4 Conclusion

This chapter presents the key findings and contributions of the research. The main points can be summarized as follows:

A hierarchical tree A2C- reinforcement learning model is proposed for online packing of airline baggage, considering practical constraints such as direction, volume, non-overlap, and stability during the loading process. To handle the large action-state space of airline baggage, a heuristic state-action space expression method based on a hierarchical tree search algorithm is introduced. By combining the hierarchical tree search algorithm with the deep reinforcement learning algorithm, the hierarchical tree structure effectively

512 P. Zhang et al.

represents baggage and stacking information. The A2C algorithm is employed to train the model, enhancing the algorithm's trainability and performance.

In consideration of the online loading scenario of airline baggage, the reward function incorporates collision interference of the robotic arm during the real loading process. Additionally, the concept of Family Unity is utilized to classify airline baggage based on size ranges, preventing excessive gaps between baggage and optimizing space utilization. These measures contribute to improved learning of the optimal strategy by the model and aid in preventing robotic arm collisions during actual loading, enhancing the algorithm's practicality.

Experimental results on a real airline baggage dataset demonstrate that the proposed algorithm achieves an average palletizing filling rate of 71.2%, surpassing existing advanced packing algorithms by 8.5%. The algorithm exhibits robustness and stability, adapting well to various baggage specifications. Field experiments are conducted using a real manipulator platform to validate the algorithm's effectiveness.

It is acknowledged that the current constraints on airline baggage loading do not account for baggage deformation caused by weight or compression during loading. Future work will address the center of gravity issue related to pile type baggage in real airline baggage loading scenarios. Additionally, the algorithm will be further developed to model and accommodate baggage with special shapes and materials, enhancing its adaptability.

Acknowledgements. The authors would like to express their gratitude to all who provided helps for this research. This study was co-supported by the National Natural Science Foundation of China and Civil Aviation Administration of China jointly funded project (No. U2033208), the Fundamental Research Funds for the Central Universities (No. 3122023018) and Tianjin Research Innovation Project for Postgraduate Students (No. 2022SKYZ247).

References

1. Xu, M.: Thinking on the development of smart airport in the new era. Civ. Aviat. Manage. **06**, 39–42 (2020)
2. Wäscher, G., Haußner, H., Schumann, H.: An improved typology of cutting and packing problems. Eur. J. Oper. Res. **183**(3), 1109–1130 (2007)
3. Ali, S., Ramos, A.G., Carravilla, M.A., Oliveira, J.F.: On-line three-dimensional packing problems: a review of off-line and on-line solution approaches. Comput. Ind. Eng. **168**, 108–122 (2022)
4. Xing, Z., Hou, X., Li, B., Zhang, T., Wen, T.: Dynamic quadtree search algorithm for luggage cart placement in Civil Aviation. J. Beijing Univ. Aeronaut. Astronaut. **48**(12), 2345–2355 (2022)
5. Hong, Z., Zhang, C.: Airport Baggage stowage strategy based on support vector machine model. J. Comput. Appl. Softw. **39**(7), 44–51+66 (2022)
6. Ha, C.T., Nguyen, T.T., Bui, L.T., Wang, R.: An online packing heuristic for the three-dimensional container loading problem in dynamic environments and the physical internet. In: Applications of Evolutionary Computation: 20th European Conference, Amsterdam, The Netherlands, April, pp. 19–21 (2017)
7. Zhang, D., Peng, Y., Zhang, L.: Multi-layer heuristic search algorithm for three-dimensional packing problem. Chin. J. Comput. **35**(12), 2553–2561 (2012)

8. Zhang, C., Zhang, Q., Zhai, Y., Liu, J.: Online airline baggage loading optimization based on improved particle swarm optimization. Packag. Eng. **42**(21), 200–206 (2021)

9. Zhang, W., Chai, S., Wang, W., Chen, Y.: Heuristic algorithm for solving the online multi-size luggage packing problem. Packag. Eng. **42**(21), 213–221 (2021)

10. Bello, I., Pham, H., Le, Q.V., Norouzi, M., Bengio, S.: Neural combinatorial optimization with reinforcement learning. arXiv preprint arXiv:1611.09940 (2016)

11. Kool, W., Van Hoof, H., Welling, M.: Attention, learn to solve routing problems! In: 7th International Conference on Learning Representations, ICLR 2019, New Orleans, LA, USA, 6–9 May (2019)

12. Yang, Y., Shen, H.: Deep reinforcement learning enhanced greedy algorithm for online scheduling of batched tasks in cloud in cloud HPC systems. IEEE Trans. Parallel Distrib. Syst. **01**, 1–12 (2021)

13. Hu, H., Zhang, X., Yan, X., Wang, L., Xu, Y.: Solving a new 3d bin packing problem with deep reinforcement learning method. arXiv preprint arXiv:1708.05930 (2017)

14. Jiang, Y., Cao, Z., Zhang, J.: Learning to solve 3-D bin packing problem via deep reinforcement learning and constraint programming. IEEE Trans. Cybernet. **53**(5), 2864–2875 (2023)

15. Verma, R., et al.: A generalized reinforcement learning algorithm for online 3d bin-packing. arXiv preprint arXiv:2007.00463 (2020)

16. Zhao, H., Zhu, C., Xu, X., Huang, H., Xu, K.: Learning Practically feasible policies for online 3D bin packing. SCIENCE CHINA Inf. Sci. **65**(112105), 1–17 (2022)

17. Erbayrak, S., Özkır, V., Yıldırım, U.M.: Multi-objective 3D bin packing problem with load balance and product family concerns. Comput. Ind. Eng. **159**, 107518 (2021)

18. Zhao, X., Xia, L., Zhang, L., Ding, Z., Yin, D., Tang, J.: Deep reinforcement learning for page-wise recommendations. arXiv preprint arXiv:1805.02343 (2018)

Author Index

Printed in the United States
by Baker & Taylor Publisher Services

Printed in the United States
by Baker & Taylor Publisher Services